AF411451

Selected Papers from the 1st International Electronic Conference on Biosensors (IECB 2020)

Selected Papers from the 1st International Electronic Conference on Biosensors (IECB 2020)

Editors

Giovanna Marrazza
Sara Tombelli

MDPI • Basel • Beijing • Wuhan • Barcelona • Belgrade • Manchester • Tokyo • Cluj • Tianjin

Editors
Giovanna Marrazza
Department of Chemistry "Ugo
Schiff", University of Florence
Italy

Sara Tombelli
Institue of Applied Physics
(IFAC), National Research
Council (CNR)
Italy

Editorial Office
MDPI
St. Alban-Anlage 66
4052 Basel, Switzerland

This is a reprint of articles from the Special Issue published online in the open access journal *Biosensors* (ISSN 2079-6374) (available at: https://www.mdpi.com/journal/biosensors/special_issues/IECB2020_SI).

For citation purposes, cite each article independently as indicated on the article page online and as indicated below:

LastName, A.A.; LastName, B.B.; LastName, C.C. Article Title. *Journal Name* **Year**, *Volume Number*, Page Range.

ISBN 978-3-0365-3401-5 (Hbk)
ISBN 978-3-0365-3402-2 (PDF)

Contents

About the Editors

Giovanna Marrazza is a full professor in Analytical Chemistry at the Department of Chemistry "Ugo Schiff" of the University of Florence, Italy. She was the President of the Course of Study in Chemistry Degree at the same university from 2017 to 2021. She has been a distinguished Visiting Professor of the Faculty of Pharmacy "Iuliu Haţieganu" University of Medicine and Pharmacy, Cluj-Napoca (Romania) since 2017.

She is a member of the steering board of Italian Sensors and Microsystems Group (AISEM) and Italian Sensor Group, Italian Chemistry Society (SCI). Her research is focused on new biosensing principles containing nanomaterials and modified interfaces with nucleic acids, enzymes, antibodies, bacteria, and molecular imprinted polymers. She is an expert in the design of procedures suitable for biosensor devices such as microflow systems, thick-film technology, and nanodispensing technologies. She has contributed more than 138 papers in leading international journals and book chapters (H-index = 42, Scopus). She has been a unit leader in national and international projects, and has participated in further national and international projects.

Sara Tombelli Institute of Applied Physics, National Research Council (CNR), Italy. Sara Tombelli is a senior researcher at the Institute of Applied Physics (CNR) in Florence, Italy. Her research activity is focused on analytical chemistry, biosensor development and surface modifications with biomolecules, such as antibodies, enzymes, aptamers, and nucleic acid probes. She has experience in intracellular nanosensors and nanoparticle manipulation, as well as in the design and application of optical nanoprobes such as molecular beacons and oligonucleotidic optical switches. On the above topics, she has contributed more than 130 publications in international refereed journals, books, and conference proceedings (H-index = 38, Scopus). She is member of the Editorial Board of Sensors; she has been a Special Issue Editor for several MDPI journals and for the Elsevier journal Sensors and Actuators Reports. She has been part of international commissions as international expert for several international PhD theses (Spain, Switzerland, Italy). She participated in several European projects and in other national and international projects; she has been a unit leader in national projects and is responsible for several measurement campaigns in the frame of European projects.

Preface to "Selected Papers from the 1st International Electronic Conference on Biosensors (IECB 2020)"

The birth of biosensors dates back to the early 1960s, when Clark and Lyons (1962) introduced the concept of using an enzyme coupled with an electrode as a reagent. After decades of intense research, biosensors attracted the attention of a large scientific community, and they currently have a very wide range of applications due to their simplicity, low cost, and precision, aiming at improving quality of life. This range covers their use for environmental monitoring, disease detection, food safety, defense, drug discovery, and many more applications.

The scope of this Special Issue is to collect some of the contributions to the First International Electronic Conference on Biosensors, which was held to bring together well-known experts currently working in biosensor technologies from around the globe, and to provide an online forum for presenting and discussing new results.

The world of biosensors is definitively a versatile and universally applicable one, as demonstrated by the wide range of topics which were addressed at the Conference, such as: bioengineered and biomimetic receptors; microfluidics for biosensing; biosensors for emergency situations; nanotechnologies and nanomaterials for biosensors; intra- and extracellular biosensing; and advanced applications in clinical, environmental, food safety, and cultural heritage fields.

Giovanna Marrazza, Sara Tombelli
Editors

 *biosensors*

 MDPI

Article

Light-Addressable Actuator-Sensor Platform for Monitoring and Manipulation of pH Gradients in Microfluidics: A Case Study with the Enzyme Penicillinase

Rene Welden [1,2,†], Melanie Jablonski [1,3,†], Christina Wege [4], Michael Keusgen [3], Patrick Hermann Wagner [2], Torsten Wagner [1,5,*] and Michael J. Schöning [1,5,*]

[1] Institute of Nano- and Biotechnologies, Aachen University of Applied Sciences, 52428 Jülich, Germany; welden@fh-aachen.de (R.W.); m.jablonski@fh-aachen.de (M.J.)
[2] Laboratory for Soft Matter and Biophysics, KU Leuven, 3001 Leuven, Belgium; patrickhermann.wagner@kuleuven.be
[3] Institute of Pharmaceutical Chemistry, Philipps University Marburg, 35032 Marburg, Germany; michael.keusgen@staff.uni-marburg.de
[4] Institute of Biomaterials and Biomolecular Systems, University of Stuttgart, 70569 Stuttgart, Germany; christina.wege@bio.uni-stuttgart.de
[5] Institute of Biological Information Processing (IBI-3), Forschungszentrum Jülich GmbH, 52425 Jülich, Germany
[*] Correspondence: torsten.wagner@fh-aachen.de (T.W.); schoening@fh-aachen.de (M.J.S.)
[†] These authors contributed equally to this work.

Abstract: The feasibility of light-addressed detection and manipulation of pH gradients inside an electrochemical microfluidic cell was studied. Local pH changes, induced by a light-addressable electrode (LAE), were detected using a light-addressable potentiometric sensor (LAPS) with different measurement modes representing an actuator-sensor system. Biosensor functionality was examined depending on locally induced pH gradients with the help of the model enzyme penicillinase, which had been immobilized in the microfluidic channel. The surface morphology of the LAE and enzyme-functionalized LAPS was studied by scanning electron microscopy. Furthermore, the penicillin sensitivity of the LAPS inside the microfluidic channel was determined with regard to the analyte's pH influence on the enzymatic reaction rate. In a final experiment, the LAE-controlled pH inhibition of the enzyme activity was monitored by the LAPS.

Keywords: light-addressable potentiometric sensor; light-addressable electrode; actuator-sensor system; enzyme kinetics; microfluidics

Citation: Welden, R.; Jablonski, M.; Wege, C.; Keusgen, M.; Wagner, P.H.; Wagner, T.; Schöning, M.J. Light-Addressable Actuator-Sensor Platform for Monitoring and Manipulation of pH Gradients in Microfluidics: A Case Study with the Enzyme Penicillinase. *Biosensors* **2021**, *11*, 171. https://doi.org/10.3390/bios11060171

Received: 30 April 2021
Accepted: 25 May 2021
Published: 27 May 2021

1. Introduction

Lab-on-a-chip systems, microfluidic bioreactors and organ-on-chip platforms with integrated sensors and actuators for the monitoring of crucial parameters (e.g., flow rate, temperature and pH) are of great interest to maintain micro-environmental conditions [1–3]. On the other hand, inducing perturbations of these parameters leads to new perceptions of such systems, e.g., by changing the extracellular pH during cell culturing [4–7]. Often, due to geometrical restrictions inside microfluidic channels, the flexible integration of "conventional" sensing devices is not easy to accomplish [1]. The sensor information is mostly obtained at a fixed position, predefined during fabrication of the usually rigid sensor geometries. At the same time, a spatially resolved mapping of the molecular species functionalized areas are defined inside the microstructure. In addition, actuation functionalities should be flexible as well, enabling manipulation of e.g., local pH changes without affecting neighboring elements. Therefore, a precise addressability of both the actuator (here, a light-addressable electrode, LAE) and the sensor (here, a light-addressable potentiometric sensor, LAPS) is required.

LAPS is a semiconductor-based chemical sensor which was first proposed by Hafeman et al., in 1988 [8]. LAPS belongs to the group of field-effect-based electrochemical sensors with an electrolyte-insulator-semiconductor (EIS) structure [9]. LAPS offers (depending on its transducer layer) the spatially resolved monitoring of concentration-dependent surface-potential changes, e.g., induced by (bio)chemical/biological molecules or living cells in the analyte solution. LAPS can be designed as multiwell- and multianalyte-sensor devices and can serve for chemical imaging, where the distribution of the analyte concentration is visualized on its chip surface [10–14]. Moreover, LAPS provides a broad range of possible applications and has been utilized for various (bio)chemical and biotechnological approaches, such as monitoring the metabolic activity of bacteria in fermentation broth [14], for on-sensor cryopreservation of cells [15], multi-ion and penicillin detection [16,17], and DNA sensing [18].

In contrast to LAPS, a LAE exhibits no insulating layer, having a direct charge transfer with the analyte. By spatially resolved illumination, conductive areas inside the semiconducting chip can be defined resulting in, e.g., photoelectrocatalytical water oxidation. Furthermore, the LAE can be used for photoelectrochemical material deposition, adjustment of a pH gradient or cell stimulation [19–21]. Advantageously, the LAE offers a high flexibility without the need for patterning complicated electrode arrays [22,23].

The combination of LAPS and LAE as a sensor-actuator system would enable the simultaneous, spatially resolved pH manipulation and monitoring inside a microfluidic system. This way, reaction processes taking place inside microchannels could be further analyzed and optimized, which is helpful for e.g., studying the response characteristics of immobilized bioreceptors in a microfluidic channel, such as enzymes.

The proposed experimental approach of a light-addressable actuator-sensor platform (consisting of a LAE and a LAPS) elaborates for the first time the mutual reaction of local pH gradients inside a microfluidic channel and the enzyme-triggered sensor signal. Penicillinase has been selected as a model enzyme to detect penicillin, since it is a robust enzyme making it advantageous for experimental application. The enzymatic reaction induces a pH change, which can be detected with the LAPS. Si_3N_4 was used as pH-sensitive transducer layer for the LAPS, and TiO_2 as photoelectrocatalytical material inducing pH-value manipulations.

The surface morphology of the LAE and the LAPS surface where physically characterized by means of scanning electron microscopy (SEM). The penicillin sensitivity of the LAPS and the effect of pH changes on the enzyme activity, provoked by the LAE, were evaluated by photocurrent-voltage and chemical-image measurements. Dynamic pH variations induced by the LAE can be further used to control the enzymatic reaction rate and to adjust the biosensor response.

2. Materials and Methods

2.1. Fabrication Process of Light-Addressable Electrodes (LAEs)

The LAE employed in this study consists of a glass/SnO_2:F/TiO_2 heterostructure. The SnO_2:F glass substrate (7 $\Omega \cdot sq^{-1}$) was purchased from Sigma Aldrich (Darmstadt, Germany). The SnO_2:F glass substrate was cleaned in an ultrasonic bath with acetone, 2-isopropanol and deionized water for 5 min, respectively, and dried with nitrogen. Afterwards, the TiO_2 layer was deposited by pulsed laser deposition (PLD). During the PLD process, a TiO_2 target (MaTeck Material-Technologie and Kristalle GmbH, Jülich, Germany) was vaporized with a KrF-excimer laser (λ = 248 nm) using a power density of 5.0 $J \cdot cm^{-2}$ with a repetition frequency of 10 Hz at a pressure of 2.0×10^{-2} hPa O_2 for 700 s. Hereby, the SnO_2:F glass substrate was heated during the PLD process to 400 °C to achieve a rutil crystal structure. To achieve an inlet and outlet for the later prepared microfluidic structure, two holes with a diameter of 1.2 mm were drilled in the LAE.

2.2. Preparation of Light-Addressable Potentiometric Sensor (LAPS) Chips

The utilized LAPS chips, consisting of a n-Si/SiO$_2$/Si$_3$N$_4$-multilayer structure, were acquired from SEIREN KST Corp. (Fukui, Japan). The thickness of the n-Si, SiO$_2$ and Si$_3$N$_4$ layer was 100 μm, 50 nm and 50 nm, respectively. To remove the surface-oxide layer, the rear side was treated by wet-chemical etching with hydrofluoric acid (HF). Afterwards, a 300 nm thick aluminum (Al) film was deposited by electron-beam evaporation with a deposition rate of 2 nm·s^{-1} to contact the n-Si substrate electrically. The wafer was diced into 20 × 20 mm^2 chips and an optical window (ca. 15 × 15 mm^2) was made by etching an inner rectangle of the Al layer with 5% HF, leaving an outer Al frame.

2.3. Enzyme Immoblization with Tobacco Mosaic Virus Particles as Enzyme Nanocarriers on LAPS Chips

For enzyme immobilization, *tobacco mosaic virus* (TMV) particles were utilized as enzyme nanocarriers. A TMV variant (S3C) that exposes a cystein residue on each coat protein was used [24,25]. To functionalize the TMV particles, bifunctional biotin-linker molecules (EZ-Link Maleimide-PEG11-Biotin, ThermoScientific, Rockford, IL, USA) were covalently bound to the thiol groups located on the surface of each coat protein, as described in [26–28]. The biotinylated TMV particles were suspended in 10 mM sodium-potassium-phosphate (SPP) buffer (pH 7.0) and stored at 4 °C until use. As a model enzyme, penicillinase from *Bacillus cereus* (Sigma-Aldrich, Darmstadt, Germany) was utilized. For specific enzyme immobilization to the biotin linkers on the surface of the TMV nanotubes, the enzyme was conjugated with streptavidin molecules using a commercial streptavidin conjugation kit (LNK162STR, Bio-Rad, Feldkirchen, Germany) [26,27]. The streptavidin-conjugated penicillinase ([SA]-penicillinase) was stored in 10 mM phosphate-buffered saline (PBS) buffer (1000 Units·mL^{-1}, pH 7.0) at 4 °C until further use.

The LAPS chips were cleaned in an ultrasonic bath for 5 min in acetone, 2-isopropanol, ethanol and deionized water, respectively. After drying with nitrogen, 10 μL TMV solution (0.1 mg·mL^{-1}) were drop-coated on the Si$_3$N$_4$ surface of the later-on microfluidic channel, and incubated for 1 h at room temperature (RT) in a humid chamber. Afterwards, the solution with unbound TMV was washed away with deionized water and the sensor chip was dried with nitrogen. In the next step, 5 μL [SA]-penicillinase solution were drop-coated on the immobilized TMV particles and incubated for 1.5 h at RT in a humid chamber. After the incubation time, the sensor chip was rinsed again with deionized water to remove unbound enzyme molecules and dried with nitrogen.

2.4. LAPS-LAE Microfluidic Assembly

To prepare the LAPS/microfluidic foil/LAE sandwich structure (schematically depicted in Figure 1a, a ~86 μm thick double-sided adhesive microfluidic tape (3M^{TM}, St. Paul, MN, USA) was patterned by laser cutting using a ProtoLaser U3 (LPKS Laser and Electronics AG, Garbsen, Germany). A 20 × 20 mm^2 rectangle with a 1.0 mm wide channel was cut out of the tape. The LAE was cleaned in an ultrasonic bath with acetone, 2-isopropanol, ethanol and deionized water and finally dried with nitrogen. Afterwards, the laser-cut microfluidic foil was stuck onto the TiO$_2$ surface of the LAE, positioning the drilled holes of the LAE at the inlet and outlet of the microfluidic channel. In the final step, the TMV- and penicillinase-functionalized LAPS chip was placed below the LAE-microfluidic structure, immobilizing the enzyme-loaded TMV particles at the bottom of the microfluidic channel. For tube connection, ferrules were attached to the inlet and outlet with the help of a photopolymer.

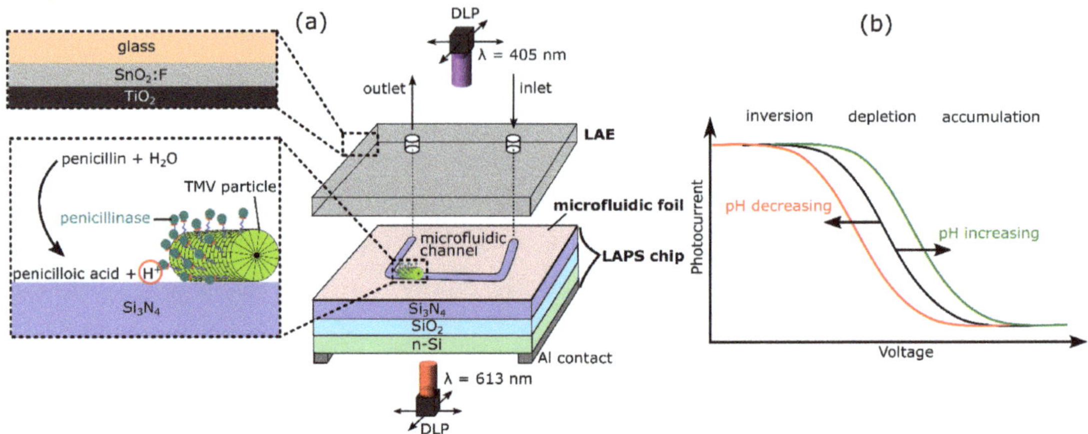

Figure 1. (**a**) Schematic of the microfluidic setup with a light-addressable potentiometric sensor (LAPS)/microfluidic foil/ light-addressable electrode (LAE)-sandwich structure. *Tobacco mosaic virus* (TMV) particles functionalized with the enzyme penicillinase are immobilized inside the microchannel. (**b**) Typical shape of a photocurrent-voltage curve for a n-type LAPS with characteristic regions of inversion, depletion and accumulation.

2.5. Measurement Setup and Characterization Methods

The inlet tube of the microfluidic setup was connected to a syringe-driven pump system (neMESYS 290N, Cetoni GmbH, Korbussen, Germany) to control the flow inside the channel. In the outlet tube, a Pt-counter electrode and a reference electrode (DRIREF-2SH, World Precision Instruments, Sarasota, FL, USA) for the electrical connection of the LAE and LAPS were inserted. A clamp connected the Al rear side contact of the LAPS to a transimpedance amplifier (gain = 10^7 V·A^{-1}, AMP100, Thorlabs GmbH, Bergkirchen, Germany) to convert the alternating photocurrent into a measurable voltage. The voltage was recorded by a measurement card (USB 7855R, NI, Austin, TX, USA). The same card also provided the bias voltage to the LAPS with respect to the counter electrode. A potential was directly applied to the LAE and Pt-counter electrode by a source measurement unit (2600b, Keithley Instruments, Solon, OH, USA).

The LAE and LAPS rear side were illuminated by a digital light processing (DLP) projector (STAR-07, ViALUX Messtechnik + Bildverarbeitung GmbH, Chemnitz, Germany) with a 405 nm and a 613 nm light-emitting diode (LED) light source. Both DLPs are modified with a lens system to focus each of the 1024×768 micromirrors to a size of 10×10 μm^2. All measurements were performed in a dark Faraday cage at room temperature (RT). For all experiments, 0.33 mM PBS buffer was used. The pH was adjusted by titration with NaOH and HCl. For penicillin detection, varying concentrations of penicillin G (Sigma Aldrich, Darmstadt, Germany) were added to the measurement solution.

For photoelectrocatalytically induced pH changes, a constant potential of 300 mV was applied to the LAE with respect to the Pt-counter electrode. After reaching steady- state conditions, the rear side was illuminated with spots of various sizes.

For LAPS characterization, three measurement modes were applied: photocurrent-voltage (I-V), chemical-image and photocurrent-time mode. The illumination of the LAPS-DLP projector was modulated with a frequency of 512 Hz to achieve an alternating photocurrent for all measurements. The voltage from the transimpedance amplifier was sampled with a frequency of 50 kHz and further processed. For I-V curves, the measurement time for each bias voltage was 400 ms. Figure 1b depicts a theoretical I-V curve of a n-type silicon LAPS. In the I-V mode, the applied voltage was swept from 0 to -3.0 V while measuring the photocurrent for a fixed illumination spot. The typical output curve shows the three characteristic regions of inversion, depletion and accumulation. A pH increase or decrease shifts the I-V curve to more positive or negative voltage, respectively, which is particularly evident in the depletion region. As slight changes of the photocurrent amplitude can occur when replacing the measurement solutions, the I-V curves were normalized with respect to the inversion region.

To obtain spatially resolved images from the microfluidic channel, the chemical image mode was utilized. A constant bias voltage, chosen close to the inflection point of the I-V curve, was applied and the rear side was scanned sequentially with a moving light beam and a sampling time of 200 ms for each spot ($250 \times 250 \ \mu m^2$). In the results section, differential chemical images are visualized. For that, the chemical image after the enzymatic reaction was subtracted from the initial reference chemical image (before enzymatic catalysis of penicillin by penicillinase). On the basis of the exemplary I-V curve in Figure 1b, at a fixed bias voltage, a pH decrease results in a photocurrent drop, while it increases with rising pH values.

Similarly, during the photocurrent-time mode, the temporal change of the photocurrent is measured for a fixed illumination spot and a fixed bias voltage (from the I-V inflection point), enabling the dynamic detection of pH changes. The sampling time was 1 s.

With all three measurement modes, changes of the pH value in the solution can be monitored. In this work, the resulting pH changes are caused by two different proton generation mechanisms, being photoelectrocatalysis and enzymatic conversion. The induced pH change from the LAE originates mostly from the water oxidation reaction,

$$2 \, H_2O \rightarrow 4 \, e^- + 4 \, H^+ + O_2, \tag{1}$$

where water is split into H^+ ions and oxygen. The second pH change is due to the immobilized enzyme penicillinase, where penicillin is converted to penicilloic acid and H^+ ions through the enzyme's ß-lactamase activity,

$$penicillin + H_2O \rightarrow penicilloic \ acid + H^+. \tag{2}$$

3. Results

3.1. Scanning Electron Microscopy (SEM) Characterization of the TiO$_2$- and Tobacco Mosaic Virus (TMV)-Modified Si$_3$N$_4$ Surface

To characterize the surface morphology of the fabricated TiO_2 and the enzyme-modified Si_3N_4, scanning electron microscopy (SEM) images were taken with a Schottky field-emission microscope (JSM-7800F, JEOL GmbH, Freising, Germany). For higher conductivity, a ~5 nm platinum-palladium layer was sputtered onto the Si_3N_4 surface before SEM images were taken.

To achieve a high spatial resolution of the LAE, a low current in the absence of illumination is required. This can be achieved with a dense and non-porous TiO_2 layer to avoid short circuits with the SnO_2:F glass [29]. Additionally, it is important to exclude a direct contact of the analyte with the highly-doped SnO_2:F layer, to circumvent unexpected surface reactions. A representative SEM image of the TiO_2 surface is given in Figure 2a. The image shows a homogeneous and dense surface structure without visible cracks. The granularity (200–250 nm) is induced by the SnO_2:F glass on which the ~200 nm TiO_2 layer is deposited.

Exemplary SEM images of the TMV-modified (carrying the penicillinase molecules) LAPS-Si_3N_4 surface with part of the microfluidic channel are shown in Figure 2b (left and right). The channel boundary is cleanly cut with no visible fringes. On the Si_3N_4, the white cloud-like area is indicating the TMV-modified surface spot (left image). A zoom into this spot (right image) shows homogeneously distributed TMV particles with immobilized penicillinase. The TMV particles appear as typical 300 nm long nanotubes, as well as in shorter particle fractions (~50–200 nm) or elongated "end-to-end"—multimer structures (up to ~600 nm) as described in previous works as enzyme nanocarriers on a Ta_2O_5-sensor surface [24,26].

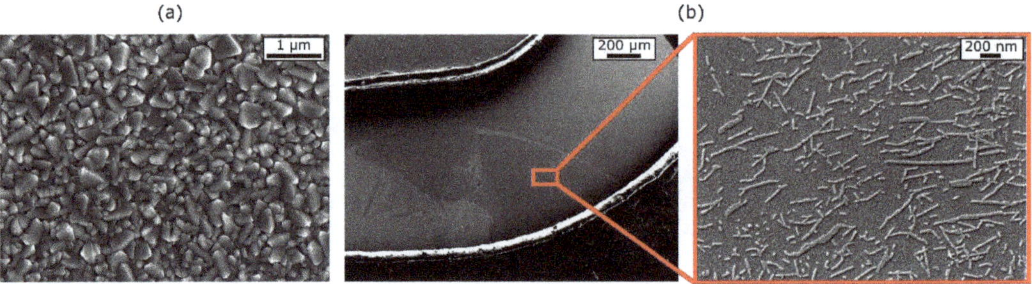

Figure 2. (**a**) Scanning electron microscope (SEM) image (magnification of 20,000× of the LAE showing the TiO_2 film surface on the SnO_2:F glass substrate. (**b**) SEM image depicting the part of the microfluidic channel where the enzyme-modified Si_3N_4 surface of the LAPS is located with a magnification of 60× (left) and with a zoom-in, showing the adsorbed TMV particles carrying the immobilized penicillinase with a magnification of 35,000× (right).

3.2. Penicillin Detection with Penicillinase-Modified LAPS

TMV particles were loaded inside the microchannel by drop-coating with subsequent penicillinase coupling by affinity binding of SA-penicillinase conjugates to the biotinylated TMV. Because modified TMV particles have been utilized for the first time inside a microfluidic channel for enzyme immobilization, chemical images and photocurrent-voltage curves were recorded by the LAPS to control the layout's functionality for penicillin detection. During the enzymatic conversion of penicillin to penicilloic acid, H^+ ions are generated resulting in a local pH change in the solution. As a first experiment, the pH change resulting from varying penicillin concentrations was studied as chemical images inside the microchannel. The rear side of the LAPS is therefore scanned sequentially, with the resulting photocurrent depicted in Figure 3a. Each chemical image represents (from top to bottom) a different penicillin concentration (from 0.1 mM to 5.0 mM) as differential image. This differential image is obtained by subtracting the particular chemical image from the reference chemical image of the microfluidic structure recorded at an applied potential of -1.65 V in 0.33 mM PBS buffer at pH 7.0. For the reference, (Figure S1 (Supplementary Information)), a high flow rate of the analyte of $1.0\ \mu L \cdot s^{-1}$ was chosen to suppress any pH changes inside the channel. During the enzymatic experiments, the flow was stopped, allowing an accumulation of enzymatically produced H^+ ions. The resulting differential chemical images (Figure 3a) show a section of the microfluidic channel with 192 measurement points, visualizing a total area of $8.0 \times 1.5\ mm^2$. The applied potential of -1.65 V was selected to be close to the inflection point of the photocurrent–voltage curve (Figure 3b).

In Figure 3a, the bottom image shows the result for a penicillin concentration of 5 mM. Due to the H^+ ion generation, the ΔPhotocurrent (ΔI_{photo}) decreased in the area with immobilized enzyme. The diameter of the pH-change region (~4.0 mm), corresponds to the drop-coated area of the TMV particles with immobilized penicillinase. In this area, ΔI_{photo} changed (decreased) by 7.3 ± 0.9 nA for a penicillin concentration of 5.0 mM. For lower penicillin concentrations, the ΔI_{photo} variations have been 6.0 ± 0.7 nA (1.0 mM) and 4.0 ± 0.5 nA (0.5 mM). For the lowest penicillin concentration (0.1 mM), a small photocurrent ($\Delta I_{photo} = 0.5 \pm 0.0$ nA) was detected. Here, due to the ΔI_{photo} scaling of the depicted chemical image, the change is hardly visible. Interestingly, the spatial pH change expansion for all concentrations was—more or less—in the same local area at 4.0 mm width (*x*-axis) and equally distributed indicating a rather low H^+ ion diffusion away from the enzyme into the surrounding medium; only the photocurrent intensity changed by varying penicillin concentrations. Besides offering a mapping of the enzyme activity inside the microchannel, the chemical image mode can be used to determine the exact spots of immobilized enzymes. Moreover, possible enzyme detachment (e.g., due to shear stress induced by high flow rates) would be directly recognized.

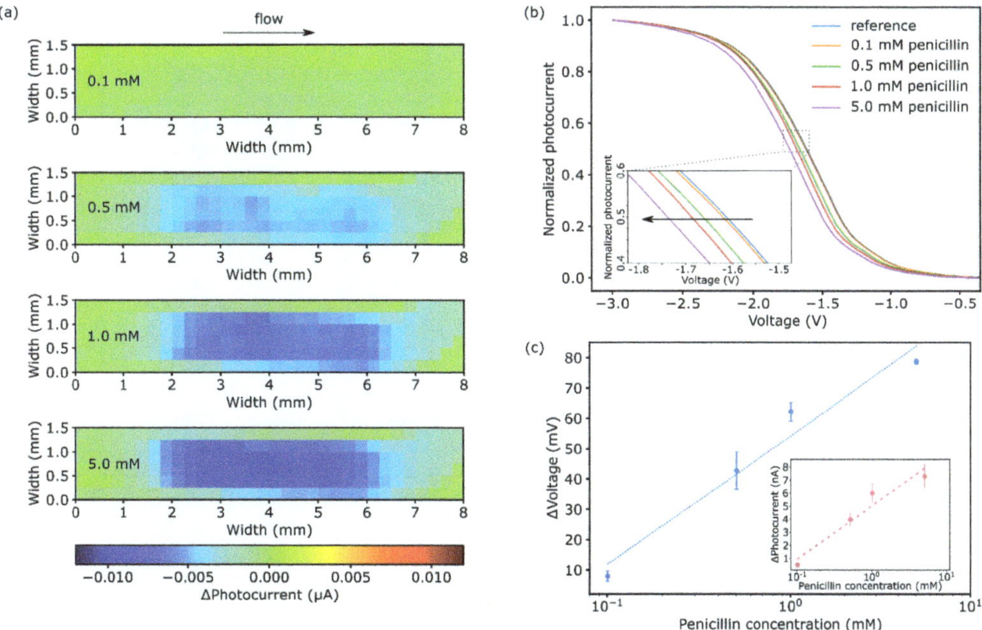

Figure 3. (**a**) Chemical images and (**b**) photocurrent-voltage curves for penicillin concentrations ranging from 0.1 to 5.0 mM after 5 min of enzymatic reaction in phosphate buffered saline (PBS) buffer, pH 7.0. (**c**) Mean calibration curve evaluated from the photocurrent-voltage curves ($n = 4$) with an average penicillin sensitivity of 42.3 mV/dec. The inlet represents the photocurrent change in dependence of the penicillin concentration, evaluated from Figure 3a.

In addition to the chemical images, for photocurrent–voltage curves, the photocurrent is recorded at a defined location inside the microchannel, while sweeping the applied bias potential from 0 to −3.0 V. The measurement spot with an illumination size of 250×250 µm^2 was located in the center of the determined pH-change area (x-axis 4.0 mm, y-axis 0.75 mm). Figure 3b displays the normalized I-V curves related to the previously discussed chemical images for different penicillin concentrations. The I-V curves exhibit the characteristic regions with inversion, depletion and accumulation. For example, the blue I-V curve represents the previously described reference measurement during vigorous flushing of the channel with PBS buffer solution (1.0 µL·s^{-1}). In the diagram, from −3.0 V to −2.4 V, the n-type semiconductor is in the inversion state. In the depletion region, the photocurrent decreases until an applied voltage of −1.0 V and reaches its minimum due to charge accumulation for further increasing bias potentials. During the enzymatic reaction, the H$^+$ ion generation leads to a surface protonation of the Si$_3$N$_4$, followed by a shift of the photocurrent-voltage curve to more negative potentials. The potential changes were taken from the inflection point at the normalized photocurrent of 0.5. In contrast to the chemical image, a potential change (ΔU) of 8.0 ± 1.7 mV with respect to the reference I-V curve was detected even for 0.1 mM penicillin. For higher penicillin concentrations, the signal shift increased to 42.8 ± 6.2 mV, 62.3 ± 3.1 mV and 78.8 ± 0.8 mV for 0.5 mM, 1.0 mM and 5.0 mM, respectively. The evaluated calibration curve is depicted in Figure 3c. A mean penicillin sensitivity in the concentration range from 0.1 to 5.0 mM of 42.3 mV/dec was achieved.

The experiments highlight, that the combination of LAPS and penicillinase-functionalized TMV can be used for the detection of penicillin inside a microfluidic channel: here, a two-dimensional mapping in x- and y-directions is possible.

3.3. Impact of pH Changes on Penicillinase Activity

A main aim of this study is to control the rate of enzymatic conversion by locally induced pH changes with the LAE. Therefore, the activity of penicillinase for pH values ranging from pH 4.0 to pH 7.0 in 0.33 mM PBS buffer with a constant penicillin concentration of 1.0 mM was characterized. The differential chemical images in Figure 4a visualize typical pH changes due to H^+ ion accumulation after stopping the enhanced flow of 1.0 $\mu L \cdot s^{-1}$. All results were obtained 5 min after the flow stopped. For pH 4.0, there is nearly no pH change, and thus no change in photocurrent (ΔI_{photo} = 0.4 nA $\pm$ 0.0) detected, which can be attributed to the inhibition of the enzyme at a such low pH value (see also activity behavior of penicillinase [30]. For pH 5.0, a slight variation in photocurrent of 2.1 $\pm$ 0.1 nA occurs, indicating a low enzymatic activity. For higher pH values of pH 6.0 and pH 7.0, ΔI_{photo} is increased to 4.3 $\pm$ 0.4 nA and 5.9 $\pm$ 0.6 nA, respectively. Here, the pH values are closer to the penicillase's activity optimum of approximately pH 7.5 (for immobilized enzyme) [31], resulting in a higher catalytic conversion of penicillin.

These results are confirmed by the associated photocurrent-voltage (I-V) curves, recorded in the center of the area with immobilized enzyme (x-axis 4.0 mm, y-axis 0.75 mm). The corresponding I-V curve for pH 4.0 PBS buffer solution is shown in Figure 4b. Between the reference (blue) and 1.0 mM penicillin I-V curve (orange), there is only a marginal shift of 6.8 $\pm$ 3.6 mV. Based on the original pH sensitivity of the sensor of 40 $mV \cdot pH^{-1}$ (data not shown), this refers to an additional pH drop of 0.17. For pH 7.0 (Figure 4c), the I-V curve shifted by 61.0 $\pm$ 0.7 mV, which is equivalent to a pH change of 1.5 from pH 7.0 to pH 5.5. Lowering the pH values in relation to pH 7.0, the photocurrent–voltage curves shifted to more negative voltages by 49.8 $\pm$ 2.8 mV at pH 6.0 and 27.0 $\pm$ 3.0 mV for pH 5.0 (not shown). It should be mentioned that, beside the inhibited enzyme's activity at lower pH values, the PBS buffer capacity is also reduced at lower pH values. By that, the enzymatically produced H^+ ions at pH 5 and pH 4 have a higher impact on the resulting pH change, than at higher pH values of pH 7 and pH 6.

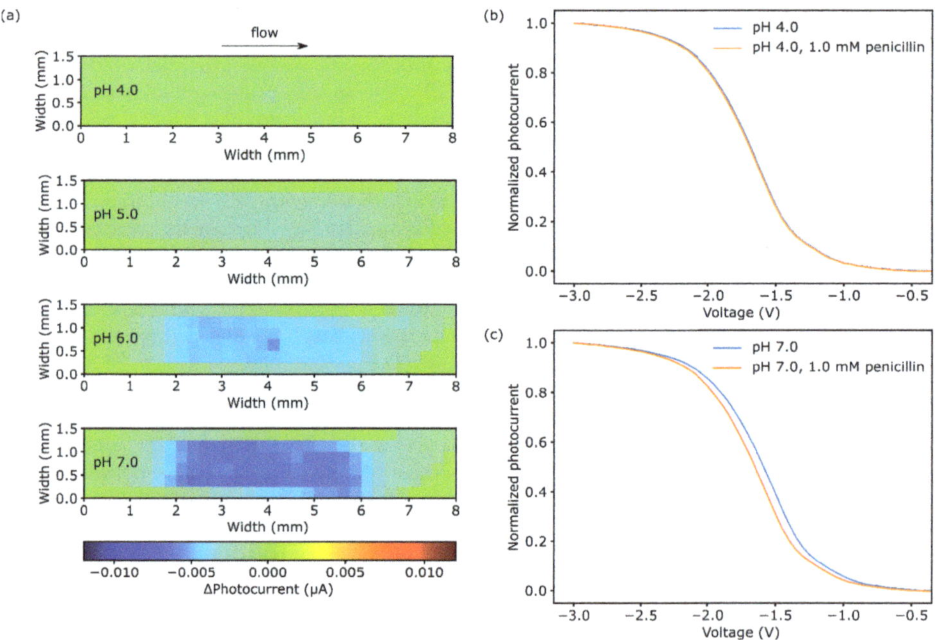

Figure 4. (**a**) Chemical images for 1.0 mM penicillin in pH 4 to pH 8 PBS buffer after 5 min of enzymatic reaction. Photocurrent-voltage curve for (**b**) pH 4 and (**c**) pH 8 PBS buffer with and without 1.0 mM penicillin.

Both, the chemical images and the I-V curve reveal the possibility to inhibit the enzymatic reaction by lowering the pH in a microfluidic setup, where the chemical imaging allows determination of 2-dimensional distribution of pH-triggered enzyme activity.

3.4. pH Manipulation with LAE

By utilizing a LAE, a direct charge transfer at the semiconductor/electrolyte interface between generated holes and species in the solution is possible. Depending on the applied LAE potential, e.g., photoelectrocatalytic water splitting takes place, allowing a flexible pH-value adjustment inside the channel. First, a potential of 0.3 V is applied to the LAE against the Pt-counter electrode. A typical transient current response is rendered in Figure 5a. Without illumination, the current equilibrates at 18 nA (3 s). During this condition, no surface reactions are triggered. When illuminating the rear side of the LAE with an area of 0.25×1.0 mm^2, a current peak occurs (4–5 s), which can be assigned to accumulated holes perturbing the surface, resulting in a capacitive discharge [32]. Afterwards, the current equilibrates at 1.04 µA after 50–60 s. Most of the current occurring during illumination can be assigned to the photoelectrocatalytic oxygen-evolution reaction of water where, besides oxygen, H$^+$ ions are produced, resulting in a pH change. After switching-off the illumination, the current decreases again to its dark current value.

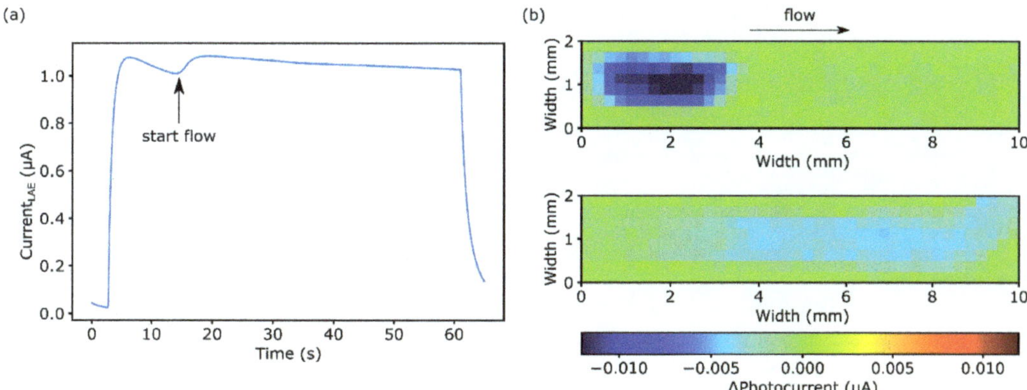

Figure 5. (**a**) Transient photocurrent signal for 60 s of illumination. (**b**) Chemical images of static (top) and dynamic (bottom) pH changes inside the microfluidic channel induced by the LAE.

Similarly to the H$^+$ ions generated by the enzymatic reaction, it is possible to visualize the photoelectrocatalytically produced protons with differential chemical images by the LAPS. In Figure 5b, differential chemical images of a 2.0×10.0 mm^2 area and an applied potential of −1.45 V were recorded in PBS buffer, pH 7.1. The top image shows a pH change induced by the LAE without flow. It can be seen that in the illuminated area, the LAPS photocurrent changes by 10.7 ± 1.9 nA. The corresponding I-V curve reveals a shift of 134.4 mV to more negative voltages, which is equal to a pH decrease by ~3.5.

To exclude that the LAE illumination wavelength of 405 nm affected the functionality of the enzyme, the LAE illumination spot was positioned 3–4 mm downwards (in flow direction) from the enzyme. After 10 s of photoelectrocatalysis, a steady flow of 0.05 µL·s^{-1} was applied, moving the generated protons upwards the channel to the location of the immobilized enzyme. Hereby, an equilibrium between the proton generation and transport is formed, resulting in a consistent pH change inside the channel. The result is depicted in the bottom image in Figure 5b. The differential chemical image shows an equally distributed pH variation with an average change of the LAPS photocurrent of 3.4 ± 0.6 nA. From the related I-V curve, a pH decrease of 1.0 ($\Delta U = 41.2$ mV) was obtained. Additionally, the start of the flow can be seen in the transient current measurement in Figure 5a. After 10 s of illumination, there is a small current increase. Since the volume above the illuminated area of 1.0×0.25 mm^2 is only 0.02 µL, due to the fluidic dimensions and no-flow conditions during the first 10 s, a reduced mass transfer can lead to a depletion of reaction partners, decreasing the current [33]. Hence, providing fresh solution with starting the flow after 10 s, the current increases to a nearly constant value of 1.04 µA, as the reaction rate between generated electron-hole pairs and reaction partners in solution equilibrate.

3.5. Regulation of Enzyme Activity by the LAE

In the final experiment, the enzymatic activity of penicillinase was directly regulated by the LAE inside the microfluidic channel. Here, photocurrent-time measurements (called as constant potential LAPS measurements) were performed to study the temporal photocurrent change during H$^+$ ion generation of the enzymatic reaction. The applied potential was -1.45 V.

Figure 6a shows the photocurrent change for 1.0 mM penicillin in PBS buffer, pH 7.1, without manipulation of the pH with the LAE. The microchannel was rinsed with a pump rate of 1.0 µL·s^{-1} for the first 60 s. After stopping the dosage, the photocurrent starts to decrease due to the accumulation of enzymatically produced H$^+$ ions and reaches an equilibrium after around 300 s. The measured photocurrent drop of 7.0 nA corresponds to a pH change of 1.75. After 300 s, the flow was started again and the channel was rinsed with fresh solution, whereby the photocurrent increased again to its initial value.

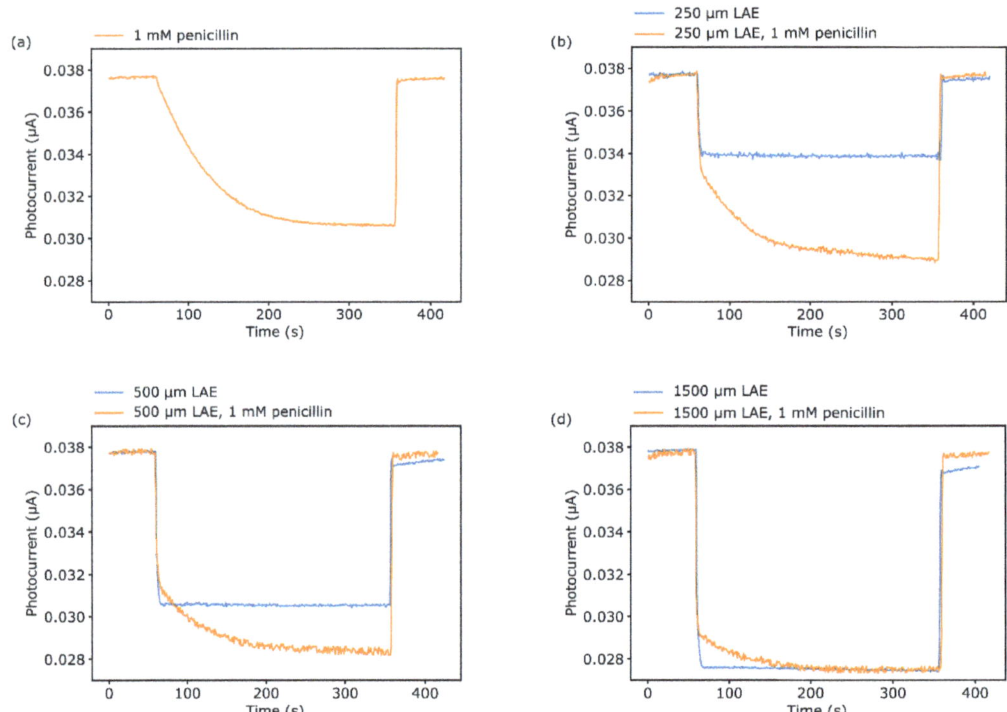

Figure 6. Constant potential LAPS measurements. (**a**) Photocurrent response for 1.0 mM penicillin in PBS buffer, pH 7.1. In (**b–d**) the blue curves depict the transient photocurrent decrease due to pH changes induced by the LAE with an illumination width of 250 µm, 500 µm and 1500 µm, respectively. The orange curves show the concatenated, additional change in photocurrent when 1.0 mM penicillin is added to the PBS buffer.

In the following measurements, the pH was regulated by the LAE and the enzymatic response was determined by the LAPS. First, the LAE induced a pH change by photoelectrocatalytic water oxidation in PBS buffer without penicillin. Since a simultaneous operation between LAPS and LAE is not possible, due to the influence of the 405 nm light beam of the DLP projector on the LAPS chip during measurements, the illustrated curves in Figure 6b–d are concatenated and normalized: The photocurrent was measured during the 60 s of rinsing, stopped while the LAE pH changed, and directly started after the LAE illumination was switched off. During the LAE illumination, the output flow of the pump was reduced to 0.05 $\mu L \cdot s^{-1}$. In Figure S2 (Supplementary Information), the effect of the lower flow velocity on the enzyme activity without LAE is analyzed. After changing the flow rate from 1.0 $\mu L \cdot s^{-1}$ to 0.05 $\mu L \cdot s^{-1}$, a small drop of ΔI_{photo} = 0.5 nA occurred. Nevertheless, the delta photocurrent after stopping the flow reached again 7.0 nA, which is identical to the photocurrent change without the decreased flow rate.

As the change of the pH is defined by the number of generated H^+ ions during the photoelectrocatalytical water oxidation, this can be influenced by varying the reactive area of the LAE with differently sized illumination spots. These have been changed with illumination lengths inside the microchannel between 250 μm and 1500 μm with the help of the DLP projector.

The pH change for an illuminated area of 1.0×0.25 mm^2 is depicted in Figure 6b. The blue curve shows the constant photocurrent shift of 3.8 nA that corresponds to a pH shift from pH 7.1 to pH 6.3. Subsequently, the measurement was repeated with 1.0 mM penicillin in the PBS buffer (orange curve). Here, after the LAE-induced pH drop, the photocurrent further decreased until it reached an equilibrium after approximately 300 s. The total photocurrent change is 8.6 nA. This is equivalent to a pH change of 2.1. As the LAE altered the pH value to 6.3, the additional pH drop by the enzymatic reaction was 1.3. In Figure 6c, the LAE was illuminated with a beam width of 500 μm, leading to a photocurrent drop of 7.2 nA. With 1.0 mM penicillin, it changed by 9.4 nA. The pH therefore decreases from pH 7.1 to pH 5.4 and pH 4.8 after the LAE and penicillin reaction, respectively. The largest illumination width of 1500 μm resulted in a ΔI_{photo} of 10.3 nA, which corresponds to a pH change of 2.6 (Figure 6d). Since this change already results in pH 4.4 inside the microchannel, no further change in the photocurrent was observed while adding 1.0 mM penicillin in the solution. This indicates that the enzymatic catalysis of penicillin was inhibited. These results underline the high potential of the proposed combination of LAPS and LAE. The flexible generation of pH gradients, using a LAE by changing the illumination spot, offers the spatially resolved control of the enzyme activity inside the microfluidic channel. Furthermore, the triggered enzymatic reaction rate can be label-free monitored by the enzyme-LAPS to validate the resulting impact on the enzymatic inhibition.

4. Conclusions

In this work, a LAE/microfluidic foil/LAPS sandwich structure was utilized for the detection and manipulation of pH gradients inside a microfluidic system. The LAPS offers the ability to detect spatially resolved pH changes inside the microfluidic channel. In contrast, locally induced pH changes can be triggered using the LAE, whereby the location may be controlled by the illuminated area. To study this sensing-actuating interplay, as a model bioreceptor, the enzyme penicillinase was immobilized inside the microchannel using plant viral (TMV) particles as enzyme nanocarriers. The enzymatic cleavage of penicillin to penicilloic acid, yielding H^+ ions, leads to local pH changes, which can be detected by the LAPS. By inducing a further pH shift via the LAE, the enzymatic activity can be inhibited.

The novel actuator-sensor platform was characterized performing photocurrent-voltage-, photocurrent-time measurements and chemical imaging with the LAPS and by transient current measurements with the LAE. The surface morphology of the LAE and TMV-modified LAPS was analyzed by means of SEM.

In the concentration range from 0.1 to 5.0 mM penicillin, the TMV-penicillinase-modified LAPS sensor achieved a penicillin sensitivity of 42.3 mV/dec, proofing the functionality as penicillin sensor inside the microfluidic setup. Additionally, the chemical images visualize, that the TMV-immobilized enzymes were confined to the area, predefined through drop-coating during assembly of the system. This extends the use of beneficial plant viral enzyme nanocarriers to a further type of microsystem. For solutions of varying pH, the inhibition of the enzymatic reaction was demonstrated at pH 4.0, whereas enzyme activity increased in LAPS measurements when pH is shifted towards the enzyme's pH optimum. Furthermore, a strategy for a spatially resolved photoelectrocatalytical pH manipulation induced by the LAE was developed. Such a pH gradient inside the microchannel was utilized to control the enzymatic reaction.

By this exemplary application, the feasibility and potential of combining two light-addressable technologies, LAPS and LAE, was highlighted to be of great benefit for further integration in lab-on-a-chip systems. The advantage of this system lies in the adaptability of both technologies, as the region of interest inside, e.g., the microfluidic channel, can be regulated in time and geometrical locus by changing the illuminated area.

In future studies, such as e.g., enzyme arrays inside a microfluidic channel, each particular enzyme might be controlled individually. Dependent on the adjusted pH value by the LAE, the enzyme activity can be either shifted to the enzyme's pH optimum (leading to increased reaction rates) or to pH values, where inhibition of enzyme takes place.

A further interesting approach for such actuator-sensing platform lies in the field of enantioselective enzymes which catalyse multiple reactions (e.g., acetoin reductase for acetoin and diacetyl determination [34,35]). Here, local pH variations triggered by the LAE could shift the pH optima corresponding to the respective substrate molecule of interest. A separation by different microchannels will address individual enzymes having a two-dimensional monitoring of each single reaction.

Supplementary Materials: The following are available online at https://www.mdpi.com/article/10.3390/bios11060171/s1, Figure S1: Reference chemical image of the microfluidic structure recorded at an applied potential of −1.65 V in 0.33 mM PBS buffer, pH 7.0. Figure S2: Photocurrent-time curve for 1.0 mM penicillin in PBS buffer, pH 7.1, for different flow rates.

Author Contributions: Conceptualization, R.W., M.J., T.W. and M.J.S.; methodology, R.W., M.J., C.W.; T.W. and M.J.S.; validation, R.W., M.J., T.W. and M.J.S.; formal analysis, R.W., M.J., T.W. and M.J.S.; investigation, R.W., M.J.; writing, R.W., M.J., C.W., M.K., P.H.W., T.W. and M.J.S.; supervision, M.K., P.H.W., T.W. and M.J.S.; All authors have read and agreed to the published version of the manuscript.

Funding: Part of this research project was funded by the German Federal Ministry of Education and Research (BMBF)–13N12585. Part of this work was funded by the Deutsche Forschungsgemeinschaft (DFG, German Research Foundation)–446507449.

Institutional Review Board Statement: Not applicable.

Informed Consent Statement: Not applicable.

Acknowledgments: The authors would like to thank Arshak Poghossian for valuable discussions, and Heiko Iken, David Rolka, Benno Schneider, Rebecca Hummel and Jürgen Schubert for technical support. Part of this work was funded by the Deutsche Forschungsgemeinschaft (DFG, German Research Foundation)–446507449. Part of this research project was funded by the German Federal Ministry of Education and Research (BMBF) within the research frame of "Nano-MatFutur": 13N12585. R. Welden thanks Aachen University of Applied Sciences for financial support.

Conflicts of Interest: The authors declare no conflict of interest.

References

1. Mousavi Shaegh, S.A.; de Ferrari, F.; Zhang, Y.S.; Nabavinia, M.; Mohammad, N.B.; Ryan, J.; Pourmand, A.; Laukaitis, E.; Banan Sadeghian, R.; Nadhman, A.; et al. A microfluidic optical platform for real-time monitoring of pH and oxygen in microfluidic bioreactors and organ-on-chip devices. *Biomicrofluidics* **2016**, *10*, 044111. [CrossRef] [PubMed]
2. Buchenauer, A.; Hofmann, M.C.; Funke, M.; Büchs, J.; Mokwa, W.; Schnakenberg, U. Micro-bioreactors for fed-batch fermentations with integrated online monitoring and microfluidic devices. *Biosens. Bioelectron.* **2009**, *24*, 1411–1416. [CrossRef] [PubMed]

3. Meller, K.; Szumski, M.; Buszewski, B. Microfluidic reactors with immobilized enzymes—Characterization, dividing, perspectives. *Sens. Actuators B* **2017**, *244*, 84–106. [CrossRef]

4. Wu, M.H.; Urban, J.P.; Cui, Z.F.; Cui, Z.; Xu, X. Effect of extracellular pH on matrix synthesis by chondrocytes in 3D agarose gel. *Biotechnol. Prog.* **2007**, *23*, 430–434. [CrossRef]

5. Kaysinger, K.K.; Ramp, W.K. Extracellular pH modulates the activity of cultured human osteoblasts. *J. Cell. Biochem.* **1998**, *68*, 83–89. [CrossRef]

6. Ayuso, J.M.; Virumbrales-Munoz, M.; McMinn, P.H.; Rehman, S.; Gomez, I.; Karim, M.R.; Trusttchel, R.; Wisinski, K.B.; Beebe, D.J.; Skala, M.C. Tumor-on-a-chip: A microfluidic model to study cell response to environmental gradients. *Lab Chip* **2019**, *19*, 3461–3471. [CrossRef]

7. Wlodkowic, D.; Cooper, J.M. Tumors on chips: Oncology meets microfluidics. *Curr. Opin. Chem. Biol.* **2010**, *14*, 556–567. [CrossRef]

8. Hafeman, D.G.; Parce, J.W.; McConnell, H.M. Light-addressable potentiometric sensor for biochemical systems. *Science* **1988**, *240*, 1182–1185. [CrossRef] [PubMed]

9. Poghossian, A.; Schöning, M.J. Capacitive field-effect EIS chemical sensors and biosensors: A status report. *Sensors* **2020**, *20*, 5639. [CrossRef]

10. Yoshinobu, T.; Miyamoto, K.I.; Werner, C.F.; Poghossian, A.; Wagner, T.; Schöning, M.J. Light-addressable potentiometric sensors for quantitative spatial imaging of chemical species. *Annu. Rev. Anal. Chem.* **2017**, *10*, 225–246. [CrossRef] [PubMed]

11. Yoshinobu, T.; Schöning, M.J. Light-addressable potentiometric sensors (LAPS) for cell monitoring and biosensing. *Curr. Opin. Electrochem.* **2021**, *28*, 100727. [CrossRef]

12. Wu, F.; Zhang, D.W.; Wang, J.; Watkinson, M.; Krause, S. Copper contamination of self-assembled organic monolayer modified silicon surfaces following a "Click" reaction characterized with LAPS and SPIM. *Langmuir* **2017**, *33*, 3170–3177. [CrossRef] [PubMed]

13. Wagner, T.; Werner, C.F.; Miyamoto, K.I.; Schöning, M.J.; Yoshinobu, T. Development and characterisation of a compact light-addressable potentiometric sensor (LAPS) based on the digital light processing (DLP) technology for flexible chemical imaging. *Sens. Actuators B* **2012**, *170*, 34–39. [CrossRef]

14. Dantism, S.; Röhlen, D.; Dahmen, M.; Wagner, T.; Wagner, P.; Schöning, M.J. LAPS-based monitoring of metabolic responses of bacterial cultures in a paper fermentation broth. *Sens. Actuators B* **2020**, *320*, 128232. [CrossRef]

15. Özsoylu, D.; Isık, T.; Demir, M.M.; Schöning, M.J.; Wagner, T. Cryopreservation of a cell-based biosensor chip modified with elastic polymer fibers enabling ready-to-use on-site applications. *Biosens. Bioelectron.* **2021**, *177*, 112983. [CrossRef] [PubMed]

16. Yoshinobu, T.; Iwasaki, H.; Ui, Y.; Furuichi, K.; Ermolenko, Y.; Mourzina, Y.; Wagner, T.; Näther, N.; Schöning, M.J. The light-addressable potentiometric sensor for multi-ion sensing and imaging. *Methods* **2005**, *37*, 94–102. [CrossRef] [PubMed]

17. Yoshinobu, T.; Ecken, H.; Poghossian, A.; Simonis, A.; Iwasaki, H.; Lüth, H.; Schöning, M.J. Constant-current-mode LAPS (CLAPS) for the detection of penicillin. *Electroanalysis* **2001**, *13*, 733–736. [CrossRef]

18. Wu, C.; Poghossian, A.; Bronder, T.S.; Schöning, M.J. Sensing of double-stranded DNA molecules by their intrinsic molecular charge using the light-addressable potentiometric sensor. *Sens. Actuators B* **2016**, *229*, 506–512. [CrossRef]

19. Vogel, Y.B.; Gonçales, V.R.; Gooding, J.J.; Ciampi, S. Electrochemical microscopy based on spatial light modulators: A projection system to spatially address electrochemical reactions at semiconductors. *J. Electrochem. Soc.* **2017**, *165*, H3085. [CrossRef]

20. Suzurikawa, J.; Nakao, M.; Kanzaki, R.; Takahashi, H. Microscale pH gradient generation by electrolysis on a light-addressable planar electrode. *Sens. Actuators B* **2010**, *149*, 205–211. [CrossRef]

21. Suzurikawa, J.; Nakao, M.; Jimbo, Y.; Kanzaki, R.; Takahashi, H. A light addressable electrode with a TiO_2 nanocrystalline film for localized electrical stimulation of cultured neurons. *Sens. Actuators B* **2014**, *192*, 393–398. [CrossRef]

22. Welden, R.; Schöning, M.J.; Wagner, P.H.; Wagner, T. Light-addressable electrodes for dynamic and flexible addressing of biological systems and electrochemical reactions. *Sensors* **2020**, *20*, 1680. [CrossRef]

23. Vogel, Y.B.; Gooding, J.J.; Ciampi, S. Light-addressable electrochemistry at semiconductor electrodes: Redox imaging, mask-free lithography and spatially resolved chemical and biological sensing. *Chem. Soc. Rev.* **2019**, *48*, 3723–3739. [CrossRef] [PubMed]

24. Jablonski, M.; Poghossian, A.; Severins, R.; Keusgen, M.; Wege, C.; Schöning, M.J. Capacitive field-effect biosensor studying adsorption of *tobacco mosaic virus* particles. *Micromachines* **2021**, *12*, 57. [CrossRef]

25. Koch, C.; Poghossian, A.; Schöning, M.J.; Wege, C. Penicillin detection by *tobacco mosaic virus*-assisted colorimetric biosensors. *Nanotheranostics* **2018**, *2*, 184. [CrossRef]

26. Poghossian, A.; Jablonski, M.; Koch, C.; Bronder, T.S.; Rolka, D.; Wege, C.; Schöning, M.J. Field-effect biosensor using virus particles as scaffolds for enzyme immobilization. *Biosens. Bioelectron.* **2018**, *110*, 168–174. [CrossRef]

27. Koch, C.; Wabbel, K.; Eber, F.J.; Krolla-Sidenstein, P.; Azucena, C.; Gliemann, H.; Eiben, S.; Geiger, F.; Wege, C. Modified TMV particles as beneficial scaffolds to present sensor enzymes. *Front. Plant Sci.* **2015**, *6*, 1137. [CrossRef]

28. Koch, C.; Poghossian, A.; Wege, C.; Schöning, M.J. TMV-based adapter templates for enhanced enzyme loading in biosensor applications. In *Virus-Derived Nanoparticles for Advanced Technologies*, 1st ed.; Wege, C., Lomonossoff, G.P., Eds.; Humana Press: New York, NY, USA, 2018; Volume 1776, pp. 553–568. [CrossRef]

29. Welden, R.; Scheja, S.; Schöning, M.J.; Wagner, P.; Wagner, T. Electrochemical evaluation of light-addressable electrodes based on TiO_2 for the integration in lab-on-chip systems. *Phys. Status Solidi A* **2018**, *215*, 1800150. [CrossRef]

30. Imsande, J.; Gillin, F.D.; Tanis, R.J.; Atherly, A.G. Properties of penicillinase from *Bacillus cereus* 569. *J. Biol. Chem.* **1970**, *245*, 2205–2212. [CrossRef]
31. do Prado, T.M.; Foguel, M.V.; Goncalves, L.M.; Maria del Pilar, T.S. β-Lactamase-based biosensor for the electrochemical determination of benzylpenicillin in milk. *Sens. Actuators B* **2015**, *210*, 254–258. [CrossRef]
32. Le Formal, F.; Sivula, K.; Graetzel, M. The transient photocurrent and photovoltage behavior of a hematite photoanode underworking conditions and the influence of surface treatments. *J. Phys. Chem. C* **2012**, *116*, 26707–26720. [CrossRef]
33. Obata, K.; van de Krol, R.; Schwarze, M.; Schomäcker, R.; Abdi, F.F. In situ observation of pH change during water splitting in neutral pH conditions: Impact of natural convection driven by buoyancy effects. *Energy Environ. Sci.* **2020**, *13*, 5104–5116. [CrossRef]
34. Molinnus, D.; Muschallik, L.; Gonzalez, L.O.; Bongaerts, J.; Wagner, T.; Selmer, T.; Siegert, P.; Keusgen, M.; Schöning, M.J. Development and characterization of a field-effect biosensor for the detection of acetoin. *Biosens. Bioelectron.* **2018**, *115*, 1–6. [CrossRef] [PubMed]
35. Jablonski, M.; Münstermann, F.; Nork, J.; Molinnus, D.; Muschallik, L.; Bongaerts, J.; Wagner, T.; Keusgen, M.; Siegert, P.; Schöning, M.J. Capacitive field-effect biosensor applied for the detection of acetoin in alcoholic beverages and fermentation broths. *Phys. Status Solidi A* **2021**, 2000765. [CrossRef]

 *biosensors*

Article

Porous Silicon Biosensor for the Detection of Bacteria through Their Lysate [†]

Roselien Vercauteren [1,*], Audrey Leprince [2], Jacques Mahillon [2] and Laurent A. Francis [1]

[1] Electrical Engineering Department, Institute of Information and Communication Technologies Electronics and Applied Mathematics, UCLouvain, 1348 Louvain-la-Neuve, Belgium; laurent.francis@uclouvain.be

[2] Laboratory of Food and Environmental Microbiology, Earth and Life Institute, UCLouvain, 1348 Louvain-la-Neuve, Belgium; audrey.leprince@uclouvain.be (A.L.); jacques.mahillon@uclouvain.be (J.M.)

* Correspondence: roselien.vercauteren@uclouvain.be

† This paper is an extended version of our paper published in: Vercauteren, R.; Leprince, A.; Mahillon, J.; Francis, L.A. Porous Silicon Biosensor for the Detection of Bacteria Through Their Lysate. In Proceedings of the 1st International Electronic Conference on Biosensors, 2–17 November 2020.

Abstract: Porous silicon (PSi) has been widely used as a biosensor in recent years due to its large surface area and its optical properties. Most PSi biosensors consist in close-ended porous layers, and, because of the diffusion-limited infiltration of the analyte, they lack sensitivity and speed of response. In order to overcome these shortcomings, PSi membranes (PSiMs) have been fabricated using electrochemical etching and standard microfabrication techniques. In this work, PSiMs have been used for the optical detection of *Bacillus cereus* lysate. Before detection, the bacteria are selectively lysed by PlyB221, an endolysin encoded by the bacteriophage Deep-Blue targeting *B. cereus*. The detection relies on the infiltration of bacterial lysate inside the membrane, which induces a shift of the effective optical thickness. The biosensor was able to detect a *B. cereus* bacterial lysate, with an initial bacteria concentration of 10^5 colony forming units per mL (CFU/mL), in only 1 h. This proof-of-concept also illustrates the specificity of the lysis before detection. Not only does this detection platform enable the fast detection of bacteria, but the same technique can be extended to other bacteria using selective lysis, as demonstrated by the detection of *Staphylococcus epidermidis*, selectively lysed by lysostaphin.

Keywords: porous silicon membrane; bacterial detection; selective lysis; endolysins; lysostaphin; flow-through

Citation: Vercauteren, R.; Leprince, A.; Mahillon, J.; Francis, L.A. Porous Silicon Biosensor for the Detection of Bacteria through Their Lysate. *Biosensors* **2021**, *11*, 27. https://doi.org/doi:10.3390/bios11020027

Received: 24 December 2020
Accepted: 16 January 2021
Published: 20 January 2021

Publisher's Note: MDPI stays neutral with regard to jurisdictional claims in published maps and institutional affiliations.

1. Introduction

This paper is the extended version of the proceedings paper presented at the 1st International Electronic Conference on Biosensors, 2–17 November 2020 [1].

A biosensor allows for the fast detection and quantification of a biological analyte, without pre-enrichment steps. It is characterized by two components: a biological element, which is often key to the specificity, and a transducer [2]. The biological element frequently takes the form of a bioreceptor, which is bound to the surface of the transducer. This binding requires several steps of surface modification or functionalization, which can be complex, time-consuming and/or expensive. On top of that, functionalization can shorten the lifespan of the biosensor and often puts heavy requirements on the storage conditions. These bottlenecks can be avoided by adding the biological element into the sample volume instead of binding it to the transducer. Among these biological elements are endolysins: produced by bacteriophages or bacterial cells, they have the capability to specifically digest the cell wall of specific bacterial strains. They have proven to be powerful specificity means for the detection of bacteria [3–5].

The transducer can rely on electrical, optical, thermal, or magnetic signals. Optical transducers however outperform most physical transducers as there is no influence of the

15

nature of the sample or of external disturbances on the signal [6]. Optical biosensors can take many forms: waveguides [7], ring resonators [8], refractometers [9,10], surface plasmon resonators [11], optical fibers [12], or even photonic crystals [13]. Porous silicon (PSi) is one widely used optical transducer for biosensors [14]. Its benefits include a large surface area, unique optical properties and a low production cost. While PSi photonic crystal [15], ring resonators [16,17], and photoluminescent sensors [18] have been demonstrated, most PSi-based optical transducers act as interferometers and rely on changes in the effective optical thickness (EOT) of the porous layer as means of detection. The EOT depends on both the effective refractive index and the thickness of the porous layer and can be quantified using reflective interferometric Fourier transform spectroscopy (RIFTS) [19,20]. The RIFTS method consists in shining a halogen light perpendicularly to the sensor surface and measuring its reflection. The reflection takes the form of Fabry-Pérot fringes due to the interferences between the light reflections from the top and the bottom interfaces of the porous film. By applying a Fourier transform to this fringe pattern, it is possible to extract their frequency, which takes the shape of a peak. The position of this peak translates the EOT. When an analyte penetrates the porous matrix, it affects the refractive index of the layer medium and induces a wavelength shift in the interference pattern; this is translated by a shift of the EOT value.

Recent works using the RIFTS as transducing mechanism of PSi-based transducers target the detection of, for instance, bacterial surface proteins [21], heat shock protein 70 [22], or bovine mastitis biomarkers [23,24], but this method has also been applied for the detection of bacteria [25–27]. These bacterial detectors however lacked sensitivity, with detection limits unable to go below 10^3 colony forming units per mL (CFU/mL). Current bacteria detection techniques, such as polymerase chain reaction (PCR) or enzyme-linked immunosorbent assay ELISA, while time-consuming, can easily reach detection limits of 10 CFU/mL [28]. This insufficient sensitivity is attributed to the hindered diffusion of bacteria into the porous matrix. Solutions, such as the electrokinetic transport for the preconcentration of analyte, have been proposed but remain to be tested on bacteria [29]. Recently, porous silicon interferometers have also been combined with gold nanoparticles for localized surface plasmon spectroscopy (LSPS), enhancing the fringe pattern contrast and increasing the sensitivity of the porous layer [30,31].

Another approach to increase the sensitivity of PSi biosensor is the fabrication of PSi membranes (PSiMs). Instead of a close-ended PSi layer over which the analyte must flow, an open-ended PSi membrane allows the analyte to flow through the porous matrix [32]. PSi membranes have been fabricated as early as the 1990s [33], but the interest in PSiMs for sensing applications only arose recently [34–38]. PSiMs can be self-supported, meaning still partly attached to the silicon substrate, or freestanding. Recently, lateral PSiMs have also been fabricated [39–41]. For biosensing applications, PSiMs have been found to increase both the response time and amplitude of the sensors [34–38].

In this work, we combine the benefits of selective endolysins and PSiM-based transducers for the fabrication of an innovative biosensor which enables the fast and label-free detection of bacteria through their lysate. The biosensing platform operates in two steps: a first step is the selective lysis of the targeted bacteria by an endolysin in a vial; and secondly, the optical monitoring of the bacterial lysate filtering through a PSiM using the RIFTS method. We demonstrate this concept with a selective optical detection of *Bacillus cereus* lysate in PBS using the recently characterized PlyB221 endolysin, encoded by the Deep-Blue phage targeting *B. cereus* [42]. The specificity is confirmed by replacing the targeted bacteria with *Staphylococcus epidermidis*, for which no optical detection was observed. To illustrate the versatility of our detection platform, the targeted bacteria strain was then switched to *S. epidermidis*, using lysostaphin as selective lytic agent.

2. Materials and Methods

2.1. Materials

Double-side polished, boron-doped silicon wafers ($\langle 100 \rangle$, 0.8–0.9 mΩ cm, 380–400 μm) were purchased from Sil'tronix Silicon Technologies (France). Aqueous hydrofluoric acid (HF, 49%) was acquired from Chem-Lab, NV (Belgium), and absolute ethanol was obtained from VWR Chemicals (France). Phosphate buffered saline (PBS, 0.01 M phosphate, pH 7.4) and lysostaphin were purchased from Sigma-Aldrich (USA).

2.2. Fabrication of Porous Silicon Layers

The PSi layer samples were prepared by the electrochemical etch of a heavily doped *p*-type silicon substrate. The etching was carried out in a custom-made Teflon$^{®}$ single bath etch-cell, with a platinum coil as the counter-electrode and a potentiostat/galvanostat (PGSTAT302N from Metrohm Belgium) as the current source. The porosification was performed in HF:ethanol (3:1, in volume) electrolyte. The first step of the anodization consisted in etching a sacrificial layer at 200 mA/cm^2 for 30 s and removing it with a 2 M solution of KOH until no more reaction was visible. This sacrificial layer removes the transitional layer and obtains a more homogeneous pore size with depth. The sample was then rinsed once in deionized water and twice in 2-propanol before being etched again at 200 mA/cm^2 for 50 s. The porous samples were thermally oxidized in an oven for 30 min at 350 °C under an oxygen flow (1.2 L/min).

2.3. Fabrication and Characterization of Porous Silicon Membranes

Figure 1 sketches the process flow, which was inspired by the work from Zhao et al. [34]. First, 3-in. highly doped *p*-type silicon wafers were cleaned in a freshly prepared piranha solution (H_2O_2:H_2SO_4, 2:5), followed by two immersions in continuously flowing deionized (DI) water during 20 min. Afterwards, 500 nm of silicon nitride (Si_3N_4) was deposited using Plasma Enhanced Chemical Vapor Deposition (PECVD). To improve the chemical resistance of the nitride layer to HF, the wafers were annealed at 900 °C in ambient air for 3 h. A first i-line optical lithography with positive resist (AZ$^{®}$ MiR^{TM} 701, MicroChemicals GmbH) provided masking for the subsequent Reactive Ion Etching (RIE) of the silicon nitride layer. The patterned nitride layer itself served as a mask during the electrochemical etch of the silicon. The porosification followed the same protocol as explained in Section 2.2, but used three different current densities in order to obtain different porosities: the sensing layer was etched at 200 mA/cm^2 for 50 s. This was followed by a 1500 s-etch at 50 mA/cm^2, making a thick optical contrast layer characterized by lower porosity, enabling the reflection of the light required for the RIFTS method; finally, a thick mechanical support layer was etched at 100 mA/cm^2. The current densities chosen for each layer have been optimized such as to guarantee a good mechanical stability and optical signal, while retaining pores large enough for the flow-through operation. The porous multilayers were passivated by thermal oxidation, for 30 min at 350 °C under an oxygen flow of 1.2 L/min. A second optical lithography was then performed on the backside of the wafers, where a thick positive resist (AZ$^{®}$ 9260, MicroChemicals GmbH) was patterned in alignment with the frontside. The thick resist served as a mask during the final step of the process, the deep reactive ion etching (DRIE) of the backside of the wafer, until the porous silicon was visible and the membranes were open.

The entire fabrication process was performed in less than a week, and could be repeated several times. Slight variations in the porous structures can happen, which can be linked to the manual positioning of the platinum electrode.

The membranes were characterized using scanning electron microscopy (SEM), both in cross section and in top view. Based on the top views, the pore size distribution could be analyzed using the ImageJ software. The porosity of each layer was determined using the spectroscopic liquid infiltration method (SLIM). In brief, the optical spectrum of a porous layer was recorded both in air and in ethanol. Using the RIFTS method described above, the EOT was calculated. Knowing the refractive indices of air, ethanol and silicon, these data were then fitted using a two-component Bruggeman effective medium approximation in order to obtain an approximation of the open porosity and the layer thickness. The experimental set up used for the SLIM method consisted in a fiber-coupled Ocean Optics JAZ spectrometer and a halogen light source.

Using the EOT measured in both air and ethanol allowed to approximate the theoretical sensitivity of the biosensor [10,31].

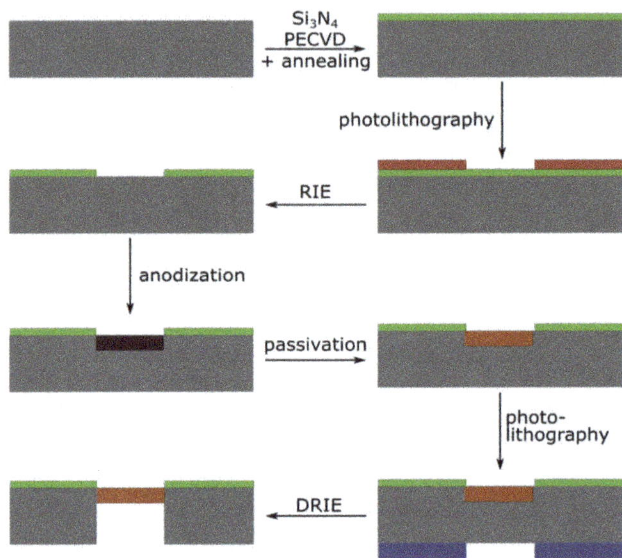

Figure 1. Schematic illustration of the process flow for the fabrication of a porous silicon membrane. Starting from a cleaned 3″ highly doped silicon wafer, the process goes through the following steps: deposition of Si₃N₄ layer using Plasma Enhanced Chemical Vapor Deposition (PECVD) and annealing; positive photolithography on the frontside; opening of the nitride layer using Reactive Ion Etching (RIE); formation of the porous silicon layer by anodization; passivation by thermal oxidation; positive photolithography on the backside; and finally opening of the membrane using Deep Reactive Ion Etching (DRIE).

2.4. PlyB221 Endolysin Expression and Purification

A detailed description of the expression and purification of PlyB221 endolysin can be found elsewhere [42]. The protein concentration was adjusted to 1 mg/mL.

2.5. Bacterial Strains, Growth Conditions

B. cereus ATCC 10987 was used as reference strain and *S. epidermidis* ATCC 35984 as negative control in the PlyB221 endolysin experiments. *S. epidermidis* ATCC 35984 was also used as target when lysostaphin was applied as selective lytic agent. Bacteria were grown overnight (O/N) in Lysogeny Broth (LB) or LB-agar plates at 30 °C for *B. cereus* and in Tryptic Soy Broth (TSB) or Tryptic Soy Agar (TSA) plates at 37 °C for *S. epidermidis*. In brief, 20 mL of LB or TSB were inoculated with 200 µL of each culture and incubated for 3 h at 30 °C (*B. cereus*) or 37 °C (*S. epidermidis*). The cultures were then centrifuged at 10,000× g for 5 min at room temperature and the supernatants were resuspended in 20 mL of PBS. This washing step was repeated once over and the optical density (OD$_{600}$) was adjusted to OD$_{600}$ = 0.2 (~10^6 CFU/mL) for *B. cereus* and OD$_{600}$ = 0.02 (~10^6 CFU/mL) for *S. epidermidis*. For the determination of the limit of detection, the *B. cereus* suspension was diluted 10 times twice, in order to obtain concentrations of ~10^5 CFU/mL and ~10^4 CFU/mL.

2.6. Lysate Observation and Characterization

B. cereus suspension and lysate were captured on a silicon surface to enable their observation using SEM. *A B. cereus* suspension was prepared as described above and adjusted to the concentrations of ~10^6 CFU/mL. The lysate was prepared by adding 600 µL of PlyB221 endolysin (1mg/ml) to 5.4 mL of bacterial suspension and by incubating this mixture and 30 °C for 30 min. Silicon dies of dimension 1 cm × 1 cm were place inside a 12 wells plate. The wells were filled with 2 mL of one of three solutions: a control solution consisting of PBS, the *B. cereus* suspension or the *B. cereus* lysate. The 12 wells plate was then incubated at 30 °C for 1 h, after which each die was rinsed 5 times in PBS by removing and adding 1 mL of PBS. After the last PBS wash, 1 mL of solutions remained in each well. A glutaraldehyde solution was added to each well in order to reach a final concentration of 2.5 %. The 12 wells plate was left at room temperature for 1 h, enabling the crosslinking of the bacterial cell walls. The samples were then washed three times with PBS using the same technique as explained before, making sure that the dies were never exposed to air. For the dehydration, the samples were then immersed in DI water solution of increasing ethanol content (25%, 50%, 75%, and finally 99.9%). Each immersion lasted 10 min. The silicon dies were then dried overnight at 58 °C. Directly before SEM observation, the samples were covered with a ~10 nm-thick layer of gold to prevent charging effects. The images were then analyzed with ImageJ to obtain information about the number and size of the bacteria and bacteria lysate.

2.7. Experimental Setup and Optical Reflectivity Measurements

PSiM samples were integrated in a custom-built polycarbonate fluidic cell. A fiber-coupled Ocean Optics JAZ spectrometer and a 10-mW halogen light source were used to record reflectivity spectra. Data were recorded every 10 s, with a spectral acquisition time of 1s over a wavelength range of 500–800 nm. Analytes were injected at flow speed of 15 to 20 µL/min using a Fluigent LINEUP™ fluidic set up. The obtained optical data were analyzed using the RIFTS method in order to obtain the effective optical thickness, EOT = 2 *nL*, with *n* being the refractive index and *L* the porous layer thickness. The relative change in EOT overtime was computed as a percentage, such that

$$\frac{\Delta EOT}{EOT_0} = \frac{EOT_t - EOT_0}{EOT} \times 100[\%].$$

The significance of the relative EOT shift was then established using a Student's t-test with a 5% confidence level, with a negative control test in PBS as reference.

2.8. Real-Time Detection of B. cereus in PBS on PSi Layer and PSi Membranes

The protocol for bacteria detection on a PSi membrane is illustrated in Figure 2. First 500 µL of purified PlyB221 endolysin were added to 4.5 mL of exponential phase *B. cereus* resuspended in PBS, so as to reach a final protein concentration of 100 µg/mL. The 5 mL final volume was sufficient for at least 4 detections. The suspension was then incubated for 30 min at 30 °C. Before flowing the bacterial lysate, PBS solution was injected at 15 to 20 µL/min for 60 min and reference measurements were performed. *B. cereus* lysate suspensions were injected at the same flow speed. Optical measurements were carried out every 10 s for 60 min. The relative EOT was then extracted from these measurements using the method described above. A control test, with only the PlyB221 endolysin at the same concentration was also performed on both types of sensors, following the same protocol described previously. For all tests, measurements were performed at least 3 times.

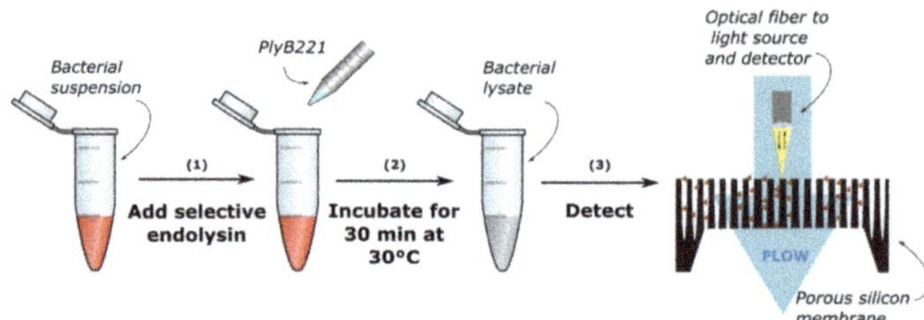

Figure 2. Protocol of bacteria detection through their lysate: (**1**) lysis of the bacteria using a selective endolysin, (**2**) incubation for 30 min and (**3**) optical detection on a porous silicon membrane.

2.9. Specificity Testing: Detection of S. epidermidis in PBS with the PlyB221 Endolysin on PSi Membranes

For this experiment, two tests were performed: a negative one using a *S. epidermidis* suspension and a positive control using a complex sample containing both *S. epidermidis* and *B. cereus*. For each test, 4.5 mL of bacterial suspension were incubated for 30 min at 30 °C with the 500 µL of PlyB221 endolysin (1 mg/mL). This final volume was sufficient for at least 4 detections. Reference measurements in PBS were performed for 5 to 15 min. The bacterial suspension was injected at a flow speed of 15 to 20 µL/min and optical measurements were performed every 10 s for 60 min. The relative EOT shift was computed as described previously.

2.10. Versatility of the Platform: Detection of S. epidermidis in PBS with Lysostaphin on PSi Membranes

In this experiment, a different lytic enzyme-bacteria pair was tested. *S. epidermidis* was selected as the targeted strain and lysostaphin was chosen as selective agent. This glycyl-glycine endopeptidase is specific of the pentaglycine bridges present in the cell wall of certain Staphylococci. For the detection experiments, 1 mg of commercialized lysostaphin was diluted in 1 mL of PBS supplemented with 30% of glycerol to reach a concentration of 20 µM. For the lysis, 150 µL of this 20 µM lysostaphin was added to 2.850 mL of bacterial suspension, yielding a final endolysin concentration of 1 µM [4]. The suspension was then incubated at 37 °C for 30 min. Reference measurements in PBS and in lysostaphin suspensions were performed for 1 h each. The bacterial suspension was injected at a flow speed of 15–20 µL/min and optical measurements were performed every minute for 60 min. The relative EOT shift was computed as described previously.

3. Results

3.1. PSi-Based Biosensor Characterization

The effective optical thickness of Porous Silicon is strongly dependent on its refractive index. When an analyte penetrates the pores, the refractive index of the porous layer increases, therefore inducing a shift in the EOT. In the case of a membrane, to make sure that the analyte remains inside the porous matrix, two approaches are possible: size exclusion or binding to the pore wall. For this project, size exclusion was chosen by accordingly selecting different pore size for each layer of the membrane. The first layer, also called the sensing layer, has an average pore size 41.05 nm, as described in Table 1 and illustrated in Figure 3. Larger pores could be etched, but the resulting layer was too easily damaged, as an increase in pore size induced a decrease of the pore wall thickness. The selected pore size is a trade-off between pore opening and mechanical integrity. The bacterial lysates are composed, among others, of cell wall fragments, DNA and RNA molecules, and cytoplasmic liquid and ribosomes, which are assumed to penetrate the sensing layer. In order to keep the bulkier ones in the top layer, the contrast layer was designed to have a smaller pore size, as indicated in Table 1. These choices in pore size also have two other motivations: (1) they allow the PlyB221 endolysin to flow through the membrane and not be retained in the sensing layer and (2) they prevent the non-lysed bacteria from penetrating the membrane, therefore enabling a selective detection. Figure 4 depicts a PSi membrane, with an up-close view of the transition between the top sensing layer and the contrast layer.

Table 1. Pore diameter, thickness and porosity measured for each of the three layers of the porous silicone membrane (PSiM).

Layer	Current Density [mA/cm^2]	Time [s]	Pore diameter [nm]	Thickness [μm]	Porosity [%]
Sensing layer	200	50	41.05 ± 20.4	4.09 ± 0.7	75.4
Contrast layer	50	1500	14.6 ± 7.8	22.8 ± 6.8	48.5
Support layer	100	2000	25.5 ± 10.4	- *	- *

* The thickness and porosity could not be accurately measured, as part of the layer was etched away during the DRIE step.

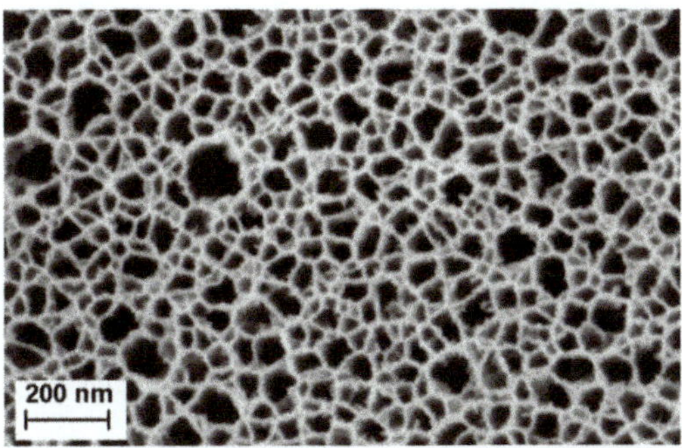

Figure 3. Scanning electron microscopy (SEM) image of the top layer porous silicon membrane. The average pore size is ~41 nm, with a standard deviation of ~20 nm.

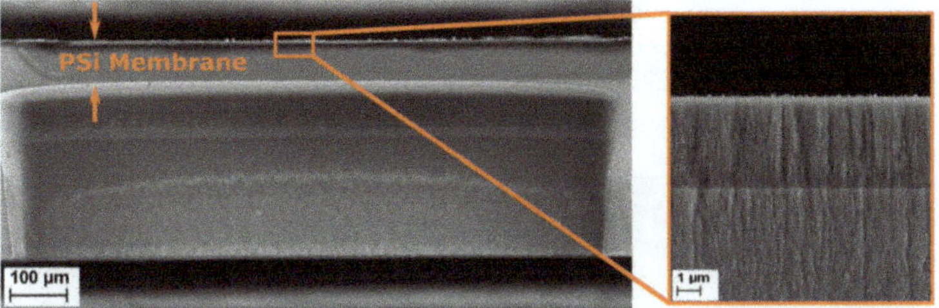

Figure 4. Scanning electron microscopy (SEM) cross-section image of a multilayered porous silicon membrane. The inset shows a close-up of the top of the porous membrane, in which the sensing and the contrast layers can be identified.

The theoretical sensitivity of the porous membrane sensor was also approximated by calculating the EOT in both air and in ethanol, applying a Fourier transform to the Fabry Perot fringes of the optical spectra, as illustrated in Figure 5. By plotting the EOT versus refractive index variation (Figure 5c), an approximation of the sensitivity can be made, which amounts to 5745.8 ± 847.7 nm·RIU^{-1} or, expressed in relative changes of EOT, as 54.4 ± 8.3 %·RIU^{-1}.

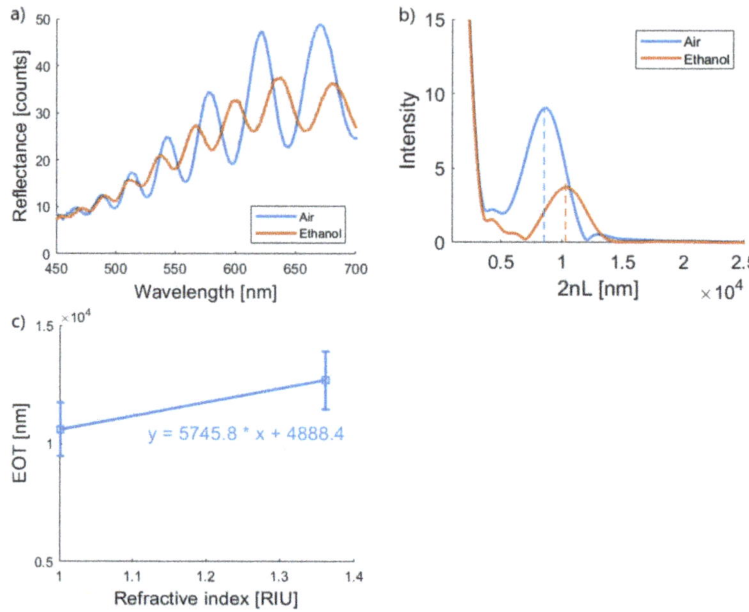

Figure 5. Porous silicon membrane characterization. (**a**) Reflection spectra of a porous silicon membrane in air (blue) and ethanol (orange). (**b**) Representative Fourier transform of the reflectance spectra in panel. (**c**). Calibration curve of the EOT versus the refractive index variation.

3.2. B. cereus Lysate Observation and Characterization

Endolysins are encoded and used by bacteriophages at the end of their replication cycle. They degrade the peptidoglycan of the targeted bacteria, creating an opening in the cell wall for the phage. When endolysins are recombinantly produced and added exogenously to bacteria, they lead to cell lysis by breaking down the exposed peptidoglycan layer which make them promising antimicrobial agents [42]. To observe the effect of the PlyB221 endolysin, *B. cereus* was observed using SEM before and after the lysis. Close up images of an intact versus a lysed bacterium are depicted in Figure 6.

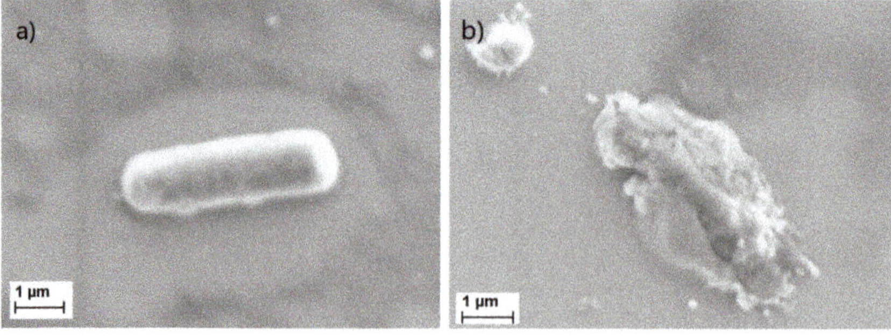

Figure 6. Scanning electron microscopy (SEM) image of (**a**) an intact *B. cereus* bacterium and (**b**) a *B. cereus* lysed bacterium.

To further characterize the bacterial lysate, the average number and size of bacterial clusters were compared before and after lysis, using images of the same magnification. Due to a lack of contrast and sharpness, only qualitative observations could be made. Over the same area of inspection, there are nearly 20 times more clusters of bacterial lysate than of bacteria; the average size of each cluster is also decrease by more than a 10-fold.

As illustrated in Figure 6, some larger clusters of bacterial lysate remain, but these are surrounded by much smaller clusters, whose area goes down to the nanometer range (<50 nm). These smaller clusters are assumed to penetrate the first porous layer, but are unable to diffuse to the second porous layer because of a size exclusion effect.

3.3. B. cereus Lysate Detection PSi Layers and PSi Membranes

In order to establish the added-value of a flow-through approach to the optical detection with respect to the traditional flow-over approach, the performances of PSi membranes were compared to those measured on PSi layers.

The average relative EOT shift measured on PSi layers, in a flow-over approach, is presented in Figure 7a. No distinction could be made between the bacterial lysate detection and the control tests using either the buffer or the endolysin suspension. Both control tests induced minute decreases of the relative EOT: on average −0.24% and −0.01% after 1 h for PBS and the PlyB221 endolysin, respectively. A small increase of relative EOT was measured in the presence of bacteria lysate, namely 0.05%, but this decrease was not significant when compared to the noise level. This noise level was calculated based on the overall standard deviation of the signal in PBS and was expressed as $3\sigma = 1.08\%$.

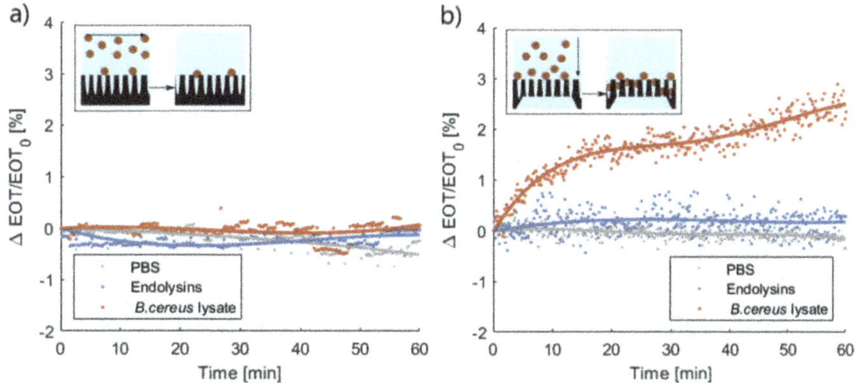

Figure 7. Characteristic relative effective optical thickness (EOT) shift measured on (**a**) a porous silicone (Psi) layer or (**b**) a PSi membrane for 1 h in phosphate buffered saline (PBS), in a PlyB221 endolysin suspension and in a *B. cereus* lysate ($n \geq 3$). In (**a**), the inset illustrates the flow-over approach; there was no significant shift visible during the detection of either the PlyB221 endolysin or bacterial lysate. In (**b**), the inset illustrates the flow-through approach; there was a significant shift in the relative EOT during the detection of bacterial lysate.

The performances of PSi membranes for the same three tests in flow-through operation are illustrated in Figure 7b. The buffer control test induced a decrease in relative EOT of −0.32% after 1 h. The control test with only the PlyB221 endolysin gave rise to a 0.28% increase of relative EOT. This increase was however not significant and remained below the noise level, which was calculated in the same manner described previously and was equal to 0.84% in the case of PSi membranes. Upon the penetration of bacterial lysate inside the membrane, a significant increase of relative EOT was measured, exceeding the noise level after 6 min and reaching an average +2.43% after 1 h.

3.4. Determination of the Limit of Detection of B. cereus

The limit of detection was determined using decreasing concentration of *B. cereus*. Results are displayed in Figure 8. As presented above, the detection of 10^6 CFU/mL is significant compared to the negative control tests in PBS and in endolysin only. For a concentration of 10^5 CFU/mL, the relative EOT shift amounted 0.96%, which is just above the noise level but still significantly different to both control tests. For 10^4 CFU/mL, the relative EOT shift decreased below the noise level to 0.64%. While this difference remains significant with respect to the control test in PBS, it is not with respect to the endolysin control test.

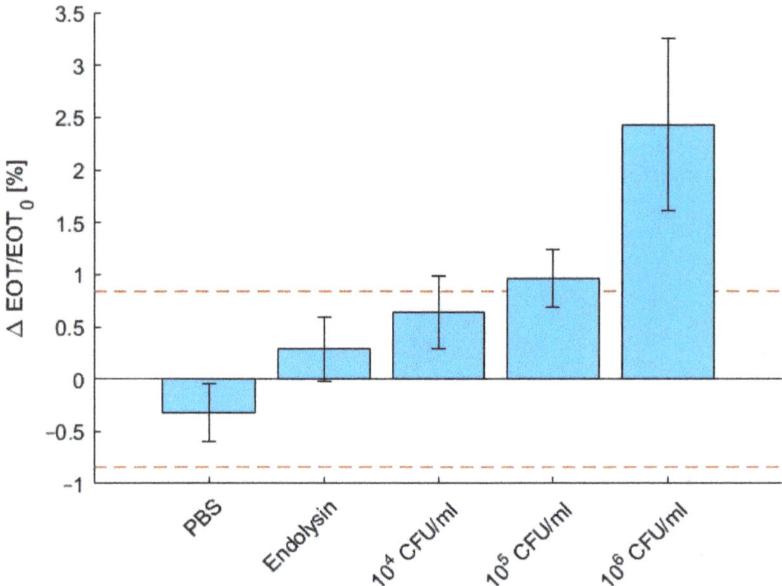

Figure 8. Characteristic relative effective optical thickness (EOT) shift measured on a PSi membrane after 1 h in PBS, in a PlyB221 endolysin suspension and in increasing concentrations of *B. cereus* lysate ($n \geq 3$). The dashed red line represents the noise level, fixed as 3σ of the signal measured in PBS. The detection limit is 10^5 CFU/ml of *B. cereus*.

3.5. Specificity Testing with S. epidermidis

To illustrate the specificity of the sensing platform, a negative control test was performed using a *S. epidermidis* and a positive control test was carried out using a mixture of *B. cereus* and *S. epidermidis*. The relative EOT shifts observed for both tests are depicted in Figure 9, where it is compared to controls in PBS and with the PlyB221 endolysin only, as well as to the detection of *B. cereus* lysate only. The detection using *S. epidermidis* induced no significant shift in relative EOT. After 1 h, the relative EOT was increased by 0.29%, which is comparable to the control test with only the PlyB221 endolysin. The positive control test showed a significant relative EOT increase of 1.59%.

3.6. Versatility of the Platform: Detection of S. epidermidis in PBS with Lysostaphin

To illustrate that the biosensor can be used for the detection of other bacteria/lytic enzyme pairs, the same detection protocol was repeated using *S. epidermidis* as targeted strain and lysostaphin as selective lytic agent. The relative EOT shift was measured overtime and compared to control tests in PBS and lysostaphin only. As depicted on Figure 10, the relative EOT shift induced after 1 h by the penetration of lysostaphin into the porous membrane is comparable to the one measure for the PlyB221 endolysin and amounts 0.26%. The penetration of *S. epidermidis* lysate causes a significant shift in relative EOT after 30 min, reaching a value of 1.83% after 1 h.

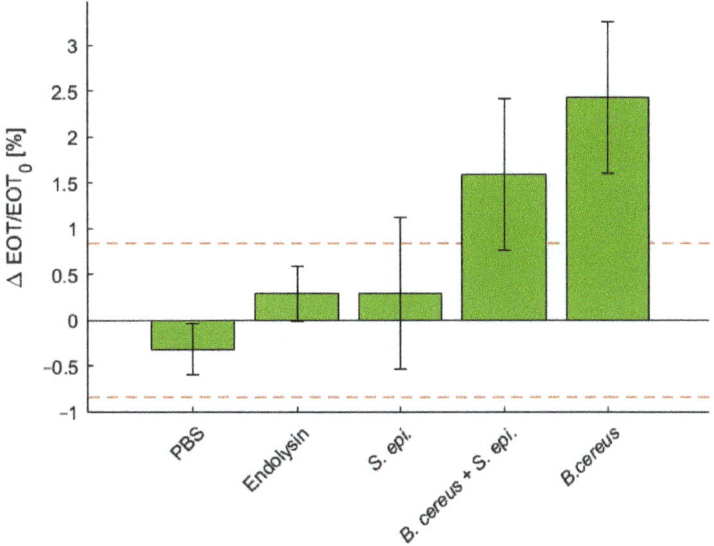

Figure 9. Characteristic relative EOT shift measured on a PSi membrane after 1 h in PBS, in a PlyB221 endolysin suspension, in a *S. epidermidis* suspension, in a mixture of *B. cereus* lysate and *S. epidermidis* and in *B. cereus* lysate only ($n \geq 3$). The dashed red line represents the noise level, fixed as 3σ of the signal measured in PBS. Only the detection of mixture of *B. cereus* lysate and *S. epidermidis* induced a significant shift.

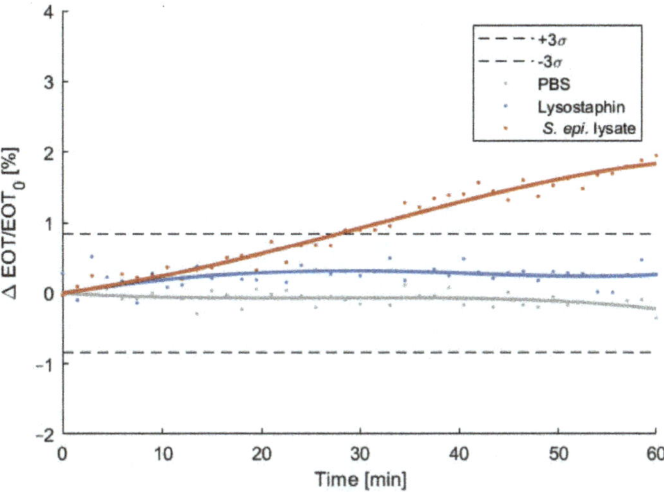

Figure 10. Relative effective optical thickness (EOT) shift measured on a PSi membrane for 1 h in PBS (grey), in a lysostaphin suspension (blue) and in a *S. epidermidis* lysate (red). The bacterial lysate is detected after 30 min.

4. Discussion

Porous silicon membranes are promising biosensors: with their flow-through operation, they overcome the lack of sensitivity of flow-over PSi-based biosensors and are characterized by short response times [34–38]. Moreover, they do not require any stratagems to transport and concentrate the analyte on the transducer. In this work, we demonstrate once again the potential of these detection platforms and focus on their use for the detection of bacteria. PSiMs were fabricated using standard microfabrication techniques and electrochemical etching. The full fabrication process took less than a week, and dozens of samples could be produced in one attempt. No functionalization was applied to the sensors, as the specificity was based on the use of endolysins. Endolysins are phage-encoded enzymes that induce bacterial lysis for certain targeted strains. In this work, the PlyB221 endolysin was used, targeting *B. cereus,* whose efficiency is demonstrated in [42]. The physical effect of PlyB221 endolysin on the bacteria was observed using SEM. While large clusters of bacterial lysate remained, they were surrounded by nanoscale-sized clusters. It is assumed that these small clusters are able to penetrate the porous membrane and enable a detection.

Combining PSiMs and the use of endolysins produced an innovative biosensing platform that was able to detect *B. cereus* lysate, with an initial concentration of 10^6 CFU/mL, in less than 10 min. In flow-over PSi-based biosensors, similar concentrations could not be detected. Two negative control tests were carried out: one in PBS and one with the PlyB221 endolysin only. The PBS control induced a slight decrease of relative EOT. This can be explained by the slow oxidation and dissolution that PSi undergoes in aqueous media [43,44]. This effect can be minimized by chemically modifying the pore surface using either hydrosilylation [45], thermal hydrocarbonization [46] or atomic layer deposition (ALD) of oxides [47]. Adequate passivation may also help reducing the noise level of the optical sensing, as demonstrated by Rasson et al. with the use of ALD [47]. The second control test, which consisted in flowing a PlyB221 endolysin suspension through the sensor for 1 h, resulted in a slight increase of the relative EOT. While it was expected that the endolysins pass through the membrane, this increase indicates that they were partly retained. We believe two hypotheses might explain this effect: (1) a minor size exclusion effect, which is understandable since there is a large pore size distribution visible in Figure 3 and (2) the binding of proteins to the pore walls.

In order to determine the detection limit, the initial concentration of bacteria was reduced. The sensor was able to detect concentrations as low as 10^5 CFU/mL after 1 h. While concentration of 10^4 CFU/mL induce a significant increase in signal compared to the PBS control test, this increase was however not significant with respect to the endolysin control test. Once again, adequate passivation may help to reduce the deviation between measurements and enable the lowering of the detection limit to 10^4 CFU/mL or lower.

While the specificity of the PlyB221 endolysin has already been demonstrated [42], the specificity of the sensor was still exemplified by adding *S. epidermidis* to the PlyB221 endolysin suspension. The observed shift in relative EOT is similar to the one observed with the PlyB221 endolysin only. This confirms that bacteria, when not lysed, are not able to penetrate the membrane, thus demonstrating the specificity of the sensing platform. A second test was performed adding both *B. cereus* and *S. epidermidis* to the endolysins and the results showed a significant increase in relative EOT, but lower than the one measured for *B. cereus* lysate only. We believe this might be explained by the accumulation of *S. epidermidis* on top of the sensor, blocking part of the pore and preventing the penetration of *B. cereus* lysate. This accumulation of intact bacteria may also explain the larger deviation between measurements when in the presence of *S. epidermidis*: part of the light is scattered by the bacteria, therefore decreasing the optical signal and adding deviations during the fit of the EOT. A solution to minimise this effect is the addition of gold nanoparticles to the porous matrix, enhancing the optical signal by increasing surface reflectivity and reducing EOT fitting deviations [30].

While most PSiM-based sensors rely on functionalization to capture the analyte [34,37,38], the biosensor presented in this work bases the capture on a size exclusion effect. This novelty has several benefits: the lack of functionalization considerably shortens the production time of the sensors and puts no requirements on the storage and detection conditions. This choice of design was made with future industrial specifications in mind: without the constraints of functionalization, a silicon-based device such as the one presented, which is compatible with all standard fabrication techniques used in microelectronic cleanrooms today, could be easily mass produced and packaged, for a very low cost. The lack of functionalization also enables the use of sensors from the same batch for the detection of different strains of bacteria, as long as a selective endolysin is available. This versatility was illustrated by the detection of *S. epidermidis* using lysostaphin as selective lytic agent.

Besides specificity, the performance of our optical biosensor can be discussed in other terms: response time, sensitivity and limit of detection [10]. The response time for this study is comparable to functionalized PSiMs [34,37,38] and since no rinsing is required to remove unbound species, the total detection time (which includes both lysis and optical monitoring) can be reduced to less than 2 h. Another major advantage is the reduced volume of analyte that is needed: only 1 mL is required for the detection. The approximated theoretical sensitivity of 5745.8 ± 847.7 nm·RIU^{-1} is high, but due to the noisy signal, the theoretical limit of detection is also quite high; this value was extrapolated as the RIU value at which the relative EOT shift is equal to the noise level 3σ and amounts to 1.5×10^{-2} RIU [30]. In terms of bacterial concentration, a limit of detection of 10^5 CFU/mL remains high compared to other PSi bacteria sensors [25,26,48,49], yet there is much room for improvement. Further studies should therefore aim to lower the detection limit to competitive levels ($<10^3$ CFU/mL) by improving the stability of the sensor and reducing the noise level of the optical signal. Once an optimized prototype is available, real samples may be analysed. A complex microfluidic integration may also enable the optical monitoring of several membranes in parallel, adding the possibility for a blank control test (with the analysed sample only) to exclude false positives, as well as for the simultaneous screening of several lytic agents.

5. Conclusions

In conclusion, we demonstrated the proof-of-concept of an innovative biosensor for the detection of bacteria through their lysate, by combining the use of both optically monitored porous silicon membranes and selective enzymatic lytic agents. We showed promising results in terms of sensitivity, specificity and speed of response by confirming the targeted detection of 10^5 CFU/ml of *B. cereus* lysate in only one hour. Furthermore, we illustrated the versatility of the detection platform by detecting selectively lysed *S. epidermidis*. These results, added to the easy fabrication and low design cost, pave the way for the development of a widespread multi-strain bacteria sensor.

Author Contributions: Conceptualization, R.V. and L.A.F.; methodology, R.V. and A.L.; software, R.V.; validation, R.V.; formal analysis, R.V.; investigation, R.V. and A.L.; resources, J.M. and L.A.F.; data curation, R.V.; writing—original draft preparation, R.V.; writing—review and editing, A.L., J.M., and L.A.F.; visualization, R.V.; supervision, J.M. and L.A.F.; project administration, J.M. and L.A.F.; funding acquisition, R.V., J.M., and L.A.F. All authors have read and agreed to the published version of the manuscript.

Funding: This research was funded by National Fund for Scientific Research (FRIA grant to R.V. and FNRS grant to J.M.) and the Research Department of the Communauté française de Belgique (Concerted Research Action, ARC no.17/22-084, grant to A.L.).

Institutional Review Board Statement: Not applicable.

Informed Consent Statement: Not applicable.

Data Availability Statement: The data presented in this study are openly available in Zenodo at 10.5281/zenodo.4446911, reference number data.xlsx.

Acknowledgments: We would like to thank Jonathan Rasson for his help during the conceptualization stages of this project.

Conflicts of Interest: The authors declare no conflict of interest.

References

1. Vercauteren, R.; Leprince, A.; Mahillon, J.; Francis, L.A. Porous Silicon Biosensor for the Detection of Bacteria through Their Lysate. *Proceedings* **2020**, *60*, 36. [CrossRef]
2. Lazcka, O.; Campo, F.J.D.; Muñoz, F.X. Pathogen Detection: A Perspective of Traditional Methods and Biosensors. *Biosens. Bioelectron.* **2007**, *22*, 1205–1217. [CrossRef] [PubMed]
3. Zourob, M.; Ripp, S. Bacteriophage-Based Biosensors. In *Recognition Receptors in Biosensors*; Zourob, M., Ed.; Springer: New York, NY, USA, 2010; pp. 415–448. ISBN 978-1-4419-0919-0.
4. Couniot, N.; Vanzieleghem, T.; Rasson, J.; Van Overstraeten-Schlögel, N.; Poncelet, O.; Mahillon, J.; Francis, L.A.; Flandre, D. Lytic Enzymes as Selectivity Means for Label-Free, Microfluidic and Impedimetric Detection of Whole-Cell Bacteria Using ALD-Al_2O_3 Passivated Microelectrodes. *Biosens. Bioelectron.* **2015**, *67*, 154–161. [CrossRef] [PubMed]
5. Paczesny, J.; Richter, Ł.; Hołyst, R. Recent Progress in the Detection of Bacteria Using Bacteriophages: A Review. *Viruses* **2020**, *12*, 845. [CrossRef] [PubMed]
6. Chen, C.; Wang, J. Optical Biosensors: An Exhaustive and Comprehensive Review. *Analyst* **2020**, *145*, 1605–1628. [CrossRef]
7. Kozma, P.; Kehl, F.; Ehrentreich-Förster, E.; Stamm, C.; Bier, F.F. Integrated Planar Optical Waveguide Interferometer Biosensors: A Comparative Review. *Biosens. Bioelectron.* **2014**, *58*, 287–307. [CrossRef]
8. Steglich, P.; Hülsemann, M.; Dietzel, B.; Mai, A. Optical Biosensors Based on Silicon-On-Insulator Ring Resonators: A Review. *Molecules* **2019**, *24*, 519. [CrossRef]
9. Chen, Y.; Liu, J.; Yang, Z.; Wilkinson, J.S.; Zhou, X. Optical Biosensors Based on Refractometric Sensing Schemes: A Review. *Biosens. Bioelectron.* **2019**, *144*, 111693. [CrossRef]
10. Chiavaioli, F.; Gouveia, C.A.J.; Jorge, P.A.S.; Baldini, F. Towards a Uniform Metrological Assessment of Grating-Based Optical Fiber Sensors: From Refractometers to Biosensors. *Biosensors* **2017**, *7*, 23. [CrossRef]
11. Prabowo, B.A.; Purwidyantri, A.; Liu, K.-C. Surface Plasmon Resonance Optical Sensor: A Review on Light Source Technology. *Biosensors* **2018**, *8*, 80. [CrossRef]
12. Zhao, Y.; Tong, R.; Xia, F.; Peng, Y. Current Status of Optical Fiber Biosensor Based on Surface Plasmon Resonance. *Biosens. Bioelectron.* **2019**, *142*, 111505. [CrossRef] [PubMed]
13. Chiappini, A.; Tran, L.T.N.; Trejo-García, P.M.; Zur, L.; Lukowiak, A.; Ferrari, M.; Righini, G.C. Photonic Crystal Stimuli-Responsive Chromatic Sensors: A Short Review. *Micromachines* **2020**, *11*, 290. [CrossRef] [PubMed]
14. Shtenberg, G.; Segal, E. Porous Silicon Optical Biosensors. In *Handbook of Porous Silicon*; Canham, L., Ed.; Springer International Publishing: Cham, Switzerland, 2014; pp. 857–868. ISBN 978-3-319-05743-9.
15. Arshavsky-Graham, S.; Massad-Ivanir, N.; Segal, E.; Weiss, S. Porous Silicon-Based Photonic Biosensors: Current Status and Emerging Applications. *Anal. Chem.* **2019**, *91*, 441–467. [CrossRef] [PubMed]
16. Caroselli, R.; Ponce-Alcántara, S.; Quilez, F.P.; Sánchez, D.M.; Morán, L.T.; Barres, A.G.; Bellieres, L.; Bandarenka, H.; Girel, K.; Bondarenko, V.; et al. Experimental Study of the Sensitivity of a Porous Silicon Ring Resonator Sensor Using Continuous In-Flow Measurements. *Opt. Express* **2017**, *25*, 31651–31659. [CrossRef] [PubMed]
17. Rodriguez, G.A.; Hu, S.; Weiss, S.M. Porous Silicon Ring Resonator for Compact, High Sensitivity Biosensing Applications. *Opt. Express* **2015**, *23*, 7111–7119. [CrossRef] [PubMed]
18. Jenie, S.N.A.; Plush, S.E.; Voelcker, N.H. Recent Advances on Luminescent Enhancement-Based Porous Silicon Biosensors. *Pharm. Res.* **2016**, *33*, 2314–2336. [CrossRef]
19. Sailor, M.J. *Porous Silicon in Practice: Preparation, Characterization and Applications*; Wiley: Weinheim, Germany, 2012; ISBN 978-3-527-64191-8.
20. Pacholski, C.; Sartor, M.; Sailor, M.J.; Cunin, F.; Miskelly, G.M. Biosensing Using Porous Silicon Double-Layer Interferometers: Reflective Interferometric Fourier Transform Spectroscopy. Available online: https://pubs.acs.org/doi/pdf/10.1021/ja0511671 (accessed on 5 January 2021).
21. Zhuo, S.; Xun, M.; Li, M.; Kong, X.; Shao, R.; Zheng, T.; Pan, D.; Li, J.; Li, Q. Rapid and Label-Free Optical Assay of S-Layer Protein with High Sensitivity Using TiO2-Coated Porous Silicon-Based Microfluidic Biosensor. *Sens. Actuators B Chem.* **2020**, 128524. [CrossRef]
22. Maniya, N.H.; Srivastava, D.N. Fabrication of Porous Silicon Based Label-Free Optical Biosensor for Heat Shock Protein 70 Detection. *Mater. Sci. Semicond. Process.* **2020**, *115*, 105126. [CrossRef]
23. Kumar, D.N.; Pinker, N.; Shtenberg, G. Inflammatory Biomarker Detection in Milk Using Label-Free Porous SiO_2 Interferometer. *Talanta* **2020**, *220*, 121439. [CrossRef]
24. Kumar, D.N.; Pinker, N.; Shtenberg, G. Porous Silicon Fabry–Pérot Interferometer for *N*-Acetyl-β-D-Glucosaminidase Biomarker Monitoring. *ACS Sens.* **2020**, *5*, 1969–1976. [CrossRef]
25. Massad-Ivanir, N.; Shtenberg, G.; Raz, N.; Gazenbeek, C.; Budding, D.; Bos, M.P.; Segal, E. Porous Silicon-Based Biosensors: Towards Real-Time Optical Detection of Target Bacteria in the Food Industry. *Sci. Rep.* **2016**, *6*. [CrossRef] [PubMed]
26. Tang, Y.; Li, Z.; Luo, Q.; Liu, J.; Wu, J. Bacteria Detection Based on Its Blockage Effect on Silicon Nanopore Array. *Biosens. Bioelectron.* **2016**, *79*, 715–720. [CrossRef] [PubMed]
27. Gongalsky, M.B.; Koval, A.A.; Schevchenko, S.N.; Tamarov, K.P.; Osminkina, L.A. Double Etched Porous Silicon Films for Improved Optical Sensing of Bacteria. *J. Electrochem. Soc.* **2017**, *164*, B581–B584. [CrossRef]

28. Rajapaksha, P.; Elbourne, A.; Gangadoo, S.; Brown, R.; Cozzolino, D.; Chapman, J. A Review of Methods for the Detection of Pathogenic Microorganisms. *Analyst* **2019**, *144*, 396–411. [CrossRef]
29. Arshavsky-Graham, S.; Massad-Ivanir, N.; Paratore, F.; Scheper, T.; Bercovici, M.; Segal, E. On Chip Protein Pre-Concentration for Enhancing the Sensitivity of Porous Silicon Biosensors. *ACS Sens.* **2017**, *2*, 1767–1773. [CrossRef]
30. Mariani, S.; Paghi, A.; La Mattina, A.A.; Debrassi, A.; Dähne, L.; Barillaro, G. Decoration of Porous Silicon with Gold Nanoparticles via Layer-by-Layer Nanoassembly for Interferometric and Hybrid Photonic/Plasmonic (Bio)Sensing. *ACS Appl. Mater. Interfaces* **2019**, *11*, 43731–43740. [CrossRef]
31. Balderas-Valadez, R.F.; Schürmann, R.; Pacholski, C. One Spot—Two Sensors: Porous Silicon Interferometers in Combination With Gold Nanostructures Showing Localized Surface Plasmon Resonance. *Front. Chem.* **2019**, *7*. [CrossRef]
32. Vercauteren, R.; Scheen, G.; Raskin, J.-P.; Francis, L.A. Porous Silicon Membranes and Their Applications: Recent Advances. *Sens. Actuators A Phys.* **2020**, 112486. [CrossRef]
33. Canham, L. *Handbook of Porous Silicon*; Springer: New York, NY, USA, 2014; ISBN 978-3-319-05743-9.
34. Zhao, Y.; Gaur, G.; Retterer, S.T.; Laibinis, P.E.; Weiss, S.M. Flow-Through Porous Silicon Membranes for Real-Time Label-Free Biosensing. *Anal. Chem.* **2016**, *88*, 10940–10948. [CrossRef]
35. Kumar, N.; Froner, E.; Guider, R.; Scarpa, M.; Bettotti, P. Investigation of Non-Specific Signals in Nanoporous Flow-through and Flow-over Based Sensors. *Analyst* **2014**, *139*, 1345. [CrossRef]
36. Martín-Sánchez, D.; Ponce-Alcántara, S.; García-Rupérez, J. Sensitivity Comparison of a Self-Standing Porous Silicon Membrane Under Flow-Through and Flow-Over Conditions. *IEEE Sens. J.* **2019**, *19*, 3276–3281. [CrossRef]
37. Zhao, Y.; Gaur, G.; Mernaugh, R.L.; Laibinis, P.E.; Weiss, S.M. Comparative Kinetic Analysis of Closed-Ended and Open-Ended Porous Sensors. *Nanoscale Res. Lett.* **2016**, *11*. [CrossRef] [PubMed]
38. Yu, N.; Wu, J. Rapid and Reagentless Detection of Thrombin in Clinic Samples via Microfluidic Aptasensors with Multiple Target-Binding Sites. *Biosens. Bioelectron.* **2019**, *146*, 111726. [CrossRef] [PubMed]
39. Leïchlé, T.; Bourrier, D. Integration of Lateral Porous Silicon Membranes into Planar Microfluidics. *Lab Chip* **2015**, *15*, 833–838. [CrossRef] [PubMed]
40. He, Y.; Vasconcellos, D.S.D.; Bardinal, V.; Bourrier, D.; Imbernon, E.; Salvagnac, L.; Laborde, A.; Dollat, X.; Leichlé, T. Lateral Porous Silicon Interferometric Transducer for Sensing Applications. In Proceedings of the 2018 IEEE SENSORS, New Delhi, India, 28–31 October 2018; pp. 1–3.
41. He, Y.; Leïchlé, T. Fabrication of Lateral Porous Silicon Membranes for Planar Microfluidics by Means of Ion Implantation. *Sens. Actuators B Chem.* **2017**, *239*, 628–634. [CrossRef]
42. Leprince, A.; Nuytten, M.; Gillis, A.; Mahillon, J. Characterization of PlyB221 and PlyP32, Two Novel Endolysins Encoded by Phages Preying on the Bacillus Cereus Group. *Viruses* **2020**, *12*, 1052. [CrossRef]
43. Janshoff, A.; Dancil, K.-P.S.; Steinem, C.; Greiner, D.P.; Lin, V.S.-Y.; Gurtner, C.; Motesharei, K.; Sailor, M.J.; Ghadiri, M.R. Macroporous P-Type Silicon Fabry–Perot Layers. Fabrication, Characterization, and Applications in Biosensing. *J. Am. Chem. Soc.* **1998**, *120*, 12108–12116. [CrossRef]
44. Dancil, K.-P.S.; Greiner, D.P.; Sailor, M.J. A Porous Silicon Optical Biosensor: Detection of Reversible Binding of IgG to a Protein A-Modified Surface. *J. Am. Chem. Soc.* **1999**, *121*, 7925–7930. [CrossRef]
45. Stewart, M.P.; Robins, E.G.; Geders, T.W.; Allen, M.J.; Choi, H.C.; Buriak, J.M. Three Methods for Stabilization and Functionalization of Porous Silicon Surfaces via Hydrosilylation and Electrografting Reactions. *Phys. Status Solidi Appl. Res.* **2000**, *182*, 109–115. [CrossRef]
46. Salonen, J.; Björkqvist, M.; Laine, E.; Niinistö, L. Stabilization of Porous Silicon Surface by Thermal Decomposition of Acetylene. *Appl. Surf. Sci.* **2004**, *225*, 389–394. [CrossRef]
47. Rasson, J.; Francis, L.A. Improved Stability of Porous Silicon in Aqueous Media via Atomic Layer Deposition of Oxides. *J. Phys. Chem. C* **2018**, *122*, 331–338. [CrossRef]
48. Tenenbaum, E.; Segal, E. Optical Biosensors for Bacteria Detection by a Peptidomimetic Antimicrobial Compound. *Analyst* **2015**, *140*, 7726–7733. [CrossRef] [PubMed]
49. Yaghoubi, M.; Rahimi, F.; Negahdari, B.; Rezayan, A.H.; Shafiekhani, A. A Lectin-Coupled Porous Silicon-Based Biosensor: Label-Free Optical Detection of Bacteria in a Real-Time Mode. *Sci. Rep.* **2020**, *10*, 16017. [CrossRef] [PubMed]

 *biosensors*

Article

Newly Developed System for the Robust Detection of *Listeria monocytogenes* Based on a Bioelectric Cell Biosensor

Agni Hadjilouka [1,2,*], Konstantinos Loizou [1], Theofylaktos Apostolou [1], Lazaros Dougiakis [1], Antonios Inglezakis [1] and Dimitrios Tsaltas [2]

[1] EMBIO Diagnostics Ltd., Athalassas Avenue 8, Strovolos, Nicosia 2018, Cyprus; k.loizou@embiodiagnostics.eu (K.L.); theo.apo@embiodiagnostics.eu (T.A.); l.dougiakis@embiodiagnostics.eu (L.D.); a.inglezakis@embiodiagnostics.eu (A.I.)

[2] Department of Agricultural Sciences, Biotechnology and Food Science, Cyprus University of Technology, 30 Archbishop Kyprianos, Limassol 3036, Cyprus; dimitris.tsaltas@cut.ac.cy

* Correspondence: agnixatz@gmail.com; Tel.: +35-7967-58608

Received: 12 October 2020; Accepted: 14 November 2020; Published: 17 November 2020

Abstract: Human food-borne diseases caused by pathogenic bacteria have been significantly increased in the last few decades causing numerous deaths worldwide. The standard analyses used for their detection have significant limitations regarding cost, special facilities and equipment, highly trained staff, and a long procedural time that can be crucial for foodborne pathogens with high hospitalization and mortality rates, such as *Listeria monocytogenes*. This study aimed to develop a biosensor that could detect *L. monocytogenes* rapidly and robustly. For this purpose, a cell-based biosensor technology based on the Bioelectric Recognition Assay (BERA) and a portable device developed by EMBIO Diagnostics, called B.EL.D (Bio Electric Diagnostics), were used. Membrane engineering was performed by electroinsertion of *Listeria monocytogenes* homologous antibodies into the membrane of African green monkey kidney (Vero) cells. The newly developed biosensor was able to detect the pathogen's presence rapidly (3 min) at concentrations as low as 10^2 CFU mL^{-1}, demonstrating a higher sensitivity than most existing biosensor-based methods. In addition, lack of cross-reactivity with other *Listeria* species, as well as with *Escherichia coli*, was shown, thus, indicating biosensor's significant specificity against *L. monocytogenes*.

Keywords: *Listeria monocytogenes*; cell-based biosensor; bioelectric recognition assay; membrane-engineering

1. Introduction

Foodborne diseases are of great concern worldwide, as they cause thousands of deaths each year, as well as significant malfunctions in health care systems, national economies, and global trade. Bacteria, viruses, parasites, and chemicals that contaminate food at any stage of food production are the causative agents for these infections. It is estimated that 600 million people (i.e., 10% of the global population) get sick and 420,000 people die every year due to the consumption of contaminated food [1]. *Listeria monocytogenes* is an intracellular pathogenic bacterium widely distributed in the environment that has been determined as a causative agent of serious epidemic and sporadic food-borne illnesses in humans. Listeriosis can lead to gastroenteritis, meningitis, or other severe symptoms with high hospitalization and mortality rates (20–30%), especially in vulnerable populations. In 2016, listeriosis was the most severe illness with the highest hospitalization and mortality rate in Europe. In the meantime, it is estimated that 1600 people get listeriosis and about 260 people die each year in the USA [2].

Despite the intensified efforts to improve the hygiene conditions and prevent cross-contamination in production processes, it is difficult to eliminate *L. monocytogenes* from all products. Hence, food safety authorities have made pathogen detection a priority. Currently, the detection of *L. monocytogenes* in food is mainly performed following the ISO 11290-1:2017 [3]. This culture method is based on the pathogen's physical and chemical characteristics and is a gold standard method, however, it is also time-consuming since it requires four days to provide the first indications for either presumptive presence or absence of the pathogen and five to seven days to undoubtedly confirm the pathogen's presence. Furthermore, DNA analysis based on quantitative polymerase chain reaction (qPCR) or microarray methods have been developed. Even though these methods are precise, they have significant limitations regarding cost, special facilities, long procedural time, and highly trained staff requirements [4]. Rapid detection methods based on molecular techniques (e.g., real-time qPCR) and immunology-based methods (e.g., VIDAS (Vitek Immuno Diagnostic Assay System) for *L. monocytogenes*) have also been developed. However, molecular techniques are still restricted by high-cost equipment and reagents, as well as complicated nucleic acid extraction procedures, while immunology-based methods are restricted by lower detection sensitivity [5]. Developing rapid, low-cost methods with high sensitivity and specificity to detect bacteria in food and reduce the risk to public health remains a challenge.

Biosensors have drawn major interest as alternative methods for food contamination analysis. Their mode of action is based on their ability to detect specific targets and transform this information into a detectable signal [6]. According to the transducing elements, biosensors are classified as electrochemical, thermal, optical, and piezoelectric sensors, with a noteworthy rapid development of electrochemical biosensors for pathogen detection in the food and agricultural sector in the last decade [7–10]. According to the type of the biomolecule recognition element, biosensors are distinguished in enzyme-based (immobilized enzymes and proteins), microorganism-based biosensors; DNA biosensors (nucleic acids); immunosensors (an immobilized antibody or antibodies coupled with an enzyme or pigment); and cellular biosensors (immobilized cells or tissues) [10,11].

Living cell-based biosensors are techno-scientific systems that use cells as sensors to detect the status of the cellular environment and physiological parameters. The innovation of these biosensors is that, unlike other types of biosensors containing only living-extracted elements, they use live cells as receptors. This type of biosensor consists of the following: (a) living cells, the biological identification element, that acts as the primary signal transducer element and is used as the primary element for the collection and transmission of signals, and (b) the secondary transducer element that converts these responses in electrical signals. When cells interact with a stimulus, they produce changes in molecules or ions, changes in potential, or changes in impedance due to cell metabolism, etc. A secondary transducer element can detect these responses and convert them into electrical signals [12]. Cell-based biosensors have the advantages of increased stability and high biocatalytic activity, and their main feature is their ability to provide relevant cellular information in response to the sample and finally measure its activity. Furthermore, their robustness, high selectivity, specificity, and evolvability have raised cell-based biosensors as a significant revolution in analytical science [13]. However, only a few studies have been performed on *L. monocytogenes* detection using cell-based biosensors [14–16].

In combination with methods such as the BERA (Bioelectric Recognition Assay), the cell-based biosensors have been used in numerous environmental, chemical, and medical applications with remarkable results [17–22]. The BERA method is based on the insertion, by electroporation, of a large number of receptor molecules (antibodies, enzymes, etc.) on the cell membrane, increasing their selectivity for recognizing target analytes [23,24]. The methodology is based on measuring the change in membrane potential caused by the binding of the target molecule to the receptors previously inserted into the cell membrane. In the beginning, the cell membrane potential is stable due to the ions flowing through the ion channels. Subsequently, and after the target molecule binds to the receptor, its structure changes, resulting in its molecular charge being displaced within the cell membrane. As a result, a large number of ions concentrate on one side of the membrane. Opening the ion channel creates an ionic current that can be measured as a corresponding current.

In the present study, we report the development of a portable BERA-type sensor based on mammalian cells for the rapid detection of *L. monocytogenes*. The biosensor was developed performing a membrane-engineering process, where anti-L. monocytogenes antibodies were electroinserted in the membrane of the Vero cells to achieve a selective response against the pathogenic bacterium. The response was measured according to the principles of BERA [18,23–26] and results were obtained based on a novel sophisticated algorithm embedded in user-friendly software that connects via Bluetooth with an android device, thus, the end-user was instantly informed of the results of each analysis performed.

2. Materials and Methods

2.1. Materials and Reagents

Monkey African green kidney (Vero) cell cultures were provided from LGC Promochem (Teddington, UK). Fetal bovine serum (FBS), antibiotics (streptomycin-penicillin), L-glutamine and L-alanine, and trypsin/EDTA (ethylenediaminetetraacetic acid) were purchased from Sigma-Aldrich (Taufkirchen, Germany). Monoclonal antibodies against *L. monocytogenes* were purchased from antibodies-online.com and *L. monocytogenes* NCTC 11,994 and *L. innocua* NCTC 11,288 from Merck (Darmstadt, Germany). Sodium chloride was purchased from Merck (Darmstadt, Germany) and Brain Heart Infusion from Biolife (Milan, Italy).

2.2. Cell Culture and Antibody Electroinsertion

Cell culture was performed according to Apostolou et al. [26]. Briefly, Vero cells were cultured in Dulbecco's medium with 10% fetal bovine serum (FBS), 10% antibiotics (streptomycin-penicillin), and 10% l-glutamine and l-alanine (nutrient medium). Cell detachment from the culture vessel was performed by adding trypsin/EDTA for 10 min at 37 °C and cells were collected by centrifugation (2 min/1200 rpm) at a final density of 2.5×10^6 mL^{-1}.

Membrane-engineered cells were created by electroinserting either the anti-L. *monocytogenes* p60 protein antibody clone p6007 (p60-biosensor) or the anti-L. *monocytogenes* actA antibody clone 3a15 (actA-biosensor) into the membrane of the Vero cells, based on a modified protocol of Zeira et al. [27]. In brief, cells were detached and collected after centrifuge (6 min/1000 rpm/25 °C). The cell pellet was resuspended in 400 µL PBS (phosphate-buffered saline) containing three different antibody concentrations (1, 5, and 10 µg mL^{-1}) and incubated on ice for 20 min. After incubation, the cell-antibody mixture was transferred into electroporator cuvettes (4 mm) and electroinsertion was performed in the Eppendorf Eporator (Hamburg, Germany) by the application of two square electric pulses at 1800 V/cm. Subsequently, the mixture was transferred in a petri dish (60×15 mm^2) containing 3 mL of nutrient medium and incubation took place at 37 °C and 5% CO$_2$ for 24 h. Then, the medium was discarded from the petri dish and Vero/anti-L. monocytogenes cells were mechanically detached and collected with the nutrient medium in Eppendorf tubes.

It has been previously indicated that, based on the above electroinserting method, the membrane-engineered cells incorporate the specific antibodies in the correct orientation and show selective interaction against the target molecules [24,28].

2.3. Bacteria Culturing and Sample Preparation

L. monocytogenes NCTC 11994, *L. ivanovii* NCTC 11846, *L. innocua* NCTC 11288, and *Escherichia coli* ATCC 10,536 were used throughout this study. The strains were kept at −20 °C in nutrient broths supplemented with 50% glycerol, then, prior to their utilization, each strain was grown twice in Brain Heart Infusion broth at 37 °C for 24 h. Experimental assays were performed using overnight bacteria cultures (*L. monocytogenes*, *L. ivanovii*, *L. innocua*: 9 log CFU (colony forming units) mL^{-1} and *E. coli*

8 log CFU mL^{-1}) that were centrifuged (10 min/3500 rpm/4 °C), washed twice with sterile saline solution, resuspended in the same diluent, and serially diluted to 2 log CFU mL^{-1}. Then, diluents with the desired concentrations (2, 4, 6, and 9 log CFU mL^{-1} for *Listeria* species and 2, 4, 6, 8 log CFU mL^{-1} for *E. coli*) were tested with the newly developed biosensor system.

For bacteria counting, the overnight cultures were serially diluted with sterile saline solution. Then, triplicate diluents were surface plated in brain heart infusion (BHI) agar plates and incubation took place at 37 °C for 24 h. Subsequently, the colonies were counted, and the concentration of the bacteria was calculated in CFU mL^{-1}.

2.4. Experimental Design and Assay Procedure

The present study was conducted in three experimental assays. The first assay was performed to compare and evaluate two different *Listeria monocytogenes* antibodies in the presence of the pathogen at various population levels. In the second assay, the effect of the concentration (1, 5, and 10 µg mL^{-1}) of the electroinserted antibody on the response of the membrane-engineered cells was investigated. Finally, in the third experimental assay, the cross-reactivity of the newly developed biosensor system was evaluated against *Escherichia coli* and other *Listeria* species.

All tests were conducted using a portable device developed by EMBIO DIAGNOSTICS (EMBIO DIAGNOSTICS Ltd., Cyprus). EMBIO's device is a multichannel potentiometer with high accuracy A/D converters that measure electric signals from various biorecognition elements, such as cells. The studied elements are added on the surface of eight screen-printed electrodes (SPE) that are connected to the underside of a replaceable connector (Figure 1A). EMBIO's device in combination with the bioelectric recognition assay (BERA), an established cell-based biosensor technology, has created a biosensor system that allows high throughput screening and rapid testing. Furthermore, the system connects via Bluetooth 4.0 with a smartphone or tablet, allowing the end-user to be instantly informed of each analysis result (Figure 1B). The system has been previously reported by Apostolou et al. [26]. The screen-printed electrodes were purchased from Zimmer and Peacock (Horten, Norway). Each single gold electrode was comprised of a 625 µm thick ceramic substrate (alumina) with three screen-printed electrodes (working electrode, reference electrode, and counter electrode). The working electrode was made of gold, and the reference electrode was made of silver/silver chloride. The counter electrode was canceled out by the measuring system (Figure 1C).

First, the samples (20 µL) were added on the top of each gold screen-printed electrode (first peak on the two-dimensional (2D) line graph) and the membrane-engineered cells (20 µL ≈ 5 × 10^4 cells) were added after 60 s (120 values)(second peak on the 2D line graph).

The cell response was recorded as a time series of potentiometric measurements (in Volts), consisting of 360 values per sample. Each measurement lasted 3 min and the sampling rate was set at 2 Hz. After each measurement, cell responses were uploaded into a cloud server (Google Firebase) and calculations were conducted (Section 2.5), based on a newly developed algorithm that provided results regarding *Listeria monocytogenes* presence on the smartphone screen (Figure 1B). Every single sample was tested eight times by using a set of eight individual sensors and each experiment was performed in duplicate.

One-hundred tests were initially conducted on sodium chloride solution (NaCl 0.85%), a diluent used for preparing microbial suspensions (blank samples). Subsequently, 100 tests of *L. monocytogenes* diluted in NaCl 0.85% at 4 different concentrations (2, 4, 6, and 9 log CFU mL^{-1}) were performed. Results obtained from the biosensors (p60-biosensor and actA-biosensor with 1, 5, or 10 µg mL^{-1} antibody concentration) were evaluated and the biosensor with the best accuracy and performance characteristics was determined and used in the subsequent tests. Performance indices used for the evaluation were sensitivity (Se), specificity (Sp), positive predictive value (PPV), and negative predictive value (NPV). An explanation of these terms can be found in the respective reference [29]. Furthermore, a total of 150 tests were performed on *L. ivanovii*, *L. innocua*, and *Escherichia coli* broth samples to evaluate biosensor's cross-reactivity with other bacteria. Finally, to demonstrate that the obtained response of the tested samples was associated with the presence of the electroinserted antibody, 40 tests (control, *L. monocytogenes*, *L. ivanovii*, *L. innocua*, and *Escherichia coli* samples) were conducted with non-engineered cells as well.

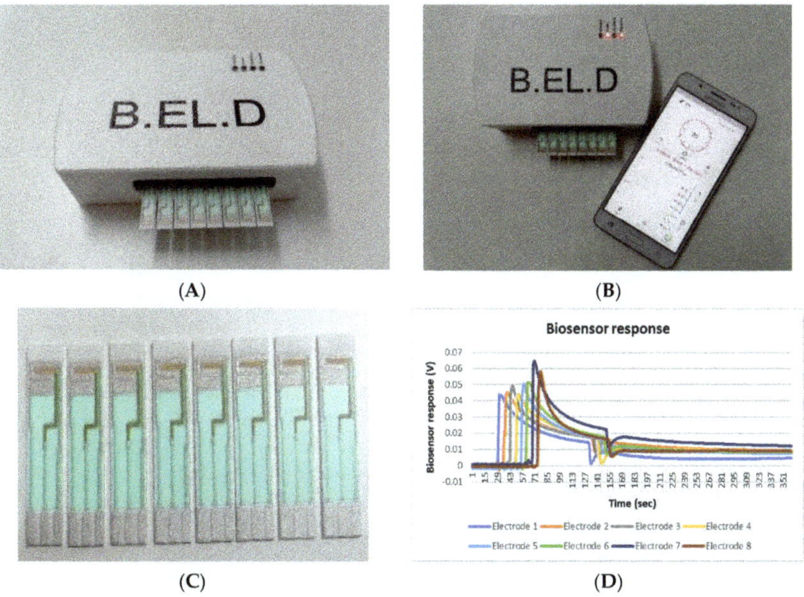

Figure 1. (**A**) Loading of the cells and samples on the electrode surface; (**B**) Results appear on the smartphone screen; (**C**) Screen-printed electrode sensors used for the assay, Zimmer and Peacock gold single electrodes; (**D**) Visualization of the electric signal via a voltage (Volts) vs. time graph.

2.5. Algorithm for Response Processing

The process of the data analysis is summarized in Figure 2.

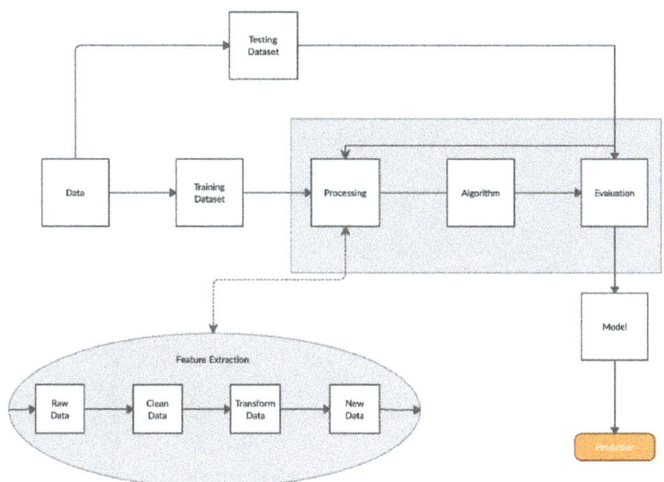

Figure 2. Bio Electric Diagnostics (B.EL.D) data analysis process on cloud functions. The analysis is being done in real time after the completion of the tests.

More analytically:

a. Dataset: The dataset contained 200 measurements from 100 negative and 100 positive samples. Each measurement consisted of the data obtained by the 8 screen-printed electrodes that recorded the cell response as a time series of potentiometric measurements (in Volts) and comprised 360 values per electrode (sampling rate of 2 Hz). The detected measurements were visualized through a voltage/time graph (Figure 1D).

b. Training/Testing Dataset: The obtained dataset was split into training and testing dataset as follows: 30% was used for training (60 measurements) and 70% was used for testing (140 measurements). The training dataset was utilized to determine the algorithm thresholds and the testing dataset was utilized for the algorithm evaluation.

c. Processing/Feature Extraction: The processing of the dataset was performed in a two-step procedure. In the first step, peaks were determined and starting noise was cleaned to smooth and calibrate the signal before the first peak. Then, all values were shifted to start at y = 0 and from each experimental dataset values from 120 to 200 were shifted at x = 0 and kept for further analysis (Figure 3). In the second step, feature vectors were extracted from the cleaned data and used as input to develop an algorithm able to detect *L. monocytogenes* in sterile saline samples. Each feature vector was calculated based on the following: (a) the average values (mean) for each cleaned dataset and (b) the rolling average with rolling window size 50 (minimum sums) [30]. This procedure was applied in each electrode channel (channel 1, channel 2, etc.) and the overall test dataset (mean average and minimum sums average for all 8 electrodes) (Figure 4). Hence, from the initial experimental raw dataset with 360 × 8 values, only 1 × 18 (1 values for each channel (8 values in total) + 1 overall value/(a) and (b)) feature values were used for the sample discrimination and the algorithm development.

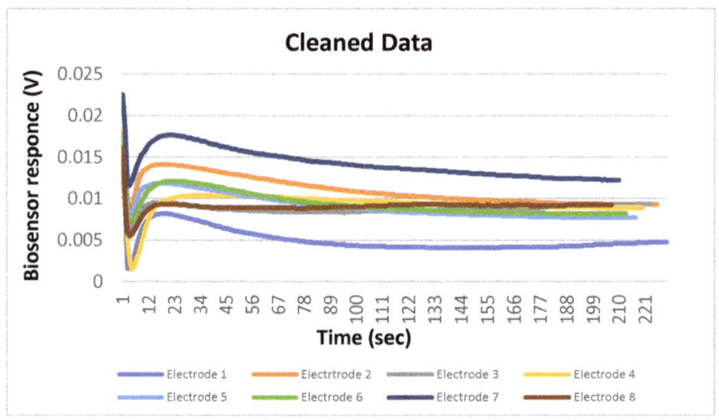

Figure 3. Visualization of the electric signal via a voltage (Volts) vs. time graph of the cleaned data.

| Category | | | | | | | | | Mean | | | | | | | | | Min Sums |
| Subcategory | channel_1 | channel_2 | channel_3 | channel_4 | channel_5 | channel_6 | channel_7 | channel_8 | Average | channel_1 | channel_2 | channel_3 | channel_4 | channel_5 | channel_6 | channel_7 | channel_8 | Average |
Level / Name																		
10^4 — 10^4 p60 5 µg (4).csv	0.007306	0.085611	-0.004549	0.007809	0.109261	-0.009979	0.031828	0.024542	0.031475	-0.0079	-0.0131	-0.0112	-0.0056	-0.0058	-0.0087	-0.0113	-0.0122	-0.008850
10^6 — 10^6 p60 5 µg (1).csv	0.027805	0.005556	0.005556	-0.009005	0.015525	-0.029536	0.039565	0.024323	0.009973	-0.0028	-0.0030	-0.0030	-0.0129	-0.0039	-0.0033	-0.0077	-0.0115	-0.006013
NaCl — NaCl & p60 (10).csv	0.035331	0.032100	0.043375	0.039822	0.028149	0.022830	0.065386	0.038510	0.038186	-0.0110	-0.0110	-0.0111	-0.0150	-0.0116	-0.0112	-0.0099	-0.0066	-0.010925
10^2 — 10^2 p60 5 µg (4).csv	0.052876	0.130632	0.119487	0.121075	0.163909	0.014258	0.053965	0.037690	0.086738	-0.0009	-0.0087	-0.0049	-0.0055	-0.0158	-0.0155	-0.0080	-0.0085	-0.008475
10^9 — 10^9 p60 5µg (2).csv	0.025354	-0.002571	-0.002571	-0.014281	0.029643	-0.026743	-0.008255	0.012370	0.001618	-0.0253	-0.0193	-0.0193	-0.0222	-0.0147	-0.0120	-0.0214	-0.0063	-0.017563

Figure 4. Feature vectors for five random samples (10^2, 10^4, 10^6, 10^9, and NaCl).

Algorithm: The algorithm used the feature vectors from the previous step as input to produce/calculate the final results. Two thresholds were set for mean values and minimum sums values and were compared to the respective values from each channel and from the average of the channels. Four primary results were extracted based on the following: (i) threshold comparing to each channel mean values, (ii) threshold comparing to the average mean value, (iii) threshold comparing to each channel minimum sums values, and (iv) threshold comparing to the average minimum sums values. In the case of each channel comparison (i and iii), the obtained result was the predominant one (e.g., if 5 electrodes gave "positive" result and 3 gave "negative"' result based on the threshold, the result would be "positive"). The final result was the predominant result obtained from the above calculations (e.g., if (i) channel mean values, "positive"; (ii) average mean value, "positive", (iii) channel minimum sums values, "negative", and (iii) average minimum sums values: "positive", the final result would be "positive"). In the case of two "positive" and two "negative", the final result was the one obtaining from the average minimum sums value (Figure 5).

Level	Category Name	Count Channels Mean	Count Channels Min Sums	Channels Mean	Mean Total	Channels Min Sums	Min Sums Total	BELD Results	Real Results	Final BELD Results
10^6	10^6 p60 5 µg (1).csv	7	5	POSITIVE	POSITIVE	POSITIVE	NEGATIVE	POSITIVE	POSITIVE	POSITIVE
NaCl	NaCl & p60 (10).csv	4	0	NEGATIVE	NEGATIVE	NEGATIVE	NEGATIVE	NEGATIVE	NEGATIVE	NEGATIVE
10^2	10^2 p60 5 µg (4).csv	5	7	POSITIVE	POSITIVE	POSITIVE	POSITIVE	POSITIVE	POSITIVE	POSITIVE
10^9	10^9 p60 5µg (2).csv	8	6	POSITIVE	POSITIVE	POSITIVE	POSITIVE	POSITIVE	POSITIVE	POSITIVE
NaCl	NaCl & p60 (15).csv	5	2	POSITIVE	POSITIVE	NEGATIVE	NEGATIVE	REPEAT	NEGATIVE	NEGATIVE

Figure 5. Calculation of the final results based on the feature vectors.

Evaluation: A testing dataset was used for the evaluation of the model. Different thresholds were tested and determination of the one providing the most accurate results was achieved.

2.6. Statiatical Analysis

One-way analysis of variance (ANOVA) was used to assess the statistical differences among the feature values obtained from the biosensors.

3. Results and Discussion

3.1. Antibody Selection and Biosensor Response in the Presence of L. monocytogenes

This study aimed to develop a biosensor that could perform rapid and robust detection of the pathogenic bacterium *L. monocytogenes*, using a modified protocol of a cell-based biosensor technology that has been previously reported by Apostolou et al. [26]. To develop the method, two different *Listeria monocytogenes* antibodies were initially used for the membrane engineering, creating two different biosensors. The two biosensors were created using either the anti-*L. monocytogenes* p60 protein antibody clone p6007 (p60-biosensor) or the anti-*L. monocytogenes* actA antibody clone 3a15 (actA-biosensor) at three different concentrations (1, 5, and 10 µg mL^{-1}). The first antibody recognizes natural and recombinant *L. monocytogenes* extracellular p60 protein, an invasion associated protein (iap) essential for cell viability encoded by *iap* gene, and the second antibody recognizes *L. monocytogenes* protein actA, a surface protein necessary for intra- and intercellular motility that appears to be a multifunctional virulence factor. The *iap* and *actA* genes also exist in other *Listeria* species but their produced proteins differ from that of *L. monocytogenes* [31].

The results indicated that when the antibody concentration was at 1 µg mL^{-1}, the actA-biosensor could not discriminate samples with and without the pathogen (0.047–0.051 mV), while the p60-biosensor was able to produce a significantly different signal only when the pathogen's population was at 9 log CFU mL^{-1} (0.027–0.031 mV in NaCl, 10^2, 10^4, 10^6 CFU mL^{-1}, and 0.041 mV in 10^9 CFU mL^{-1}) (Figure 6A). More accurately, no pattern was visible for the different antibodies at this concentration, and this was attributed to the low number of the electoinserted antibodies into the membrane-engineered cells.

The best discrimination was achieved by p60-biosensor with 5 µg mL^{-1} antibody concentration. At this antibody concentration, increasing biosensor response was observed against the increasing *L. monocytogenes* population, with a 50% higher voltage response between the control and the inoculated samples with the higher bacterial concentrations (6 and 9 log CFU mL^{-1}). Contrary to this, the p60-biosensor with 10 µg mL^{-1} antibody concentration could distinguish blank samples (0.116 mV) from samples with *L. monocytogenes* at 4, 6, and 9 log CFU mL^{-1}, but not from samples with *L. monocytogenes* at 2 log CFU mL^{-1} (0.117 mV). Hence, it was demonstrated that the p60-biosensor with 5 µg mL^{-1} antibody concentration had the best discrimination ability, setting the method's limit of detection at 10^2 CFU mL^{-1}. This is the lowest LOD being reported in studies that have developed various biosensors (optical, piezoelectric, amperometric, impedimetric, cell-based) for the detection of *L. monocytogenes* in culture and food samples [32–36].

From the conducted experiments, it was observed that the p60-biosensor with 5 µg mL^{-1} antibody concentration produced an increasing pattern against the increasing analyte (CFU mL^{-1}). However, when the concentration of the antibody was augmented at a density of 10 µg mL^{-1}, a decreasing pattern against the increasing analyte (CFU mL^{-1}) was revealed. Although the sensitivity of a biosensor is largely dependent on the physical properties of the transducer system, the amount of the biological receptor molecules immobilized on the sensor surface remains an issue.

The signal obtained from the actA-biosensors with 5 and 10 µg mL^{-1} antibody concentration could distinguish blank samples from samples with *L. monocytogenes* at 6 and 9 log CFU mL^{-1} with values ranging from 0.035 to 0.056 mV and from 0.028 to 0.037 mV, respectively. Nevertheless, the biosensors were not able to discriminate blank samples from samples with *L. monocytogenes* at 2 and 4 log CFU mL^{-1}, since similar voltage responses were observed (Figure 6B,C).

The binding of the antigens to antibodies adsorbed on a surface is a complex phenomenon and can be described as a two-step process. Firstly, the surface-adsorbed antibodies bind to the antigens through one of their fragment antigen-binding (F(ab)) domains. If they remain bounded, the second F(ab) domain has an opportunity to bind to any other antigen. Consequently, the equilibrium surface concentration of bounded IgG molecules, which increases with the surface concentration of the antigens, is the sum of two contributions. Therefore, antibodies which bound through a single F(ab) lead to a lower surface equilibrium.

De Michele et al. [37] recently, showed that the large size of IgGs played an instrumental role. Antigens diffusing in the bulk, see a few available IgGs on the surface which is reduced concerning the number of sites that are already occupied. In fact, a significant number of the unbounded antigens are unavailable due to the large size of IgGs, which overlap the unbounded sites of neighboring IgGs, thus, making them invisible for other antigens. In this way, antibodies that are immobilized onto the surfaces in a relatively higher density will induce locally crowded areas. Cho et al. [38] showed that such cluster formation could negatively affect the performance of the immunoassays, reporting that the binding capability was approximately 10 times lower than the maximum.

Additionally, Kwon et al. [39], in one of their studies, found that the immobilized antibodies that were active for binding antigens decreased from 150% to 60% as the density of immobilized antibodies was increased from 0.3 to 2.5 ng/mm^2. In either case, the data clearly showed that as the density of the antibody increased, the steric accumulation of antigen-antibody complexes decreased the activity of the immobilized antibody. This demonstrates that the amount of antigen that binds depends on the density of the immobilized antibody.

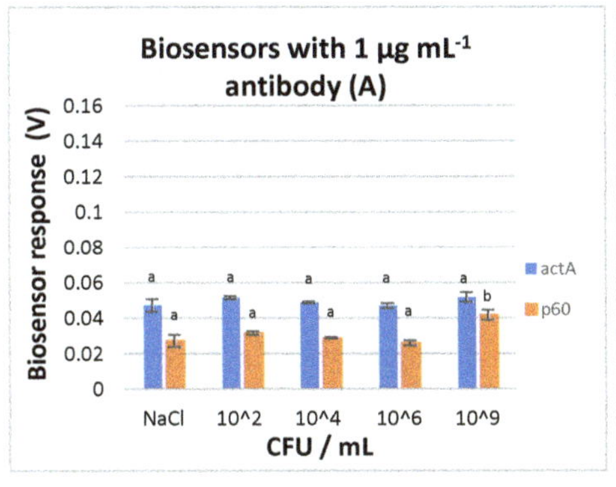
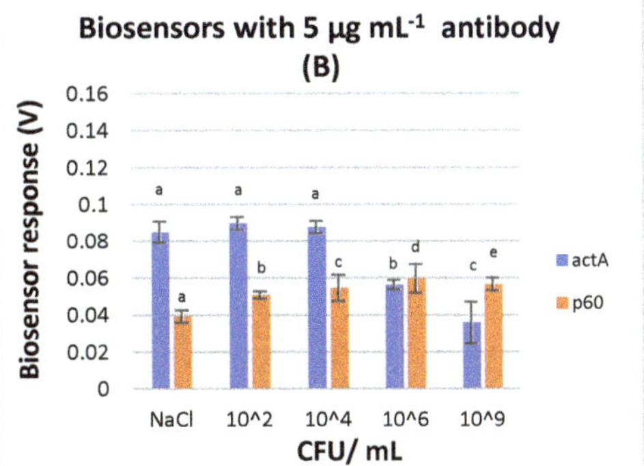
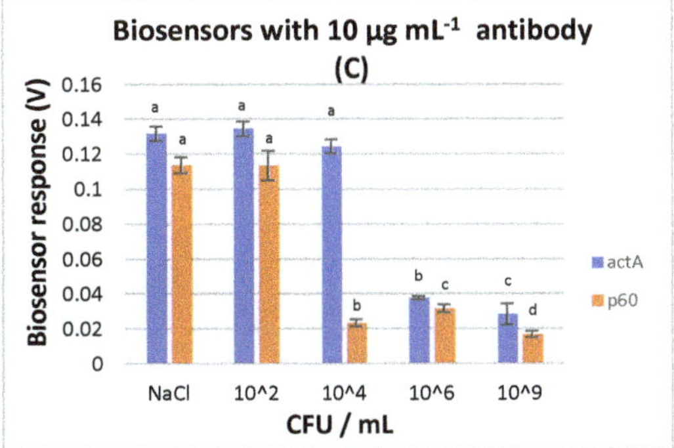

Figure 6. Response of actA and p60-biosensors with (**A**) 1 µg mL^{-1}, (**B**) 5 µg mL^{-1}, and (**C**) 10 µg mL^{-1} antibody concentration in the absence (NaCl 0.85%) and presence of *L. monocytogenes* at 4 different final concentrations (2, 4, 6, and 9 log CFU mL^{-1}). Error bars represent the standard errors of the mean value of all replications. Columns marked with different letters indicate that response was significantly ($p < 0.05$) different from the respective of the blank samples (NaCl) for each experimental assay.

Ideally, the antibody should specifically recognize and bind its antigen at the lowest possible concentration. Since the biosensor surfaces are in the µm scale, the optimum density of immobilized antibodies onto the surface result in enhanced detection sensitivity.

On the basis of the above, and since the p60-biosensor with 5 µg mL^{-1} antibody concentration had the best ability to separate blank from inoculated samples, especially when the pathogen was present at high concentrations, the p60 protein antibody clone p6007 was selected as the best antibody for the detection of the pathogen and the 5 µg mL^{-1} antibody concentration was selected as the optimum density for the *L. monocytogenes* biosensor.

Calculation of Method's Performance Characteristics

As mentioned above (Section 2.5), 70% of the samples was used for testing (i.e., 140 samples, 77 positive and 63 negative). Parallel testing of the samples with the ISO and the newly developed method resulted in true-positive (TP), true-negative (TN), false-positive (FP), and false-negative (FN) results. On the basis of these results, the performance indices of the newly developed method were calculated and are summarized in the following Table 1:

Table 1. Performance indices of the p-60 biosensor with 5 µg mL^{-1} antibody concentration.

Results [1]		Performance Indices [2]	
TP	75	Se	97.4%
FP	2	Sp	84.13%
TN	53	PPV	88.24%
FN	10	NPV	96.36%

[1] TP, true positive; TN, true negative; FP, false positive; FN, false negative. [2] Se, sensitivity; Sp, specificity; PPV, positive predictive value; NPV, negative predictive value.

Furthermore, the p60 biosensor with 5 µg mL^{-1} antibody concentration demonstrated perfect discrimination among samples with the highest inoculum level (9 log CFU mL^{-1}) and the rest of the samples with 100% accuracy, sensitivity, specificity, positive predictive value, and negative predictive value.

3.2. Selectivity Assay

The p60 protein that exists in other *Listeria* species differs from the *L. monocytogenes* p60 protein. However, *Listeria ivanovii*, a pathogenic bacterium with extremely rare cases of human disease, and *Listeria innocua*, a non-pathogenic bacterium widely distributed in the environment, were also included in the experimental assay to evaluate biosensor's cross-reactivity with other *Listeria* species. In addition, the biosensor's response in the presence of *Escherichia coli*, a microorganism that is often present in several food categories indicating faecal contamination and poor hygiene practices, was also evaluated. Evaluation of the biosensor's cross-reactivity with *E. coli*, *L. ivanovii*, and *L. innocua* indicated that the *L. monocytogenes* p60-biosensor with 5 µg mL^{-1} antibody concentration was able to distinguish *L. monocytogenes* from the other *Listeria* species, as well as from *Escherichia coli*, infallibly. To evaluate biosensor's selectivity, all bacteria were diluted in sterile saline solution and tested at the same concentration range (Figure 7). In every case examined, the biosensor's response was statistically different ($p < 0.05$) from the observed response against the similar concentrations of *L. monocytogenes*.

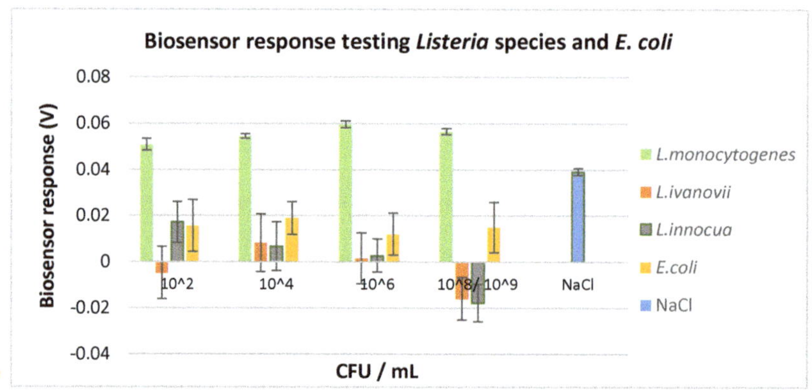

Figure 7. Response of p60-biosensor with 5 µg mL^{-1} antibody concentration at broth samples of *L. monocytogenes*, *L. ivanovii*, *L. innocua*, and *Escherichia coli* at four different population levels (*L. monocytogenes*, *L. ivanovii*, *L. innocua*: 2, 4, 6, and 9 log CFU mL^{-1}; *E. coli*: 2, 4, 6, 8 log CFU mL^{-1}).

More accurately, in the presence of *L. ivanovii* and *L. innocua*, the biosensor's response was significantly different with substantial variations at every population level. The observed response was attributed to bacterial membrane potential dynamics [40]. Hence, it was indicated that the newly developed biosensor had no cross-reactivity with other *Listeria* species. A similar response was observed in the presence of *Escherichia coli*, thus, indicating no reaction between the samples and the biosensor. The response was less variable among different concentrations but significantly different ($p < 0.05$) to the ones observed for the similar concentrations of *L. monocytogenes* (Figure 7). Furthermore, the presence of *L. ivanovii*, *L. innocua*, and *Escherichia coli* revealed different cell responses as compared with the control values. However, the response was not a result of a specific reaction, since no detection pattern was demonstrated. Finally, based on the conducted experiments the non-engineered membrane cells seem to react in the presence of *L. monocytogenes*, *L. ivanovii*, *L. innocua*, and *Escherichia coli*, but the differences on response were statistically non-significant (data not shown). Hence, the biosensor's selectivity assay was determined for *L. monocytogenes*. Contrary to these results, the p60 biosensor with 1 μg mL^{-1} antibody concentration was not able to distinguish samples with *L. monocytogenes* from samples with the other *Listeria* species (data not shown).

From the selectivity assay, it was validated that the bioelectric recognition assay was combined with the molecular identification through membrane engineering for the detection of the pathogenic bacterium *Listeria monocytogenes* with successful results. The presence of *L. monocytogenes* and attachment to its respective antibody caused a measurable change in the cell membrane structure that led to the discrimination between positive and negative samples. Hence, membrane engineering [19] was a critical step in the successful development of the biosensor system, since the electroinsertion of the anti-L. monocytogenes antibody on the surface of the Vero cells fortified the selectivity of the system against the pathogenic bacterium.

4. Conclusions

The present study demonstrates a rapid, high throughput, and portable screening system for *L. monocytogenes* detection, based on membrane-engineered cells. The newly developed biosensor system is combined with a sophisticated algorithm embedded in user-friendly software that allows the end-user to be instantly informed of the analysis results. The aim of this study was the proof-of-concept of the biosensor system for *L. monocytogenes* detection and the optimization of parameters that affect its performance, such as the type and the concentration of the electroinserted antibody. The newly developed biosensor was proven to be a robust and selective tool for *L. monocytogenes* detection, with a limit of detection as low as 10^2 CFU mL^{-1}. Therefore, in future research, the essay will be optimized in different food substrates and validation of its ability to detect pathogen's presence on actual food samples will be performed, identifying and eradicating potential impediments due to the matrix effect.

Author Contributions: Conceptualization, D.T. and K.L.; Data curation, A.H., T.A., L.D., K.L. and A.I.; Investigation, A.H. and T.A.; Methodology, A.H., T.A. and K.L; Supervision, D.T.; Validation, A.H. and T.A.; Visualization, K.L.; Writing—original draft, A.H. and T.A.; Writing—review and editing, D.T. All authors have read and agreed to the published version of the manuscript.

Funding: This research was funded by RESTART 2016–2020 Programmes for Research, Technological Development and Innovation of the Research and Innovation Foundation (RIF) in Cyprus. (project "Post-Doc/0718/0003", Post Doctoral Researchers Programme).

Conflicts of Interest: The authors declare no conflict of interest.

References

1. World Health Organization (WHO). Available online: https://www.who.int/NEWS-ROOM/FACT-SHEETS/DETAIL/FOOD-SAFETY/ (accessed on 29 June 2020).
2. Centers for Disease Control and Prevention (CDC). Listeria (Listeriosis). 2020. Available online: https://www.cdc.gov/listeria/index.html/ (accessed on 29 June 2020).
3. International Organization for Standardization (ISO). *ISO 11290-1:2017: Microbiology of the Food Chain—Horizontal Method for the Detection and Enumeration of Listeria Monocytogenes and of Listeria spp.—Part 1: Detection Method*; ISO: Geneva, Switzerland, 2017; Available online: https://www.iso.org/standard/60313.html (accessed on 1 November 2020).
4. Ali, S.; Hassan, A.; Hassan, G.; Eun, C.; Bae, J.; Lee, C.H.; Kim, I. Disposable all-printed electronic biosensor for instantaneous detection and classification of pathogens. *Sci. Rep.* **2018**, *8*, 5920. [CrossRef] [PubMed]

5.   Kanayeva, D.A.; Wang, R.; Rhoads, D.; Erf, G.F.; Slavik, M.F.; Tung, S.; Li, Y. Efficient separation and sensitive detection of Listeria monocytogenes using an impedance immunosensor based on magnetic nanoparticles, a microfluidic chip, and an interdigitated microelectrode. *J. Food Prot.* **2012**, *75*, 1951–1959. [CrossRef] [PubMed]

6.   Scognamiglio, V.; Pezzotti, G.; Pezzotti, I.; Cano, J.; Buonasera, K.; Giannini, D.; Giardi, M.T. Biosensors for effective environmental and agrifood protection and commercialization: From research to market. *Microchim. Acta* **2010**, *170*, 215–225. [CrossRef]

7.   Thevenot, D.R.; Toth, K.; Durst, R.A.; Wilson, G.S. Electrochemical Biosensors: Recommended Definitions and Classification. *Pure Appl. Chem.* **1999**, *7*, 2333–2348. [CrossRef]

8.   Damborský, P.; Švitel, J.; Katrlík, J. Optical biosensors. *Essays Biochem.* **2016**, *60*, 91–100.

9.   Bhalla, N.; Jolly, P.; Formisano, N.; Estrela, P. Introduction to biosensors. *Essays Biochem.* **2016**, *60*, 1–8.

10.   Zhang, Z.; Cong, Y.; Huang, Y.; Du, X. Nanomaterials-based Electrochemical Immunosensors. *Micromachines* **2019**, *10*, 397. [CrossRef]

11.   Asal, M.; Özen, Ö.; Şahinler, M.; Polatoğlu, İ. Recent developments in enzyme, DNA and immuno-based biosensors. *Sensors* **2018**, *18*, 1924. [CrossRef]

12.   Gupta, N.; Renugopalakrishnan, V.; Liepmann, D.; Paulmurugan, R.; Malhotra, B.D. Cell-based biosensors: Recent trends, challenges and future perspectives. *Biosens. Bioelectron.* **2019**, *141*, 111435. [CrossRef]

13.   Inda, M.E.; Mimee, M.; Lu, T.K. Cell-based biosensors for immunology, inflammation, and allergy. *J. Allergy Clin. Immunol.* **2019**, *144*, 645–647. [CrossRef]

14.   Curtis, T.; Naal, R.M.Z.G.; Batt, C.; Tabb, J.; Holowka, D. Development of a mast cell-based biosensor. *Biosens. Bioelectron.* **2008**, *23*, 1024–1031. [CrossRef] [PubMed]

15.   Banerjee, P.; Lenz, D.; Robinson, J.P.; Rickus, J.L.; Bhunia, A.K. A novel and simple cell-based detection system with a collagen-encapsulated B-lymphocyte cell line as a biosensor for rapid detection of pathogens and toxins. *Lab. Investig.* **2008**, *88*, 196–206. [CrossRef] [PubMed]

16.   Banerjee, P.; Bhunia, A.K. Cell-based biosensor for rapid screening of pathogens and toxins. *Biosens. Bioelectron.* **2010**, *26*, 99–106. [CrossRef] [PubMed]

17.   Kintzios, S. Cell-based sensors in clinical chemistry. *Mini Rev. Med. Chem.* **2007**, *7*, 1019–1026. [CrossRef]

18.   Mavrikou, S.; Moschopoulou, G.; Tsekouras, V.; Kintzios, S. Development of a portable, ultra-rapid and ultra-sensitive cell-based biosensor for the direct detection of the SARS-CoV-2 S1 spike protein antigen. *Biosensors* **2020**, *20*, 3121.

19.   Moschopoulou, G.; Valero, T.; Kintzios, S. Molecular Identification through Membrane Engineering as a revolutionary concept for the construction of cell sensors with customized target recognition properties: The example of superoxide detection. *Procedia Eng.* **2011**, *25*, 1541–1544. [CrossRef]

20.   Moschopoulou, G.; Vitsa, K.; Bem, F.; Vassilakos, N.; Perdikaris, A.; Blouhos, P.; Yialouris, C.; Frossiniotis, D.; Anthopoulos, I.; Maggana, O.; et al. Engineering of the membrane of fibroblast cells with virus-specific antibodies: A novel biosensor tool for virus detection. *Biosens. Bioelectron.* **2008**, *24*, 1033–1036. [CrossRef]

21.   Perdikaris, A.; Alexandropoulos, N.; Kintzios, S. Development of a novel, ultra-rapid biosensor for the qualitative detection of hepatitis b virus-associated antigens and anti-HBV, based on "membrane-engineered" fibroblast cells with virus-specific antibodies and antigens. *Sensors* **2009**, *9*, 2176–2186. [CrossRef]

22.   Perdikaris, A.; Vassilakos, N.; Yiakoumettis, I.; Kektsidou, O.; Kintzios, S. Development of a portable, high throughput biosensor system for rapid plant virus detection. *J. Virol. Methods* **2011**, *177*, 94–99. [CrossRef]

23.   Moschopoulou, G.; Kintzios, S. Application of "membrane-engineering" to bioelectric recognition cell sensors for the detection of picomole concentrations of superoxide radical: A novel biosensor principle. *Anal. Chim. Acta* **2006**, *573–574*, 90–96. [CrossRef]

24.   Moschopoulou, G.; Valero, T.; Kintzios, S. Superoxide determination using membrane engineered cells: An example of a novel concept for the construction of cell sensors with customized target recognition properties. *Sens. Actuators B Chem.* **2012**, *175*, 88–94. [CrossRef]

25.   Kintzios, S.; Pistola, E.; Konstas, J.; Bem, F.; Matakiadis, T.; Alexandropoulos, N.; Bem, F.; Ekonomou, G.; Biselis, J.; Levin, R. Bioelectric recognition assay (BERA). *Biosens. Bioelectron.* **2001**, *16*, 467–480. [CrossRef]

26.   Apostolou, T.; Loizou, K.; Hadjilouka, A.; Inglezakis, A.; Kintzios, S. Newly developed system for acetamiprid residue screening in the lettuce samples based on a bioelectric biosensor. *Biosensors* **2020**, *10*, 8. [CrossRef]

27. Zeira, M.; Tosi, P.F.; Mouneimne, Y.; Lazarte, J.; Sneed, L.; Volsky, D.L.; Nikolau, C. Full-length CD4 electroinserted in the erythrocyte membrane as a long-lived inhibitor of infection by human immunodeficiency virus. *Proc. Natl. Acad. Sci. USA* **1991**, *88*, 4409–4413. [CrossRef]

28. Kokla, A.; Blouchos, P.; Livaniou, E.; Zikos, C.; Kakabakos, S.E.; Petrou, P.S.; Kintzios, S. Visualization of the membrane-engineering concept: Evidence for the specific orientation of electroinserted antibodies and selective binding of target analytes. *J. Mol. Recognit.* **2013**, *26*, 627–632. [CrossRef]

29. Glas, A.S.; Lijmer, J.G.; Prins, M.H.; Bonsel, G.J.; Bossuyt, P.M.M. The diagnostic odds ratio: A single indicator of test performance. *J. Clin. Epidemiol.* **2003**, *56*, 1129–1135. [CrossRef]

30. Van der Walt, S.; Colbert, S.C.; Varoquaux, G. The NumPy Array: A Structure for Efficient Numerical Computation, Computing in Science & Engineering. *Comput. Sci. Eng.* **2011**, *13*, 22–30.

31. Paramithiotis, S.; Hadjilouka, A.; Drosinos, E.H. Listeria pathogenicity island 1. Structure and function. In *Listeria Monocytogenes: Food Sources, Prevalence and Management Strategies*; Hambrick, E.C., Ed.; Nova Publishers: New York, NY, USA, 2014; pp. 265–282.

32. Koubova, V.; Brynda, E.; Karasova, L.; Skvor, J.; Homola, J.; Dostalek, J.; Tobiska, P.; Rosicky, J. Detection of foodborne pathogens using surface plasmon resonance biosensors. *Sens. Actuators B Chem.* **2001**, *74*, 100–105. [CrossRef]

33. Leonard, P.; Hearty, S.; Quinn, J.; O'Kennedy, R. A generic approach for the detection of whole Listeria monocytogenes cells in contaminated samples using surface plasmon resonance. *Biosens. Bioelectron.* **2004**, *19*, 1331–1335. [CrossRef]

34. Sharma, H.; Mutharasan, R. Rapid and sensitive immunodetection of Listeria monocytogenes in milk using a novel piezoelectric cantilever sensor. *Biosens. Bioelectron.* **2013**, *45*, 158–162. [CrossRef]

35. Cheng, C.; Peng, Y.; Bai, J.; Zhang, X.; Liu, Y.; Fan, X.; Ning, B.; Gao, Z. Rapid detection of *Listeria monocytogenes* in milk by self-assembled electrochemical immunosensor. *Sens. Actuators B Chem.* **2014**, *190*, 900–906. [CrossRef]

36. Wang, D.; Chen, Q.; Huo, H. Efficient separation and quantitative detection of *Listeria monocytogenes* based on screen-printed interdigitated electrode, urease and magnetic nanoparticles. *Food Control* **2017**, *73*, 555–561. [CrossRef]

37. De Michele, C.; De Los Rios, P.; Foffi, G.; Piazza, F. Simulation and Theory of Antibody Binding to Crowded Antigen-Covered Surfaces. *PLoS Comput. Biol.* **2016**, *12*, e1004752. [CrossRef]

38. Cho, I.H.; Paek, E.H.; Lee, H.; Kang, J.Y.; Kim, T.S.; Paek, S.H. Site-directed biotinylation of antibodies for controlled immobilization on solid surfaces. *Anal. Biochem.* **2007**, *365*, 14–23. [CrossRef]

39. Kwon, Y.; Han, Z.; Karatan, E.; Mrksich, M.; Kay, B.K. Antibody Arrays Prepared by Cutinase-Mediated Immobilization on Self-Assembled Monolayers. *Anal. Chem.* **2004**, *76*, 5713–5720. [CrossRef]

40. Benarroch, J.M.; Asally, M. The Microbiologist's Guide to Membrane Potential Dynamics. *Trends Microbiol.* **2020**, *28*, 304–314. [CrossRef]

Publisher's Note: MDPI stays neutral with regard to jurisdictional claims in published maps and institutional affiliations.

Article

A Comparative Study of Approaches to Improve the Sensitivity of Lateral Flow Immunoassay of the Antibiotic Lincomycin

Kseniya V. Serebrennikova [1], Olga D. Hendrickson [1], Elena A. Zvereva [1], Demid S. Popravko [1], Anatoly V. Zherdev [1], Chuanlai Xu [2] and Boris B. Dzantiev [1,*]

[1] Research Center of Biotechnology of the Russian Academy of Sciences, A.N. Bach Institute of Biochemistry, Leninsky Prospect 33, 119071 Moscow, Russia; ksenijasereb@mail.ru (K.V.S.); odhendrick@gmail.com (O.D.H.); zverevaea@yandex.ru (E.A.Z.); dspopravko@mitht.ru (D.S.P.); zherdev@inbi.ras.ru (A.V.Z.)

[2] School of Food Science and Technology, Jiangnan University, Wuxi 214122, China; xcl@jiangnan.edu.cn

* Correspondence: dzantiev@inbi.ras.ru; Tel.: +7-495-954-31-42

Received: 6 November 2020; Accepted: 1 December 2020; Published: 3 December 2020

Abstract: This study provides a comparative assessment of the various nanodispersed markers and related detection techniques used in the immunochromatographic detection of an antibiotic lincomycin (LIN). Improving the sensitivity of the competitive lateral flow immunoassay is important, given the increasing demands for the monitoring of chemical contaminants in food. Gold nanoparticles (AuNPs) and CdSe/ZnS quantum dots (QDs) were used for the development and comparison of three approaches for the lateral flow immunoassay (LFIA) of LIN, namely, colorimetric, fluorescence, and surface-enhanced Raman spectroscopy (SERS)-based LFIAs. It was demonstrated that, for colorimetric and fluorescence analysis, the detection limits were comparable at 0.4 and 0.2 ng/mL, respectively. A SERS-based method allowed achieving the gain of five orders of magnitude in the assay sensitivity (1.4 fg/mL) compared to conventional LFIAs. Therefore, an integration of a SERS reporter into the LFIA is a promising tool for extremely sensitive quantitative detection of target analytes. However, implementation of this time-consuming technique requires expensive equipment and skilled personnel. In contrast, conventional AuNP- and QD-based LFIAs can provide simple, rapid, and inexpensive point-of-care testing for practical use.

Keywords: lateral flow immunoassay; antibiotics; lincomycin; gold nanoparticles; quantum dots; surface-enhanced Raman spectroscopy

---

1. Introduction

The lateral flow immunoassay (LFIA) is a common analytical platform for the point-of-care testing of medical diagnostics and environmental monitoring because of its rapidity and simplicity. The LFIA provides clear advantages, including the availability of results within a few minutes, the small volume of an analyzed sample, and inexpensive and user-friendly point-of-care testing [1]. The LFIA combines immunochemical reactions with a chromatography principle. It relies on interactions between an analyte and pre-immobilized recognition elements initiated by the addition of a liquid sample. The LFIA result is a signal at the test line generated by a nanodispersed reporter used. Despite all the advantages mentioned above, the widespread use of LFIAs has been limited by their insufficient sensitivity. Significant effort has been devoted to improving LFIA sensitivity, including the use of alternative labels and detectors, as well as the addition of amplification stages [2,3].

To improve the sensitivity of the immunoassay, integration of the LFIA and surface-enhanced Raman spectroscopy (SERS) was proposed. Because of a simple and cost-effective synthesis, gold and silver nanoparticles are the most common SERS substrates [4]. Typically, nanostructured substrates are functionalized with Raman reporter molecules to produce strong and characteristic peaks in SERS spectra, thus enabling quantitative detection of target analytes. The effectiveness of the SERS-based LFIA technique has been confirmed in numerous recent studies [5–8]. In addition to common AuNPs or latex beads, magnetic and fluorescent particles are used as labels in LFIAs. QDs are used as labels because of their unique optical properties, such as high fluorescence, broad and continuous distributed excitation, photostability, and proven immunoassay effectiveness [9,10]. LFIAs with magnetic and photoluminescent labels showed improved sensitivity for a wide range of analytes [11–14]. Among other markers applied in LFIA, carbon nanoparticles can be mentioned [15,16]. Compared to other labels, carbon nanoparticles are easily detected visually, which contributes to reducing the detection limit of the analyte.

A survey of the literature shows there have been many works published on new immunoassay markers, but they do not go beyond the description of the effectiveness at detecting a particular analyte or report a comparison of the results with conventional gold nanoparticle-based LFIAs. These regularities are poorly transformed into other objects of research. Therefore, the assessment of the test systems with the same reagents that vary according to the kind of marker and readout technique applied will provide more information.

During the study, we explored three approaches to improving LFIA sensitivity. To verify the effectiveness of the proposed methods, we selected the antimicrobial lincomycin (LIN), which is a product of *Streptomyces lincolnensis* bacteria. The known varieties of methods for quantitative detection of LIN include mainly microbiological and chromatographic techniques [17]. The use of accurate chromatographic methods is a common practice to identify and quantify antibiotics in different matrices. Although chromatography–mass spectrometry is a highly sensitive and efficient method, its use requires sample pretreatment, costly equipment, and specially trained personnel [18,19]. Recently, other techniques have also been reported for the determination of LIN in foodstuffs [20,21]. Numerous studies have reported the use of the enzyme-linked immunosorbent assay (ELISA) and LFIA for monitoring LIN residues [22–24]. However, despite the availability of the techniques to control antibiotics, there is great demand for the development of highly sensitive alternative ways of (a) achieving simple pretreatment procedures (reduce it to dilution eliminating the matrix effect) and (b) minimizing the risk of long-term consumption of contaminants at concentrations below threshold levels.

In this study, the same bioreagents were used to compare different labels and readout systems in a competitive LFIA for LIN. An increase in competitive LFIA sensitivity is possible by reducing the concentration of immunoreagents; however, this decrease is limited by the ability to detect the analytical signal. Beyond the optimization of reagent concentrations, improving the signal-generating elements and readout techniques are other effective strategies to achieve increased assay sensitivity. Moreover, the integration of sensitive detection techniques with LFIA allows for a reduction in immunoreagent consumption.

The current study is a systematic investigation using LFIA integrated with different labels (AuNPs and QDs) and readout techniques (colorimetry, fluorescence, and SERS) to detect LIN. AuNPs were implemented both for traditional colorimetric detection and for coupling to SERS readouts. The quantitative detection of LIN was performed by registering the colorimetric or fluorescence intensity of AuNPs or QDs, respectively, captured on the test line. To design a SERS-based LFIA, AuNPs functionalized with 4-mercaptobenzoic acid (4-MBA) and coupled with anti-LIN monoclonal antibodies (AuNPs–MBA–Ab) were used as a SERS reporter bioprobe. In this case, a conventional LFIA procedure was followed by registration of Raman spectra from the test line.

2. Materials and Methods

2.1. Reactants

Lincomycin hydrochloride monohydrate (LIN), $HAuCl_4$, sodium azide, sodium citrate, Tween-20, Triton X-100, and 4-MBA were obtained from Sigma-Aldrich (St. Louis, MO, USA). N-(3-dimethylaminopropyl)-N′-ethylcarbodiimide hydrochloride (EDC) and sulfo-N-hydroxysuccinimide (NHS) were supplied from Fluka (Buchs, Switzerland). Goat antibodies against mouse immunoglobulins (GAMI) were purchased from Arista Biologicals (Allentown, PA, USA). Bovine serum albumin (BSA) was supplied from Eximio Biotec (Wuxi, China). The CdSe/ZnS QDs with an emission peak at 625 nm were obtained from Invitrogen (Catalog No A10200, Thermo Fisher Scientific, Waltham, MA, USA). All other reagents were of analytical grade.

Ultrapure water (Millipore Corporation, Burlington, MA, USA) with resistivity of 418.2 MΩ was used to prepare the AuNPs and their conjugates as well as LIN stock solutions (100 µg/mL). The LFIAs were carried out in 96-well transparent Costar 9018 polystyrene microplates provided by Corning Costar (Tewksbury, MA, USA). Amicon Ultra-0.5 mL Centrifugal Filter (100 K) was purchased from Millipore (Billerica, MA, USA).

2.2. Preparation of Monoclonal Anti-LIN Antibodies

A synthesis of the LIN–BSA conjugate and a preparation of anti-LIN antibodies were carried out in accordance with the procedure described in the study by Cao et al. [25].

2.3. Synthesis and Characterization of AuNPs

AuNPs with an average diameter of 30 nm and 40 nm were prepared according to the citrate-reduction method [26]. To obtain 30 nm AuNPs, 1 mL of 1% $HAuCl_4$ was added to 97.5 mL of ultrapure water and heated to boiling. After that, 1.5 mL of 1% sodium citrate was added immediately to the boiling solution during vigorous stirring. The mixture was left to boil for 25 min and then cooled. The colloidal AuNPs were stored at 4 °C.

To obtain AuNPs with an average diameter of 40 nm, 1.5 mL of 1% sodium citrate was added to 100 mL of boiling 0.01% $HAuCl_4$ aqueous solution under rapid agitation. The solution was then boiled for another 15 min and cooled to room temperature.

The transmission electron microscopic (TEM) images were recorded with a JEM-100C electron microscope (JEOL, Tokyo, Japan) operating at 80 kV. The AuNP preparations were applied to 300-mesh grids (Pelco International, Redding, CA, USA) coated with formvar film. The images obtained were analyzed using Image Tool software (University of Texas Health Science Center, San Antonio, TX, USA). UV–vis absorption spectra were obtained through spectrophotometer UV-2450 (Shimadzu, Kyoto, Japan).

2.4. Conjugation of Antibodies to AuNPs

Antibody–AuNPs conjugates were prepared according to the previously described technique [27]. Anti-LIN antibodies were dialyzed against a Tris-HCl buffer (10 mM, pH 8.5), and added to AuNPs at a concentration of 10 µg/mL (OD520 = 1). The mixture was incubated for 45 min while stirring at room temperature. BSA in the final concentration of 0.25% was further added to this preparation, followed by stirring for 15 min. The excess reagents were removed by centrifugation at 9500× g for 15 min, followed by resuspension of the antibody–AuNPs pellet in Tris buffer (10 mM, pH 8.5) with 1% BSA, 1% sucrose, and 0.1% sodium azide (TBSA).

2.5. Conjugation of Antibodies with QDs

Anti-LIN antibodies were dialyzed against a borate buffer (50 mM, pH 8.7). The molar ratio of QDs to anti-LIN antibodies during synthesis was 1:2. Antibodies (300 μL, 0.2 mg/mL), QDs (25 μL, 8 μM), and freshly prepared EDC and NHS solutions (50 μL, 0.8 mM each) were mixed. After incubation for 90 min in a dark place at room temperature, the resulting mixture was purified by centrifugation at 10,000× g for 15 min using Amicon Ultra 100 kDa tubes (Billerica, MA, USA).

The centrifugation was repeated four times, and, finally, 14 μL of QDs with a concentration of 4.26 mg/mL was obtained.

2.6. Synthesis of the Raman Reporter Bioprobe

To a solution of 40 nm diameter AuNPs (10 mL), 10 μL of 1 mM 4-MBA in ethanol was added [28]. The mixture was incubated for 3 h, followed by centrifugation at 5000× g for 15 min. The resulting pellet was resuspended in water.

To prepare the AuNPs–MBA–Ab bioprobe, Au–MBA conjugate and anti-LIN antibodies were adjusted to pH 8.9 with 0.1 M K_2CO_3. Anti-LIN antibodies at a concentration of 10 μg/mL were added to 2 mL of Au–MBA and incubated for 3.5 h at room temperature. Then, 50 μL of 10% BSA was added and incubated overnight at 4 °C. After that, the mixture was centrifuged at 9000× g, for 10 min. The pellet was resuspended in an equal volume of water and stored at 4 °C.

2.7. Preparation of Test Strips

The schemes of three LFIA formats are shown in Figure 1. Test strips were assembled using MdiEasypack membrane sets (Advanced Microdevices, Ambala Cantt, India) comprising the following elements: a plastic support, a CNPC nitrocellulose working membrane with a pore size of 15 μm, a PT–R7 conjugate fiberglass pad (in case of conventional AuNP- and QD-based LFIAs), a GFB-R4 sample pad, and an AP045 absorbent pad. The control line was formed by applying 0.5 mg/mL GAMI in a K-phosphate buffer (PBS, 50 mM, pH 7.4, with 0.1 M NaCl) by an Iso-Flow automatic dispenser (Imagene Technology, Hanover, NH, USA). To form a test line, LIN–BSA conjugate (0.5 mg/mL—for AuNPs-based, 0.15 mg/mL—for QD-based, and 0.2 mg/mL—for SERS-based LFIAs, in PBS) was applied. After that, the test strips were dried at 37 °C for 2 h. For AuNP-based LFIA, the antibody-AuNPs conjugate in TBSA containing 0.05% Tween-20 was applied to the conjugate pad and dried at room temperature overnight. For QD-based LFIA, 1 μL of antibody–QDs conjugate (0.09 μM) in a borate buffer (BB, 0.05 M with 1% BSA, 0.1% sucrose, and 0.1% sodium azide, 0.05% Tween-20) was applied to the interface of the sample pad and nitrocellulose membrane and dried at room temperature overnight. Finally, the assembled multimembrane composites were cut into individual test strips 3 mm wide using an automatic guillotine cutter (Index Cutter-1, A-Point Technologies, Gibbstown, NJ, USA).

2.8. LFIA Procedures

2.8.1. Colorimetric and Fluorescent LFIAs

Solutions of LIN (1 μg/mL–1 pg/mL) in PBST (100 μL) were dripped onto the microplate wells. The test strips were vertically placed into the well and left to react for 15 min. The color intensity (in the case of AuNP-based LFIA) of the formed bands was scanned by the CanoScanLiDE 90 (Canon, Tokyo, Japan). The fluorescence intensity (for QD-based LFIA) was recorded under UV light excitation. The obtained images were then digitized using the TotalLab program (Nonlinear Dynamics, Newcastle upon Tyne, UK).

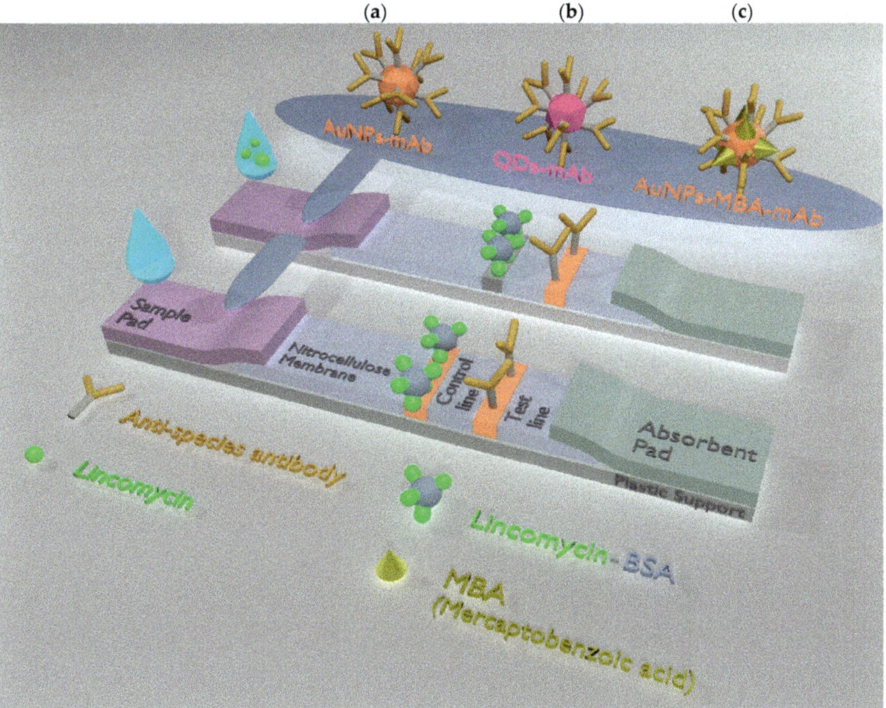

Figure 1. Schemes of lateral flow immunoassay (LFIA) formats developed in the study: conventional colorimetric gold nanoparticle (AuNP)-based LFIA (**a**); fluorescent quantum dot (QD)-based LFIA (**b**); AuNP-based LFIA with surface-enhanced Raman scattering (SERS) detection (**c**).

2.8.2. SERS-Based LFIA

To perform the SERS-based LFIA, 2 μL of AuNPs–MBA–Ab bioprobe was pipetted onto the sample pad approximately 1 cm below the nitrocellulose membrane, and 100 μL of LIN $(100,000–1 \times 10^{-8}$ ng/mL) in PBS containing 0.05% Triton X-100 (PBST) was added into the microplate wells. The test strips were vertically inserted into the wells and left to react for 15 min. Then, the Raman spectra from 10 points along the middle of the test line were collected using a DXR Raman microscope (Thermo Fisher Scientific, Madison, WI, USA). The SERS settings were selected with the identical registering technique [29,30]. All spectra were obtained under the same conditions: The excitation source was tuned at 780 nm and laser power of 20 mW; the exposure time was 10 s. A 10× objective lens (NA = 0.25) was used to focus a laser spot on the surface of the test strip.

3. Results and Discussion

3.1. Synthesis and Characterization of Signal Markers

AuNPs were used as a reporter label in conventional and SERS-based LFIAs. AuNPs of a diameter close to 30 nm were reported to be optimal for traditional immunochromatography [31], whereas larger particles are preferable in SERS-based LFIAs. The optimal size of AuNPs for the preparation of a SERS-active probe was previously found to be no more than 50 nm [32]. Therefore, to achieve a desirable sensitivity and high reproducibility in SERS-based LFIAs, AuNPs with an average size of 40 nm were preferred. To prepare AuNPs with diameters of 30 and 40 nm, a simple method of sodium citrate-associated reduction of chloroauric acid was applied. According to this method, the size of the resulting AuNPs is varied by adding different amounts of the reducing agent: to obtain larger particles, a smaller volume of reducing agent is required. The selection of 4-MBA as a Raman reporter molecule stems from the widespread use of thiol-containing aromatic molecules because of their ability to conjugate directly to the gold surface and provide surface carboxyl groups for biomolecule binding [33]. To optimize the composition of the AuNPs–MBA–Ab bioprobe, the amount of added 4-MBA was varied in the range of 10–50 µL per 10 mL of AuNPs. It was demonstrated that an excess of 4-MBA could cause the aggregation of AuNPs, as evidenced by a red-shifted absorbance peak in UV–vis spectra and a color change of the AuNPs–MBA–Ab probe (data not shown). Therefore, 10 µL of 1 mM MBA was proven to be sufficient for preparation of a stable AuNPs–MBA conjugate.

The size and shape of AuNPs were estimated by TEM and UV–vis spectroscopy (Figure 2). The as-prepared AuNPs showed localized surface plasmon resonance at 523 and 527 nm for AuNPs of 30 and 40 nm, respectively. After the conjugation process, the slight redshift of the maximum peak of AuNPs–MBA–Ab was observed, which indicates a successful conjugation of Au–MBA and anti-LIN antibodies. The TEM images revealed spherical morphology and homogeneity with a size distribution in the range of 29.5 ± 7.4 nm and 39.5 ± 5.0 nm, and a degree of ellipticity of 1.3 for two AuNPs preparations (Figure 2b,c). According to the manufacturer, the size of carboxyl quantum dots varies from 15 to 20 nm.

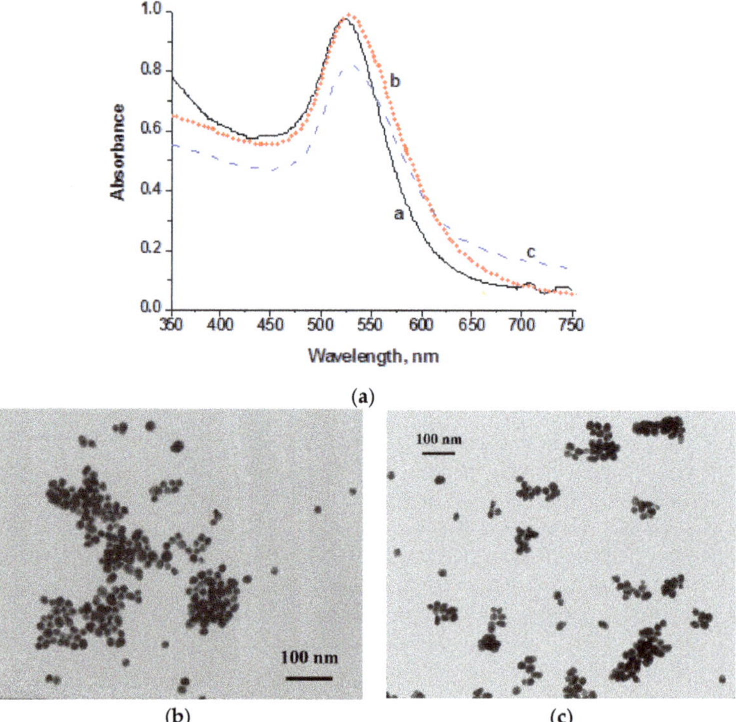

Figure 2. (a) UV–vis spectra of AuNPs of 30 nm (**a**) and 40 nm (**b**) diameter, and AuNPs functionalized with 4-mercaptobenzoic acid (4-MBA) and coupled with anti-lincomycin (LIN) monoclonal antibodies (AuNPs–MBA–Ab) (**c**); (**b**) microphotograph of AuNPs for colorimetric LFIA. The average diameter is 29.5 ± 7.4 nm and the degree of polydispersity is 1.3; (**c**) microphotograph of AuNP–MBA for SERS-based LFIA. The average diameter is 39.5 ± 5.0 nm and the degree of polydispersity is 1.3.

3.2. AuNP-Based LFIA

Given the need to detect a low molecular weight compound in this study, a direct competitive LFIA format was performed (Figure 1). In this assay, free LIN present in the sample competed with the immobilized LIN–BSA conjugate in regard to binding with specific anti-LIN antibodies. The binding sites of the specific anti-LIN antibodies labeled with different nanodispersed markers were first occupied with the target analyte. And thereafter the excess labeled antibodies were captured by the LIN-BSA conjugate, which in turn was detected by employing different detection techniques. Thus, the signal intensity on the test line of the strip was inversely proportional to the concentration of LIN in the sample. The preliminary characterization of the immune properties of monoclonal anti-LIN antibodies used by ELISA confirmed their high affinity (Figure S1) and allowed for the development of LFIAs.

The scheme of the conventional AuNP-based LFIA is presented in Figure 1a. For the LFIA, LIN–BSA conjugate, and GAMI were applied to form test and control lines on the working membrane, respectively. The specific antibody-labeled AuNPs was immobilized on the fiberglass pad. The assay conditions were optimized to achieve the lowest detection limit at a high amplitude of the analytical signal. As a result, the following conditions were found to be optimal for three formats of assay: 0.5 mg/mL—for AuNP-based, 0.15 mg/mL—for QD-based, and 0.2 mg/mL—for SERS-based LFIAs (the concentration varies from 0.2 to 1 mg/mL) and 0.5 mg/mL for GAMI (the concentration varies from 0.15 to 0.5 mg/mL). The AuNPs-anti-LIN antibodies solution was then applied to the conjugate pad at the concentration corresponding to $OD_{520} = 1$ (we tested OD_{520} in the range from 0.5 to 2.5). The overall performance of the LFIA was explored by varying the concentration of the analyte (from 1000 to 0.001 ng/mL). Under optimal experimental conditions, the AuNP-based LFIA exhibits linearity over the range of 0.7–7.2 ng/mL with an instrumental detection limit of 0.4 ng/mL (Figure 3). The cutoff was 10 ng/mL with the assay duration of 15 min.

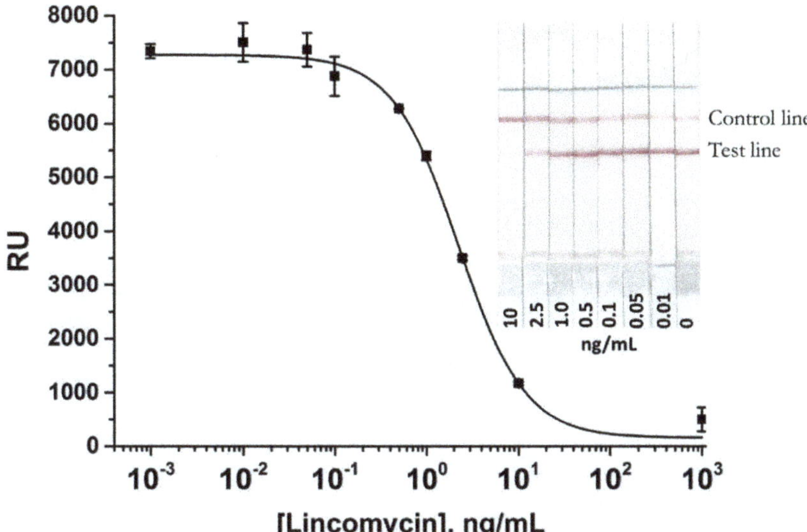

Figure 3. Calibration curve of LIN in the AuNP-based LFIA and the digital photographs of the LFA strips after conventional AuNP-based LFIA procedure. The LIN-BSA conjugate was applied at the test line at a concentration of 0.5 mg/mL. The AuNPs-anti-LIN antibodies solution was applied at a concentration corresponding $OD_{520} = 1$. LIN concentrations are given at the bottom of the test strips. The error bars indicate the standard deviations for three measurements.

3.3. QD-Based LFIA

The scheme of QD-based LFIA is demonstrated in Figure 1b. For QD-based LFIA, the selection of working membranes aimed to decrease background fluorescence was carried out together with the optimization of specific reagent concentrations described above. For this purpose, CNPC SS12 12/15 μ (Advanced Microdevices), HF120, and HF180 (Millipore) membranes differing in pore size and flow rates were tested. Figure 4 indicates that the use of CNPC SS12 (of a 12 and 15 μm pore size, respectively) leads to the formation of background coloration over its entire surface. Testing of Millipore membranes with different pore sizes and flow rates demonstrated that the application of the Millipore HF180 membrane facilitated achieving the maximum analytical signal intensity (as opposed to a 20% reduction in intensity when using a HF120 membrane), eliminating nonspecific binding, and ensuring uniform movement of samples.

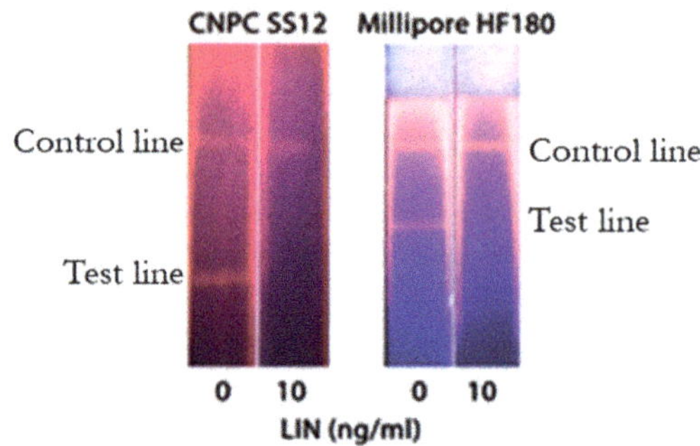

Figure 4. Images of the test zones after the LFIA performed using CNPC SS12 and Millipore HF180 membranes. The LIN-BSA conjugate concentration was 0.15 mg/mL.

The next stage of assay optimization was to select the optimal reaction medium that would decrease nonspecific binding and provide a higher signal intensity. The use of PBST as a buffer solution for QD-based LFIA led to nonspecific binding of the antibody–QDs conjugate and background staining of the working membrane. For LIN detection, significantly higher signal intensities were obtained with BB. It is acknowledged that, for the effective elution of the antibody–QDs conjugate and its movement along the membrane, detergents must be added to the buffer [26]. It was shown that the addition of Tween-20 (0.05%), BSA (1%), and sucrose (0.1%) to BB eliminated the nonspecific sorption of QD-labeled antibodies in the test zone and increased the intensity of the analytical signal by 15%. BSA and sucrose were added to the buffer to reduce the flow rate (due to a viscosity increase) and, hence, to maximize the contact time of the sample with the labeled antibodies. Furthermore, the use of BSA allows blocking the sites of nonspecific sorption of the conjugate [34].

An antibody–QDs conjugate solution was applied to the interface of the sample pad and working membrane at a volume of 1 μL and a concentration range of 0.09 to 0.28 μM. The 0.09 μM conjugate concentration was shown to give the optimal fluorescence intensity. A further decrease in the concentration of the antibody–QDs conjugate led to a drop in the signal amplitude and a decrease in the reproducibility of the test results.

Figure 5 shows the calibration curve for LIN detection in the optimized LFIA. The instrumental LOD was 0.2 ng/mL and the dynamic linear range was 0.6–10.4 ng/mL. The visual LOD was 20 ng/mL. QD-based LFIA can provide results in 15 min.

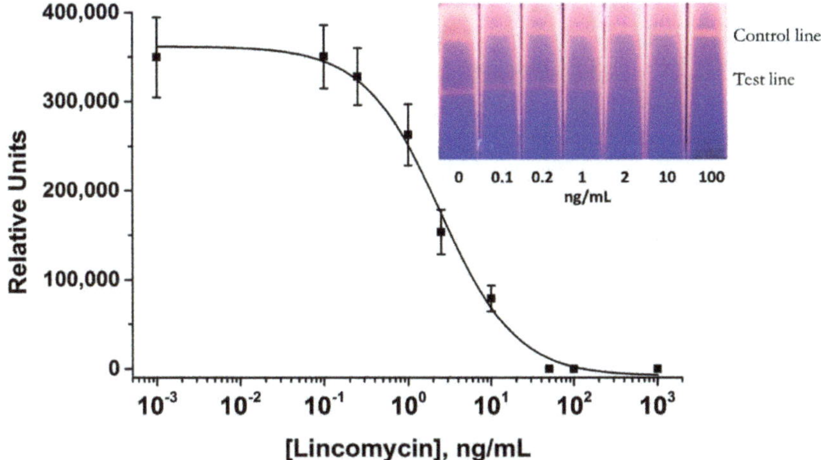

Figure 5. Calibration curve of LIN in the QD-based LFIA and image of the test strips with increasing concentration of LIN ranging from 0.1 to 100 ng/mL under following optimal conditions: The LIN-BSA conjugate was applied at the test line at a concentration of 0.15 mg/mL; 1 µL of antibody–QDs conjugate (0.09 µM) was applied to the interface of the sample pad and nitrocellulose membrane. The error bars indicate the standard deviations for three measurements.

3.4. SERS-Based LFIA

The principle of the SERS-based LFIA is illustrated in Figure 1c. In this study, the bioconjugates of the anti-LIN antibody and AuNPs functionalized with 4-MBA were applied as both a Raman reporter bioprobe and a detection probe. In this study, 4-MBA was chosen as the reporter molecule because of its ability to provide a strong binding to the AuNP surface and high SERS-signal while being in proximity with a metal surface (enhancement factor up to 1×10^7) [35]. To obtain the optimal characteristics of the test-system, the following parameters were optimized: The amount of the LIN–BSA conjugate immobilized on the test line; the amount of the Raman reporter bioprobe. After immersing the test strips in 100 µL of the LIN solutions, visual staining was detected on the test line after 15 min. According to the competitive format of assay, the color intensity and, consequently, the SERS signal provided by the AuNPs-MBA-Ab probe is inversely proportional to the LIN concentration. As shown in Figure 6, the SERS spectra of the Raman reporter bioprobe are characterized by two intense peaks at 1077 cm^{-1} and 1580 cm^{-1}, which in conformity with spectral data for the MBA molecule [36] correspond to vibrations in the aromatic ring. Therefore, this confirms the specific binding of reporter bioprobe to the LIN-BSA conjugate on the test line. The highest peak at 1077 cm^{-1} was used further to quantify the antibiotic content. As follows from the spectra corresponding to different LIN concentrations, the SERS intensity gradually decreases at 1077 cm^{-1} with an increase in the concentration of antibiotics.

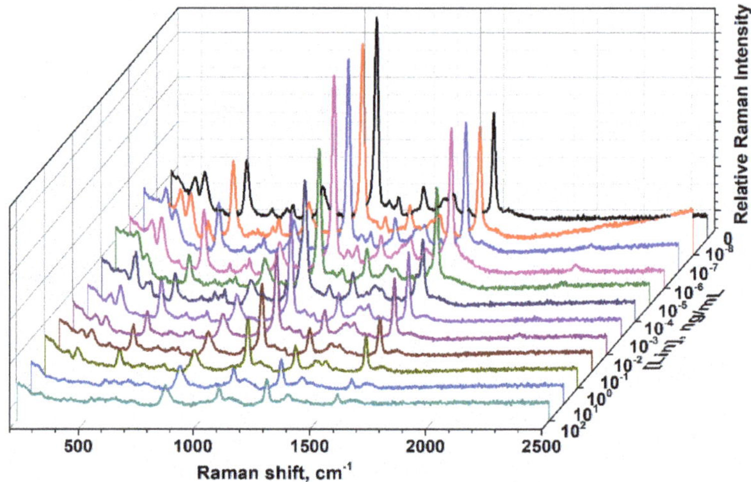

Figure 6. SERS spectra arising from MBA on a test line for various LIN concentrations following the LFIA procedures under optimal conditions: The amount of AuNPs–MBA–Ab bioprobe was 2 µL; the amount of LIN–BSA conjugate was 0.2 mg/mL.

To investigate the impact of the Raman reporter bioprobe, AuNPs–MBA–Ab in amounts ranging from 1 to 5 µL were dotted on a sample pad of the strip (Figure 7a). For the coating antigen immobilized on the test line, LIN–BSA with concentrations ranging from 0.2 to 0.5 mg/mL were investigated, respectively (Figure 7b). The amount of these components is shown to have no significant influence on assay sensitivity (Figure 7).

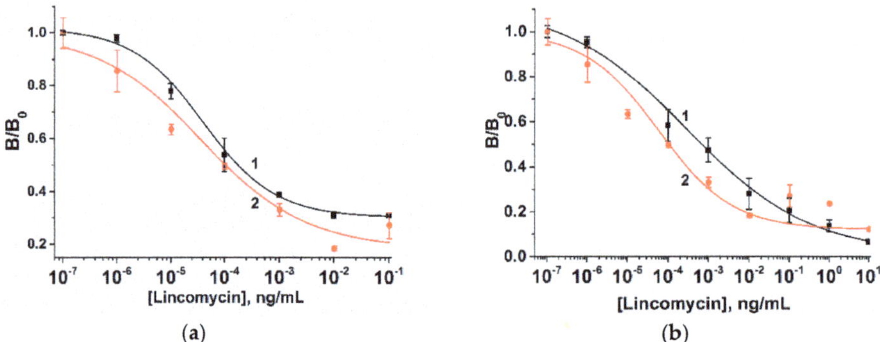

Figure 7. Calibration curves of the SERS-based LFIA of LIN for various LIN–BSA conjugate and AuNPs–MBA–Ab bioprobe amounts. (**a**) The amount of AuNPs–MBA–Ab bioprobe was 1 µL (1) and 4 µL (2); (**b**) the amount of LIN–BSA conjugate was 0.2 mg/mL (1) and 0.5 mg/mL (2). B and B_0 ($B_0 \approx 1000$ a.u.) correspond to the SERS intensities of MBA at 1077 cm^{-1}, when standard and zero LIN solutions were applied to the sample pad, respectively. The error bars indicate the standard deviations for three measurements.

When the LIN–BSA concentration and a loading of an AuNPs–MBA–Ab bioprobe increased, the Raman intensities decreased. In contrast, a decrease in the amount of Raman reporter bioprobe and the concentration of the immobilized LIN–BSA conjugate allowed for the expansion of the dynamic range of the detected LIN concentrations. Therefore, the optimal amount of LIN–BSA conjugate was 0.2 mg/mL and the amount of AuNPs–MBA–Ab bioprobe added to the strip was 2 µL. The results shown in Figure 8 indicate that the dynamic linear range of the SERS-based LFIA varies from 2.8×10^{-6} to 10 ng/mL with a detection limit of 1.4×10^{-6} ng/mL. Such improved analytical characteristics can be explained by the high sensitivity of SERS detection toward a reporter molecule, which allows for simultaneous reduction in the amount of immobilized LIN-BSA conjugate and the Au-MBA-Ab reporter bioprobe. Not enough attention is being paid to the development of LFIA for LIN. According to the data shown in Table 1, most studies are devoted to the development of conventional colorimetric LFIAs for the detection of LIN, including instances where LIN is part of a panel of antibiotics (multiplex assay format).

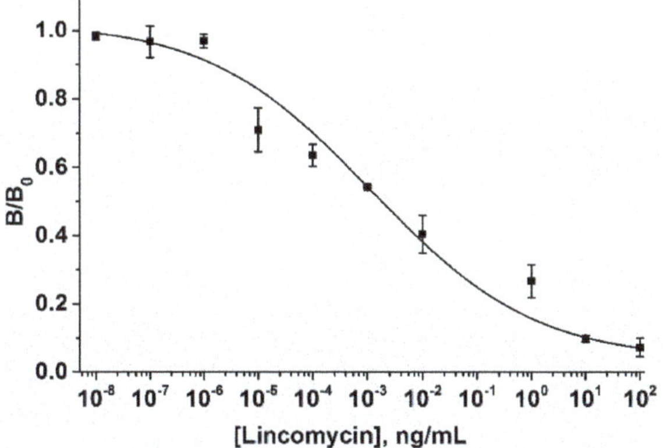

Figure 8. The calibration curve of the SERS-based LFIA for LIN detection according to the following optimal assay conditions: The amount of AuNPs–MBA–Ab bioprobe was 2 µL; the LIN–BSA conjugate concentration was 0.2 mg/mL. The error bars indicate the standard deviation of the Raman intensities from MBA reporter molecule at 1077 cm^{-1} measured from 10 points along the middle of the test line.

Table 1. Lateral flow immunoassay (LFIA) tests for lincomycin detection.

Target Analyte	LFIA Formats	Signal Marker	Limit of Detection	References
Lincomycin (and chloramphenicol, tetracycline)	Multiplex LFIA	30 nm AuNPs	0.4 ng/mL	[37]
Lincomycin (and gentamicin, kanamycin, streptomycin, neomycin)	Multiplex LFIA	15 nm AuNPs	2.5 ng/mL	[38]
Lincomycin	Fluorescence LFIA	Fluorescent microspheres	0.69 ng/mL	[23]
Lincomycin (and clindamycin, pirlimycin)	Conventional LFIA	20 nm AuNPs	10 ng/mL	[39]
Lincomycin	Indirect LFIA	30 nm AuNPs	8 pg/mL	[24]

Abbreviations: AuNPs—gold nanoparticles.

Replacing AuNPs with fluorescent microspheres made it possible to achieve a small gain in the assay sensitivity [23]. However, a significant decrease in the detection limit of LIN was achieved in our previous study (up to 8 pg/mL) when an indirect LFIA was implemented. In this study, the implementation of SERS readout technique in AuNP-based LFIA using the same immunoreagents revealed an approximately three orders of magnitude improvement in assay sensitivity. It should, however, be pointed out that such high sensitivity is resulted not only from the implementation of effective readout techniques but also from the excellent characteristics of the applied antibodies, the affinity of which was 1.15×10^{9} M^{-1} (according to information provided by the manufacturer).

To date, the integration of immunoassay, in particular LFIA, with the SERS detection technique for the development of highly sensitive quantitative test systems is just getting started. Nevertheless, a number of studies prove the effectiveness of this method for the determination of target analytes at low concentrations [30,40–42]. The current study demonstrates excellent performance of the test system reached by the integration of the SERS technique with LFIA, and this integration may be considered as a potential tool for sensitive screenings of antibiotics. However, SERS technique, which is attractive because of its ability to detect extremely low analyte concentrations, is difficult to classify as point-of-care testing unless it is a handheld Raman reader format. As a result of the steps taken to design simple and portable SERS-based LFIA readers, the commercial availability of such devices may be expected in the future [43,44]. On the contrary, AuNP- and QD-LFIAs are fast and cheap out-of-laboratory techniques available today, the results of which can be quantified even using smartphones or handheld readers [45,46]. Therefore, the choice of appropriate technique is influenced by the object of study.

4. Conclusions

In the current study, several LFIA approaches using the antibiotic LIN as the relevant contaminant of food products were performed and compared, including conventional AuNP-based LFIA, fluorescent QD-based LFIA, and SERS-based LFIA. AuNP- and QD-based LFIAs are confined to the limit of detection of the nanodispersed label used and the analyses were carried out in the direct competitive format with use of anti-LIN antibodies labeled with AuNPs or QDs. The colorimetric AuNPs-based LFIA was characterized by the detection limit of 0.4 ng/mL. The replacement of the colorimetric marker with a fluorescent one resulted in a slight enhancement in sensitivity (the detection limit was 0.2 ng/mL). To address current challenges of LFIA biosensors associated with the lack of sensitivity and limits in quantitative analysis, the novel SERS-based LFIA for LIN was developed. The limit of detection determined by SERS experiments was 1.4×10^{-6} ng/mL. Notably, the sensitivity of AuNP- and QD-based LFIAs are defined by the detection limit of the nanodispersed marker on the test strip, while in the case of SERS-based LFIA an indirect registration of the signal from the Raman reporter molecule using a highly sensitive device is performed. Therefore, the ordinary comparison of the detection limits achieved using the considered three approaches is not quite legitimate and the choice of a nanodispersed marker and a signal detection technique should be determined by several parameters, in particular, the aim of the study, facilities of the laboratory, the nature of the target analyte and requirements to its maximum residue limits. The proposed SERS-based LFIA, which possesses both high sensitivity and quantitative evaluation capabilities, confirms the effectiveness of the SERS technique for the sensitive detection of target analytes. This implies that handheld Raman readers for quantitative LFIA could potentially facilitate sensitive point-of-care tests. To our knowledge, ours is the first report of quantitative LIN detection by fluorescence and SERS-based LFIA, which also presents a promising tool for other contaminants.

Supplementary Materials: The following are available online at http://www.mdpi.com/2079-6374/10/12/198/s1. Figure S1: Competitive curve for the ELISA of LIN. The detection limit of LIN (IC10) is 0.08 ng/mL, its concentration causing 50% inhibition of the antibody binding (IC50) is 0.69 ng/mL. The error bars indicate the standard deviations for three measurements.

Author Contributions: Conceptualization, B.B.D.; methodology, A.V.Z.; formal analysis, K.V.S., O.D.H. and E.A.Z.; investigation, K.V.S., O.D.H., E.A.Z. and D.S.P.; resources, C.X.; writing—original draft preparation, K.V.S.; writing—review and editing, O.D.H. and A.V.Z.; visualization, K.V.S., O.D.H. and E.A.Z.; supervision, B.B.D.; project administration, A.V.Z. All authors have read and agreed to the published version of the manuscript.

Funding: This research was financially supported by the Russian Science Foundation (Project 19-14-00370).

Acknowledgments: The authors are grateful to S.M. Pridvorova (Research Center of Biotechnology of the Russian Academy of Sciences) for obtaining electronic microphotographs of the AuNPs, and to D.V. Kuznetsov (National University of Science and Technology (MISiS), Moscow, Russia) for assistance in using the DXR Raman Microscope.

Conflicts of Interest: The authors declare no conflict of interest.

References

1. Quesada-González, D.; Merkoçi, A. Nanoparticle-based lateral flow biosensors. *Biosens. Bioelectron.* **2015**, *73*, 47–63. [CrossRef]

2. Urusov, A.E.; Zherdev, A.V.; Dzantiev, B.B. Towards Lateral flow quantitative assays: Detection approaches. *Biosensors* **2019**, *9*, 89. [CrossRef]

3. Qu, Z.; Wang, K.; Alfranca, G.; de la Fuente, J.M.; Cui, D. A plasmonic thermal sensing based portable device for lateral flow assay detection and quantification. *Nanoscale Res. Lett.* **2020**, *15*, 10. [CrossRef] [PubMed]

4. Mosier-Boss, P.A. Review of SERS substrates for chemical sensing. *Nanomaterials* **2017**, *7*, 142. [CrossRef] [PubMed]

5. Hwang, J.; Lee, S.; Choo, J. Application of a SERS-based lateral flow immunoassay strip for the rapid and sensitive detection of staphylococcal enterotoxin B. *Nanoscale* **2016**, *8*, 11418–11425. [CrossRef] [PubMed]

6. Khlebtsov, B.N.; Bratashov, D.N.; Byzova, N.A.; Dzantiev, B.B.; Khlebtsov, N.G. SERS-based lateral flow immunoassay of troponin I by using gap-enhanced Raman tags. *Nano Res.* **2019**, *12*, 413–420. [CrossRef]

7. Qian, J.; Xing, C.; Ge, Y.; Rui, L.; Aitong, L.; Yan, W. Gold nanostars-enhanced Raman fingerprint strip for rapid detection of trace tetracycline in water samples. *Spectrochim. Acta A* **2020**, *232*, 118146. [CrossRef]

8. Wang, Y.; Sun, J.; Hou, Y.; Zhang, C.; Li, D.; Li, H.; Yang, M.; Fan, C.; Sun, B. A SERS-based lateral flow assay biosensor for quantitative and ultrasensitive detection of interleukin-6 in unprocessed whole blood. *Biosens. Bioelectron.* **2019**, *141*, 111432. [CrossRef]

9. Berlina, A.N.; Taranova, N.A.; Zherdev, A.V.; Vengerov, Y.Y.; Dzantiev, B.B. Quantum dot-based lateral flow immunoassay for detection of chloramphenicol in milk. *Anal. Bioanal. Chem.* **2013**, *405*, 4997–5000. [CrossRef]

10. Tripathi, P.; Upadhyay, N.; Nara, S. Recent advancements in lateral flow immunoassays: A journey for toxin detection in food. *Crit. Rev. Food Sci. Nutr.* **2018**, *58*, 1715–1734. [CrossRef]

11. Zangheri, M.; Di Nardo, F.; Anfossi, L.; Giovannoli, C.; Baggiani, C.; Roda, A.; Mirasoli, M. A multiplex chemiluminescent biosensor for type B-fumonisins and aflatoxin B1 quantitative detection in maize flour. *Analyst* **2014**, *140*, 358–365. [CrossRef] [PubMed]

12. Morales-Narváez, E.; Naghdi, T.; Zor, E.; Merkoçi, A. Photoluminescent lateral-flow immunoassay revealed by graphene oxide: Highly sensitive paper-based pathogen detection. *Anal. Chem.* **2015**, *87*, 8573–8577. [CrossRef] [PubMed]

13. Han, G.R.; Kim, M.G. Highly Sensitive chemiluminescence-based lateral flow immunoassay for cardiac troponin i detection in human serum. *Sensors* **2020**, *20*, 2593. [CrossRef] [PubMed]

14. Yan, L.; Dou, L.; Bu, T. Highly sensitive furazolidone monitoring in milk by a signal amplified lateral flow assay based on magnetite nanoparticles labeled dual-probe. *Food Chem.* **2018**, *261*, 131–138. [CrossRef] [PubMed]

15. Noguera, P.; Posthuma-Trumpie, G.A.; van Tuil, M.; van der Wal, F.J.; de Boer, A.; Moers, A.P.H.A.; van Amerongen, A. Carbon nanoparticles in lateral flow methods to detect genes encoding virulence factors of Shiga toxin-producing Escherichia coli. *Anal. Bioanal. Chem.* **2011**, *399*, 831–838. [CrossRef] [PubMed]

16. Wiriyachaiporn, N.; Sirikett, H.; Maneeprakorn, W.; Dharakul, T. Carbon nanotag based visual detection of influenza A virus by a lateral flow immunoassay. *Microchim. Acta* **2017**, *184*, 1827–1835. [CrossRef]

17. Parthasarathy, R.; Monette, C.E.; Bracero, S.; Saha, M.S. Methods for field measurement of antibiotic concentrations: Limitations and outlook. *FEMS Microbiol. Ecol.* **2018**, *94*, fiy 105. [CrossRef]

18. Benetti, C.; Piro, R.; Binato, G.; Angeletti, R.; Biancotto, G. Simultaneous determination of lincomycin and five macrolide antibiotic residues in honey by liquid chromatography coupled to electrospray ionisation mass spectrometry. *Food Addit. Contam.* **2006**, *23*, 1099–1108. [CrossRef]

19. Jank, L.; Martins, M.; Bazzan, J.; Magalhães, T.; Motta, C.; Hoff, R.; Barreto, F.; Pizzolato, T. High-throughput method for macrolides and lincosamides antibiotics residues analysis in milk and muscle using a simple liquid–liquid ex- traction technique and liquid chromatography–electrospray–tandem mass spectrometry analysis (LC–MS/MS). *Talanta* **2015**, *144*, 686–695. [CrossRef]

20. Leng, Y.; Hu, F.; Ma, C.; Du, C.; Ma, L.; Xu, J.; Lin, Q.; Sang, Z.; Luand, Z. Simple, rapid, sensitive, selective and label-free lincomycin detection by using $HAuCl_4$ and NaOH. *RSC Adv.* **2019**, *9*, 28248–28252. [CrossRef]

21. Li, S.; Liu, C.; Yin, G.; Zhang, Q.; Luo, J.; Wu, N. Aptamer-molecularly imprinted sensor base on 2916 electrogenerated chemiluminescence energy transfer for detection of lincomycin. *Biosens. Bioelectron.* **2017**, *91*, 687–691. [CrossRef] [PubMed]

22. Hendrickson, O.D.; Zvereva, E.A.; Zherdev, A.V.; Godjevargova, T.; Xu, C.; Dzantiev, B.B. Development of a double immunochromatographic test system for simultaneous determination of lincomycin and tylosin antibiotics in foodstuffs. *Food Chem.* **2020**, *318*, 126510. [CrossRef] [PubMed]

23. Zhou, J.; Zhu, K.; Xu, F.; Wang, W.; Jiang, H.; Wang, Z.; Ding, S. Development of a microsphere-based fluorescence immunochromatographic assay for monitoring lincomycin in milk, honey, beef, and swine urine. *J. Agric. Food Chem.* **2014**, *62*, 12061–12066. [CrossRef] [PubMed]

24. Hendrickson, O.D.; Zvereva, E.A.; Popravko, D.S.; Zherdev, A.V.; Xu, C.; Dzantiev, B.B. An immunochromatographic test system for the determination of lincomycin in foodstuffs of animal origin. *J. Chromatogr. B* **2020**, *1141*, 122014. [CrossRef] [PubMed]

25. Cao, S.; Song, S.; Liu, L.; Kong, N.; Kuang, H.; Xu, C. Comparison of an enzyme-linked immunosorbent assay with an immunochromatographic assay for detection of lincomycin in milk and honey. *Immunol. Investig.* **2015**, *44*, 438–450. [CrossRef]

26. Frens, G. Controlled nucleation for the regulation of the particle size in monodisperse gold suspensions. *Nat. Phys. Sci.* **1973**, *241*, 20–22. [CrossRef]

27. Hendrickson, O.D.; Zvereva, E.A.; Shanin, I.A.; Zherdev, A.V.; Tarannum, N.; Dzantiev, B.B. Highly sensitive immunochromatographic detection of antibiotic ciprofloxacin in milk. *Appl. Biochem. Microbiol.* **2018**, *54*, 670–676. [CrossRef]

28. Li, X.; Yang, T.; Song, Y.; Zhu, J.; Wang, D.; Li, W. Surface-enhanced Raman spectroscopy (SERS)-based immunochromatographic assay (ICA) for the simultaneous detection of two pyrethroid pesticides. *Sens. Actuators B* **2019**, *283*, 230–238. [CrossRef]

29. Serebrennikova, K.; Samsonova, J.; Osipov, A.; Senapati, D.; Kuznetsov, D. Gold nanoflowers and gold nanospheres as labels in lateral flow immunoassay of procalcitonin. *Nano Hybrids Compos.* **2017**, *13*, 47–53. [CrossRef]

30. Fu, X.; Chu, Y.; Zhao, K.; Li, J.; Deng, A. Ultrasensitive detection of the β-adrenergic agonist brombuterol by a SERS-based lateral flow immunochromatographic assay using flower-like gold-silver core-shell nanoparticles. *Microchim. Acta* **2017**, *184*, 1711–1719. [CrossRef]

31. Alasel, M.; Keusgen, M. Two-protein modified gold nanoparticles for one-step serological diagnosis. *Phys. Status Solidi A* **2018**, *215*, 1700700. [CrossRef]

32. Karnorachai, K.; Sakamoto, K.; Laocharoensuk, R.; Suwussa, B.; Dharakul, T.; Miki, K. SERS-based immunoassay on 2D-arrays of Au@Ag core–shell nanoparticles: Influence of the sizes of the SERS probe and sandwich immunocomplex on the sensitivity. *RSC Adv.* **2017**, *7*, 14099–14106. [CrossRef]

33. Li, R.; Lv, H.; Zhang, X.; Liu, P.; Chen, L.; Cheng, J.; Zhao, B. Vibrational spectroscopy and density functional theory study of 4-mercaptobenzoic acid. *Spectrochim. Acta Part. A* **2015**, *148*, 369–374. [CrossRef] [PubMed]

34. O'Farrell, B. Evolution in lateral flow–based immunoassay systems. In *Lateral Flow Immunoassay*; Wong, R.C., Tse, H.Y., Eds.; Humana Press: New York, NY, USA, 2009; Chapter 1; p. 236.

35. Orendorff, C.J.; Gole, A.; Sau, T.K.; Murphy, C.J. Surface-enhanced raman spectroscopy of self-assembled monolayers: sandwich architecture and nanoparticle shape dependence. *Anal. Chem.* **2005**, *77*, 3261–3266. [CrossRef]

36. Smith, G.; Girardon, J.S.; Paul, J.F.; Berrier, E. Dynamics of a plasmon-activated p-mercaptobenzoic acid layer deposited over Au nanoparticles using time-resolved SERS. *Phys. Chem. Chem. Phys.* **2016**, *18*, 19567–19573. [CrossRef]

37. Bartosh, A.V.; Sotnikov, D.V.; Hendrickson, O.D.; Zherdev, A.V.; Dzantiev, B.B. Design of multiplex lateral flow tests: A case study for simultaneous detection of three antibiotics. *Biosensors* **2020**, *10*, 17. [CrossRef]

38. Peng, J.; Wang, Y.; Liu, L.; Kuang, H.; Li, A.; Xu, C. Multiplex lateral flow immunoassay for five antibiotics detection based on gold nanoparticle aggregations. *RSC Adv.* **2016**, *6*, 7798–7805. [CrossRef]

39. Guo, L.; Wu, X.; Liu, L.; Kuang, H.; Xu, C. Gold Immunochromatographic assay for rapid on site detection of lincosamide residues in milk, egg, beef, and honey samples. *Biotechnol. J.* **2019**, *15*, 1900174. [CrossRef]

40. Fu, X.; Cheng, Z.; Yu, J.; Choo, P.; Chen, L.; Choo, J. A SERS-based lateral flow assay biosensor for highly sensitive detection of HIV-1 DNA. *Biosens. Bioelectron.* **2016**, *78*, 530–537. [CrossRef]

41. Shu, L.; Zhou, J.; Yuan, X.; Petti, L.; Chen, J.; Jia, Z.; Mormile, P. Highly sensitive immunoassay based on SERS using nano-Au immune probes and a nano-Ag immune substrate. *Talanta* **2014**, *123*, 161–168. [CrossRef]

42. Wang, Y.; Chen, S.; Wei, C.; Xu, M.; Yao, J.; Li, Y.; Deng, A.; Gu, R. A femtogram level competitive immunoassay of mercury (II) based on surface-enhanced Raman spectroscopy. *Chem. Commun. (Camb. Engl.)* **2014**, *50*, 9112–9114. [CrossRef] [PubMed]

43. Xiao, R.; Lu, L.; Rong, Z.; Wang, C.; Peng, Y.; Wang, F.; Wang, J.; Sun, M.; Dong, J.; Wang, D.; et al. Portable and multiplexed lateral flow immunoassay reader based on SERS for highly sensitive point-of-care testing. *Biosens. Bioelectron.* **2020**, *168*, 112524. [CrossRef] [PubMed]

44. Tran, V.; Walkenfort, B.; König, M.; Salehi, M.; Schlücker, S. Rapid, quantitative, and ultrasensitive point-of-care testing: A portable SERS reader for lateral flow assays in clinical chemistry. *Angew. Chem. (Int. Ed. Engl.)* **2019**, *58*, 442–446. [CrossRef] [PubMed]

45. Hou, Y.; Wang, K.; Xiao, K.; Qin, W.; Lu, W.; Tao, W.; Cui, D. Smartphone-based dual-modality imaging system for quantitative detection of color or fluorescent lateral flow immunochromatographic strips. *Nanoscale Res. Lett.* **2017**, *12*, 291. [CrossRef]

46. Ruppert, C.; Phogat, N.; Laufer, S.; Kohl, M.; Deigner, H.-P. A smartphone readout system for gold nanoparticle-based lateral flow assays: Application to monitoring of digoxigenin. *Microchim. Acta* **2019**, *186*, 119. [CrossRef]

Publisher's Note: MDPI stays neutral with regard to jurisdictional claims in published maps and institutional affiliations.

biosensors

MDPI

Article

Aptamer–Target–Gold Nanoparticle Conjugates for the Quantification of Fumonisin B1

Vicente Antonio Mirón-Mérida *, Yadira González-Espinosa, Mar Collado-González, Yun Yun Gong, Yuan Guo and Francisco M. Goycoolea *

School of Food Science and Nutrition, University of Leeds, Leeds LS2 9JT, UK;
y.gonzalezespinosa@leeds.ac.uk (Y.G.-E.); m.d.m.colladogonzalez@leeds.ac.uk (M.C.-G.);
y.gong@leeds.ac.uk (Y.Y.G.); y.guo@leeds.ac.uk (Y.G.)
* Correspondence: fsvamm@leeds.ac.uk (V.A.M.-M.); f.m.goycoolea@leeds.ac.uk (F.M.G.)

Abstract: Fumonisin B1 (FB1), a mycotoxin classified as group 2B hazard, is of high importance due to its abundance and occurrence in varied crops. Conventional methods for detection are sensitive and selective; however, they also convey disadvantages such as long assay times, expensive equipment and instrumentation, complex procedures, sample pretreatment and unfeasibility for on-site analysis. Therefore, there is a need for quick, simple and affordable quantification methods. On that note, aptamers (ssDNA) are a good alternative for designing specific and sensitive biosensing techniques. In this work, the assessment of the performance of two aptamers (40 and 96 nt) on the colorimetric quantification of FB1 was determined by conducting an aptamer–target incubation step, followed by the addition of gold nanoparticles (AuNPs) and NaCl. Although $MgCl_2$ and Tris-HCl were, respectively, essential for aptamer 96 and 40 nt, the latter was not specific for FB1. Alternatively, the formation of Aptamer (96 nt)–FB1–AuNP conjugates in $MgCl_2$ exhibited stabilization to NaCl-induced aggregation at increasing FB1 concentrations. The application of asymmetric flow field-flow fractionation (AF4) allowed their size separation and characterization by a multidetection system (UV-VIS, MALS and DLS online), with a reduction in the limit of detection from 0.002 µg/mL to 56 fg/mL.

Keywords: Fumonsin B1; aptamers; gold nanoparticles; UV/VIS spectroscopy; asymmetric flow field-flow fractionation

Citation: Mirón-Mérida, V.A.;
González-Espinosa, Y.;
Collado-González, M.; Gong, Y.Y.;
Guo, Y.; Goycoolea, F.M.
Aptamer–Target–Gold Nanoparticle
Conjugates for the Quantification of
Fumonisin B1. *Biosensors* **2021**, *11*, 18.
https://doi.org/10.3390/bios
11010018

Received: 23 December 2020
Accepted: 5 January 2021
Published: 8 January 2021

Publisher's Note: MDPI stays neutral
with regard to jurisdictional claims in
published maps and institutional affiliations.

1. Introduction

Exposure to FB1 occurs not only in African [1] and Latin-American [2] countries but also in several regions of Asia [3] and Europe [4]. For that reason, monitoring and controlling contamination of food commodities with fumonisins becomes a highly pressing matter for protecting human health worldwide. Fumosinins are a group of toxins generated by diverse fungi including *Fusarium verticillioides* [5], *Alternaria alternata* [6], *Aspergillus niger* [7], *Tolypocladium cylindrosporum*, *Tolypocladium geodes* and *Tolypocladium inflatum* [8]. Their chemical structure commonly consists of alkylamines, whose substitution of up to seven side chains allows the formation of different analogs, group B being the most common in nature [9]. From the latter, Fumonisin B1($C_{34}H_{59}NO_{15}$) has been reported to be a latent risk in several food products, such as cereals and beverages.

Classified as group 2B hazard, FB1 (Figure S1) has a possible carcinogenic effect on human health [10], whose toxicity has been related to the disruption of sphingolipid metabolism, oxidative stress and epigenetic changes [11], along with the interruption of barrier functions [12]. Nevertheless, current conventional methods for mycotoxin detection, including enzyme-linked immunosorbent assay (ELISA), high-performance liquid chro-matography with fluorescence detection (HPLC-FLD) and liquid chromatography-mass spectrometry (LC-MS), are costly, time consuming, are difficult to be applied on site and require experienced users [13]. Therefore, there is a need for developing sensitive, yet quick and affordable methods for quantifying mycotoxins.

Biosensors are a suitable alternative to conventional methods by means of their general mechanism, where any target detection is carried out by a biological receptor and transduced into a signal (optical, electrochemical, mechanical, etc.). From the different biorecognition receptors (enzymes, antibodies, nuclei acids, cells, etc.), the performance of aptamers has been exceptional [14]. Aptamers are single-stranded DNA or RNA molecules with high molecular recognition toward different types of molecules, distinct binding affinities, target selectivity and high capability to discriminate slight chiral differences [15]. Their selection technique called Systematic Evolution of Ligands by Exponential enrichment (SELEX), involves incubating a DNA library with the specific target or other relevant molecules, followed by the amplification of potential binders after several selection and discrimination rounds [16]. In addition to their good performance, compared to antibodies, aptamers are easy to modify, as well as reproducible by solid-phase chemical synthesis, which represents a reduction in cost and time. To date, two aptamers specific for FB1 composed of 96 and 80 nucleotides (nt) have been selected by SELEX and used in up to 31 different biosensing approaches [15,16], through aptamer modifications, hybridization with complementary strands or reduction in the sequence length (Figure S2). Those applications have involved optical, chemiluminescence, electrochemical, surface-enhanced Raman spectroscopy (SERS), mass spectrometry (MS) and bending related signals, where the most sensitive methods were correlated with fluorescent and SERS read outs (Figure S2).

A decisive step during the design of aptamer-based biosensors is the selection of the target-aptamer incubation conditions (buffer, time and temperature). In such approaches, Tris, Tris-HCl, phosphate buffer, NaCl, $CaCl_2$, KCl and $MgCl_2$ are normally added during the binding stage, which normally results in a 3D conformational change of aptamers upon binding [16]. Apart from the sensitive response obtained through aptamer dehybridization from complementary sequences at specific binding sites, mycotoxin detection has been carried out by immobilization of aptamers onto the surface of different platforms such as graphene and gold nanoparticles. In this regard, gold nanoparticles (AuNPs) are suitable for developing colorimetric methods, which are still relevant due to its feasible application for on-site analysis and reduction in engineering costs. The thermodynamic instability of colloidal gold suspensions potentiates their flocculation. This aggregation, however, can be prevented when gold nanoparticles are coated with negatively charged citrate molecules, or at low ionic strength and pH above the isoelectric point of citrate. Particle aggregation in this case occurs when an aqueous dispersant medium containing the nanoparticles has a high enough ionic strength to screen their electrostatic repulsion charges caused by their citrate stabilization (e.g., salt-induced aggregation) [17]. AuNPs are commonly analyzed in terms of their absorbance by using surface plasmon resonance related to their color modification (red to blue) by aggregation with cationic compounds, changes in the suspension pH and ionic strength [18,19]. The SPR spectra of AuNPs are closely related to their particle size, which plays an important role in their absorption, scattering and excitation behavior [20]. Functionalization of AuNPs can be promoted through aptamer (ssDNA) uncoiling, whose bases are exposed to the negative surface of AuNPs, thus enabling their interaction by van der Waals forces [21].

In addition to the spectrophotometric analysis of AuNPs, particle size characterization can be achieved by more robust methods capable of probing the interaction between aptamers and AuNPs in the presence of a target molecule. For instance, asymmetric flow field-flow fractionation (AF4) is a technique that allows the separation of particles and aggregates in a size range from 1 nm to 1 μm, based on their diffusion coefficient in aqueous media. This analytical method offers several advantages, such as minimal sample alteration and efficient quantitative analysis of the physico-chemical parameters of the study materials such as their concentration and particle size by multidetection (UV/VIS, fluorescence, multiangle light scattering (MALS) and differential refractive index (dRI)) [22]. The general AF4 operation principle is based on the generation of a parabolic laminar flow profile within the separation channel transporting the particles. The action of a perpendicular flow, known as cross flow (CF) across the semipermeable membrane, drives separation according to the diffusion coefficient of particles. Some AF4 methods have been developed for studying the properties of AuNPs as either standard samples [22] or additives [23], as well as to study the target-binding relation between aptamers and their specific targeted proteins [24].

In this paper, we developed a bulk colorimetric method to examine the efficiency of two aptamers, namely, the first 96 nt sequence and a shortened version (40 nt) from the 80 nt aptamer, to quantify FB1 in different binding buffers and assessed its potential as a robust biosensing method. The general procedure involved incubating FB1 with aptamers, followed by another incubation stage with AuNPs, addition of NaCl and subsequent analysis by UV/VIS spectroscopy (λ = 400–800 nm). In addition, the characterization of the formation of Aptamer (96 nt)–FB1–AuNP conjugates was carried out by multidetection AF4, thus uncovering a promising application on the sensitive detection of this mycotoxin.

2. Materials and Methods

2.1. Materials

Fumonisin B1 (FB1, CAT FB1147), aflatoxin B1(AFB1, CAT A6636) from *Aspergillus flavus*, ochratoxin A (OTA, CAT 32937) and sodium azide (CAT 71290) were obtained from Sigma-Aldrich (St. Louis, MO, USA). Tris-HCl Buffer (UltraPure™ 1M pH 7.5, CAT 15567027) was acquired from Invitrogen™(USA). Sodium chloride (CAT BP358-1), methanol (CAT A454-1) and phosphate buffered saline (PBS, pH 7.4, CAT BP2944-100) tablets were purchased from Fisher Scientific (Loughborough, UK). Magnesium chloride (CAT J364) and potassium chloride (CAT 1.04936) were both bought from VWR (Lutterworth, UK) and BioChemica (Barcelona, Spain), respectively. Novachem surfactant 100 (C-SUR-100, lot 162167) was purchased from Postnova Analytics (Landsberg am Lech, Germany). All experiments were conducted using Mili-Q water (MQ water). Synthesized, dried and HPLC-purified aptamers specific for FB1 (Aptamer 40 nt: 5′-C GAT CTG GAT ATT ATT TTT GAT ACC CCT TTG GGG AGA CAT-3′ and Aptamer 96 nt: 5′-ATA CCA GCT TAT TCA ATT AAT CGC ATT ACC TTA TAC CAG CTT ATT CAA TTA CGT CTG CAC ATA CCA GCT TAT TCA ATT AGA TAG TAA GTG CAA TCT-3′) were purchased from Biomers.net (Germany) and diluted in sterile Milli-Q water.

2.2. Synthesis and Characterization of Gold Nanoparticles (AuNP)

Two stock solutions (stock 1 and 2) of gold nanoparticles (AuNP) were synthesized by citrated reduction [25]. The concentration of AuNP was determined according to the Lambert–Beer equation, Equation (1), based on a wavelength scan (200–800 nm) performed using a Specord 210 Plus Analytic Jena spectrophotometer (Jena, Germany).

$$A = C * \varepsilon * L \tag{1}$$

where A is the absorbance, C the molar concentration, ε is the molar extinction coefficient (3.67×10^8 M^{-1}cm^{-1}-specific for AuNP with a surface plasmon resonance (SPR) peak wavelength of 520 nm) and L the path length (1 cm).

Particle size distribution (nm) of AuNPs was determined in triplicate by dynamic light scattering with non-invasive back scattering (DLS-NIBS) at a measurement angle of 173° at 25 °C, in a Malvern Zetasizer NanoZS (Malvern Instruments, UK) fitted with a red laser (λ = 632.8 nm), and software in automatic mode.

2.3. Functionalization of Gold Nanoparticles (AuNP) with Aptamers

To perform the functionalization of AuNPs with aptamer, first, AuNPs (30 μL) were separately mixed with different concentrations of NaCl (1:1 *v/v*) to find their SPR peak shifting point (aggregation point). Then, different aptamer:AuNP mol ratios were mixed and incubated for 90 min at room temperature (RT~ 22 °C), from which 30 μL was mixed in a 96-well microplate with the selected NaCl molar concentration (1:1 *v/v*). Measurements to find the point of aptamer functionalization (until non-aggregation was observed) were conducted through a 400–800 nm scan using a Tecan Spark 10 M plate reader (Tecan, Reading, UK).

2.4. Assays with Aptamer 40 nt

2.4.1. Effect of Tris-HCl, PBS, and Its Combination on the Binding Effect of Aptamer 40 nt

Three concentrations of FB1 (0, 10 and 100 μg/mL), including two high values to secure and observable effect, were dissolved in either Tris-HCl buffer 31.1 mM, PBS 12.79 mM (NaCl equivalents) or a combination of both, and were then incubated with aptamer 40 nt in a vial (calculated to a final volume of 5 μL per microplate-well), for 60 min at 37 °C. After this step, a volume of AuNPs (stock 1) necessary to reach the selected aptamer:AuNP molar ratio (117:1) was added and incubated at 37 °C for 2 h. From this solution, 30 μL was placed on a microplate well and combined with NaCl 0.4 M (1:1 v:v) and subsequently subjected to a wavelength scan (λ = 400–800 nm) on a Tecan microplate reader. The absorbance ratio at 650 and 520 nm ($A_{650/520}$) was then calculated. The general procedure applied as a biosensing technique in this work is outlined in Figure 1a.

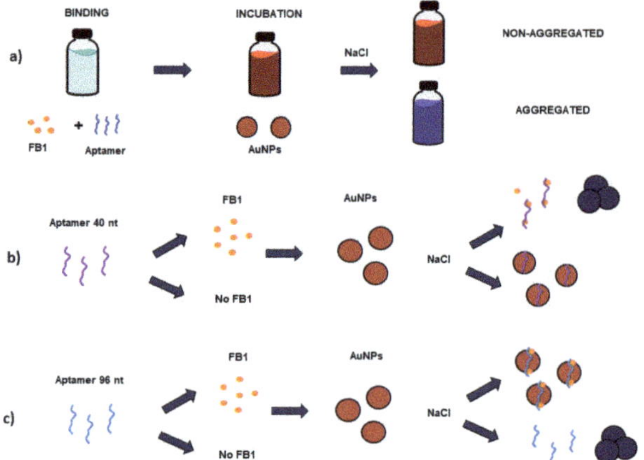

Figure 1. (**a**) Bulk aptasensor for the colorimetric determination of FB1with both aptamers through the binding mechanism proposed for the quantification of FB1 with (**b**) aptamer 40 nt and (**c**) aptamer 96 nt.

2.4.2. Effect of Different Buffers on the Performance of Aptamer 40 nt

An increasing concentration of FB1 dissolved in different binding buffers was combined with aptamer 40 nt (final volume: 5 μL per well), for 60 min at 37 °C. The addition of AuNPs (stock 1) to achieve a molar ratio of 117:1 was conducted according to Table 1. The $A_{650/520}$ ratio values after the addition of NaCl 0.4 M (1:1 v:v) were used to calculate the limit of detection (*LOD*) for each assay, according to Equation (2) [26].

$$LOD = Blank + 3\sigma_{blank} \tag{2}$$

where the Blank accounts for the signal exhibited by the blank (sample with no FB1), and $3\sigma_{blank}$ is three times the standard deviation of the blank. The proposed mechanism for aptamer 40 nt is displayed in Figure 1b.

Table 1. Different binding conditions for aptamer-based quantification of FB1 (aptamer 40 nt) [1].

Assay	Binding Buffer	Incubation with AuNP Time/ Temperature (min/°C)	Equation A_{650}/A_{520} =	r^2	Range of Tested Concentrations (µg/mL)	Limit of Detection (LOD) (µg/mL)
1	MgCl$_2$ 0.19 mM	120/37	$-0.008 \ln [X] + 0.2977$	0.6542	0.008–8.0	4.16
2	NaCl 0.06M + MgCl$_2$ 0.1mM	120/37	$0.0164 \ln [X] + 0.3025$	0.6795	0.0074–7.4	0.35
3	PBS (eq. 12 mM NaCl) + Tris-HCl buffer 17 mM	120/37	$0.0442 \ln [X] + 0.6709$	0.737	0.0086–8.6	0.11
4	Tris HCl buffer 14.06 mM	105/20	$0.0419 \ln [X] + 0.9072$	0.8311	0.0096–9.66	0.03

[1] X = FB1 concentration in µg/mL. Assay conditions: Aptamer:AuNP (mol:mol):117:1; FB1–aptamer incubation: 60 min (37 °C).

2.4.3. Reduction in the Aptamer: AuNP Molar Ratio

A folding step was integrated by placing a vial containing Aptamer 40 nt (dissolved in the binding buffer for this section), in a water bath at 94 °C for 5 min, followed by 15 min on ice. FB1 was also dissolved in binding buffer (Tris-HCl buffer 15 mM, NaCl 85 mM, CaCl$_2$ 1mM, KCl 5mM and MgCl$_2$ 2mM), and then incorporated (final volume: 5 µL per well) and incubated for 60 min at RT. Stock 1 of AuNPs was added to achieve the required aptamer:AuNPs molar ratio (47:1 to 117:1) and left in a shaking incubator (Titramax 1000, Heidolph, UK) at 300 RPM for another 60 min at RT, followed by the addition of NaCl 0.4 M (1:1 ratio v:v) and calculation of its $A_{650/520}$ ratio and LODs.

2.4.4. Specificity of Aptamer 40 nt

The specificity of aptamer 40 nt was tested on the addition of FB1, OTA or AFB1 (13.8 µM), with an aptamer:AuNPs molar ratio of 117:1 in Tris-HCl 14.06 mM with 1 h binding at 37 °C and 105 min of AuNP functionalization at RT. In addition, an aptamer:AuNPs ratio of 47:1 in buffer from Section 2.4.3 (folding included) was used for testing the specificity against OTA after 1 h binding (RT) and 1 h functionalization (RT).

2.5. Assays with Aptamer 96 nt

2.5.1. AuNP Functionalization with Aptamer 96 nt

Aptamer 96 nt was dissolved in MgCl$_2$ 1 mM and folded through a 5 min incubation in a water bath (94 °C), followed by 15 min immersion on ice. Then, different concentrations of FB1 (0.01–10 µg/mL) in MgCl$_2$ 1 mM were added and incubated at 37 °C for 30 min, followed by the addition of stock 2 AuNPs (30:1 molar ratio) for 1h at RT. A wavelength scan (λ = 200–800 nm) was performed on 30 µL of the mixture combined in a 96-well microplate with NaCl 0.2 M (1:1 v:v) for calculating the $A_{650/520}$ ratio. Additionally, the difference in absolute area was calculated in Origin Pro v. 8.6 software, between the curves of each sample and the respective blank (without FB1). The proposed mechanism for aptamer 96 nt is displayed in Figure 1c.

2.5.2. Specificity of Aptamer 96 nt

The specificity of aptamer 96 nt was also tested on the addition of FB1, OTA or AFB1 (1.38 µM), by following the incubation conditions in Section 2.5.1. Such mycotoxins were selected due to their relevance and simultaneous occurrence in some food products (also related to some synergistic effects).

2.5.3. Asymmetric Flow Field-Flow Fractionation (AF4)

After the functionalization of AuNPs with aptamer 96 nt at different FB1 concentrations (0.001–10 µg/mL), as indicated in Section 2.5.1, NaCl 0.2 M was added. The suspensions were subjected to AF4 conducted in an AF2000 Multiflow system from Postnova Analytics UK Ltd. (Malvern, UK). The method for the size separation of AuNPs stabilized with the aptamer–FB1 complex occurred within the channel (Postnova Z-AF4-CHA-611) having a 350 µM spacer and was performed in a 10 kDa cut-off regenerated cellulose membrane (Z-AF4-MEM-612-10KD). The temperature of the channel was controlled by a thermostat (PN4020) and set at 30 °C for all experiments. A solution of 0.05% Novachem® with sodium azide (3 mM) to avoid bacterial growth in the channel was used as the carrier liquid and prepared in Milli-Q water filtered through a 0.1 µm membrane filter (VCWP Millipore). The autosampler was equipped with a 500 µL loop allowing the injection of a 100 µL sample. All the samples were measured using the following optimized AF4 method that first consisted in an injection step at a flow of 0.2 mL/min; the sample was then focused for 6 min at a rate of 1.30 mL/min with a cross-flow (CF) set at 1 mL/min. After the focusing step and a transition period of 0.2 min, the CF was decreased with an exponent decay as follows: CF was kept first constant at 1 mL/min for 0.2 min, then decreased with an exponent decay of 0.2 to 0.1 mL/min over a 40 min period and finally kept constant at 0.1 mL/min for a further 20 min. The detector flow was kept along the process constant at 0.5 mL/min to ensure detectors baseline stability. Eluted samples were finally passed and analyzed through a series of online multiple detectors: first through a dual UV/VIS detector (PN3211) set at λ = 520 and 600 nm, a refractive index detector, RI (PN3150), a 21 angle multiangle light scattering detector, MALS (PN3621) and finally through an online dynamic light scattering detector (Zetasizer Nano ZS). All recorded signals were analyzed at increasing concentrations of FB1. Specifically, the areas under UV and MALS signals were determined in Origin Pro v. 8.6 software, with normalization of the base line from each fractogram.

2.5.4. Prediction of the Aptamer Folded Structure

The folded structure of nucleic acids was predicted using MFold Web Server, according to the folding conditions for aptamer 40 nt and aptamer 96 nt.

2.5.5. Circular Dichroism Spectroscopy

Far-UV circular dichroism (CD) spectroscopy was conducted in a Jasco J715 spectropolarimeter with a 6-cell changer and Peltier temperature control, from λ = 200 to 340 nm. The incubation was performed as described in Section 2.5.1, at a concentration of 10 µg/L for FB1.

3. Results and Discussion

3.1. Effect of Buffer Incubation on the Quantification of FB1 with Aptamer 40 nt

The UV/VIS absorption spectra obtained for the citrate-stabilize gold nanoparticles (AuNPs) in stock 1 (6.93 nM) had a maximum peak at a wavelength of 520 nm. The aggregation of colloidal gold nanoparticles, due to charge screening, is produced by salts and cationic compounds, and can be visually observed by a change in the dispersion color from red to blue, and spectrophotometrically confirmed by a peak shift from the absorbance from λ = 520 to ~650 nm. As denoted in Figure S3a, the properties for stock 1 (Size = 18.49 ± 0.4 nm, Pdl = 0.199 ± 0.017) indicated an aggregation point after the addition of NaCl 0.4 M in a 1:1 v:v ratio (Figure S3b). Hence, this point served as the main reference for functionalization with aptamer 40 nt.

Before any incubation with aptamers, the effect of different buffers was tested on both AuNP stock solutions. A key finding was uncovering the effect that Tris-HCl buffer 50 mM (pH 7.5) exerted on the resulting aggregation of both AuNP stocks (1 and 2). Therefore, we conducted a more detailed study to assess the effect of Tris-HCl and PBS ionic strength. This study revealed that 33 mM was the maximum buffer concentration capable of inducing particle aggregation. On the other hand, the concentration of NaCl in PBS to reach the aggregation point was established at 0.4 M (data not shown). In addition, the possible aggregation effects from such buffers were diminished by an initial aptamer–

target incubation, and the later addition of AuNPs. Unlike the approach using AuNPs reported here, binding buffer formulations including concentrations as high as 20 mM $CaCl_2$, 20 mM $MgCl_2$ and 120 Mm NaCl have been documented with silica spheres [21,27] and fluorescence detection methods [28].

The study of the effect of different buffers was driven by the variety of buffers applied for aptamer-based detection of FB1 [29]. Therefore, all the tested buffer conditions in this work, were selected based on their previously reported inclusion during the binding step of several aptasensors for FB1. As shown in Figure 2a, the effect of incubation in Tris-HCl buffer (pH 7.5, 31.1 mM), PBS (pH 7.4, 12.79 mM) and its combination on the binding effect of aptamer 40 nt toward FB1 was assessed. It was observed that under the same binding conditions, Tris-HCl buffer and a Tris-HCl/PBS mixture provided optimal performance at an increasing concentration of FB1 (0–100 μM). The trend of such increments was similar between the incubation with Tris-HCl and that in the mixed buffer. In contrast, PBS resulted in an opposite effect on the $A_{650/520}$ value. A non-significant difference was found between Tris-HCl and mixed buffer ($p = 0.33$) at the highest toxin concentration (10 μM), which denoted their similar effect on FB1 binding. Yet, as a more linear trend can be observed when employing the mixture of both buffers (Figure 2a), this was selected for exploring a wider range of mycotoxin concentrations (0.0086–8.6 μg/mL). As suggested by the linear curves shown in Figure 2b, the incubation of aptamer 40 nt with increasing FB1 concentrations, in the presence of Tris-HCl and its combination with PBS, decreased the number of available aptamer strands due to aptamer binding, so that fewer AuNPs were functionalized and thus they were not protected from NaCl-induced aggregation corresponding to the visible appearance of a blue color (Figure 1b) (Figure S4a), and an increment on the $A_{650/520}$ value.

The result of incubating aptamer 40 nt and FB1 in the presence of four different buffers is displayed in Figure 2b. As noted, the incubation with $MgCl_2$ (Assay 1) and its combination with NaCl (Assay 2) were unfavorable for the quantification of FB1, which was confirmed by the determination coefficients (r^2) and the high LODs in Table 1. In accordance with Figure 2a, Tris-HCl and its mixture with PBS (Assays 3 and 4) afforded greater r^2 and lower LODs, from which the sole application of Tris-HCl resulted in an enhancement effect on the method sensitivity, which resulted in an overall greater performance by means of the NH_3^+ group in Tris-HCl, compared to the ions (Cl^-, Na^+, K^+) from PBS. Despite such confirmed effect, incubation under the presence of Tris-HCl buffer with further AuNP functionalization (Assay 4) was not specific for FB1, as corroborated in Figure 2c, where the addition of the same concentration of FB1, OTA and AFB1 did not show significant differences between the signals for OTA and FB1 ($p = 0.065$). In both cases, the aptamer–mycotoxin incubation step resulted in less unbound aptamer strands at increasing target concentrations, which diminished the number of functionalized AuNPs, hence more particle aggregation was observed. Aptamer 40 nt is a shortened version of an 80 nt sequence (Kd = 62 nM), selected in 100 mM NaCl, 20 mM Tris-HCl, 2 mm $MgCl_2$, 5 mM KCl, 1mM $CaCl_2$ and 0.02% Tween 20 [16]. The application of shorter sequences corresponds to an attempt to reduce synthesis costs, while increasing its affinity by selecting specific binding regions [16,30].

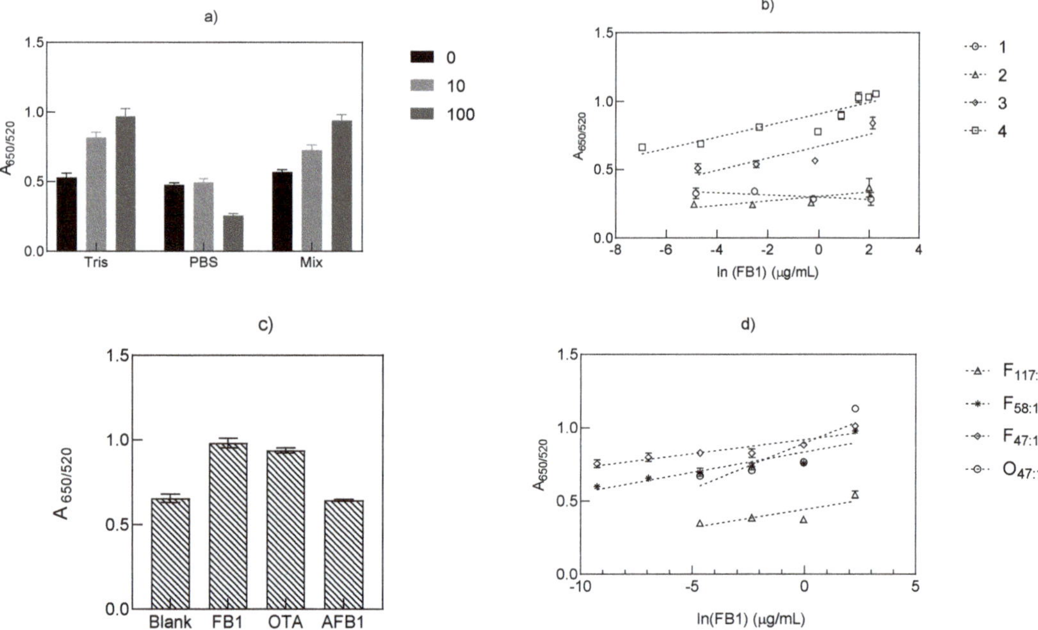

Figure 2. Effect of the incubation of aptamer 40 nt with FB1 on the $A_{650/520}$ ratio in (**a**) Tris-HCl buffer 31.1 mM, PBS 12.79 mM (NaCl equivalence) and a combination of both (Mix) to a final concentration of 31.1 (Tris-HCl) and 12.79 mM (PBS in NaCl equivalence), under the same binding conditions ($n = 6$), and (**b**) different buffers at the conditions outlined in Table 1 (the numbers correspond to each assay, $n = 3$). (**c**) Specificity of assay 4 incubated with other mycotoxins (13.8 μM, $n = 4$) and (**d**) the effect of the reduction in the aptamer: AuNP molar ratio (as shown in the legend subscripts) in the incubation with FB1 (F) and OTA (O)($n = 4$).

To apply the favorable effects of Tris-HCl buffer and reduce its lack of specificity, more cationic compounds, such as NaCl, $CaCl_2$, KCl and $MgCl_2$, were included, along with a reduction in the aptamer:AuNP ratio (117:1 to 47:1), and examined the improvement of the LODs as indicated in Table 2. The inclusion of counterions such as Na^+ is intended in this case to reduce the repulsion between negative charges from the DNA backbone; calcium promotes the formation of coordination complexes with carboxyl groups present in FB1, whereas the use of monovalent ions (K^+) is commonly applied to steady guanine tetrads [15]. Even when this strategy enhanced the sensitivity in terms of the increment of $A_{650/520}$, from 0.54 to 0.0007 μg/mL, while simultaneously increasing the r^2 values of the curves from Figure 2d, the incubation with OTA at a molar ratio of 47:1 (aptamer:AuNP) was highly correlated (R = 0.998, $p = 0.002$) to the values obtained through the determination of FB1 at a 58:1 molar ratio, as observed in the overlapping curves (F58:1 and O47:1) from Figure 2d.

Likewise, the equation parameters and LOD reported for OTA (0.06 μg/mL) revealed the absence of specificity from aptamer 40 nt to FB1, under the selected conditions and the biosensing technique presented in this section. The latter was opposite to the specificity reported for aptamer 40 nt through an electrochemical approach in the presence of OTA and thrombin incubated in PBS and Tris buffers [31]. Although Tris-HCl was confirmed as an ideal buffer for aptamer 40 nt, the lack of specificity could explain the existence of only two aptamer 40 nt-based methods since their disclosure in 2017 [28], and the role of the incubation conditions and sensing platform on the variable specificity of a certain sequence. In keeping with this argument, a recent study based on in silico docking assays evaluated the affinity of the 80 nt aptamer, from which aptamer 40 nt was obtained. Unlike the high affinity to free FB1, no binding was observed when FB1 was immobilized in magnetic beads [32].

Table 2. Equations and LODs for the aptamer-based quantification of FB1 with aptamer 40 nt at different molar ratios.

Assay [1]	Equation [2] $A_{650/520} =$	r^2	Range of Tested Concentrations (μg/mL)	LOD (μg/mL)
$F_{117:1}$	$0.0248 \ln[X] + 0.4434$	0.695	0.0096–9.66	0.54
$F_{58:1}$	$0.0276 \ln [X] + 0.8332$	0.833	0.00096–9.66	0.001
$F_{47:1}$	$0.0192 \ln [X] + 0.9169$	0.84	0.000096–9.66	0.0007
$O_{47:1}$	$0.0624 \ln [X] + 0.8937$	0.76	0.0096–9.66	0.06

[1] F: Assays with FB1; O: Assays with Ochratoxin A; Numbers as subscript indicate the aptamer: AuNP molar ratio; [2] X = FB1concentration in μg/mL. Note: Buffer: Tris-HCl buffer 15 mM + NaCl 85 mM + $CaCl_2$ 1 mM + 5 mM KCl + $MgCl_2$ 2 mM; FB1 aptamer incubation: 60 min (RT); incubation with AuNP: 60 min RT.

Such unspecific binding can also be addressed in terms of buffer pH (7.4–7.5) in which FB1 appears as zwitterion due to the pKa values of its trycarballylic acid (3.49, 4.56, 5.83) and the amine group (pKa > 9) resulting in non-specific electrostatic interactions with aptamer 40 nt, similar to the adsorption on several materials [33], which was also observed through the role of the ionic forms of OTA from its carboxyl (pKa = 4.3–4.4) and phenolic hydroxyl groups (pKa = 7–7.3) [34]. Based on these results, we confirmed that the specificity and good performance of aptamers depends on the binding buffer and binding conditions, along with the selected sensing mechanism. Although Tris-HCl combined with PBS buffer indicated a good sensing capability from aptamer 40 nt, its lack of specificity was revealed under the presence of OTA.

3.2. Quantification of FB1 with Aptamer 96 nt

Stock 2 of AuNPs also had a maximum peak at a λ = 520 nm; nevertheless, the aggregation profile induced by NaCl was described by a slight reduction in the absorbance at λ = 520 nm and a pronounced peak increment from λ = 550 to 650 nm (Figure S5b). As previously mentioned, the characteristics of AuNPs are relevant to describe and understand their resulting signals. Hence, the properties of stock 2 (Size = 21.65 $\pm$ 0.22 nm, Pdl = 0.087 $\pm$ 0.016) are displayed in Figure S5a,b, where NaCl 0.2 M 1:1 ratio (v:v) indicated the main constrain for functionalization with aptamer 96 nt at an aptamer:AuNP molar ratio of 30:1 (Figure S5c). In this regard, when compared to stock 1, the application of a lower concentration of NaCl on AuNPs functionalized with aptamer 96 nt corresponds to the lower concentration of stock 2 (2.21 nM) when compared to stock 1 (6.93 nM).

From all the buffers previously tested for aptamer 40 nt, only $MgCl_2$ exhibited a positive performance for the proposed bulk technique with aptamer 96 nt; therefore, it was selected for detecting FB1. The addition of 5 mM $MgCl_2$ showed aggregation of stock 2 AuNPs by a colorimetric shift from red (stable AuNPs) to purple (i.e., an indicative of a certain degree of aggregation). A reduced concentration of $MgCl_2$ (1 mM) was, therefore, used to diminish those negative effects. Adding Mg^{2+} ions in binding buffers contributes to their blocking effect toward the repulsion of the negative charges from the DNA backbone, which allows a more compact folding [15]. Unlike other aptasensors, it seemed that the incubation of aptamer 96 nt with FB1 in $MgCl_2$ did not lead to a conformational change in its structure, thus resulting in the formation of an aptamer–FB1–AuNP complex (Figure 1c), stable against NaCl-induced aggregation at increasing concentrations of FB1 (Figures S5d and S6d). The formation of an aptamer–target–AuNP conjugate was also observed for the determination of serotonin with AuNPs and aptamers dissolved in 2 mM $MgCl_2$ (pH 7.4), where a serotonin–aptamer complex was capable of protecting AuNPs from salt-induced aggregation, especially at high serotonin concentrations [35].

As shown in Table 3, a reduction in the LOD value was achieved with aptamer 96 nt, where the value for the $A_{650/520}$ ratio (0.003 μg/mL), calculated from the curve in Figure 3a, was lowered when the absolute area was analyzed instead (0.002 µg/mL). To increase the sensitivity of the analysis as well as minimize reproducibility issues, calculation of the absolute area was conducted. The estimation of the differential area under the curve between the blank and its corresponding standard curve points improved the linearity and slope. This mathematical comparison, where each experiment had a reference point (blank), could reduce the variability of the results when changing the stock solutions, using AuNPs with varying shapes and particle sizes, or when observing stock aggregation upon storage. Therefore, before its subtraction, the spectrum of the corresponding blank must be acquired during each batch run. Contrary to the issues observed for aptamer 40 nt, the proposed method evinced high specificity from aptamer 96 nt toward FB1 as displayed in Figure 3b,c for the $A_{650/520}$ ratio and absolute area, respectively. In both cases, AFB1 and OTA had similar signals to the blank values, where FB1 displayed a distinct result. In different biosensing approaches, aptamer 96 nt has been confirmed as specific to FB1 in the presence of almost 19 compounds, which justifies its use in 24 biosensors [29]. From these results, we observed the positive effect of $MgCl_2$ for the biorecognition of aptamer 96 nt to FB1, through the formation of an aptamer 96 nt–FB1–AuNPs conjugate. Such mechanism was specific to FB1, with a protective effect to salt-induced aggregation at increasing target concentrations.

Table 3. Equations and LOD for aptamer-based quantification of FB1 with aptamer 96 nt through different signals [1].

Signal		Equation	r^2	Range of Tested Concentrations (µg/mL)	LOD (µg/mL)
$A_{650/520}$		$A_{650/520} = -0.074 \ln [X] + 0.5153$	0.9179	0.01–10	0.003
Absolute Area		$AA = 1.34 \ln [X] + 8.298$	0.9243	0.01–10	0.002
Peak Area 520 nm	Peak 1	$Area = 0.0002 \ln [X] + 0.0036$	0.9763	0.001–10	1.68
	Peak 2	$Area = - 0.0005 \ln [X] + 0.0065$	0.8735	0.001–10	0.0001
Peak Area 600 nm	Peak 1	$Area = 0.0017 \ln [X] + 0.0334$	0.872	0.001–10	7.83
	Peak 2	$Area = - 0.006 \ln [X] + 0.0946$	0.858	0.001–10	0.000000056
Peak 2 Area /Peak 1 Area	520 nm	$P1/P2 = - 0.268 \ln [X] + 1.8427$	0.7639	0.001–10	0.007
	600 nm	$P1/P2 = - 0.364 \ln [X] + 2.9186$	0.9637	0.001–10	0.0006
Peak 2 Area–Peak 1 Area	600 nm	$P–P2 = - 0.008 \ln [X] + 0.059$	0.9525	0.001–10	0.0000016
28°		$Area = - 0.07 \ln [X] + 0.4632$	0.9128	0.001–10	0.00000016
Diameter		$D = - 9.498 \ln [X] + 124.61$	0.9514	0.001–10	0.000959

[1] X = FB1concentration in µg/mL. Buffer: $MgCl_2$ 1mM; FB1 aptamer 96 nt incubation: 30 min (37 °C); incubation with AuNP: 60 min (RT); AuNP:Aptamer molar ratio 30:1.

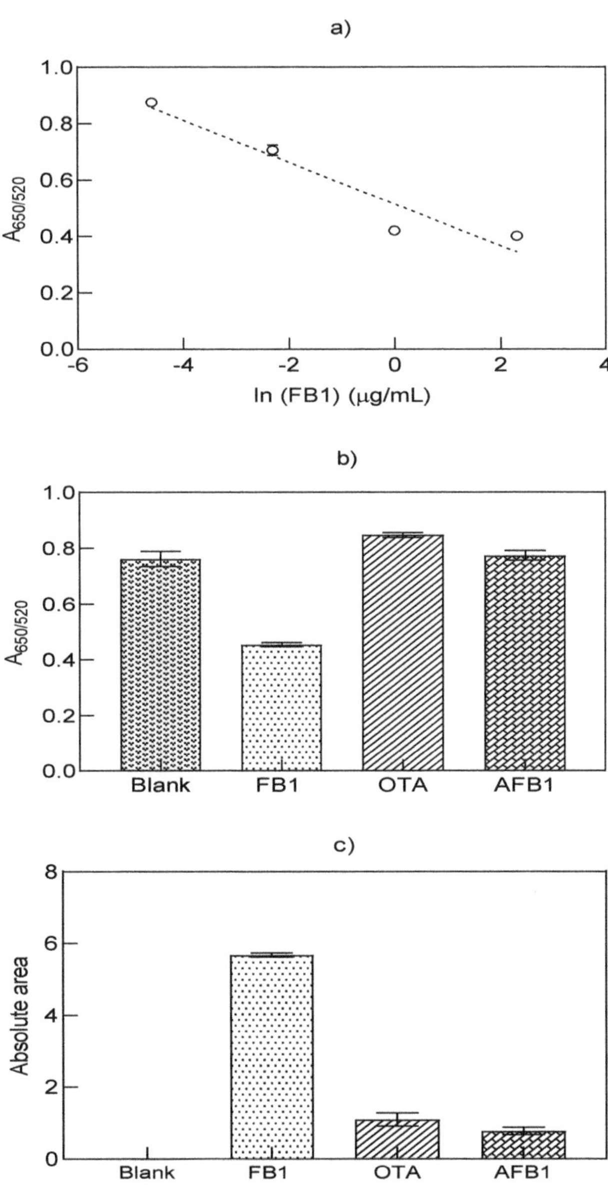

Figure 3. (a) Standard curve for the quantification of FB1 with aptamer 96 nt in $MgCl_2$ through the analysis of the $A_{650/520}$ ratio. Specificity test under the presence of FB1, AFB1 and OTA at a concentration of 1.38 µM for the (**b**) $A_{650/520}$ ratio and the (**c**) absolute area values ($n = 3$).

3.3. Asymmetric Flow Field-Flow Fractionation (AF4) of the Aptamer 96 nt–FB1–AuNP Conjugates

The characterization of different aptamer 96 nt–FB1–AuNP conjugates after the addition of NaCl 0.2 M was further examined by multidetection AF4, to confirm the combined role of the aptamer 96 nt–FB1 complex in the functionalization and protection of AuNPs against NaCl-driven aggregation at different target concentration. As a first step, it was determined both visually and spectrophotometrically that the UV/VIS signal at λ = 520 and 600 nm after AF4 separation showed higher peaks at lower FB1 contents (results not shown).

AF4 relies on the separation of particles of varying size depending on their hydrodynamic and diffusion properties, where the initial elution corresponded to small stabilized particles (Peak 1), whereas larger particles (Peak 2) eluted after, before passing through the detectors as shown in representative fractograms (Figures S6a and S6b). According to the AF4 principle, longer elution times correspond to particles of larger sizes, which in many cases could also be diagnostic of aggregation. Compared to the spectrophotometric analysis, AF4 contributed to the refinement of the detected signals, with a much greater resolution at lower target concentrations. Even when two peaks were detected at λ = 520 and 600 nm, the peak corresponding to the highest particle sizes (Peak 2) had a more favored trend for its potential application in biosensing techniques, as displayed in Figure 4a,b. The selection of this peak was consistent with the lower LODs achieved when analyzing the values of Peak 2 at both wavelengths (Table 3). The UV/VIS signal at λ = 600 nm was mainly ascribed to the aggregation profile of AuNPs; for that reason, the peaks between 40 and 60 min were larger and more noticeable among different concentrations of FB1 (Figure S6b).

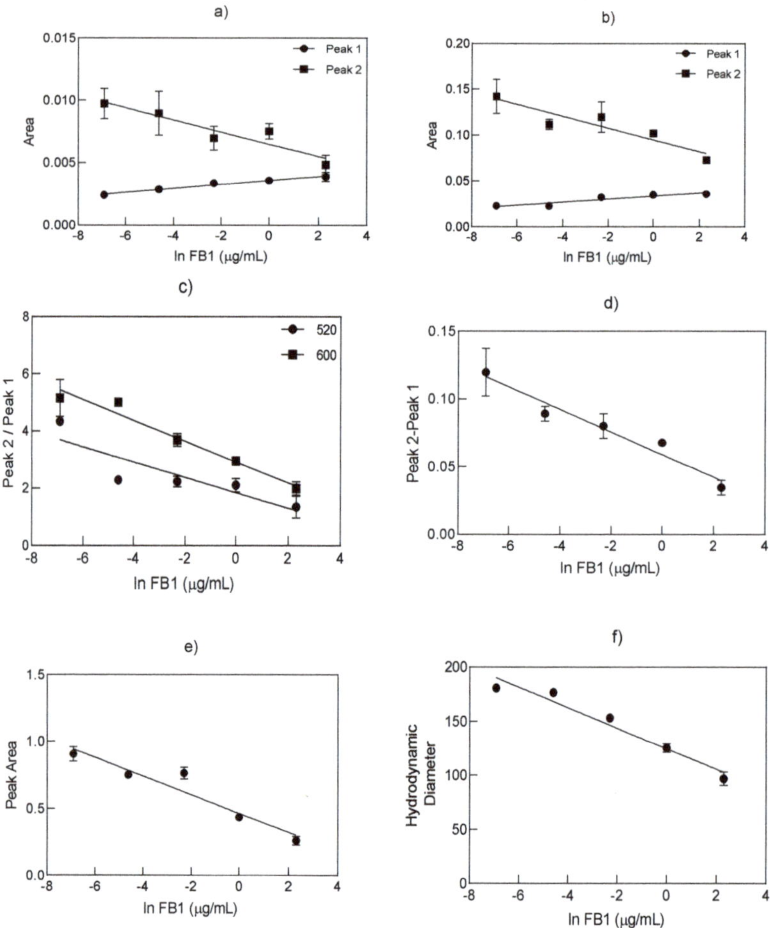

Figure 4. Standard curves for the quantification of FB1 with aptamer 96 nt through the analysis of the AF4 fractograms from the UV-Vis peak areas at (**a**) 520 nm, (**b**) 600 nm, (**c**) peak ratio between Peak 2 (larger particles) and Peak 1 (smaller particles), (**d**) peak area differences at 600 nm, (**e**) MALS peak area at 28° and (**f**) hydrodynamic diameter determined by DLS for the Aptamer (96 nt)–FB1–AuNPs conjugates in NaCl 0.2 M (*n* = 3).

The reported analytical method demonstrated the presence of stabilized and aggregated particles in the same suspension, rather than their complete aggregation or stabilization through aptamer FB1 functionalization. Such profile was regulated by the concentration of FB1 where fewer aggregated particles were detected at higher target concentrations (Figure S6b). The presence of different particle sizes could also be attributed to the native heterogeneity of the selected stock, which is a common issue in AuNPs, derived during their manufacturing/synthesis process, and previously confirmed by AF4 as an arrangement of particles with different size, shape and zeta potential [36].

To achieve a better separation resolution of different particle populations, a regenerated cellulose membrane was reported due to its high recovery of AuNPs [22], associated with its charge, which depends on the pH value of the carrier solvent. Here, as the carrier solvent has a pH = 9.5, the regenerated cellulose membrane bears a net negative charge (zeta potential < ~−20 mV) [37], which decreases the interaction between the membrane and AuNPs, thus favoring their elution. In addition, the application of a combination of ionic and non-ionic surfactants, such as Novachem® 0.05%, in the carrier solvent was considered for preventing particle aggregation and accumulation on the cellulose membrane, as reported for other surfactants, which also improved the retention profiles [38]. Similarly, surfactants also decrease particle loss, null peaks, particle separation and recovery loss [22].

Even though the retention time during AF4 fractionation can be correlated with particle size, and the MALS signal detection can be used for the determination of the radius of gyration [38,39], in our study, we focused on the analysis of the peak areas of the different signals. Therefore, the effect of FB1 on the functionalization of AuNPs with aptamers was studied as a whole system, regardless of the differences in the absolute particle size. Despite the initial characterization purpose for the application of AF4 on the conjugates at varying FB1 concentrations, the low LODs achieved, especially after analyzing the peak areas at 600 nm (56 fg/mL), unveiled its promising usage as a biosensing technique.

As highlighted in the equations and LODs (Table 3) from Figure 4c,d, neither the Peak 2/Peak 1 ratio for both wavelengths (7 and 0.6 ng/mL, respectively), nor the subtraction of Peak 1 from Peak 2 at 600 nm (1.6 pg/mL), were as sensitive as the sole analysis of Peak 2 at 600 nm. Similarly, the parameters quantified from the MALS signal at 28° (LOD = 0.16 pg/mL) and the hydrodynamic diameter obtained by DLS (LOD = 0.96 ng/mL), plotted in Figure 4e,f, respectively, resulted in lower sensitivity compared to the values determined from UV signals at 600 nm (Table 3). Although the MALS signal at 90° has been analyzed for AuNPs [38,39], in this work, more distinctive peaks between samples were found at 28°, which allowed the examination of various degrees of aggregation with a better signal-to-noise ratio. Such MALS signal (28°) was adequate for representing the effects of an increasing concentration of FB1 on the same system by analysis of its peak area, which also demonstrated that smaller angles from MALS are more sensitive to larger particles. Interestingly, the hydrodynamic diameter complied with the UV/VIS fractograms and spectrophotometric analysis, where greater hydrodynamic diameters corresponded to the presence of aggregated particles at lower FB1 contents. To the best of our knowledge, AF4 has not been yet used for the quantification of mycotoxins. One study applied this technique for quantifying the molecular weight of an aptamer–streptavidin–immunoglobulin G (IgG) complex in a biosensing technique for OTA [40]. In summary, the analysis of the aptamer 96 nt–FB1–AuNPs conjugates by AF4 revealed the presence of varied degrees of aggregated and non-aggregated particles at different target concentrations. From all the detected signals, the analysis of the UV/VIS peak area at 600 nm was the most suitable for portraying such variability, with a promising scope for the application of AF4 as a new biosensing technique in the fg/mL range.

3.4. Interaction of the Conjugate Elements (Aptamer 96 nt–FB1–AuNPs)

The biosensing approach in this work did not require any complementary strand or aptamer modification (label), especially when considering that the latter might decrease its binding affinity. Here, the selection of the incubation buffer is also a relevant step for the success of the aptamer-target binding stage [41]. As shown in Figure 5a, when compared to aptamer 40 nt (dG = −6.64), the structure of aptamer 96 nt (dG = −12.43) has elongated hairpins and loops, due to its longer sequence.

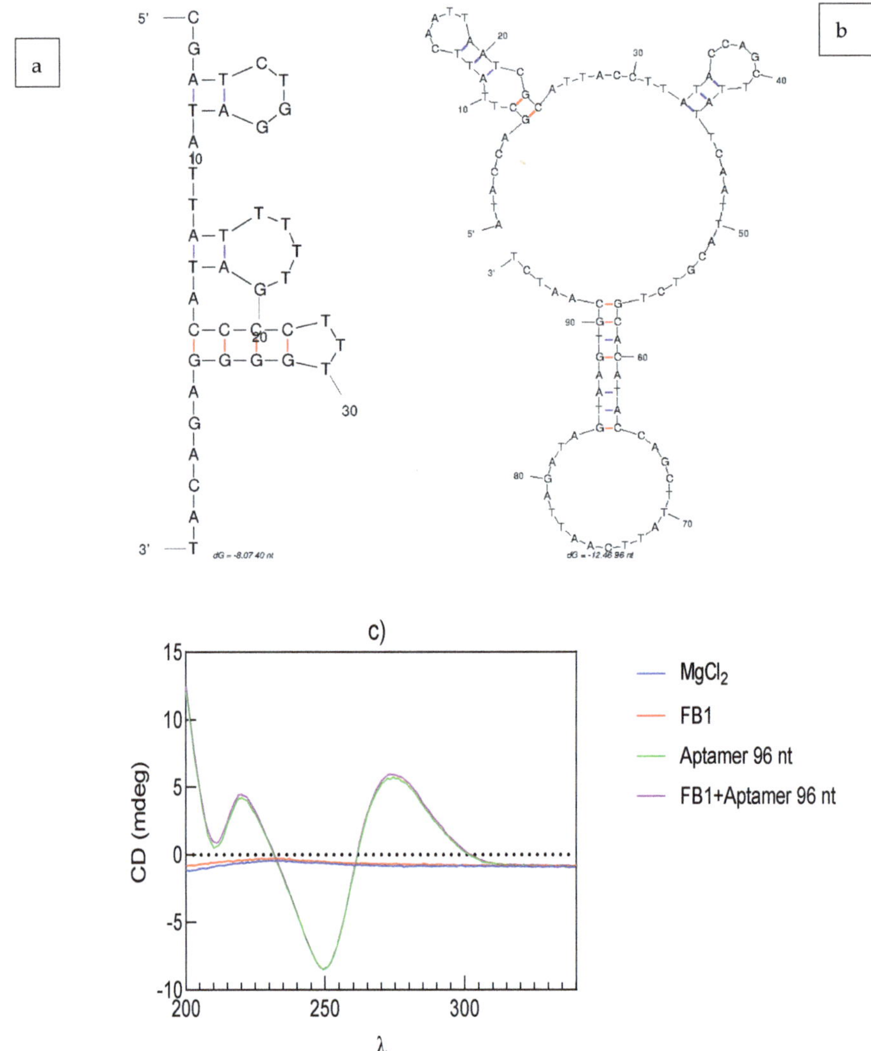

Figure 5. Predicted folded structures of (**a**) aptamer 40 nt (dG = −8.07), (**b**) aptamer 96 nt (dG = −12.46), and (**c**) circular dichroism spectrum of aptamer 96 nt in the absence and presence of 10 µg/L FB1 incubated in MgCl₂.

A study on the binding affinity of minimers from aptamer 96 nt (Kd= 100 nM) reported that the 16 nt sequence next to the 3′ primer binding region (AGATAGTAAGTGCAATCT-3′) is related to target binding. However, the regions following the 5′ end (5′-ATACCAGCTTAT TCAATT) are also necessary for the overall binding efficiencies, which correspond to the favorable dissociation constant of the sequence without both primer binding regions (60 nt, Kd = 195 nM) [30]. Nevertheless, binding assays for this minimer denoted low or null binding towards FB1 [32], which explains the majority of applications with the complete

96 nt sequence. In addition, the amine group in FB1 has been used in immobilization approaches during SELEX, motivated by the immunodominant epitope region in FB1 (close to the concurrence of tricarballylic acid to C-11 and C-20), which is distant from such functional group [16]. Even when the presence of multiple free dihedrals within the structure of FB1 requires more configurational space during binding, high binding propensity has been reported between FB1 and the backbone of different FB1 aptamers, including aptamer 96 nt [32].

Based on this, the formation of aptamer 96 nt–FB1–AuNPs conjugates first occurred by incubation of the binding region and backbone in aptamer 96 nt with its epitope zone in FB1, where a conformational change was unlikely to have occurred due to the length of the aptamer and the role of the negative backbone charges from both primer binding regions, which could have generated steric hindrance and simultaneous binding in some regions. Therefore, the conjugation was followed by immobilization of the aptamer 96 nt–FB1 complex on AuNPs by means of the NH_2 group in FB1, which was previously evaluated on the prevented aggregation from binding buffers through the sole incubation with the target molecule, but was weak against salt-induced aggregation (Figure S4b). We thus suggest that the mechanism was completed by the formation of an aptamer 96 nt–FB1 layer by non-covalent attachment of the primer regions to the surface of AuNPs. This proposal is consistent with the increased density of aptamer 96 nt–FB1 complexes on AuNPs, and the prevented aggregation upon the addition of salt.

Circular dichroism (CD) spectra of $MgCl_2$ and FB1 were very close to the solvent baseline (Figure 5c). In turn, the CD spectra for the aptamer 96 nt showed two trough bands at $\lambda = 205$ and 250 nm and a positive peak band at 280 nm region before and after binding with FB1. The CD spectrum was close to previous analysis on this sequence, which denoted helicity of its parallel arrangement [42]. The same negative (250 nm) and positive (280 nm) bands were previously described as an indicator of an A-form hairpin duplex structure, where the flat zone from 265 to 285 nm was attributed to base pair formation between A-T and G-C [16,43]. However, the 3-D representation of the most stable docked pose in aptamer 96 nt was indicated as a B-form duplex structure, as A-forms are mostly favored in RNA [32]. The peak similarity from both spectra also confirmed the absence of a conformational change upon binding. Depending on the selected immobilization and biosensing method, when a long-length aptamer is used, the FB1-modulated conformational change might not be observed, as only some regions of the aptamer display affinity. Such absence of conformational change was also confirmed by gel electrophoresis, as indicated in Figure S7, where a slight increase of 2.5% was calculated for the band intensities of aptamer 96 nt incubated with the maximum FB1 concentration (10.02 µg/mL) compared to the bands for aptamer 96 nt in $MgCl_2$ 1mM. The aforementioned was opposite to the trend observed in other reports, when an increasing target concentration enhanced the band intensities from the aptamer–target complex due to conformational change [44]. To conclude, based on the CD and gel electrophoresis results, it was understood that no conformational change was found upon target binding, which also corresponded to the long structure of aptamer 96 nt, predicted in Mfold.

As a first attempt toward the subsequent full validation of the developed method, we evaluated the matrix-matched quantification of FB1 in various liquid and solid food samples. For instance, the incubation of aptamers with its target in either buffer or a corn extract showed color differences among both samples, after the addition of AuNPs (Figure S8a). Likewise, the analysis in spiked vodka displayed different aggregation spectra to those from FB1 in binding buffer ($MgCl_2$ 1mM), as shown in Figure S8b,c. While spirits can be directly measured and injected into the AF4 system, a pretreatment step is required for the analysis of more complex samples such as corn extracts. As plotted in Figure S2, the LOD, for the specific quantification of FB1 by the reported aptamer–target–AuNP complex ($A_{650/520}$ absolute area), is comparable to that for some fluorescent and electrochemical aptasensors specific for this mycotoxin. Yet, the application of AF4 is a promising technique for the analysis of FB1, which regardless of the increased assay time resulted in a low

LOD value comparable with the most sensitive aptamer-based sensors reported for this purpose. The novelty of this approach lies in the integration of the unmodified 96 nt ssDNA sequence, without the inclusion of other complementary strands, supports or DNA modification (SH, NH_2, biotin, FAM, Cy5.5). Additionally, to the best of our knowledge, this is the first report on the use of AF4 for evaluating the performance of an aptamer on the formation of a conjugate enhanced by the presence of its target molecule, thus revealing to be a highly sensitive and specific method.

4. Conclusions

This work presents a comparison on the performances of two aptamers for the colorimetric quantification of FB1. The results indicated that, along with the aptamer sequence, the selected buffer and incubation conditions play an important role in the final sensitivity and specificity of each assay. In this regard, incubation with Tris-HCl and $MgCl_2$ was suitable for the 40 and 96-mer aptamers, respectively. Contrary to previous reports [31], the assay with a short length aptamer (40 nt) was not specific for FB1, as similar results were observed through the incubation with OTA. A different mechanism has been proposed for the long aptamer (96 nt), previously reported for several aptamer-based approaches. In this case, an aptamer–FB1–AuNP conjugate was formulated in the presence of $MgCl_2$ 1 mM, showing stability to salt-aggregation at an increasing concentration of FB1(0.001–10 µg/mL). Unlike other aptasensors, the 96 nt aptamer offered a simplified approach as an unmodified ssDNA sequence was applied without the need of end modifications or complementary strands. Analysis of the spectrophotometric signals resulted in LODs similar to other sensitive techniques; however, the exploration of the aggregation profile by AF4 with multidetection (UV/VIS, MALS, DLS) derived in a promising sensing technique with sensitivity in the fg/mL level. The characterization of the complex formation revealed the absence of DNA conformational change upon binding, yet this new mechanism might be suitable for the direct analysis of different food matrices, where there is scope for exploring other targets, such as emerging mycotoxins. To our knowledge, this is the first aptasensing technique for FB1 applying the 96 nt aptamer sequence without any end modification, label or complementary strand. Likewise, there is no evidence for the use of AF4 in the exploration of aptamer–target–AuNPs interactions.

Further validation and standardization steps are still required for the commercial application and possible scaling to paper-based techniques, which might enhance the opportunities for on-site quantifications, while decreasing the total manufacturing cost. Nevertheless, this work established a new mechanism for detecting FB1 with a 96 nt aptamer in bulk, while at the same time presents for the first time the use of a more robust method, as it is AF4, resulting in LODs with strong advantages over more complex designs.

Supplementary Materials: The following is available online at https://www.mdpi.com/article/10.3390/bios11010018/s1.

Author Contributions: Conceptualization, V.A.M.-M.; Y.G.; Y.Y.G. and F.M.G.; methodology, V.A.M.-M.; Y.G.-E. and M.C.-G.; software, V.A.M.-M.; validation, V.A.M.-M.; formal analysis, V.A.M.-M.; investigation, V.A.M.-M.; resources, Y.G., Y.Y.G. and F.M.G.; data curation, V.A.M.-M.; writing—original draft preparation, V.A.M.-M.; writing—review and editing, V.A.M.-M., Y.G.-E., M.C.-G. and F.M.G.; visualization, V.A.M.-M.; supervision, Y.Y.G. and F.M.G.; project administration, V.A.M.-M. and F.M.G.; funding acquisition, V.A.M.-M. and F.M.G. All authors have read and agreed to the published version of the manuscript.

Funding: V.A.M.-M. was awarded with a scholarship from CONACYT (Mexico) for completing his PhD studies. M.C.G. acknowledges a fellowship for postdoctoral training (20381/PD/17) funded by the Consejería de Empleo, Universidades y Empresa de la CARM, through the Fundación Séneca de la Región de Murcia.

Institutional Review Board Statement: Not applicable.

Informed Consent Statement: Not applicable.

Data Availability Statement: The data presented in this study are available on request from the corresponding author. The data are not publicly available due to its classification as intellectual property by the University of Leeds.

Acknowledgments: Help from G. Nasir Khan for the CD spectroscopy measurements at the Astbury Centre for Structural Molecular Biology, University of Leeds, is gratefully acknowledged. We are also indebted to B. Sabagh from Postnova Analytics UK for his kind advice in the development of the AF4 method.

Conflicts of Interest: The authors declare no conflict of interest.

References

1. Wangia, R.N.; Githanga, D.P.; Xue, K.S.; Tang, L.; Anzala, O.A.; Wang, J.S. Validation of urinary sphingolipid metabolites as biomarker of effect for fumonisins exposure in Kenyan children. *Biomarkers* **2019**, *24*, 379–388. [CrossRef] [PubMed]
2. Wall-Martínez, H.A.; Ramírez-Martínez, A.; Wesolek, N.; Brabet, C.; Durand, N.; Rodríguez-Jimenes, G.C.; García-Alvarado, M.A.; Salgado-Cervantes, M.A.; Robles-Olvera, V.J.; Roudot, A.C. Risk assessment of exposure to mycotoxins (aflatoxins and fumonisins) through corn tortilla intake in Veracruz City (Mexico). *Food Addit. Contam. A* **2019**, *36*, 929–939. [CrossRef] [PubMed]
3. Hu, L.; Liu, H.; Yang, J.; Wang, C.; Wang, Y.; Yang, Y.; Chen, X. Free and hidden fumonisins in raw maize and maize-based products from China. *Food Addit. Contam. B* **2019**, *12*, 90–96. [CrossRef] [PubMed]
4. Martins, C.; Vidal, A.; De Boevre, M.; De Saeger, S.; Nunes, C.; Torres, D.; Goios, A.; Lopes, C.; Assunção, R.; Alvito, P. Exposure assessment of Portuguese population to multiple mycotoxins: The human biomonitoring approach. *Int. J. Hyg. Environ. Health* **2019**, *222*, 913–925. [CrossRef]
5. Abbas, H.; Vesonder, R.F.; Boyette, C.D.; Hoagland, R.E.; Krick, T. Production of fumonisins by Fusarium moniliforme cultures isolated from jimsonweed in Mississippi. *J. Phytopathol.* **1992**, *136*, 199–203. [CrossRef]
6. Abbas, H.K.; Riley, R.T. The presence and phytotoxicity of fumonisins and AAL-toxin in *Alternaria alternata*. *Toxicon* **1996**, *34*, 503. [CrossRef]
7. Mansson, M.; Klejnstrup, M.L.; Phipps, R.K.; Nielsen, K.F.; Frisvad, J.C.; Gotfredsen, C.H.; Larsen, T.O. Isolation and NMR characterization of fumonisin B2 and a new fumonisin B6 from *Aspergillus niger*. *J. Agric. Food Chem.* **2010**, *58*, 949–953. [CrossRef]
8. Mogensen, J.M.; Møller, K.A.; Von Freiesleben, P.; Labuda, R.; Varga, E.; Sulyok, M.; Kubátová, A.; Thrane, U.; Andersen, B.; Nielsen, K.F. Production of fumonisins B 2 and B 4 in *Tolypocladium* species. *J. Ind. Microbiol. Biotechnol.* **2011**, *38*, 1329–1335. [CrossRef]
9. Rheeder, J.P.; Marasas, W.F.; Vismer, H.F. Production of fumonisin analogs by *Fusarium* species. *Appl. Environ. Microbiol.* **2002**, *68*, 2101–2105. [CrossRef]
10. Ostry, V.; Malir, F.; Toman, J.; Grosse, Y. Mycotoxins as human carcinogens—the IARC Monographs classification. *Mycotoxin Res.* **2017**, *33*, 65–73. [CrossRef]
11. Liu, X.; Fan, L.; Yin, S.; Chen, H.; Hu, H. Molecular mechanisms of fumonisin B1-induced toxicities and its applications in the mechanism-based interventions. *Toxicon* **2019**, *167*, 1–5. [CrossRef] [PubMed]
12. Yuan, Q.; Jiang, Y.; Fan, Y.; Ma, Y.; Lei, H.; Su, J. Fumonisin B1 induces oxidative stress and breaks barrier functions in pig iliac endothelium cells. *Toxins* **2019**, *11*, 387. [CrossRef] [PubMed]
13. Lee, S.; Kim, G.; Moon, J. Performance improvement of the one-dot lateral flow immunoassay for aflatoxin B1 by using a smartphone-based reading system. *Sensors* **2013**, *13*, 5109–5116. [CrossRef] [PubMed]
14. Ahmed, M.U.; Zourob, M.; Tamiya, E. *Food Biosensors*; Royal Society of Chemistry: Cambridge, UK, 2016; pp. 2–9.
15. McKeague, M.; Bradley, C.R.; Girolamo, A.D.; Visconti, A.; Miller, J.D.; DeRosa, M.C. Screening and initial binding assessment of fumonisin B1 aptamers. *Int. J. Mol. Sci.* **2010**, *11*, 4864–4881. [CrossRef] [PubMed]
16. Chen, X.; Huang, Y.; Duan, N.; Wu, S.; Xia, Y.; Ma, X.; Zhu, C.; Jiang, Y.; Ding, Z.; Wang, Z. Selection and characterization of single stranded DNA aptamers recognizing fumonisin B 1. *Microchim. Acta* **2014**, *181*, 1317–1324. [CrossRef]
17. Pamies, R.; Hernández, J.G.C.; Fernández Espín, V.; Collado-González, M.; Díaz, F.G.B.; García de la Torre, J. Aggregation behaviour of gold nanoparticles in saline aqueous media. *J. Nanopart. Res.* **2014**, *16*, 2376. [CrossRef]
18. Li, H.; Rothberg, L. Colorimetric detection of DNA sequences based on electrostatic interactions with unmodified gold nanoparticles. *Proc. Natl. Acad. Sci. USA* **2004**, *101*, 14036–14039. [CrossRef]
19. Pandey, P.C.; Pandey, G. Synthesis of gold nanoparticles resistant to pH and salt for biomedical applications; functional activity of organic amine. *J. Mater. Res.* **2016**, *31*, 3313. [CrossRef]
20. Amendola, V.; Pilot, R.; Frasconi, M.; Maragò, O.M.; Iatì, M.A. Surface plasmon resonance in gold nanoparticles: A review. *J. Phys. Condens. Matter* **2017**, *29*, 203002. [CrossRef]
21. Yue, S.; Jie, X.; Wei, L.; Bin, C.; Dou Dou, W.; Yi, Y.; QingXia, L.; JianLin, L.; TieSong, Z. Simultaneous detection of ochratoxin A and fumonisin B1 in cereal samples using an aptamer–photonic crystal encoded suspension array. *Anal. Chem.* **2014**, *86*, 11797–11802. [CrossRef]
22. Hagendorfer, H.; Kaegi, R.; Traber, J.; Mertens, S.F.; Scherrers, R.; Ludwig, C.; Ulrich, A. Application of an asymmetric flow field flow fractionation multi-detector approach for metallic engineered nanoparticle characterization–prospects and limitations demonstrated on Au nanoparticles. *Anal. Chim. Acta* **2011**, *706*, 367–378. [CrossRef] [PubMed]

23. Mudalige, T.K.; Qu, H.; Van Haute, D.; Ansar, S.M.; Linder, S.W. Capillary electrophoresis and asymmetric flow field-flow fractionation for size-based separation of engineered metallic nanoparticles: A critical comparative review. *TrAC Trends Anal. Chem.* **2018**, *106*, 202–212. [CrossRef]
24. Schachermeyer, S.; Ashby, J.; Zhong, W. Aptamer–protein binding detected by asymmetric flow field flow fractionation. *J. Chromatogr. A* **2013**, *1295*, 107–113. [CrossRef] [PubMed]
25. Derbyshire, N.; White, S.J.; Bunka, D.H.; Song, L.; Stead, S.; Tarbin, J.; Sharman, M.; Zhou, D.; Stockley, P.G. Toggled RNA aptamers against aminoglycosides allowing facile detection of antibiotics using gold nanoparticle assays. *Anal. Chem.* **2012**, *84*, 6595–6602. [CrossRef] [PubMed]
26. Quesada-González, D.; Stefani, C.; González, I.; De la Escosura-Muñiz, A.; Domingo, N.; Mutjé, P.; Merkoçi, A. Signal enhancement on gold nanoparticle-based lateral flow tests using cellulose nanofibers. *Biosens. Bioelectron.* **2019**, *141*, 111407. [CrossRef]
27. Yang, Y.; Li, W.; Shen, P.; Liu, R.; Li, Y.; Xu, J.; Zheng, Q.; Zhang, Y.; Li, J.; Zheng, T. Aptamer fluorescence signal recovery screening for multiplex mycotoxins in cereal samples based on photonic crystal microsphere suspension array. *Sens. Actuators B Chem.* **2017**, *248*, 351–358. [CrossRef]
28. Tian, H.; Sofer, Z.; Pumera, M.; Bonanni, A. Investigation on the ability of heteroatom-doped graphene for biorecognition. *Nanoscale* **2017**, *9*, 3530–3536. [CrossRef]
29. Mirón-Mérida, V.A.; Gong, Y.Y.; Goycoolea, F.M. Aptamer-based detection of fumonisin B1: A systematic comparison with conventional and other novel methods. *ChemRxiv* **2020**. (Preprint). [CrossRef]
30. Frost, N.R.; McKeague, M.; Falcioni, D.; DeRosa, M.C. An in solution assay for interrogation of affinity and rational minimer design for small molecule-binding aptamers. *Analyst* **2015**, *140*, 6643–6651. [CrossRef]
31. Cheng, Z.X.; Bonanni, A. All-in-one: Electroactive nanocarbon as simultaneous platform and label for single-step biosensing. *Chem. Eur. J.* **2018**, *24*, 6380–6385. [CrossRef]
32. Ciriaco, F.; De Leo, V.; Catucci, L.; Pascale, M.; Logrieco, A.F.; DeRosa, M.C.; De Girolamo, A. An in-silico pipeline for rapid screening of DNA aptamers against mycotoxins: The case-study of Fumonisin B1, Aflatoxin B1 and Ochratoxin A. *Polymers* **2020**, *12*, 2983. [CrossRef]
33. Baglieri, A.; Reyneri, A.; Gennari, M.; Nègre, M. Organically modified clays as binders of fumonisins in feedstocks. *J. Environ. Sci. Health Part B* **2013**, *48*, 776–783. [CrossRef] [PubMed]
34. Zhao, Z.; Liu, N.; Yang, L.; Wang, J.; Song, S.; Nie, D.; Yang, X.; Hou, J.; Wu, A. Cross-linked chitosan polymers as generic adsorbents for simultaneous adsorption of multiple mycotoxins. *Food Control* **2015**, *57*, 362–369. [CrossRef]
35. Chávez, J.L.; Hagen, J.A.; Kelley-Loughnane, N. Fast and selective plasmonic serotonin detection with aptamer-gold nanoparticle conjugates. *Sensors* **2017**, *17*, 681. [CrossRef]
36. Riley, K.R.; El Hadri, H.; Tan, J.; Hackley, V.A.; MacCrehan, W.A. High separation efficiency of gold nanomaterials of different aspect ratio and size using capillary transient isotachophoresis. *J. Chromatogr. A* **2019**, *1598*, 216–222. [CrossRef] [PubMed]
37. González-Espinosa, Y.; Sabagh, B.; Moldenhauer, E.; Clarke, P.; Goycoolea, F.M. Characterisation of chitosan molecular weight distribution by multi-detection asymmetric flow-field flow fractionation (AF4) and SEC. *Int. J. Biol. Macromol.* **2019**, *136*, 911–919. [CrossRef] [PubMed]
38. Cho, T.J.; Hackley, V.A. Fractionation and characterization of gold nanoparticles in aqueous solution: Asymmetric-flow field flow fractionation with MALS, DLS, and UV–Vis detection. *Anal. Bioanal. Chem.* **2010**, *398*, 2003–2018. [CrossRef]
39. Schmidt, B.; Loeschner, K.; Hadrup, N.; Mortensen, A.; Sloth, J.J.; Bender Koch, C.; Larsen, E.H. Quantitative characterization of gold nanoparticles by field-flow fractionation coupled online with light scattering detection and inductively coupled plasma mass spectrometry. *Anal. Chem.* **2011**, *83*, 2461–2468. [CrossRef]
40. Samokhvalov, A.V.; Safenkova, I.V.; Eremin, S.A.; Zherdev, A.V.; Dzantiev, B.B. Use of anchor protein modules in fluorescence polarisation aptamer assay for ochratoxin A determination. *Anal. Chim. Acta* **2017**, *962*, 80–87. [CrossRef]
41. Baaske, P.; Wienken, C.J.; Reineck, P.; Duhr, S.; Braun, D. Optical thermophoresis for quantifying the buffer dependence of aptamer binding. *Angew. Chem. Int. Ed.* **2010**, *49*, 2238–2241. [CrossRef]
42. Wu, S.; Duan, N.; Li, X.; Tan, G.; Ma, X.; Xia, Y.; Wang, Z.; Wang, H. Homogenous detection of fumonisin B1 with a molecular beacon based on fluorescence resonance energy transfer between NaYF4: Yb, Ho upconversion nanoparticles and gold nanoparticles. *Talanta* **2013**, *116*, 611–618. [CrossRef] [PubMed]
43. Lin, P.H.; Chen, R.H.; Lee, C.H.; Chang, Y.; Chen, C.S.; Chen, W.Y. Studies of the binding mechanism between aptamers and thrombin by circular dichroism, surface plasmon resonance and isothermal titration calorimetry. *Colloids Surf. B* **2011**, *88*, 552–558. [CrossRef] [PubMed]
44. Jing, M.; Bowser, M.T. Methods for measuring aptamer-protein equilibria: A review. *Anal. Chim. Acta* **2011**, *686*, 9–18. [CrossRef] [PubMed]

biosensors

MDPI

Article

Vibrational Spectra of Nucleotides in the Presence of the Au Cluster Enhancer in MD Simulation of a SERS Sensor [†]

Tatiana Zolotoukhina *, Momoko Yamada and Shingo Iwakura

Department of Mechanical Engineering, University of Toyama, Toyama 930-8555, Japan
* Correspondence: zolotu@eng.u-toyama.ac.jp; Tel.: +81-76-445-6739
† This paper is an extended version of our paper published in: Zolotoukhina, T.; Yamada, M.; Iwakura, S. Influence of the Au Cluster Enhancer on Vibrational Spectra of Nucleotides in MD Simulation of a SERS Sensor. In Proceedings of the 1st International Electronic Conference on Biosensors, 2–17 November 2020.

Abstract: Surface-enhanced Raman scattering (SERS) nanoprobes have shown tremendous potential in in vivo imaging. The development of single oligomer resolution in the SERS promotes experiments on DNA and protein identification using SERS as a nanobiosensor. As Raman scanners rely on a multiple spectrum acquisition, faster imaging in real-time is required. SERS weak signal requires averaging of the acquired spectra that erases information on conformation and interaction. To build spectral libraries, the simulation of measurement conditions and conformational variations for the nucleotides relative to enhancer nanostructures would be desirable. In the molecular dynamic (MD) model of a sensing system, we simulate vibrational spectra of the cytosine nucleotide in FF2/FF3 potential in the dynamic interaction with the Au20 nanoparticles (NP) (EAM potential). Fourier transfer of the density of states (DOS) was performed to obtain the spectra of bonds in reaction coordinates for nucleotides at a resolution of 20 to 40 cm^{-1}. The Au_{20} was optimized by ab initio density functional theory with generalized gradient approximation (DFT GGA) and relaxed by MD. The optimal localization of nucleotide vs. NP was defined and the spectral modes of both components vs. interaction studied. Bond-dependent spectral maps of nucleotide and NP have shown response to interaction. The marker frequencies of the Au_{20}—nucleotide interaction have been evaluated.

Keywords: vibrational spectra; molecular dynamics; nucleotides; Au nanoparticle; SERS

Citation: Zolotoukhina, T.; Yamada, M.; Iwakura, S. Vibrational Spectra of Nucleotides in the Presence of the Au Cluster Enhancer in MD Simulation of a SERS Sensor . *Biosensors* **2021**, *11*, 37. https://doi.org/10.3390/bios11020037

Received: 20 December 2020
Accepted: 25 January 2021
Published: 29 January 2021

1. Introduction

Sensing and analysis of DNA sequences and epigenetic structure modifications in the human genome are essential for the understanding of the mechanism of gene expression in cells and the development of diseases. Efforts to detect and map DNA sequences and modifications with 2D materials such as graphene and hexagonal BN have been made and have utilized a number of methods, such as nanopore-based ionic and transverse current, and bonding to complimentary molecules in nanopores and STM. The development of the surface-enhanced and tip-enhanced Raman spectroscopy (surface-enhanced Raman scattering (SERS) and TERS) [1–5] with nanoscale resolution [6–9] applicable to single biomolecule [10,11] analysis, as well as stimulated Raman scattering and ultrafast two-dimensional infrared (2D-IR) spectroscopy of small organic molecules [11–14] opens a way to identify molecular structures in a label-free way and investigate mechanisms of interaction with the environment in biomolecules. Attention to DNA molecules in manifold applications, from sequencing, epigenetic studies, disease relation, etc., to nano-biostructures, makes identification of nucleotides and bases in a variety of states highly desirable in the non-contact label-free way of nanophotonic vibrational spectroscopy [15]. SERS is a promising method for nucleic acid [16] and protein detection [17,18]. In terms of its abundant fingerprint information, anti-interference with water, and extraordinarily high sensitivity, SERS holds many attractive advantages over other techniques. Recent research discusses probing of proteins by SERS in solution using gold nanoparticles in the

Biosensors **2021**, *11*, 37. https://doi.org/10.3390/bios11020037

absence of immobilizing agents so that neither the tertiary structure of the proteins nor the surface properties of the nanoparticles are affected [17]. Methods for the rapid and sensitive detection of extended ($\geq$100-base) nucleic acids with reduced preparation are also being developed [16]. They utilize the DNA sequence-specific assembly of silver nanoparticles labelled with a Raman tag to provide molecular recognition of the target DNA with probe orientation and hybridisation procedures found to be critical for the methods.

Another path to the identification of small organic molecules, such as nucleotides, can employ interaction with a nanopore in a 2D solid film, especially graphene [19–27], though experiments and calculations with hexagonal BN were also carried out [28–31]. Detection of the DNA/RNA mononucleotides [32–34] and nucleobases [35] with SERS techniques has also been carried out. Despite the many advantages and simplicity of the 2D nanopore sequencing, a few issues [20] remain to be resolved, including controlling the DNA translocation rate, suppressing stochastic nucleobase motions, and determining the signal overlap between different nucleobases. The graphene-based nanopores suffer from significant signal noises due to graphene hydrophobicity and trapping of DNA bases during translocation [36]. The study of the influence of the interaction force between the nucleobases and graphene nanopore on translocating molecules helps to find a solution to some of the mentioned issues.

Surface-enhanced Raman scattering (SERS) using plasmonic metal nanoparticles has been applied to characterize complex biological samples, ranging from biomacromolecules such as nucleic acids and proteins, to living eukaryotic and prokaryotic cells in whole animals for several decades now. The sensitivity of the SERS signal with respect to interactions at the surface of the plasmonic nanoparticles makes it specifically useful for studies of molecular contacts of nanoparticles in biological systems and does not require chemical or biological functionalization.

Control of the 3D structure of DNA and proteins in SERS spectra acquisition still remains a complex question. In the case of DNA molecules, a translocation of the single or double strand through the nanopore substrate has been well established. For proteins, conformation and unfolding are considered. The discrimination of protein conformations is of critical importance for identifying the unfolding states. Attempts to measure dynamic SERS of proteins [37] were carried out over a long acquisition time. Characteristic protein signatures at different time points were observed and compared to conventional Raman with SERS. TG-SERS can distinguish discrete features of proteins, such as the secondary structures, and is therefore indicative of folding or unfolding of the protein [38]. Other studies [39] mention that at a low gold-nanoparticle: protein-molar ratio, significant unfolding of the protein is observed at the surface of the gold NP. When using unfolded proteins, they can be translocated through the nanostructures [40] and nanopores [41]. Therefore, acquisition of SERS spectra of unfolded proteins at the translocation through a nanopore with an attached enhancer NP should be possible in principle. The process of unfolding and conformation should involve the interaction of the building blocks with system elements, as well as between amino acids in chain molecule and tertiary structures. For secondary and tertiary structures, the amino acids closest to the NP can have slight changes in SERS spectra due to the interactions, and simulation of such changes would help to distinguish and register them in acquired spectra of proteins.

In order to design and simulate the system that would allow for single molecule selection and detection of its state by high-resolution SERS methods, we tried to predict direction and possible range of spectral changes in the interacting molecule-SERS enhancer systems. The placement or growth of the small Au nanoparticle at the edge of the graphene nanopore is considered as shown in Figure 1. Plasmonic enhancement by the Au NP will make the vibrational signal measurable, while graphene nanopores would allow for control localization and transient conformation of passing DNA nucleotides, strand fragments, or proteins. Interaction at the Van der Waals interaction distances with both components, graphene and metal NP, has high probability of changing the vibrational spectral maps of the passing molecules. The vibrational mode changes in nucleotides with respect to the interaction with graphene, which were shown in a molecular dynamic (MD) simulation to be present at short distances in our previous calculations. At present, the estimation of the influence of the small Au nanoparticle on the vibrational spectral modes of the cytosine nucleotide has been attempted to be clarified in MD calculations as a model of nucleotide–Au NP interaction. If the spectral modes of both components of interaction are studied relative to localization and molecule conformation, changes in nucleotide spectra due to interaction conditions can be mapped into a kind of library. We performed the initial steps by clarifying the possibility of such mapping. Bond-dependent spectra of nucleotide and NP were tested on their response to interactions to reveal the marker frequencies of the Au_{20}—nucleotide interaction.

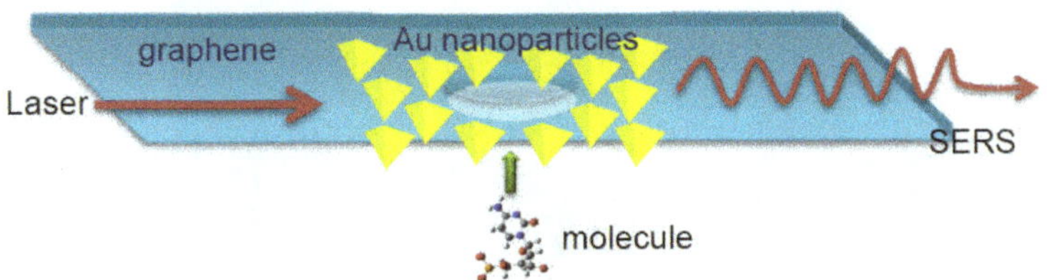

Figure 1. Schematic of the flow-through setup that allows single nucleotide to flow through in a graphene nanopore at plasmonic resonance upon the laser excitation.

2. Model and Methods

The present study considers the MD approach to the identification of nucleotides in interactions with the nanopore environment for vibrational spectroscopy. The nucleotide vibrational frequencies we obtained have been attributed to stretching, bending, or ring-breathing modes. By comparison of nucleobase spectra to the ones of nucleotides, the presence of the 2′-deoxyribose in the nucleotides can be separated into spectral mapping and mode relaxation. The vibrational density of states is calculated in the transient regime during passing time through the graphene nanopore for each atom of the molecule to resolve the difference in the structure of nucleotides. In dynamics, the vibrational spectra are being evaluated from MD propagation velocities computed in the anharmonic interaction potential, in graphene, and in the molecule, with the Lennard-Jones (LJ) potential between them. Fourier transfer $I(f)$ of the velocity autocorrelation function $G(\tau)$ is as follows:

$$G(\tau) = \frac{<v_i(t_0)\cdot v_i(t_0+\tau)>}{<v_i(t_0)\cdot v_i(t_0)>} \qquad I(f) = \int_{-\infty}^{\infty} G(\tau)\exp(-2\pi i f t)d\tau \qquad (1)$$

where τ is the duration of correlation, $v_i(t_0)$ is the velocity of the atom at time t_0, and $v_i(t_0+\tau)$ is the velocity of the atom during correlation time. According to the Wiener–Khinchin theorem, $I(f)$ defines the vibrational density of states (DOS) of the system. The potential used for DNA nucleotides is the MM2/MM3 force field potential [42]. We investigate the transient interaction with graphene modelled by REBO potential [43] and the molecule-graphene interaction for C-X (X = H, O, C, N) by the LJ potential [42]. The interaction causes a shift in some frequencies of vibrations of the nucleotide DOSs. The vibrational spectra in MD exhibit shifts and intensity changes due to interaction with the environment that causes intramolecular vibrational mode dynamics.

The calculation setup is presented in Figure 2. The graphene sheet with the 1.5 nm in diameter pore at its center is shown in two projections, a top and front view, with a location of the nucleotide molecule relative to the pore plane and center. The graphene sheet is oriented in the x-y-plane, the edges along the y-axis are fixed, and the edges along the x-axis are free. The nucleotide can move with a given fixed velocity of the center of mass ($v_{c.o.m.}$) in the positive z-direction that reproduces the motion of DNA fragments through the pores [44,45] in a driving constant electric field in experimental setups. The $v_{c.o.m.}$ is optimized to enhance the interaction force between the nucleotide and edge of graphene pore and reduce rotation of the nucleotide in the translocation as much as possible. All atoms of the system except fixed ones are thermally relaxed to the temperatures $T_{graphene}$ = 300 K and $T_{nucleotide}$ = 30 °C, as in Figure 2, before sampling. In the translocation process, the single-layer graphene sheet interacts with nucleotides in graphene nanopore. Graphene affects the interaction field of the passing molecule close to the pore edge. The graphene-molecule C-X (X = H, O, N, C) potential is considered a VdW one to avoid bond creation and nucleotide attachment to the pore.

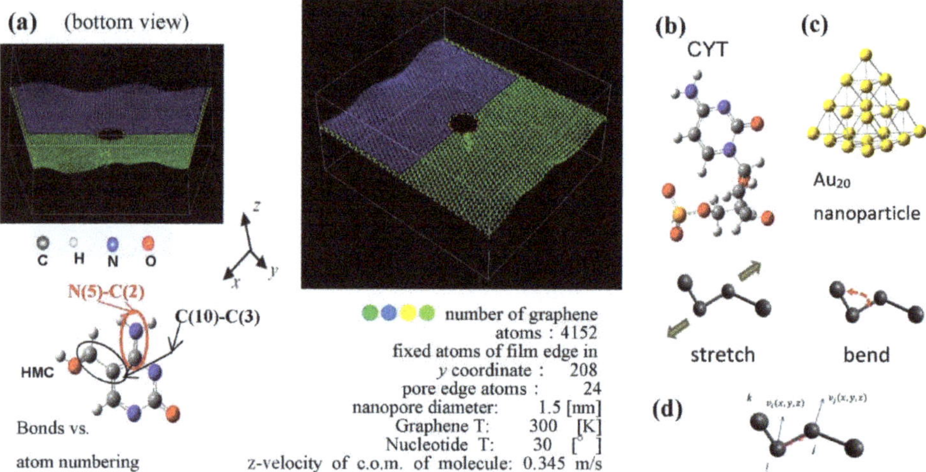

Figure 2. (**a**) The structure and initial position of cytosine nucleotide relative to the graphene pore. The spatial orientation of the nucleotide's cyclic plane is at 30° to the z-x plane. Atoms are shown: C in grey, N in blue, O in red, and H in light grey. (**b**) Initial configuration of cytosine (right). Hydroxymethylcytosine (HMC) base is shown on the left as an example of corresponding atom numbering for bonds in molecular dynamic (MD) calculations. (**c**) Optimized by density functional theory with generalized gradient approximation (DFT GGA) calculations Au_{20} nanoparticle. (**d**) Reaction coordinates for stretching and bending in velocity correlation calculation use the $\vec{v}$ projections along the bonds.

The relative size and weight of the nucleotides as compared with nucleobases reduce rotational motion inside the pore [46–49] that is further controlled by translocation velocity. For the given graphene-nucleotide setup, evaluation of the vibrational spectral maps of the DNA nucleotides, cytosine, thymine, adenine, and guanine, has been initially carried out in reactive coordinates by accumulating spectra of individual atom of bonds obtained by auto-correlation functions of Equation (1). The resulting spectral tables (Tables S1–S8, Figure S1) of each base and nucleotide are included in Supplementary Materials. It was confirmed that the nucleotides–graphene pore interaction does not generate recognizable changes in the spectral map of graphene pore atoms in our system. However, the conformation of the molecules responds to the interaction flexibly, exhibiting spectral changes. The center of mass (c.o.m.) of nucleotides was adjusted relative to the center of the pore along the y-axis to maximize force at the pore edge. The orientation of the cyclic planes of each type of nucleotide was selected to have the same tilt angle for comparison of spectra for various nucleotide interactions. An initial position of the nucleotides was chosen to be outside the interaction region in the z-direction with the tilt of the nucleobase plane taken to be 30 degrees relative to z-x-plane. The single-layer graphene was thermally equilibrated at room temperature prior to sampling. The present simulation was performed in the absence of hydration in the system and hydrophobicity of graphene; therefore, the electrostatic interactions term is not included. For correlation evaluation of nucleotide spectra, the interaction time with graphene was taken at the interaction interval of the passing time through the pore. The high-resolution spectral calculations were carried out at the step size of Δt = 0.05 fs and τ = 32,768 steps of correlation delay for the shown states of nucleotides; in the present calculations, cytosine was used to test the system. To attribute obtained frequencies to a particular type of vibrations, such as stretching, bending, or torsion, the autocorrelation function of Equation (1) was calculated over the reaction coordinates that were taken by the projection of velocities on the bond vector, and the respective angle between these vectors as can be seen in Figure 2d. The single frequency obtained in such co-ordinates can be pinpointed as corresponding to the atom of the vibrating molecule, as well as to the type of vibrational mode that can be pinpointed as corresponding to the particular bond or angle. The ring-breathing modes can be identified by the same frequency present in the spectra of all bonds forming the ring. However, low-level noise from the nearest bonds to the bond studied was also present due to the correlation's calculation method. Therefore, only relatively high intensity spectral modes were considered for evaluation.

The stability of the obtained frequencies relative to the trajectory randomization has also been tested for cytosine molecules. The 10% to 15% randomization in the initial position and velocity distributions of the cytosine nucleotide at five different calculations have shown a preliminary absence of the changes in spectra above a few percent for stretching of the X-Y (X, Y = C, N, O) bonds. Our calculation system does not include solution molecules yet.

To create a calculation model that would be close to an experimental set up in a SERS type of measurement, we decided to introduce a metal nanoparticle in the nucleotide–graphene system of Figure 1c. The Au_{20} nanoparticle was pyramid-shaped and optimized in ab initio DFT GGA (density functional theory with generalized gradient approximation) calculations with the b3pw91 basis set. The size and shape of the NP was taken to be a stable configuration [50]. The obtained configuration was further relaxed by the MD calculation with the EAM potential [51] and free boundary conditions during 3000 steps with Δt = 0.1 fs at T = 300 K to obtain stable nanoparticle samples that can be used separately or on the top of the graphene sheet. The potential parameters for LJ interaction with nucleotide and graphene were defined using the Lorentz–Berthelot mixing rule and based on the LJ parameters ε (Au−Au) = 0.039 kcal/mol, σ (Au−Au) = 2.9 Å [52]. The mixing was done using Van der Waals parameters of the MM2/MM3 force field potential for nucleotides. For the interaction with graphene, mixing used the sp3 C atom parameters from the force field potential, the same values that were applied for the nucleotide–graphene Van der Waals parameters.

The study of the vibrational spectra of the nucleotides in the dynamic interaction with metallic nanoparticles (NP) [53,54] located close to the graphene nanopore can combine translocation localization and nucleotide interaction enhancement or modification due to (1) the graphene LJ interaction and (2) the Au LJ interaction. For the initial interaction of the Au_{20} nanoparticle with cytosine nucleotide, the Au_{20} NP was localized close to the cytosine initial position as it is shown in Figure 3. As the first step, both NP and cytosine had zero c.o.m. velocity. Initial orientation and distance of NP relative to the nucleotide was initially controlled to estimate vibrational spectra of both interacting parts of the system.

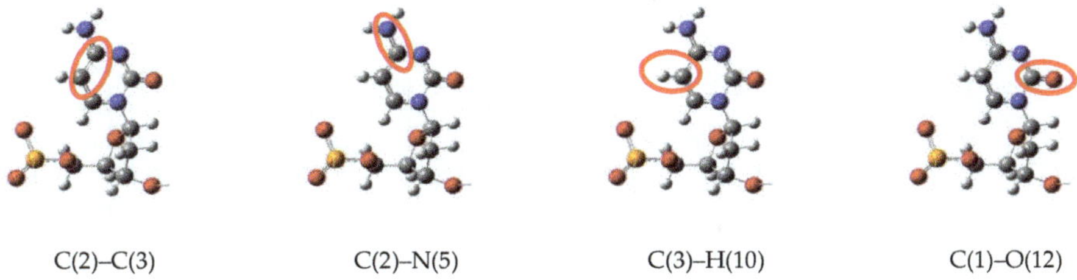

Figure 3. Bonds in cytosine nucleotide (cytidine) for which vibrational spectra are shown below. Numbers of atoms are in internal numbering order.

3. Results

The influence of the interaction in the nucleotide-metal nanoparticle system is numerically modelled in the MD system. The validation of the model was performed twofold: first, to obtain a spectral resolution sufficient to register changes due to the LJ interaction, and secondly, to evaluate the corresponding variations in the vibrational spectra of the nucleotide-nanoparticle system.

3.1. Resolution of Nucleotide Spectra

First, we evaluated the variations in the nucleotide–graphene system at translocation. Spectral variations which were examined due to interaction force in the nucleobase–graphene nanopore system pointed to the influence of the conformation on the nucleotide spectral maps [55,56]. The transient vibrational frequencies of all passing nucleotides were calculated at the same initial incident angle and shift distance from pore edge. To separate decay in the calculated data from the proper spectrum, we have extracte decay components out of calculated spectral data. The calculation of the transient autocorrelation function includes relatively short interaction time during which the correlation data are accumulated. As a result of the transience of the correlation signal, the exponential function of time decay is converted by the Fourier transfer into a decaying spectral map. In order to separate decay components from the spectra itself, we propose the two-parametric exponential fitting, as follows:

$$\begin{aligned} x_{\exp fit} &= a \times \exp(bx_1), \\ x_{amp\, fit} &= x_{amp} - x_{\exp fit} \end{aligned} \tag{2}$$

The introduced exponential function was fitted by parameters a, b either to the function's low-frequency region (head) or high-frequency (tail) region. The exponential decay component $x_{\exp fit}$ was then subtracted from the calculated x_{amp} spectra. The low-frequency fitting produced better-resolved spectra above 5 THz compared with the high-frequency fitting. All calculated spectra of nucleotides that are discussed were obtained in the low-frequency fitting. The initial spectral resolution Δf of Fourier transform that was tested was 40 cm^{-1}.

In contrast with quantum density functional theory (DFT) calculations of IR and Raman spectra, transient MD calculations are sensitive to the duration of correlation time and relative interaction of the structures studied. Therefore, obtained spectra should be compared with the other available calculations and experimental data. The spectral maps of nucleotides that we calculated were highly sensitive to interactions in the system. Since all nucleotide spectral maps were calculated in identical conditions of nanopore translocation, the obtained transient frequencies could be used as reference values to distinguish nucleotides from each other. Comparison with results of 2D-IR experiments and DFT calculations [57] shows a 50–80 cm^{-1} discrepancy between corresponding stretching C-C and C-N frequencies calculated in the MD model at 40 cm^{-1} resolution. The SERS data on the breathing mode [35] of nucleotides with the full width at half maximum of the peak wavenumber being 13 and 20 cm^{-1} show the presence of the mode in the 660–800 cm^{-1} interval for nucleotides. Our single bond spectra in Appendix A exhibit the presence of the bending type of mode in the above interval, different for each nucleotide; however, our resolution in the case is limited by 40 cm^{-1}. For guanine, we obtained a bending peak at 655 cm^{-1} as compared to the measured [35] peak at 661 cm^{-1}; thymine exhibited the calculated 860 cm^{-1} peak vs. the measured 795 cm^{-1}. All measurements were held in the aqueous solution with nucleotides in proximity to the gold nanoslits or nanoparticles for enhancement of the signal. To improve our model, the resolution was improved and the metal nanoparticle was introduced.

The employed spectral resolution in our MD calculations has been sufficient/enough to register structural dependence of the molecular spectral map on a bond-to-bond basis. However, the 2000 cm^{-1} upper limit excludes the C-H and N-H bonds from the calculated frequency region. With the increased resolution of IR and SERS/TERS SM spectroscopy [35,58] and ab initio MD (AIMD) calculations of the chemical bonding effect in SERS [59,60], the information on vibrational mode variations due to interaction with environment, which could be hydration changes or a Van der Waals interaction with graphene pore, should be available in MD calculations as well. The approximation of the MM2/MM3 potential used for nucleotides cannot allow the estimation of dipole moment and polarizability changes with the environment that are calculated in DFT and AIMD spectra. However, transient dynamics sample a sufficient number of states that constitute several periods of vibrations. The coupled anharmonic bonding is also present in the MD potentials which are used. Therefore, a partial reproduction of the changes of dipole moments and polarizabilities with interaction has a cumulative presence in velocity correlation data calculated in MD. To extend the spectral range up to 4000 cm^{-1}, we considered the sampling resolution in real space over the interaction potential $U(r(\Delta t))$ by decreasing Δt, and in reciprocal space by extending the number of time steps [61]. The testing of the proposed spectral resolution was carried out for the hydroxymethylcytosine (HMC) nucleotide that has H-X bonding sites to X=C and O atoms in hydroxymethyl group (marked in Figure 4).

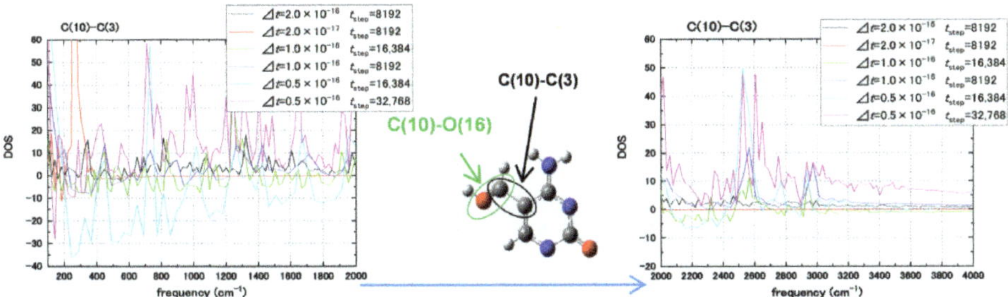

Figure 4. The spectra of a single bond marked as C(10)-C(3) in internal atom numbering of the hydroxymethylcytosine (HMC) nucleotide. The C(10) atom belongs to the hydroxymethyl group and exhibits high frequencies of the C-H bond. The spectra are collected outside of graphene pore interaction (v$_z$ = 0) at a different propagation time step (2.0 ÷ 0.5 × 10^{-16} s) and total simulation time (in step number) in two frequency domains: 0–2000 cm^{-1} at the left and 2000–4000cm^{-1} at the right. The base part of the HMC nucleotide with the marked bond adjacent to H-X (C, O) is shown in the center.

As shown in Figure 4, the time step of 0.05 fs and 32,768 steps has given the resolution of vibrational modes at Δf = 20 cm^{-1} and up to 4000 cm^{-1}. The Δf = 20 cm^{-1} spectral resolution is comparable to the 15 cm^{-1} half-width of the Lorentzian function that is used to broaden Raman spectral lines in the DFT calculation [54]. To further confirm our resolution of the Raman spectra of nucleotides, HMC was calculated at the DFT level with the 20 Å size of the supercell for the methylated nucleotide using the Quantum Espresso package [62]. Some of the H-X stretching modes in the hydroxymethylcytosine are shown in Figure 5. The C(10)-C(3) and C(10)-O(16) bonds of hydroxymethyl group are scanned for highest intensity vibration frequencies, which are compared with the Raman frequencies obtained in the DFT calculations. Correspondence, where it was obtained between the MD and DFT frequency modes, is within 10 cm^{-1}. Not all MD vibrational modes with high amplitudes correspond to the DFT normal modes. Some of the modes that have high intensity in the MD calculation (2524 and 2605 cm^{-1}) can be Raman inactive in the DFT calculation but exhibit IR activity. This relation should be further clarified. The MD calculations with 20 cm^{-1} spectral resolution have been carried out in equilibrium conditions without interaction with graphene.

As compared to the typical sampling timescale of the SERS and TERS techniques being a few microseconds, the typical time scale of the plasmon excitation of the metal enhancer system relates to the scale of 100 fs, and the coupling between the electronic and vibrational degrees of freedom is on the time scale of 0.1–0.5 ps. Therefore, interaction and conformation-dependent single molecule measurements and calculations should concentrate on the time scale of few vibrational periods that we presently cover with a 1.64 ps time interval. Experimentally, nanophotonic probing of the acoustic phonon propagation has been achieved now at the ps timescale [63]. Only the highest peaks in the spectrum are considered as modes in our case. In spectral calculations, the increase in the spectral range resolution simultaneously causes a noise connected to the time step reduction as a consequence of Fourier transfer. The source of the noise is twofold at short time step: (1) the velocity projection on the bond vectors starts to resolve traces of vibrations related to nearest bonds connected by the atom and (2) the FFT procedure in spectra calculations can produce low intensity side lobes of the signal peaks that are not smoothed out with window functions in order not to lose spectral information on essential modes.

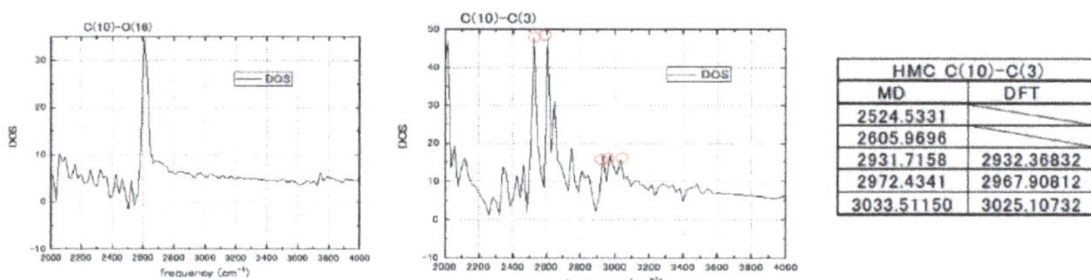

HMC C(10)–C(3)	
MD	DFT
2524.5331	
2605.9696	
2931.7158	2932.36832
2972.4341	2967.90812
3033.51150	3025.10732

Figure 5. Spectral frequencies of the C(10)-C(3) and C(10)-O(16) bonds of hydroxymethylcytosine (HMC) (see Figure 3) calculated by MD outside of graphene pore ($v_z = 0$) for the 2000–4000 cm^{-1} frequency region. Comparison to the density functional theory (DFT) calculations of Raman frequencies.

3.2. Interaction between Metal Clusters and Nucleotides

In order to confirm the effect of the LJ interaction on the vibration spectra of in the nucleotide–Au$_{20}$ nanoparticle system, the Au$_{20}$ NP was placed into an already-tested graphene-nucleotide system close to initial nucleotide position as shown in Figure 6a. Cytosine nucleotides' vibrational spectra have been estimated for base bonds circled in Figure 3 in red. To exclude the interaction with the graphene sheet at the initial stage, the distance from the graphene was out of range of the LJ interaction. The initial orientation of the NP relative to the cytosine nucleotide at the 4 Å distance between the tip of NP and the atoms of the nucleotide was selected to keep the LJ interaction at a relatively weak level. Figure 7 compares vibrational spectra of the bonds C(2)–C(3), C(2)–N(5), C(3)–H(10), and C(1)–O(12) in the absence and presence of Au$_{20}$ NP. Explicit changes in modes due to interaction are shown with vertical arrows ↓ in each case.

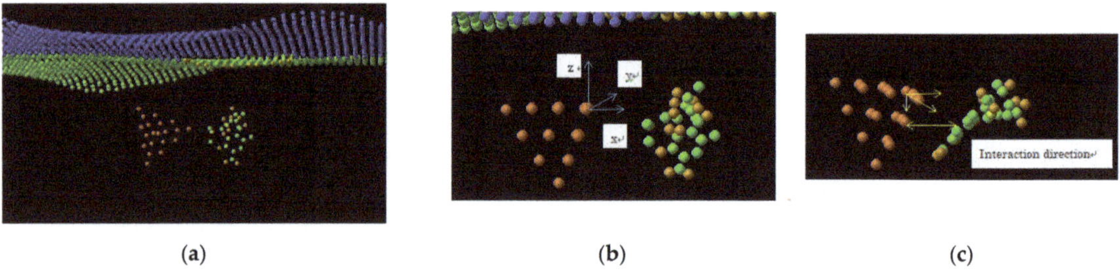

(a)   (b)   (c)

Figure 6. (**a**) System of Au$_{20}$ nanoparticle and cytosine nucleotide vs. graphene sheet in initial Au$_{20}$ NP orientation where interaction is localized primarily within the tip atom of the Au$_{20}$ pyramid; (**b**) rotated NP with the upper plane of the pyramid parallel to the graphene (x,y) plane and interaction with the edge atoms of the Au$_{20}$ pyramid enhanced; (**c**) interaction direction of nucleotide with NP changes during nucleotide conformation and alignment at the vibration time. Calculations were first carried out with the initial $v_{com} = 0$ in x,y,z directions for cytosine and NP.

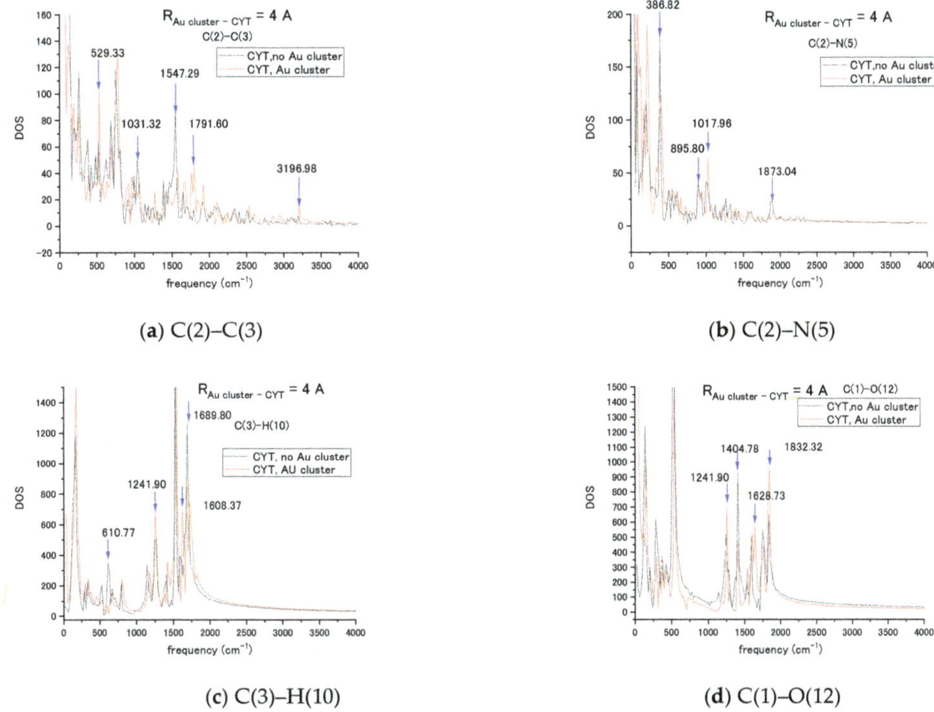

(**a**) C(2)–C(3)

(**b**) C(2)–N(5)

(**c**) C(3)–H(10)

(**d**) C(1)–O(12)

Figure 7. Vibrational spectra of the cytosine (CYT) bonds (**a**) C(2)–C(3), (**b**) C(2)–N(5), (**c**) C(3)–H(10), and (**d**) C(1)–O(12) in the absence and presence of interaction with the Au_{20} NP. Changes in spectra such as mode shifts and large amplitude changes are marked by vertical arrows.

In the C(2)–C(3) case, there were changes in bending and twisting at the 500–1000 cm^{-1} frequency range. On the 1300–2000 cm^{-1} interval, the stretch mode is blue shifted due to interaction with NP, so it is attributed to the breathing mode of the nucleotide base. In the C(2)–N(5) bond, the changes in modes related to bending and twisting can be also seen: the amplitude of the breathing mode was very weak, and the C-N stretching mode had no shift but only an increase in amplitude. The C(3)–H(10) bond had changes in C-H bending and stretching modes at the 1000–2000 cm^{-1} interval. For the C(1)–O(12) bond, the largest changes were in breathing and the stretching modes at the 1300–2000 cm^{-1} range. Interaction with the nanoparticle modified transient vibrational frequencies of cytosine at the ps interval in our MD simulation.

For the Au_{20} NP, the vibrational spectra were also estimated for individual atoms in Cartesian coordinates. Figure 8 exhibits the spectra of the tip atoms in the pyramid in the absence of interaction in x, y, and z coordinates with the basic 188 cm^{-1} cluster mode for all tip atoms that can be a mode characterising the whole Au_{20} nanoparticle. Comparing results in Figure 8, it can be seen that the closer to the nucleotide, the stronger the influence. In addition, it can be seen that there is an even greater effect between atoms 4190 and 4196. Obtained spectra relate to a single orientation of the NP vs. nucleotide. A change in NP orientation would lead to variation in LJ interaction strength and subsequent changes in vibration spectrum.

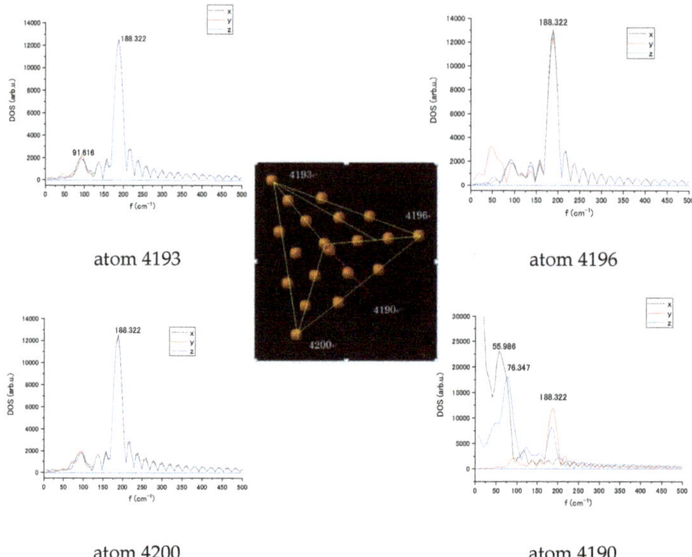

atom 4193   atom 4196

atom 4200   atom 4190

Figure 8. Vibrational spectra of the tip atoms of Au_{20} nanoparticle numbered 4190, 4193, 4196, and 4200 in x, y, and z coordinates without interaction with the cytosine.

The Au NP was then rotated as seen in Figure 6b,c. The upper plane of the pyramid was moved to be in the (x,y) plane and z-axis rotation placed the four atoms of the edge of the pyramid into LJ interaction proximity to the nucleotide. Figure 6c shows the animation snapshot after rotation, and Figure 9 shows the spectral change in 4196 atom's modes after rotation. The vibrational spectra of the interaction with the cytosine single atom (4196) of Au_{20} nanoparticle tip contrast with the spectra of the rotated NP with edge interaction (Figure 9a,b). The basic 188 cm^{-1} frequency of the NP remains intact after rotation. A strong amplitude enhancement is seen for frequencies 76 and 81 cm^{-1} in y and z-direction due to the change in the LJ interaction. The interaction with the nucleotide initiates a slight movement of the NP, which can be reflected in the appearance of the low-frequency modes. For relatively large spherically shaped Au nanoparticles, a typical Raman band over the range of 300–900 cm^{-1} was observed experimentally. For the pyramid-shaped Au_{20} NP, we obtained lower basic frequency that remains stable at different strength of Van der Waals interactions with the nucleotide.

The sensor's design often suggests a fixed attachment of the enhancer nanoparticle and transport of the molecules in the measurement process. To measure transient vibrational spectra of the nucleotides, a scope of molecule's passing velocity vz has been tested in MD simulations. For the given spectral resolution, a cytosine's c.o.m. velocity in the range vz = 0.25 m/s = 25 Å/ps has been selected by interaction time being approximately half of the nucleotide's total propagation time. Passing of the nucleotide limits interaction duration with the NP and leads to changes in vibrational bands. To reveal the connection of the transient time with the changes in the spectrum, we compared spectra of the bond C(2)-C(3) from both end atoms for cytosine alone, a cytosine at 4 Å distance without transient velocity ($v_z = 0$), and with v_z = 0.025 m/s as shown in Figure 10. The C(2) atom of the ring is bound to the amino group NH_2 and C(3) atom of the ring has a C-H bond. The spectra reflect the presence of the different set of bands for each atom in the bond and attachment to the amino group influences transient regime in Figure 10a not only by changes in band amplitude and frequency shifts, but by a velocity of DOS decay that becomes slower. The C(2) atom in Figure 10b exhibits smaller changes in the basic C-C ring vibration frequencies corresponding to ~1550 and 1690 cm^{-1} in present calculations (seen in Figure 7a,c) in stationary and transient regimes. The velocity has been added to the c.o.m. motion of nucleotide at all durations of propagation. Such a simple model reproduces, in general, the motion of partially charged molecules in a uniform electric field. The present result connects the transient velocity of a molecule, size of NP, and interaction time with the molecule's vibrational spectra. The ring remains relatively stable in the transient regime while the amino group attached to the ring structure responds to the transient interaction.

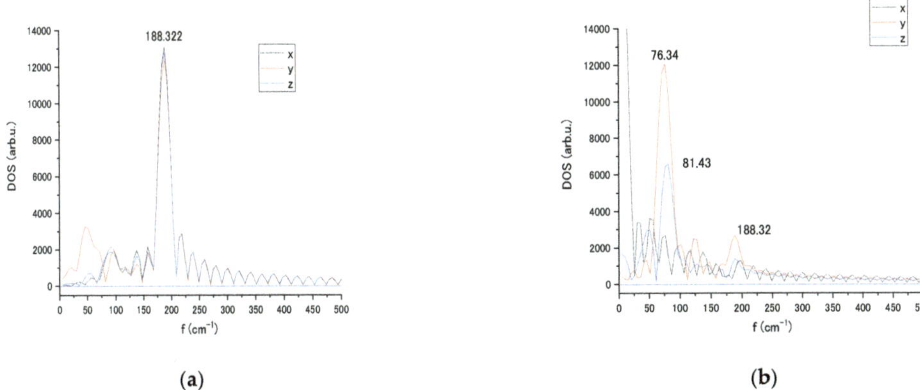

Figure 9. Vibrational spectra of the interacting tip atom (4196) of Au_{20} nanoparticle in x, y, and z coordinates at initial (**a**) and rotated (**b**) orientation relative to cytosine localization.

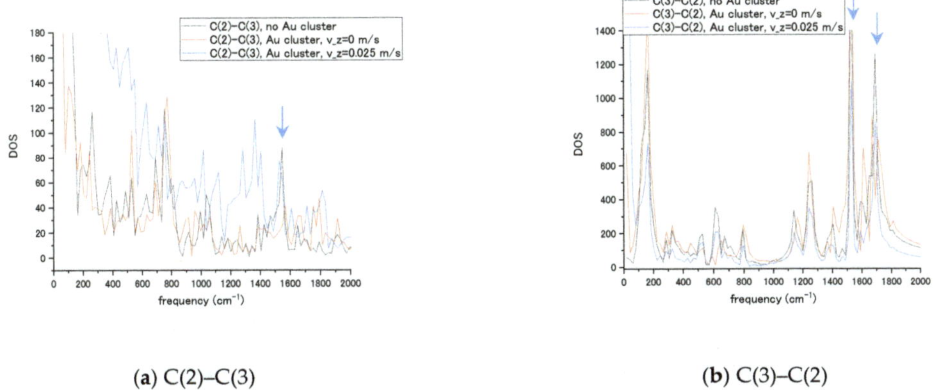

(**a**) C(2)–C(3)   (**b**) C(3)–C(2)

Figure 10. Vibrational spectra of the cytosine bonds (**a**) C(2)–C(3) and (**b**) C(3)–C(2), in the absence and presence of interaction with the Au_{20} NP. Transient velocity of the nucleotide was $v_z = 0.025$ m/s.

4. Discussion

As seen in existing experimental studies of single-molecule SERS of DNA and proteins [64,65], researchers have gradually developed methods to characterize a single biological molecule with Raman spectra at the time scale of interaction with the environment and other molecules and structures. The simulation methods can create spectral libraries for vibrations of molecular species vs. different types of interaction to specify the interaction's type and strength.

We connect the importance of the proposed theoretical calculations with the development of machine learning in various fields of biomolecular modeling and simulation.

AI algorithms are employed in protein structure prediction based on the data in the form of trajectories that can be used to train machine learning algorithms [66] or applied to the analysis of the SERS/TERS spectra measurements of the presence of drug-in-solution [67]. All such algorithms need data to learn from. Another aspect is that machine learning methods generally require extra preprocessing or knowledge of key features in data of measured structures to handle large-scale data. The proposed simulation approach can be the method to produce a sufficient amount of data to train machine learning algorithms, not only on structural differences in spectra of measured molecules, but also on conformational and interaction changes.

The data on the changes in spectra relative to the interactions in the measured system will provide knowledge of key features in spectra as a part of data sets. Such data in combination with machine learning can reduce spectra accumulation and averages over a few seconds' acquisition time applied currently, and open the way to a single pulse measurement of vibrational spectra. In the present study, the possibility of the simulation of Raman spectra as depending on the transient interactions between the parts of a SERS system is investigated. A simple system that contains the molecule itself, an Au nanoparticle as Raman enhancer and graphene sheet with nanopore that can control the localization of the measured molecule relative to enhancer if it is attached to graphene, is selected. The position-controlled agglomeration of nanoparticles can be a key to the successful collection of SERS spectra [39] of proteins and DNA as shown in the experiment. We suppose that when the positioning of nanoparticles can be controlled, and detailed knowledge about spectral changes due to interaction with the nanoparticles and substrate such as graphene with nanopore can be used for the training of machine learning algorithms, fast transient single-molecule SERS measurements can be realized.

The question of extension of the achieved accuracy and simulation' sensitivity of building block molecules to the 3D structures of biomacromolecules such as nucleic acids and proteins are also important for applications. In the case of DNA molecules, a slow translocation of the single or double strand through the nanopore has a timescale around a microsecond in the experiment. The duration time of the nucleotide presence in the vicinity of the graphene nanopore with attached Au NPs should be far below the conventional time resolution of current SERS and TERS techniques that accumulate pulse spectra. However, the single pulse measurement will get a spectrum of the nucleotide located in the vicinity and currently enhanced by agglomerated Au NPs at the translocation time. Each such single pulse spectrum is relatively weak and has a fingerprint of the interaction with the enhancer NP, therefore, preliminary knowledge of the spectral changes relative to interaction could help to isolate the signal in the spectrum from the noise level.

5. Conclusions

With the development of machine learning in biomolecular modeling and characterization, knowledge of interaction-dependent features in spectral data of SERS-measured molecules and structures leads to dynamic single-molecule SERS fingerprints. We have shown that the dynamic vibrational motion analysis of a single molecule can investigate the Van der Waals interaction.

The findings presented in the current research prove the MD simulation's applicability for transient vibrational spectra of biomolecular building blocks such as nucleotides or amino acids. We have shown that the obtained vibrational spectra are sensitive enough to reflect even weak Van der Waals interactions in the few component systems studied, such as nucleotide–Au-NP–graphene. The transient regime of nucleotide passing by NP and through the graphene was shown to be spectrally sensitive. This is confirmed by changes in the bands of amino and methyl groups attached to the rings dependent on interaction strength and length. The transient vibrational spectra record enables discriminating different interaction events with the spectral fingerprints of molecules and NP that also exhibit spectral modification. We consider expanding the scope of the method for protein fragments and conformations. The use of interaction-dependent MD and ab initio

MD simulations of transient spectra can make the measurement of SERS of unattached molecules attainable in a small number of pulses.

Supplementary Materials: The following are available online at https://www.mdpi.com/article/10.3390/bios11020037/s1. Table S1: Frequencies of cytosine nucleotide bonds, highest intensities. Table S2: Frequencies of thymine nucleotide bonds, highest intensities. Table S3: Frequencies of adenine nucleotide bonds, highest intensities. Table S4: Frequencies of guanine nucleotide bonds, highest intensities. Figure S1: The numbering of the nucleotide atoms used in Tables S1–S4 and S5–S8. Table S5: Frequencies of cytosine base bonds, highest intensities. Table S6: Frequencies of thymine base bonds, highest intensities. Table S7: Frequencies of adenine base bonds, highest intensities. Table S8: Frequencies of guanine base bonds, highest intensities.

Author Contributions: Conceptualization, methodology, software development, validation, supervision: T.Z.; data curation: M.Y. and S.I.; writing—original draft preparation, review, and editing: T.Z. All authors have read and agreed to the published version of the manuscript.

Funding: This research received no external funding.

Institutional Review Board Statement: Not applicable.

Informed Consent Statement: Not applicable.

Data Availability Statement: Not applicable.

Acknowledgments: T.Z. gratefully acknowledges the technical assistance in the nucleotide calculations and preparation of the results of Appendix A done by H. Mizuguchi and T. Kitani. All calculations were performed at Applied Mechano-Informatics Laboratory in Toyama University.

Conflicts of Interest: The authors declare no conflict of interest.

Appendix A

For several selected bonds, the stretching spectra of four nucleotides are shown below in Figure A1 for the low resolution calculations.

Each bond between atoms i and j includes the projection of the vectors $\bar{v}_i$ and $\bar{v}_j$ on the bond's $\bar{r}_{ij}$ vector. Investigated spectra for each bond are evaluated from velocity autocorrelations for i and j atoms and are shown in black and red in the figures. Since each atom participates in several bonds' vibrations, calculated spectra contain low amplitude bands of adjacent bonds. The valid frequency range is 100–2000 cm^{-1}. The calculations of spectra were done for approximately 16 ps, with the time step equal to 0.2 fs, at 40 cm^{-1} resolution. The Tables of calculated vibrational bands for the four types of nucleotides and their bases are collected in Supplementary Materials.

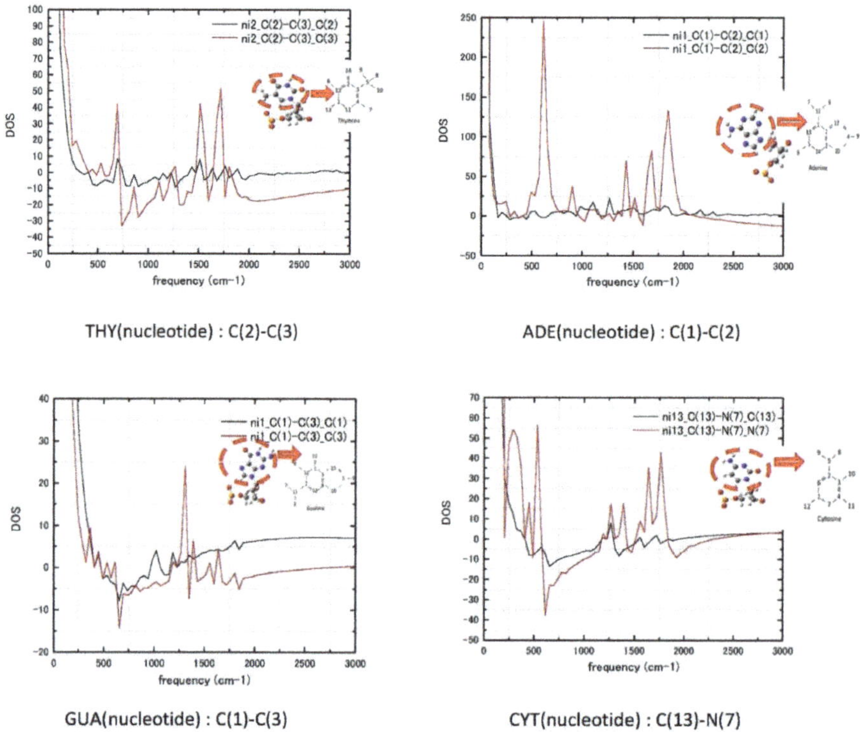

THY(nucleotide) : C(2)-C(3)

ADE(nucleotide) : C(1)-C(2)

GUA(nucleotide) : C(1)-C(3)

CYT(nucleotide) : C(13)-N(7)

Figure A1. Stretching spectra of adenine, guanine, thymine C(1)–C(2)–C(3) bonds for atoms in hexagonal ring of the base, C(13)–N(7) stretching frequencies of cytosine in the base-to-2'-deoxyribose bond, see numbering in the inserted scheme of each base. Red and black colors mark the *i-j* bond from the *i* and *j* side. Atoms at the side of the bond are written after the bond: X-Y X.

References

1. Stöckle, R.M.; Suh, Y.D.; Deckert, V.; Zenobi, R. Nanoscale chemical analysis by tip-enhanced Raman spectroscopy. *Chem. Phys. Lett.* **2000**, *318*, 131–136. [CrossRef]
2. Hayazawa, N.; Inouye, Y.; Sekkat, Z.; Kawata, S. Metallized tip amplification of near-field Raman scattering. *Opt. Commun.* **2000**, *183*, 333–336. [CrossRef]
3. Anderson, M.S. Locally enhanced Raman spectroscopy with an atomic force microscope. *Appl. Phys. Lett.* **2000**, *76*, 3130–3132. [CrossRef]
4. Knoll, B.; Keilmann, F. Near-field probing of vibrational absorption for chemical microscopy. *Nature* **1999**, *399*, 134–137. [CrossRef]
5. Taubner, T.; Keilmann, F.; Hillenbrand, R. Nanomechanical Resonance Tuning and Phase Effects in Optical Near-Field Interaction. *Nano Lett.* **2004**, *4*, 1669–1672. [CrossRef]
6. Zhu, Z.; Zhan, L.; Hou, C.; Wang, Z. Nanostructured Metal-Enhanced Raman Spectroscopy for DNA Base Detection. *IEEE Photonics J.* **2012**, *4*, 1333–1339.
7. Anker, J.N.; Hall, W.P.; Lyandres, O.; Shah, N.C.; Zhao, J.; van Duyne, R.P. Biosensing with plasmonic nanosensors. *Nat. Mater.* **2008**, *7*, 442–453. [CrossRef]
8. Saha, K.; Agasti, S.S.; Kim, C.; Li, X.; Rotello, V.M. Gold Nanoparticles in Chemical and Biological Sensing. *Chem. Rev.* **2012**, *112*, 2739–2779. [CrossRef]
9. Guerrini, L.; Krpetic, Z.; van Lierop, D.; Alvarez-Puebla, R.A.; Graham, D. Direct surface-enhanced Raman scattering analysis of DNA duplexes. *Angew. Chem. Int. Ed.* **2015**, *54*, 1144–1148. [CrossRef]
10. Kneipp, K.; Wang, Y.; Kneipp, H.; Perelman, L.T.; Itzkan, I.; Dasari, R.R.; Field, M.S. Single Molecule Detection Using Surface-Enhanced Raman Scattering (SERS). *Phys. Rev. Lett.* **1997**, *78*, 1667–1670. [CrossRef]
11. Greve, C.; Elsaesser, T. Ultrafast Two-Dimensional Infrared Spectroscopy of Guanine–Cytosine Base Pairs in DNA Oligomers. *J. Phys. Chem. B* **2013**, *117*, 14009–14017. [CrossRef] [PubMed]
12. Hamm, P.; Zanni, M. *Concepts and Methods of 2D Infrared Spectroscopy*; Cambridge University Press: Cambridge, UK, 2011.

13. Peng, C.S.; Jones, K.C.; Tokmakoff, A. Anharmonic Vibrational Modes of Nucleic Acid Bases Revealed by 2D IR Spectroscopy. *J. Am. Chem. Soc.* **2011**, *133*, 15650–15660. [CrossRef] [PubMed]

14. Sanstead, P.J.; Stevenson, P.; Tokmakof, A. Sequence-Dependent Mechanism of DNA Oligonucleotide Dehybridization Resolved through Infrared Spectroscopy. *J. Am. Chem. Soc.* **2016**, *138*, 11792–11801. [CrossRef] [PubMed]

15. Zolotoukhina, T.; Yamada, M.; Iwakura, S. Influence of the Au Cluster Enhancer on Vibrational Spectra of Nucleotides in MD Simulation of a SERS Sensor. *Proceedings* **2020**, *60*, 25. [CrossRef]

16. Huang, J.A.; Mousavi, M.Z.; Zhao, M.Z.; Hubarevich, Y.; Omeis, F.; Giovannini, G.; Schütte, M.; Garoli, D.; De Angelis, F. SERS discrimination of single DNA bases in single oligonucleotides by electro-plasmonic trapping. *Nat. Com.* **2019**, *10*, 5321. [CrossRef] [PubMed]

17. Han, X.X.; Zhao, B.; Ozaki, Y. Surface-enhanced Raman Scattering for Protein Detection. *Anal. Bioanal. Chem.* **2009**, *394*, 1719–1727. [CrossRef]

18. Fazio, B.; D'andrea, C.; Foti, A.; Messina, E.; Irrera, A.; Donato, M.G.; Villari, V.; Micali, N.; Maragò, O.M.; Gucciardi, P.G. SERS detection of biomolecules at physiological pH via aggregation of gold nanorods mediated by optical forces and plasmonic heating. *Sci. Rep.* **2016**, *6*, 26952. [CrossRef]

19. Li, J.; Gershow, M.; Golovchenko, J.A. DNA molecules and configurations in a solid-state nanopore microscope. *Nat. Mater.* **2003**, *2*, 611–615. [CrossRef]

20. Yang, N.; Jiang, X. Nanocarbons for DNA sequencing: A review. *Carbon* **2017**, *115*, 293–311. [CrossRef]

21. Franc, L.T.C.; Carrilho, E.; Kist, T.B.L. A review of DNA sequencing techniques. *Q. Rev. Biophys.* **2002**, *35*, 169–200. [CrossRef]

22. Thompson, J.F.; Milos, P.M. The properties and applications of single-molecule DNA sequencing. *Genome Biol.* **2011**, *12*, 217. [CrossRef] [PubMed]

23. Izrailev, S.; Stepaniants, S.; Isralewitz, B.; Kosztin, D.; Lu, H.; Molnar, F.; Wriggers, W.; Schulten, K. Steered Molecular Dynamics. In *Computational Molecular Dynamics: Challenges, Methods, Ideas*; Springer: Berlin/Heidelberg, Germany, 1999; pp. 39–65.

24. Shankla, M.; Aksimentiev, A. Conformational transitions and stop-and-go nanopore transport of single-stranded DNA on charged graphene. *Nat. Commun.* **2014**, *5*, 5171. [CrossRef] [PubMed]

25. Liang, L.; Shen, J.-W.; Zhang, Z.; Wang, Q. DNA sequencing by two-dimensional materials: As theoretical modeling meets experiments. *Biosens. Bioelectron.* **2017**, *89*, 280–292. [CrossRef] [PubMed]

26. Zhang, Z.; Shen, J.-W.; Wang, H.; Wang, Q.; Zhang, J.; Liang, L.; Ågren, H.; Tu, Y. Effects of Graphene Nanopore Geometry on DNA Sequencing. *J. Phys. Chem. Lett.* **2014**, *5*, 1602–1607. [CrossRef]

27. Liang, L.; Zhang, Z.; Shen, J.; Zhe, K.; Wang, Q.; Wu, T.; Ågren, H.; Tu, Y. Theoretical studies on the dynamics of DNA fragment translocation through multilayer graphene nanopores. *RSC Adv.* **2014**, *4*, 50494–50502. [CrossRef]

28. Zhou, Z.; Hu, Y.; Wang, H.; Xu, Z.; Wang, W.; Bai, X.; Shan, X.; Lu, X. DNA Translocation through Hydrophilic Nanopore in Hexagonal Boron Nitride. *Sci. Rep.* **2013**, *3*, 3287. [CrossRef]

29. De Souza, F.A.L.; Amorim, R.G.; Scopel, W.L.; Scheicher, R.H. Electrical detection of nucleotides via nanopores in a hybrid graphene/h-BN sheet. *Nanoscale* **2017**, *9*, 2207–2212. [CrossRef]

30. Zhang, L.; Wang, X. DNA Sequencing by Hexagonal Boron Nitride Nanopore: A Computational Study. *Nanomaterials* **2016**, *6*, 111. [CrossRef]

31. Gilbert, S.M.; Dunn, G.; Azizi, A.; Pham, T.; Shevitski, B.; Dimitrov, E.; Liu, S.; Aloni, S.; Zettl, A. Fabrication of Subnanometer-Precision Nanopores in Hexagonal Boron Nitride. *Sci. Rep.* **2017**, *7*, 15096. [CrossRef]

32. Lee, D.; Lee, S.; Seong, G.H.; Choo, J.; Lee, E.K.; Gweon, D.-G.; Lee, S.A. Quantitative analysis of methyl parathion pesticides in a polydimethylsiloxane microfluidic channel using confocal surface-enhanced Raman spectroscopy. *Appl. Spectrosc.* **2006**, *60*, 373–377. [CrossRef]

33. Bell, S.E.J.; Sirimuthu, N.M.S. Surface-Enhanced Raman Spectroscopy (SERS) for Sub-Micromolar Detection of DNA/RNA Mononucleotides. *J. Am. Chem. Soc.* **2006**, *128*, 15580–15581. [CrossRef] [PubMed]

34. Madzharova, F.; Heiner, Z.; Gühlke, M.; Kneipp, J. Surface-Enhanced Hyper-Raman Spectra of Adenine, Guanine, Cytosine, Thymine, and Uracil. *J. Phys. Chem. C* **2016**, *120*, 15415–15423. [CrossRef] [PubMed]

35. Chen, C.; Li, Y.; Kerman, S.; Neutens, P.; Willems, K.; Cornelissen, S.; Lagae, L.; Stakenborg, T.; van Dorpe, P. High spatial resolution nanoslit SERS for single-molecule nucleobase sensing. *Nat. Commun.* **2018**, *9*, 1733. [CrossRef] [PubMed]

36. Merchant, C.A.; Healy, K.; Wanunu, M.; Ray, V.; Peterman, N.; Bartel, J.; Fischbein, M.D.; Venta, K.; Luo, Z.; Johnson, A.T.C.; et al. DNA Translocation through Graphene Nanopores. *Nano Lett.* **2010**, *10*, 2915–2921. [CrossRef]

37. Brule, T.; Bouhelier, A.; Dereux, A.; Finot, E. Discrimination between Single Protein Conformations Using Dynamic SERS. *ACS Sens.* **2016**, *1*, 676–680. [CrossRef]

38. Kögler, M.; Itkonen, J.; Viitala, T.; Casteleijn, M.G. Assessment of recombinant protein production in E. coli with Time-Gated Surface Enhanced Raman Spectroscopy (TG-SERS). *Sci. Rep.* **2020**, *10*, 2472. [CrossRef]

39. Szekeres, G.P.; Kneipp, J. SERS Probing of Proteins in Gold Nanoparticle Agglomerates. *Front. Chem.* **2019**, *7*, 30. [CrossRef]

40. Zeng, S.; Wen, C.; Solomon, P.; Zhang, S.-L.; Zhang, Z. Rectification of protein translocation in truncated pyramidal nanopores. *Nat. Nanotechnol.* **2019**, *14*, 1056–1062. [CrossRef]

41. Chen, H.; Li, L.; Zhang, T.; Qiao, Z.; Tang, J.; Zhou, J. Protein Translocation through a MoS2 Nanopore: A Molecular Dynamics Study. *J. Phys. Chem. C* **2018**, *122*, 2070–2080. [CrossRef]

42. Cornell, W.D.; Cieplak, P.; Baryly, C.T.; Gould, I.R.; Merz, K.M., Jr.; Ferguson, F.M.; Spellmeyer, D.C.; Fox, T.; Caldwell, J.W.; Kollman, P.A. A Second Generation Force Field for the Simulation of Proteins, Nucleic Acids, and Organic Molecules. *J. Am. Chem. Soc.* **1995**, *117*, 5179–5197. [CrossRef]

43. Zayak, A.T.; Hu, Y.S.; Choo, H.; Bokor, J.; Cabrini, S.; Schuck, P.J.; Neaton, J.B. Chemical Raman Enhancement of Organic Adsorbates on Metal Surfaces. *Phys. Rev. Lett.* **2011**, *106*, 083003. [CrossRef] [PubMed]

44. Akahori, R.; Hag, T.; Hatano, T.; Yanagi, I.; Ohura, T.; Hamamura, H.; Iwasaki, T.; Yokoi, T.; Anazawa, T. Slowing single-stranded DNA translocation through a solid-state nanopore by decreasing the nanopore diameter. *Nanotechnology* **2014**, *25*, 275501. [CrossRef] [PubMed]

45. Mirsaidov, U.; Comer, J.; Dimitrov, V.; Aksimentiev, A.; Timp, G. Slowing the translocation of double-stranded DNA using a nanopore smaller than the double helix. *Nanotechnology* **2010**, *21*, 395501. [CrossRef] [PubMed]

46. El-Khoury, P.Z.; Hu, D.; Hess, W.P. Junction Plasmon-Induced Molecular Reorientation. *J. Phys. Chem. Lett.* **2013**, *4*, 3435–3439. [CrossRef]

47. El-Khoury, W.P.; Hess, W.P. Raman scattering from 1,3-propanedithiol at a hot spot: Theory meets experiment. *Chem. Phys. Lett.* **2013**, *581*, 57–63. [CrossRef]

48. El-Khoury, P.Z.; Ueltschi, T.W.; Mifflin, A.L.; Hu, D.; Hess, W.P. Frequency-Resolved Nanoscale Chemical Imaging of 4,4′-Dimercaptostilbene on Silver. *J. Phys. Chem. C* **2014**, *118*, 27525–27530. [CrossRef]

49. El-Khoury, P.Z.; Johnson, G.E.; Novikova, I.V.; Gong, Y.; Joly, A.G.; Evans, J.E.; Zamkov, M.; Laskin, J.; Hess, W.P. Enhanced Raman scattering from aromatic dithiols electrosprayed into plasmonic nanojunctions. *Faraday Discuss.* **2015**, *184*, 339–357. [CrossRef] [PubMed]

50. Wang, J.; Wang, G.; Zhao, J. Structures and electronic properties of Cu20, Ag20, and Au20 clusters with density functional method. *Chem. Phys. Lett.* **2003**, *380*, 716–720. [CrossRef]

51. Olsson, P.A.T. Transverse resonant properties of strained gold nanowires. *J. Appl. Phys.* **2010**, *108*, 34318. [CrossRef]

52. Qing, P.; Leng, Y.; Zhao, X.; Cummings, P.T. Molecular simulations of stretching gold nanowires in solvents. *Nanotechnology* **2007**, *18*, 424007.

53. Muntean, C.M.; Bratu, I.; Leopold, N.; Morari, C.; Buimaga-Iarincaa, L.; Purcaru, M.A.P. Subpicosecond surface dynamics in genomic DNA from in vitro-grown plant species: A SERS assessment. *Phys. Chem. Chem. Phys.* **2015**, *17*, 21323–21330. [CrossRef] [PubMed]

54. Latorre, F.; Kupfer, S.; Bocklitz, T.; Kinzel, D.; Trautmann, S.; Grafe, S.; Deckert, V. Spatial resolution of tip-enhanced Raman spectroscopy–DFT assessment of the chemical effect. *Nanoscale* **2016**, *8*, 10229–10239. [CrossRef] [PubMed]

55. Keyser, U.F.; Koeleman, B.N.; Dorp, S.V.; Krapf, D.; Smeets, R.M.M.; Lemay, S.G.; Dekker, N.H.; Dekker, C. Direct force measurements on DNA in a solid-state nanopore. *Nat. Phys.* **2006**, *2*, 473–477. [CrossRef]

56. Takeuchi, K.; Zolotoukhina, T. Individual DNA base identification at the transport through graphene nanopore. In Proceedings of the ASME 11th International Conference on Nanochannels, Microchannels, and Minichannels, ICNMM2013, Sapporo, Japan, 16–19 June 2013; p. ICNMM2013-73053.

57. Fornaro, T.; Biczysko, M.; Monti, S.; Barone, V. Dispersion corrected DFT approaches for anharmonic vibrational frequency calculations: Nucleobases and their dimers. *Phys. Chem. Chem. Phys.* **2014**, *16*, 10112–10128. [CrossRef] [PubMed]

58. Guchhait, B.; Liu, Y.; Siebert, T.; Elsaessera, T. Ultrafast vibrational dynamics of the DNA backbone at different hydration levels mapped by two-dimensional infrared spectroscopy. *Struct. Dyn.* **2016**, *3*, 043202. [CrossRef] [PubMed]

59. Fisher, S.A.; Apra, E.; Govind, N.; Hess, W.P.; El-Khoury, P.Z. Nonequilibrium Chemical Effects in Single-Molecule SERS Revealed by Ab Initio Molecular Dynamics Simulations. *J. Phys. Chem. A* **2017**, *121*, 1344–1350. [CrossRef]

60. Fisher, S.A.; Ueltschi, T.W.; El-Khoury, P.Z.; Miffin, A.L.; Hess, W.P.; Wang, H.-F.; Cramer, C.J.; Govind, N. Infrared and Raman Spectroscopy from Ab Initio Molecular Dynamics and Static Normal Mode Analysis: The C–H Region of DMSO as a Case Study. *J. Phys. Chem. B* **2016**, *120*, 1429–1436. [CrossRef]

61. Zolotoukhina, T.; Nitta, T.; Takeuchi, S.; Wakamatsu, D. Vibrational spectra of methylated forms of cytosine and adenine in the graphene nanopore and for regions of hydrogen binding. In Proceedings of the 19th International Conference on the Science and Application of Nanotubes and Low-Dimensional Materials (NT18), Beijing, China, 15–20 July 2018; p. PO060.

62. Giannozzi, P.; Baroni, S.; Bonini, N.; Calandra, M.; Car, R.; Cavazzoni, C.; Ceresoli, D.; Chiarotti, G.L.; Cococcioni, M.; Dabo, I. QUANTUM ESPRESSO: A modular and open-source software project for quantum simulations of materials. *J. Phys. Condens. Matter* **2009**, *21*, 395502. [CrossRef]

63. Mante, P.-A.; Belliard, L.; Perrin, B. Acoustic phonons in nanowires probed by ultrafast pump-probe spectroscopy. *Nanophotonics* **2018**, *7*, 1759–1780. [CrossRef]

64. Almehmadi, L.M.; Curley, S.M.; Tokranova, N.A.; Tenenbaum, S.A.; Lednev, I.K. Surface Enhanced Raman Spectroscopy for Single Molecule Protein Detection. *Sci. Rep.* **2019**, *9*, 12356. [CrossRef]

65. Macdonald, D.; Smith, E.; Faulds, K.; Graham, D. DNA detection by SERS: Hybridisation parameters and the potential for asymmetric PCR. *Analyst* **2020**, *145*, 1871–1877. [CrossRef] [PubMed]

66. Verkhivker, G.; Spiwok, V.; Gervasio, F.L. Editorial: Machine Learning in Biomolecular Simulations. *Front. Mol. Biosci.* **2019**, *6*, 76. [CrossRef] [PubMed]

67. Weng, S.; Yuan, H.; Zhang, X.; Li, P.; Zheng, L.; Zhao, J.; Huang, L. Deep learning networks for the recognition and quantitation of surface-enhanced Raman spectroscopy. *Analyst* **2020**, *145*, 4827–4835. [CrossRef] [PubMed]

 *biosensors*

Article

2D Nanomaterial, Ti₃C₂ MXene-Based Sensor to Guide Lung Cancer Therapy and Management †

Mahek Sadiq [1], Lizhi Pang [2], Michael Johnson [3], Venkatachalem Sathish [2], Qifeng Zhang [3,4] and Danling Wang [1,3,4,*]

[1]   Biomedical Engineering Program, North Dakota State University, Fargo, ND 58108, USA; mahek.sadiq@ndsu.edu
[2]   Department of Pharmaceutical Science, North Dakota State University, Fargo, ND 58108, USA; lizhi.pang@ndsu.edu (L.P.); s.venkatachalem@ndsu.edu (V.S.)
[3]   Materials and Nanotechnology Program, North Dakota State University, Fargo, ND 58108, USA; michael.johnson.1@ndsu.edu (M.J.); qifeng.zhang@ndsu.edu (Q.Z.)
[4]   Department of Electrical and Computer Engineering, North Dakota State University, Fargo, ND 58102, USA
*   Correspondence: danling.wang@ndsu.edu; Tel.: +1-701-231-8396
†   This paper is an extended version of our paper published in: 2D Nanomaterial, Ti3C2 MXene-Based Sensor to Guide Lung Cancer Therapy and Management. In Proceedings of the 1st International Electronic Conference on Biosensors, 2–17 November 2020.

Abstract: Major advances in cancer control can be greatly aided by early diagnosis and effective treatment in its pre-invasive state. Lung cancer (small cell and non-small cell) is a leading cause of cancer-related deaths among both men and women around the world. A lot of research attention has been directed toward diagnosing and treating lung cancer. A common method of lung cancer treatment is based on COX-2 (cyclooxygenase-2) inhibitors. This is because COX-2 is commonly overexpressed in lung cancer and also the abundance of its enzymatic product prostaglandin E2 (PGE_2). Instead of using traditional COX-2 inhibitors to treat lung cancer, here, we introduce a new anti-cancer strategy recently developed for lung cancer treatment. It adopts more abundant omega-6 (ω-6) fatty acids such as dihomo-γ-linolenic acid (DGLA) in the daily diet and the commonly high levels of COX-2 expressed in lung cancer to promote the formation of 8-hydroxyoctanoic acid (8-HOA) through a new delta-5-desaturase (D5Di) inhibitor. The D5Di does not only limit the metabolic product, PGE_2, but also promote the COX-2 catalyzed DGLA peroxidation to form 8-HOA, a novel anti-cancer free radical byproduct. Therefore, the measurement of the PGE_2 and 8-HOA levels in cancer cells can be an effective method to treat lung cancer by providing in-time guidance. In this paper, we mainly report on a novel sensor, which is based on a newly developed functionalized nanomaterial, 2-dimensional nanosheets, or Ti_3C_2 MXene. The preliminary results have proven to sensitively, selectively, precisely, and effectively detect PGE_2 and 8-HOA in A549 lung cancer cells. The capability of the sensor to detect trace level 8-HOA in A549 has been verified in comparison with the traditional gas chromatography–mass spectrometry (GC–MS) method. The sensing principle could be due to the unique structure and material property of Ti_3C_2 MXene: a multilayered structure and extremely large surface area, metallic conductivity, and ease and versatility in surface modification. All these make the Ti_3C_2 MXene-based sensor selectively adsorb 8-HOA molecules through effective charge transfer and lead to a measurable change in the conductivity of the material with a high signal-to-noise ratio and excellent sensitivity.

Keywords: 2D Ti_3C_2 MXene; PGE_2; 8-HOA; lung cancer

Citation: Sadiq, M.; Pang, L.; Johnson, M.; Sathish, V.; Zhang, Q.; Wang, D. 2D Nanomaterial, Ti₃C₂ MXene-Based Sensor to Guide Lung Cancer Therapy and Management . *Biosensors* **2021**, *11*, 40. https://doi.org/10.3390/bios11020040

Received: 1 January 2021
Accepted: 29 January 2021
Published: 4 February 2021

Publisher's Note: MDPI stays neutral with regard to jurisdictional claims in published maps and institutional affiliations.

1. Introduction

The most common cancers occur in the lungs, breasts, pancreas, colon, skin, and stomach [1]. Lung cancer is the second most common cancer in men and women and the leading cause of cancer deaths in the United States. The two major types of lung cancer are small cell lung cancer (SCLC, ~15%) [2] and non-small cell lung cancer (NSCLC, ~85%) [3]. The survival rate of both types of lung cancer is very low [4]. According to the American Cancer Society, lung cancer and asbestos-related lung cancer [5] alone were responsible for 142,670 estimated deaths in 2019, making it the number one killer and three times deadlier than breast cancer [6]. This is because most patients (~75%) are diagnosed at a late stage of the disease (stage III or IV) [7]. To increase the survival rate, major advances in lung cancer control or prevention can be facilitated by early detection and effective anti-cancer therapy. In recent years, a variety of therapeutic and adjuvant methods and nutritional approaches have been developed for lung cancer treatment such as chemotherapy, targeted therapy [8,9], cyclooxygenase (COX)-2 inhibition [10], and omega-3 fatty acid dietary manipulation [11,12].

In addition to these methods, many physical "visualization/detection" methods [13] are available for tumor detection and cancer diagnosis [14]. Some of them are positron emission tomography (PET), magnetic resonance imaging (MRI), computerized tomography (CT), ultrasonography, endoscopy, and the gas chromatography method. However, these methods have some major issues for applications in cancer diagnosis. For example, MRI is very expensive and time-consuming. Sometimes it cannot even distinguish between malignant and benign cancer [15]. In the case of PET, radioactive material is used which is combined with glucose and injected into the patient. This process might cause a health concern for diabetic patients [16]. High-dose radiation involved in CT scanning can even increase the risk of cancer [17]. Ultrasound, however, cannot provide accurate diagnosis and frequently has difficulty determining whether a mass is malignant or not [18]. Endoscopy is relatively safer but still has complications such as perforation, infection, bleeding, and pancreatitis [19]. The fundamental limitation of gas chromatography is that the substance must be volatile. It means that a finite portion of the substance needs to be distributed into the gaseous state [20], which could make it problematic to use gas chromatography–mass spectroscopy (GC–MS) in cancer detection because its sampling procedure is very complicated and the results are difficult to interpret. In addition, the GC–MS technique is very expensive and must be operated by very skilled personnel [21]. Therefore, an effective and accurate technique to diagnose cancer and assist in effective treatment is urgently needed.

In particular, studies have confirmed that cyclooxygenase (COX), typically the inducible form of COX-2, is commonly overexpressed in lung cancer and the abundance of its enzymatic product prostaglandin E_2 (PGE_2) plays an important role in influencing cancer development. Since PGE_2 is a deleterious metabolite formed from COX-2-catalyzed peroxidation of an upstream omega-6 (ω-6) fatty acid called arachidonic acid (AA), PGE_2 promotes tumor growth and metastasis [22]. Subsequently, it can be treated as an indicator of local COX activity to regulate or control lung cancer. Many efforts in lung cancer therapy have been focused on the development of COX-2 inhibitors because they can be used to suppress prostaglandinE_2 (PGE_2) formation from COX-2-catalyzed ω-6 arachidonic acid peroxidation [23]. However, most COX-2 inhibitors can severely injure the gastrointestinal tract, increase the risk of cardiovascular disease, and provide limited clinical responses [22,23]. To seek a safer and more efficient method to treat cancers, a new anti-cancer strategy [24], as shown in Figure 1, has been recently developed. This is a very different approach than the classic COX-2 inhibitors [24–26]. In detail, this is a strategy that adopts more abundant ω-6s such as dihomo-γ-linolenic acid (DGLA) in the daily diet and the commonly high level of COX expressed in most cancers to promote the formation of 8-hydroxyoctanoic acid (8-HOA) using a newly developed inhibitor, delta-5-desaturase inhibitor (D5Di). This is because the D5D is an enzyme that converts an upstream DGLA in a diet to AA. The high expression of COX-2 will promote the conversion of AA to PGE_2, while the D5Di will (1) knock down the conversion of DGLA to AA and limit the generation of metabolic product, PGE_2; and (2) promote the COX-2-catalyzed DGLA peroxidation to form 8-HOA, a novel anti-cancer free radical by-product. This strategy has proven to produce more effective and safer therapeutic outcomes in cancer treatment and has been validated in colon and pancreatic cancers [27]. Therefore, detection of the PGE_2 and 8-HOA in lung cancer should be an effective method to evaluate the efficiency of the cancer treatment. Furthermore, the relative ratio of PGE_2 and 8-HOA concentrations can become a useful adjuvant method to help diagnose cancers at an early stage. Therefore, it is very critical to develop a sensing technique or device that can track/monitor the PGE_2 and 8-HOA concentrations in cancer and provide in-time guidance and feedback for cancer treatment and prevention. However, due to the extremely low concentrations of PGE_2 and 8-HOA in cancer cells ~ ng/mL or μM, the detection of these components is quite challenging. Traditional methods of measuring low concentrations of compounds, such as PGE_2 and 8-HOA, are using gas chromatography–mass spectrometry (GC–MS) or liquid chromatography–mass spectrometry (LC–MS). These techniques, as described above, are accurate and sensitive but heavy (not portable), expensive (needing skilled personnel to operate), and time-consuming (complicated sample processing) and cannot provide in-time feedback.

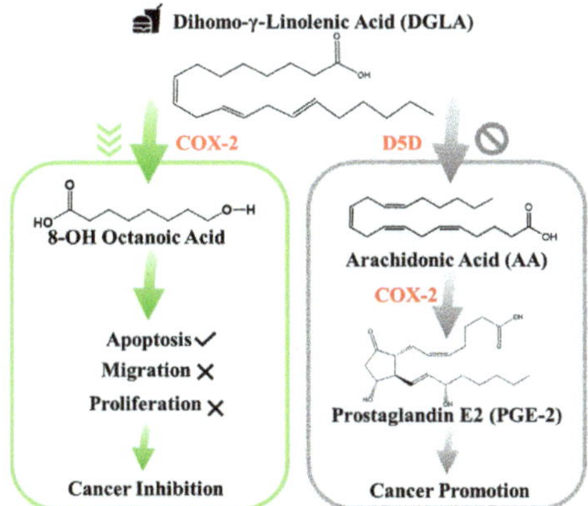

Figure 1. New anti-cancer strategy: target but do not inhibit cyclooxygenase-2 (COX-2) in cancer.

Recent advances in nanotechnology have made it possible to synthesize functionalized nanomaterials for applications in electronics, sensing, biomedicines for disease diagnosis and control, drug delivery, and the food industry [28–31]. Due to the increased surface areas and the feasibility of controllable size and surface properties, nanomaterials such as nanofibers, nanowires, and nanoparticles provide great opportunities for the development of advanced sensing systems and portable device/instrumentation with improved sensitivity and selectivity [32–38]. In particular, the use of structure-directing synthetic approaches in nanomaterials allows the tailoring of the nanomaterial crystalline phase, surface states, morphology, and facets for specific sensing application. Recently, with the development of two-dimensional (2D) nanomaterials such as graphene, these types of materials have gained tremendous attention because of their astonishing electrical and optical properties featured with an "all-surface" nature [39–43]. This all-surface nature can offer great opportunities to tune material properties through surface treatment for targetable detection.

In 2011 [44], the discovery of MXenes introduced a new family into the two-dimensional (2D) materials and further proves to be promising in the flexible and broad application due to its controllable preparation methods and fascinating properties. In essence, MXenes consist of transition metals (including Ti, V, Nb, Mo, etc.) and carbon or nitrogen, sharing a general formula of $M_{n+1}X_n$ (n = 1–3). As a new star of 2D materials, MXenes have the metallic conductivity and hydrophilic nature due to their uncommon surface terminations. Moreover, the unique accordion-like morphology (Figure 2), excellent conductivity, and rich but tailorable surface functional groups endow MXenes with attractive electronic, mechanical, physical, and chemical properties [45] for applications in energy storage [46], environmental science [47], and sensors [48]. The numerous applications of MXene as sensors in various fields are summarized in the paper [49]. In our conference paper [50], we reported that the as-synthesized Ti_3C_2 MXene-based nanosensor has effectively detected 8-HOA in cancer cells without and with using a D5D inhibitor. In that paper, we described more details about how the new sensor based on the two-dimensional nanomaterial Ti_3C_2 MXene [51] can facilitate the lung diagnosis and treatment efficiently by using the new D5D inhibitor and 8-HOA therapy on lung cancer. The preliminary data indicate that this new sensor device can sensitively detect PGE_2 and 8-HOA levels in healthy and cancerous lung cells (BEAS2B and A549 respectively) with similar accuracy to GC–MS but much faster and in an in-time manner to guide the cancer treatment.

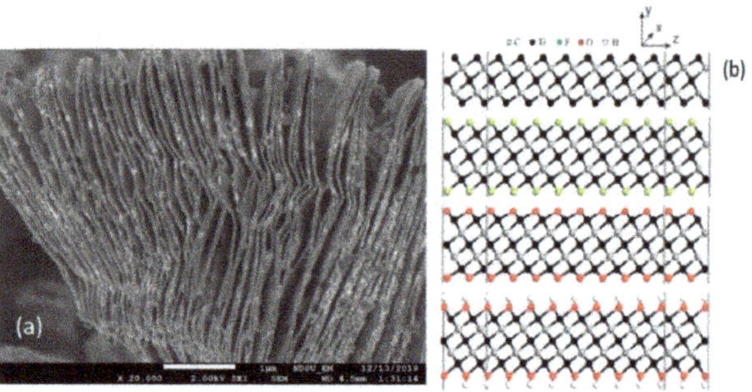

Figure 2. Newly synthesized 2D multilayered Ti_3C_2 MXene nanosheets (**a**) Scanning electron microscope (SEM) image; (**b**) pristine and surface-terminated Ti_3C_2 MXene with different functional groups.

2. Materials and Methods

2.1. Sensing Material Synthesis, Sensing Tests, and Cell Lines Preparation

2.1.1. Ti_3C_2 Nanomaterial-Based Sensor Preparation

The sensor that we used is based on a new 2D nanomaterial, Ti_3C_2 MXene. This nanomaterial was prepared using a method developed in our group and named the "hot etching method" [52]. In detail, the synthesis of Ti_3C_2 MXene followed the steps: (1) Preparing the Ti_3AlC_2 MAX phase. It was obtained through ball milling TiC, Ti, and Al powders in the molar ratio 2:1:1.2 respectively, for 5 h. Under argon flow, the resulting powder was then pressed into a pellet and sintered at 1350 °C for 4 h. The collected pellet after being milled back into powder was sieved through a 160-mesh sieve; (2) etching Al from the MAX phase to form the MXene phase. The as-prepared MAX powder was collected at an elevated temperature for etching through the "hot-etch method". Hydrofluoric (HF) acid in 25 mL Teflon line autoclave at a temperature of 150 °C was used in a Thermolyne furnace for 5 h to etch 0.5 g of the MAX phase. To remove Al from the MAX phase, 5%wt of HF was used. Materials after being sonicated for one hour using a sonicating bath were collected through centrifuge; all the materials were dried overnight in a drying oven at 65 °C; (3) synthesizing MXene powders for the sensing film. Finally, the synthesized nanomaterial was drop-casted on the gold electrode-patterned glass substrate to form a thin film. The thin film is made by first making a paste for the application of the MXene powder to the substrate. This paste is made using 0.1 g of MXene material in 0.3 mL of ethanol which is then dispersed via mixing on a stir-plate. This paste is then blade-coated onto the sensor substrate at a thickness of 0.05 mm. The morphology of the synthesized 2D multilayered nanomaterial is shown in Figure 2a as the scanning electron microscope (SEM) image, which clearly exhibits multilayered nanosheets and accordion-like morphology. Figure 2b reveals the as-synthesized Ti_3C_2 material's special surface terminations, which can lead to the unique surface and material properties of Ti_3C_2.

2.1.2. Ti$_3$C$_2$ MXene Based Sensor Device Fabrication

In order to monitor the promoting 8-HOA formation and the variation of PGE$_2$ in cancer cells during the new anti-cancer treatment, the sensor device, as shown in Figure 3, is fabricated followed by the steps: (1) Ti$_3$C$_2$ MXene sensing film fabrication: the Au electron contact patterned using photolithography and deposited on the glass wafer as the substrate, and then direct drop-casting the as-synthesized Ti$_3$C$_2$ suspension solution onto the patterned substrate to form the sensor slide; (2) the resistance change caused by the exposure of 8-HOA/PGE$_2$ on the sensing film is measured through a Keithley resistance meter and data collected via computer.

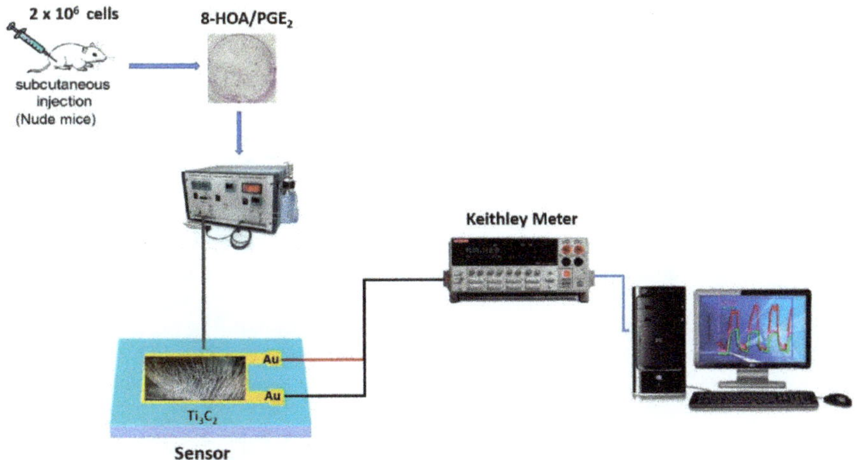

Figure 3. Sketch of Ti$_3$C$_2$ MXene-based sensing system.

2.1.3. Cancer Cell Lines and Materials

A549 (ATCC®CCL-185™), NCI-H1299 (ATCC® CRL-5803), and BEAS-2B (ATCC®CRL-9609™) were purchased from American Type Culture Collection (ATCC, VA, USA). Iminodibenzyl (CAS Number: 494-19-9) and 8-hydroxyoctanoic acid (8-HOA) were obtained from Sigma-Aldrich (St. Louis, MO, USA). PGE$_2$ and DGLA (for in vitro study) and DGLA ethyl ester (for in vivo study) were acquired from Cayman Chemical (Ann Arbor, MI, USA).

2.1.4. Preparation of Cell Samples

About 3×10^5 A549 or BEAS-2B cells were trypsinized and seeded into each well of the 6-well plates. Then, the cells were randomly assigned into different groups for the administration of DGLA (100 µM), iminodibenzyl (10 µM), or their combination accordingly. After 48 h, the cell culture medium was collected. Cells were washed with phosphate buffer solution (PBS) and collected by centrifugation after trypsinization. A 1 mL cell culture medium with collected cells was homogenized and ready for testing. Three different groups of control samples were prepared using the same preparation procedures, including (a) blank group in 1 mL cell homogenate without any treatment; (b) 8-HOA group in 1 mL cell homogenate containing 0.6 ug/mL exogenous 8-HOA; (c) PGE$_2$ group in 1 mL cell homogenate containing 6 ug/mL exogenous PGE$_2$.

2.1.5. Xenografted Lung Tumor Model on Nude Mice

Six-week-old nude mice were purchased from The Jackson Laboratory. The mice were housed in a pathogen-free IVC System with water and food ad libitum. All the animal experiments in this study were approved by the Institutional Animal Care and Use Committees at North Dakota State University. About 2×10^6 A549 or H1299 cells were injected into the hind flank of the nude mouse to induce tumors as we previously described [26]. The mice were randomly assigned to the following treatments: Control (treated with the same volume of the vehicle), DGLA (5 mg/mouse, oral gavage, every day), iminodibenzyl (15 mg/kg, intraperitoneal injection, daily), and DGLA+ iminodibenzyl. The treatment started at two weeks of injection of A549 cells in nude mice. All the administrations lasted for four weeks. At the end of the treatment, mice were sacrificed, and tumors were isolated. Tumor tissues were crushed and homogenized by using a mortar in liquid nitrogen. The blood was centrifuged for 10 min at 2000 rpm for separating serum. The supernatant of tumor tissues and serum was collected for analysis.

2.2. Methodology

To verify the roles of 8-HOA and PGE_2 in cancer development and treatment, the experiments have been designed to do testing in healthy lung cells and A549 lung cancer cells. The 8-HOA and PGE_2 concentrations were carefully calculated before applying them onto the sensing films with certain accuracy and consistency. The sensor performance was further verified in comparison with the measurement using traditional GC–MS.

2.2.1. Normal Cells

For application to test the effect of 8-HOA and PGE_2 in normal lung cells, 10^6 BEAS2B non-tumorigenic epithelial cell lines were collected. 8-HOA, PGE_2, and BSA (Bovine Serum Albumin) were applied to the samples right before measuring the resistance change. Once the samples were applied onto the Ti_3C_2 MXene-based sensors, resistances were measured immediately and repeated at regular time intervals. The experiment is listed in Table 1 and the resistance change of the MXene slides for each of the samples is measured and shown in Figure 4. The resistance increases dramatically when BEAS2B is added with PGE_2 but BEAS2B alone and BEAS2B with 8-HOA do not show obvious change of resistance.

Table 1. Table showing the composition of each sample for BEAS2B.

Sample	Cell	8-HOA	PGE_2	BSA
1	10^6 BEAS2B	none	none	None
2	10^6 BEAS2B	0.6 ug/mL	none	None
3	10^6 BEAS2B	none	6 ug/mL	None
4	None	none	none	1 mg/mL

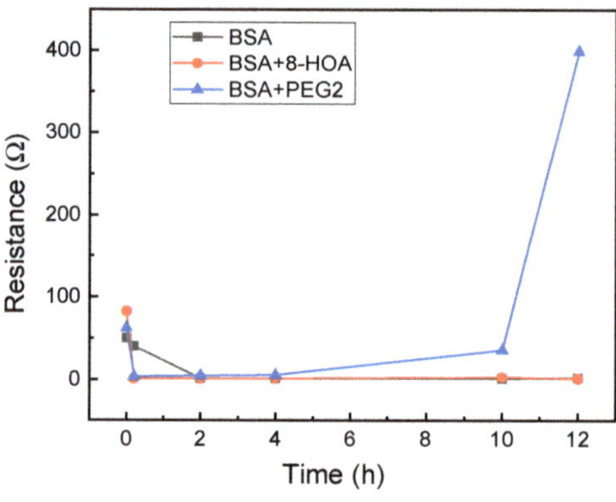

Figure 4. Resistance change measured using Ti_3C_2 MXene-based sensors for BEAS2B cells.

2.2.2. A549/H1299 Lung Cancer Cells

A549 and H1299 both are lung cancer epithelial cell lines. To study the effect of 8-HOA, both H1299 and A549 cell lines have been studied on lung cancer cell apoptosis, proliferation, and survival as shown in Figure 5. H1299 cells were treated with 1 µM 8-HOA for 48 h and afterward were subjected to flow cytometry to observe the apoptosis in the staining of Annexin V-FITC/PI. Wound-healing assay of H1299 lung cancer (non-small cell lung carcinoma cell line) treated with 8-HOA was observed and recorded having the relative area of the wound to 0 h respectively normalized to 1. Finally, the Colony formation assay of H1299 lung cancer cells treated with 8-HOA was observed. Survival fraction of different treatment groups to control were respectively normalized to 1.

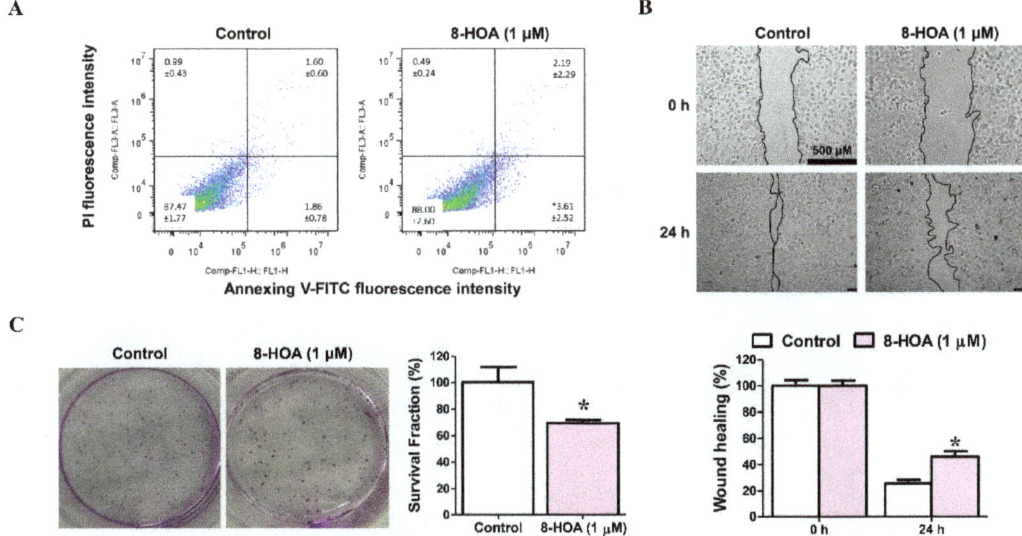

Figure 5. Effect of 8-hydroxyoctanoic acid (8-HOA) on lung cancer cell apoptosis, proliferation, and survival. (**A**) Cell apoptosis was determined by flow cytometry. (**B**) Wound-healing assay of H1299 lung cancer cells. (**C**) Colony formation assay of H1299 lung cancer. * $p < 0.05$ vs. Control group. Data represent mean ± SEM, unpaired t-test.

Figure 6 shows the effect of D5D inhibitor on lung cancer apoptosis. Cell apoptosis was determined by flow cytometry on H1299 lung cancer cells in staining of Annexin V-FITC/PI. H1299 cells were treated with DGLA (100 µM) and D5D inhibitor (10 µM) for 48 h before flow cytometry analysis. Immunofluorescence images of cleaved poly (ADP-ribose) polymerase (PARP) in lung tumor tissues after 4 weeks of treatment with DGLA (5 mg/mouse) and D5D inhibitor (15 mg/kg) were collected. The expression of cleaved PARP was stained in violet, and cell nuclei were counter-stained with 4′,6-diamidino-2-phenylindole (DAPI).

The experimental results showed a similar response to 8-HOA in A549 and H1299 cell lines. In sensing tests, A549 cells were collected after being cultured. The complete design of the experiments to verify the relative concentration of generated 8-HOA and PGE_2 with and without using the new cancer treatment via the detection of Ti_3C_2 Mxene-based sensor are listed in Table 2. Similar to the BEAS-2B cell lines, 8-HOA and PGE_2 samples were applied to the A549 cell lines just before conducting the experiment. The sensing tests of these samples are shown in Figure 7. The resistances of A549 cancer cells, A549 cells treated by DGLA, and PGE_2 are much higher than the cancer cells treated by adding 8-HOA, applying D5Di, or using the new anti-cancer treatment DGLA + D5Di.

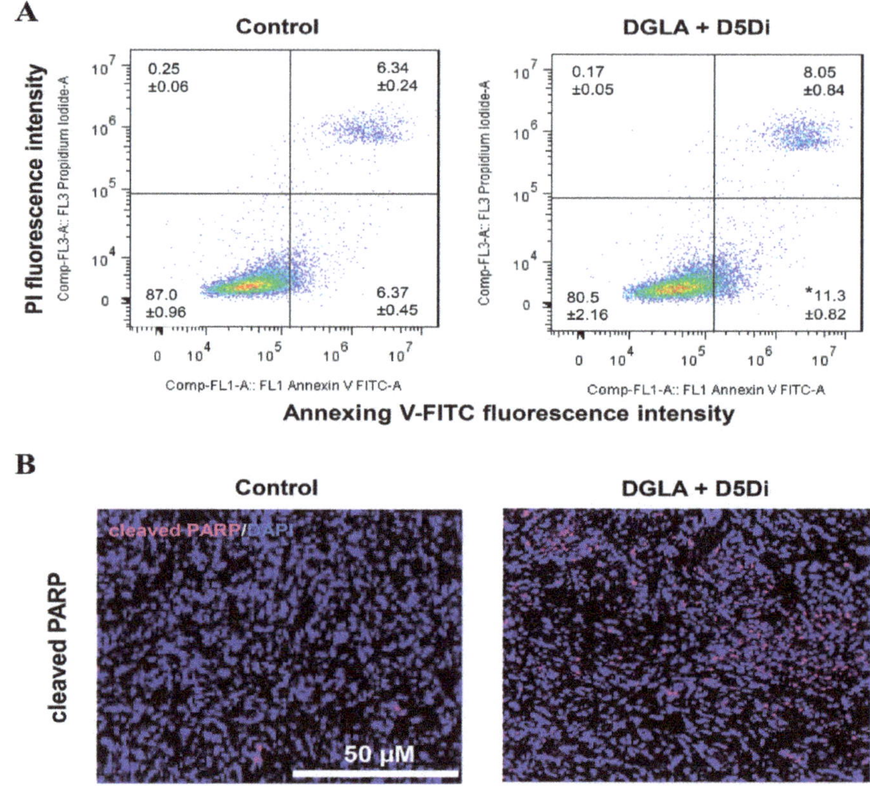

Figure 6. Effect of delta-5-desaturase (D5D) inhibitor on lung cancer apoptosis. (**A**) Cell apoptosis determined by flow cytometry. (**B**) Immunofluorescence images of cleaved poly (ADP-ribose) polymerase (PARP) in lung tumor tissues. * $p < 0.05$ vs. Control group. Data represent mean $\pm$ SEM, unpaired t-test.

Table 2. The composition of each sample for A549 cells treated by 8-hydroxyoctanoic acid (8-HOA), Prostaglandin E2 (PGE2), dihomo-γ-linolenic acid (DGLA), delta-5-desaturase inhibitor (D5Di), and DGLA + D5Di.

Sample	Cell	DGLA	D5Di	8-HOA	PGE$_2$	Estimated 8-HOA/PGE$_2$ Level
1	10^6 A549	none	none	none	none	Low 8-HOA; low PGE2
2	10^6 A549	none	none	0.6 µg/mL	none	High 8-HOA; low PGE2
3	10^6 A549	none	none	none	6 µg/mL	Low 8-HOA; high PGE2
4	10^6 A549	100 µM	none	none	none	Low 8-HOA; high PGE2
5	10^6 A549	none	10 µM	none	none	Low 8-HOA; low PGE2
6	10^6 A549	100 µM	10 µM	none	none	High 8-HOA; low PGE2

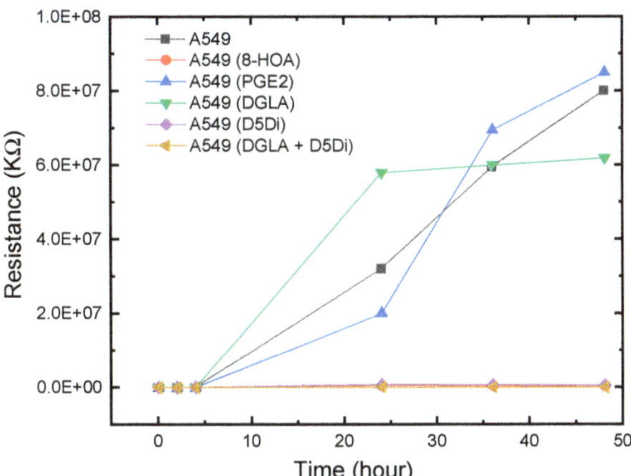

Figure 7. Resistance change measured using Ti_3C_2 MXene-based sensors for A549 cancer cells with and without using the new anti-cancer treatment.

3. Results and Discussion

3.1. Observation from the Non-Tumorigenic Sample Graph

In a healthy subject, both the concentration of PGE_2 and 8-HOA should be low. The sensing test is conducted on the normal lung cell, BEAS2B, without extra treatment and BEAS2B by treating with extra PGE_2 or 8-HOA. A significant resistance increase is observed in BEAS2B by adding 10 μM PGE_2, while the untreated normal cells and cells treated by 8-HOA do not show obvious resistance change. This result indicates a unique role of PGE_2 in healthy cells through the change of the electrical property of sensing material. Considering the elevated concentration of PGE_2 can indicate a cancer development, such a sensitive response to PGE_2 using Ti_3C_2 MXene-based sensor can be potentially used to diagnose cancer even at a very early stage.

3.2. Observation from the CARCINOGENIC Samples

As we have discussed in this paper previously, D5D inhibitor (D5Di) is used for preventing the conversion of DGLA to AA and ultimately limiting the formation of PGE_2. According to the main mechanism of the new anti-cancer strategy, D5Di along with DGLA can effectively limit the formation of PGE_2 but promote the formation of 8-HOA. The sensing test using the newly developed Ti_3C_2 MXene-based sensor, as shown in Figure 6, exhibits an interesting trend of resistance change. Similar to showing high resistance for A549, A549 with adding 10 uM PGE_2, and A549 treated by DGLA both show high resistance in the sensing test. The results indicate a higher concentration of PGE_2 generated in A549 cells while the high resistance in the sample only treated by DGLA confirms that omega-6 (DGLA) are pro-inflammatory and promote the formation of PGE_2. However, the new anti-cancer treatment using DGLA and D5Di to treat A549 cells shows a similar low resistance level to that of A549 cells with 8-HOA. This result indicates promising information: the Ti_3C_2 MXene-based sensor can be used to monitor or validate the anti-cancer effect of the new strategy: DGLA + D5Di, which should be an effective anti-cancer effect because of the generation of 8-HOA.

3.3. Correlation between the Sensing Test Results and GC–MS Results

To verify the Ti_3C_2-based sensor for PGE_2 and 8-HOA detection, both sensor and GC–MS have been used to detect very low concentrations of 8-HOA and PGE_2 in A495 lung cancer cells. The Ti_3C_2 Mxene sensors can provide the information of concentration of 8-HOA via the value of resistance while GC–MS can quantitatively provide the exact concentration of 8-HOA. As shown in Figure 8, an obvious correlation is obtained between the GC–MS measurement and resistances that the Ti_3C_2 MXene-based sensor measured. This correlation further confirmed the capability of the Ti_3C_2 MXene-based sensor to detect trace concentrations of 8-HOA. It can be a convenient, fast, and low-cost tool to help the anti-cancer strategy in lung cancer treatment.

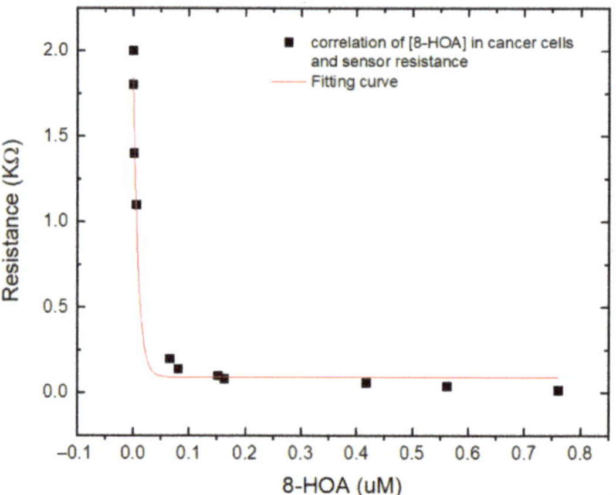

Figure 8. Correlation between different concentration of 8-HOA detected by gas chromatography–mass spectroscopy (GC–MS) and resistance measured by Ti_3C_2 MXene sensor using the same sampling conditions.

4. Conclusions and Discussion

A new sensor based on 2D nanosheets, Ti_3C_2 MXene, has been designed and used for the sensing response to 8-HOA and PGE_2 in lung cancer cells. The preliminary results indicate an important conclusion: this new Ti_3C_2-based sensor can provide a convenient and simple method for anti-cancer treatment guidance. In addition, the high sensitivity of this new sensor opens a potential application for early-stage cancer detection via monitoring variation of PGE_2 and 8-HOA in cells. Instead of using heavy, expensive, and time-consuming GC–MS to assist the anti-cancer treatment, the Ti_3C_2 MXene-based sensor can provide a fast, simple, low-cost, highly efficient, and much less invasive assistant tool to detect and cure cancer.

Author Contributions: M.S. initially wrote the paper, performed sensing experiments, and analyzed experimental data; L.P. prepared cancer cells and data analysis; M.J. performed all XRD, FT-IR, and Raman spectroscopy experiments, analyzed data, and synthesized the materials; V.S. designed cancer testing; Q.Z. designed material synthesis procedures; D.W. revised and finalized the paper and led the research team. All authors have read and agreed to the published version of the manuscript.

Funding: This work is supported in part by the Offerdahl Seed Grant, NDSU Centennial Endowment Award, FAR0029296; and ND NASA EPSCoR research grant, FAR0030154, and ND EPSCoR seed award, FAR0030452.

Institutional Review Board Statement: All the animal studies were approved by the IACUC (Institutional Animal Care and Use Committees) at the North Dakota State University (protocol code A18031 and date of approval: 7 February 2018).

Informed Consent Statement: Not applicable.

Acknowledgments: We wish to acknowledge the NDSU Core Research Facilities for providing access to microfabrication tools and materials characterization instruments.

Conflicts of Interest: The authors declare no conflict of interest.

References

1.   Cancer. Available online: https://www.who.int/news-room/fact-sheets/detail/cancer (accessed on 10 October 2020).
2.   Small Cell Lung Cancer—Cancer Therapy Advisor. Available online: https://www.cancertherapyadvisor.com/home/decision-support-in-medicine/imaging/small-cell-lung-cancer/ (accessed on 10 October 2020).
3.   Molina, J.R.; Yang, P.; Cassivi, S.D.; Schild, S.E.; Adjei, A.A. Non-Small Cell Lung Cancer: Epidemiology, Risk Factors, Treatment, and Survivorship. In *Mayo Clinic Proceedings*; Elsevier Ltd.: Amsterdam, The Netherlands, 2008; Volume 83, pp. 584–594. [CrossRef]
4.   Torre, L.A.; Siegel, R.L.; Jemal, A. Lung Cancer Statistics. *Adv. Exp. Med. Biol.* **2016**, *893*, 1–19. [CrossRef]
5.   Asbestos Lung Cancer: Causes, Diagnosis & Treatment. Available online: https://www.asbestos.com/cancer/lung-cancer/ (accessed on 10 October 2020).
6.   Deadliest Cancers Receive the Least Attention. Available online: https://www.asbestos.com/featured-stories/cancers-that-kill-us/ (accessed on 11 October 2020).
7.   Dela Cruz, C.S.; Tanoue, L.T.; Matthay, R.A. Lung Cancer: Epidemiology, Etiology, and Prevention. *CME* **2011**, *32*, 605–644. [CrossRef]
8.   Lee, S.H. Chemotherapy for Lung Cancer in the Era of Personalized Medicine. *Tuberc. Respir. Dis.* **2019**, *82*, 179–189. [CrossRef] [PubMed]
9.   Cheng, M.; Jolly, S.; Quarshie, W.O.; Kapadia, N.; Vigneau, F.D.; Kong, F.M. Modern Radiation Further Improves Survival in Non-Small Cell Lung Cancer: An Analysis of 288,670 Patients. *J. Cancer* **2019**, *10*, 168–177. [CrossRef]
10. Sandler, A.B.; Dubinett, S.M. COX-2 Inhibition and Lung Cancer. *Semin. Oncol.* **2004**, *31* (Suppl. 7), 45–52. [CrossRef] [PubMed]
11. Vega, O.M.; Abkenari, S.; Tong, Z.; Tedman, A.; Huerta-Yepez, S. Omega-3 Polyunsaturated Fatty Acids and Lung Cancer: Nutrition or Pharmacology? *Nutr. Cancer* **2020**, 1–21. [CrossRef]
12. Yin, Y.; Sui, C.; Meng, F.; Ma, P.; Jiang, Y. The Omega-3 Polyunsaturated Fatty Acid Docosahexaenoic Acid Inhibits Proliferation and Progression of Non-Small Cell Lung Cancer Cells through the Reactive Oxygen Species-Mediated Inactivation of the PI3K/Akt Pathway. In *Lipids in Health and Disease*; BioMed Central Ltd.: London, UK, 2017. [CrossRef]
13. Kouremenos, K.A.; Johansson, M.; Marriott, P.J. Advances in Gas Chromatographic Methods for the Identification of Biomarkers in Cancer. *J. Cancer* **2012**, *3*, 404–420. [CrossRef] [PubMed]
14. Cancer—Diagnosis and treatment—Mayo Clinic. Available online: https://www.mayoclinic.org/diseases-conditions/cancer/diagnosis-treatment/drc-20370594 (accessed on 10 October 2020).
15. Problems with MRI for Cancer Diagnosis. Available online: https://www.ctoam.com/precision-oncology/why-we-exist/standard-treatment/diagnostics/mri/ (accessed on 12 October 2020).
16. The Pros and Cons of PET/CT Scans | Independent Imaging. Available online: https://www.independentimaging.com/the-pros-and-cons-of-pet-ct-scans/ (accessed on 12 October 2020).
17. Fred, H.L. Drawbacks and Limitations of Computed Tomography: Views from a Medical Educator. *Tex. Heart Inst. J.* **2004**, *31*, 345–348.
18. Bitencourt, A.G.V.; Graziano, L.; Guatelli, C.S.; Albuquerque, M.L.L.; Marques, E.F. Ultrasound-Guided Biopsy of Breast Calcifications Using a New Image Processing Technique: Initial Experience. *Radiol. Bras.* **2018**, *51*, 106–108. [CrossRef]
19. Endoscopy: Purpose, Procedure, Risks. Available online: https://www.webmd.com/digestive-disorders/digestive-diseases-endoscopy#1 (accessed on 13 October 2020).
20. Gas Chromatography. Available online: http://www.chemforlife.org/teacher/topics/gas_chromatography.htm (accessed on 13 October 2020).
21. Buszewski, B.; Ligor, T.; Jezierski, T.; Wenda-Piesik, A.; Walczak, M.; Rudnicka, J. Identification of Volatile Lung Cancer Markers by Gas Chromatography-Mass Spectrometry: Comparison with Discrimination by Canines. *Anal. Bioanal. Chem.* **2012**, *404*, 141–146. [CrossRef] [PubMed]
22. Borer, J.S.; Simon, L.S. Cardiovascular and Gastrointestinal Effects of COX-2 Inhibitors and NSAIDs: Achieving a Balance. *Arthritis Res. Ther.* **2005**, *7* (Suppl. 4), S14–S22. [CrossRef]
23. Yang, P.; Chan, D.; Felix, E.; Cartwright, C.; Menter, D.G.; Madden, T.; Klein, R.D.; Fischer, S.M.; Newman, R.A. Formation and Antiproliferative Effect of Prostaglandin E3 from Eicosapentaenoic Acid in Human Lung Cancer Cells. *J. Lipid Res.* **2004**, *45*, 1030–1039. [CrossRef] [PubMed]
24. Xu, Y.; Qi, J.; Yang, X.; Wu, E.; Qian, S.Y. Free Radical Derivatives Formed from Cyclooxygenase-Catalyzed Dihomo-γ-Linolenic Acid Peroxidation Can Attenuate Colon Cancer Cell Growth and Enhance 5-Fluorouracil's Cytotoxicity. *Redox Biol.* **2014**, *2*, 610–618. [CrossRef]
25. Xu, Y.; Yang, X.; Wang, T.; Yang, L.; He, Y.Y.; Miskimins, K.; Qian, S.Y. Knockdown Delta-5-Desaturase in Breast Cancer Cells That Overexpress COX-2 Results in Inhibition of Growth, Migration and Invasion via a Dihomo-γ-Linolenic Acid Peroxidation Dependent Mechanism. *BMC Cancer* **2018**, *18*, 1–15. [CrossRef] [PubMed]
26. Pang, L.; Shah, H.; Wang, H.; Shu, D.; Qian, S.Y.; Sathish, V. EpCAM-Targeted 3WJ RNA Nanoparticle Harboring Delta-5-Desaturase SiRNA Inhibited Lung Tumor Formation via DGLA Peroxidation. *Mol. Ther. Nucleic Acids* **2020**, *22*, 222–235. [CrossRef]
27. Yang, X.; Xu, Y.; Wang, T.; Shu, D.; Guo, P.; Miskimins, K.; Qian, S.Y. Inhibition of Cancer Migration and Invasion by Knocking down Delta-5-Desaturase in COX-2 Overexpressed Cancer Cells. *Redox Biol.* **2017**, *11*, 653–662. [CrossRef]

28. Jafarizadeh-Malmiri, H.; Sayyar, Z.; Anarjan, N.; Berenjian, A.; Jafarizadeh-Malmiri, H.; Sayyar, Z.; Anarjan, N.; Berenjian, A. Nano-Sensors in Food Nanobiotechnology. In *Nanobiotechnology in Food: Concepts, Applications and Perspectives*; Springer International Publishing: Berlin/Heidelberg, Germany, 2019; pp. 81–94. [CrossRef]

29. Wang, P.; Zhang, L.; Zheng, W.; Cong, L.; Guo, Z.; Xie, Y.; Wang, L.; Tang, R.; Feng, Q.; Hamada, Y.; et al. Thermo-Triggered Release of CRISPR-Cas9 System by Lipid-Encapsulated Gold Nanoparticles for Tumor Therapy. *Angew. Chem. Int. Ed.* **2018**, *57*, 1491–1496. [CrossRef]

30. Lei, Y.; Tang, L.; Xie, Y.; Xianyu, Y.; Zhang, L.; Wang, P.; Hamada, Y.; Jiang, K.; Zheng, W.; Jiang, X. Gold Nanoclusters-Assisted Delivery of NGF SiRNA for Effective Treatment of Pancreatic Cancer. *Nat. Commun.* **2017**, *8*, 1–15. [CrossRef]

31. Li, N.; Wu, D.; Li, X.; Zhou, X.; Fan, G.; Li, G.; Wu, Y. Effective Enrichment and Detection of Plant Growth Regulators in Fruits and Vegetables Using a Novel Magnetic Covalent Organic Framework Material as the Adsorbents. *Food Chem.* **2020**, *306*, 125455. [CrossRef]

32. Wang, D.; Zhang, Q.; Hossain, R.; Johnson, M. High Sensitive Breath Sensor Based on Nanostructured $K_2W_7O_{22}$ for Detection of Type 1 Diabetes. *IEEE Sens. J.* **2018**, *18*, 4399–4404. [CrossRef]

33. Huber, F.; Riegert, S.; Madel, M.; Thonke, K. H2S Sensing in the Ppb Regime with Zinc Oxide Nanowires. *Sens. Actuators B Chem.* **2017**, *239*, 358–363. [CrossRef]

34. Wang, C.; Chu, X.; Wu, M. Detection of H2S down to Ppb Levels at Room Temperature Using Sensors Based on ZnO Nanorods. *Sens. Actuators B Chem.* **2006**, *113*, 320–323. [CrossRef]

35. Rai, P.; Majhi, S.M.; Yu, Y.T.; Lee, J.H. Noble Metal@metal Oxide Semiconductor Core@shell Nano-Architectures as a New Platform for Gas Sensor Applications. *RSC Adv.* **2015**, *5*, 76229–76248. [CrossRef]

36. Korotcenkov, G.; Brinzari, V.; Cho, B.K. Conductometric Gas Sensors Based on Metal Oxides Modified with Gold Nanoparticles: A Review. *Microchim. Acta* **2016**, *183*, 1033–1054. [CrossRef]

37. Wang, D.L.; Chen, A.T.; Zhang, Q.F.; Cao, G.Z. Room-Temperature Chemiresistive Effect of TiO_2-B Nanowires to Nitroaromatic and Nitroamine Explosives. *IEEE Sens. J.* **2011**, *11*. [CrossRef]

38. Wang, D.; Sun, H.; Chen, A.; Jang, S.H.; Jen, A.K.Y.; Szep, A. Chemiresistive Response of Silicon Nanowires to Trace Vapor of Nitro Explosives. *Nanoscale* **2012**, *4*, 2628–2632. [CrossRef]

39. Wang, T.; Huang, D.; Yang, Z.; Xu, S.; He, G.; Li, X.; Hu, N.; Yin, G.; He, D.; Zhang, L. A Review on Graphene-Based Gas/Vapor Sensors with Unique Properties and Potential Applications. *Nano-Micro Lett.* **2016**, *8*, 95–119. [CrossRef]

40. Toda, K.; Furue, R.; Hayami, S. Recent Progress in Applications of Graphene Oxide for Gas Sensing: A Review. *Anal. Chim. Acta* **2015**, *878*, 43–53. [CrossRef] [PubMed]

41. Ji, J.; Wen, J.; Shen, Y.; Lv, Y.; Chen, Y.; Liu, S.; Ma, H.; Zhang, Y. Simultaneous Noncovalent Modification and Exfoliation of 2D Carbon Nitride for Enhanced Electrochemiluminescent Biosensing. *J. Am. Chem. Soc.* **2017**, *139*, 11698–11701. [CrossRef]

42. Tan, C.; Cao, X.; Wu, X.J.; He, Q.; Yang, J.; Zhang, X.; Chen, J.; Zhao, W.; Han, S.; Nam, G.H.; et al. Recent Advances in Ultrathin Two-Dimensional Nanomaterials. *Chem. Rev.* **2017**, *117*, 6225–6331. [CrossRef]

43. Wen, W.; Song, Y.; Yan, X.; Zhu, C.; Du, D.; Wang, S.; Asiri, A.M.; Lin, Y. Recent Advances in Emerging 2D Nanomaterials for Biosensing and Bioimaging Applications. *Mater. Today* **2018**, *21*, 164–177. [CrossRef]

44. Naguib, M.; Kurtoglu, M.; Presser, V.; Lu, J.; Niu, J.J.; Heon, M.; Hultman, L.; Gogotsi, Y.; Barsoum, M.W. Two-Dimensional Nanocrystals Produced by Exfoliation of Ti_3AlC_2. *Adv. Mater.* **2011**, *23*, 4248–4253. [CrossRef]

45. Shahzad, F.; Zaidi, S.A.; Naqvi, R.A. 2D Transition Metal Carbides (MXene) for Electrochemical Sensing: A Review. *Crit. Rev. Anal. Chem.* **2020**, 1–17. [CrossRef] [PubMed]

46. Ghidiu, M.; Lukatskaya, M.R.; Zhao, M.Q.; Gogotsi, Y.; Barsoum, M.W. Conductive Two-Dimensional Titanium Carbide "clay" with High Volumetric Capacitance. *Nature* **2015**, *516*, 78–81. [CrossRef]

47. Peng, Q.; Guo, J.; Zhang, Q.; Xiang, J.; Liu, B.; Zhou, A.; Liu, R.; Tian, Y. Unique Lead Adsorption Behavior of Activated Hydroxyl Group in Two-Dimensional Titanium Carbide. *J. Am. Chem. Soc.* **2014**, *136*, 4113–4116. [CrossRef] [PubMed]

48. Kim, S.J.; Koh, H.J.; Ren, C.E.; Kwon, O.; Maleski, K.; Cho, S.Y.; Anasori, B.; Kim, C.K.; Choi, Y.K.; Kim, J.; et al. Metallic $Ti_3C_2T_x$ MXene Gas Sensors with Ultrahigh Signal-to-Noise Ratio. *ACS Nano* **2018**, *12*, 986–993. [CrossRef]

49. Sinha, A.; Dhanjai; Zhao, H.; Huang, Y.; Lu, X.; Chen, J.; Jain, R. MXene: An Emerging Material for Sensing and Biosensing. *TrAC Trends Anal. Chem.* **2018**, *105*, 424–435. [CrossRef]

50. Lorencova, L.; Sadasivuni, K.K.; Kasak, P.; Tkac, J. Ti3C2 MXene-Based Nanobiosensors for Detection of Cancer Biomarkers. *Intech* **1989**, *32*, 137–144.

51. Mahek, S.; Lizhi, P.; Michael, J.; Venkatachalem, S.; Danling, W. 2D Nanotmaterial, Ti_3C_2 MXene-Based Sensor to Guide Lung Caner Therapy and Management. *Proceedings* **2020**, *60*, 29. [CrossRef]

52. Michael, J.; Qifeng, Z.; Danling, W. Titanium Carbide MXene: Synthesis, Electrical and Optical Properties and Their Applications in Sensors and Energy Storage Devices. *Nanomater. Nanotechnol.* **2019**, *9*, 184798041882447. [CrossRef]

 *biosensors*

Article

Optimisation of an Electrochemical DNA Sensor for Measuring KRAS G12D and G13D Point Mutations in Different Tumour Types [†]

Bukola Attoye [1,*], Matthew J. Baker [2], Fiona Thomson [3], Chantevy Pou [3] and Damion K. Corrigan [1]

[1] Department of Biomedical Engineering, University of Strathclyde, 40 George Street, Glasgow G1 1QE, UK; damion.corrigan@strath.ac.uk

[2] Technology and Innovation Centre, Department of Pure and Applied Chemistry, University of Strathclyde, 99 George Street, Glasgow G1 1RD, UK; matthew.baker@strath.ac.uk

[3] Wolfson Wohl Cancer Research Centre, Institute of Cancer Sciences, University of Glasgow, Glasgow G61 1QH, UK; fiona.thomson@glasgow.ac.uk (F.T.); chantevy.pou@glasgow.ac.uk (C.P.)

* Correspondence: bukola.omolaiye@strath.ac.uk

[†] This paper is an extended version of our paper published in: Attoye, B.; Corrigan, D.K.; Baker, M.J.; Thomson, F.; Pou, C. Development of electrochemical DNA detection methods to measure circulating tumour DNA for enhanced diagnosis and monitoring cancer. In Proceedings of the 1st International Electronic Conference on Biosensors, 2–17 November 2020.

Abstract: Circulating tumour DNA (ctDNA) is widely used in liquid biopsies due to having a presence in the blood that is typically in proportion to the stage of the cancer and because it may present a quick and practical method of capturing tumour heterogeneity. This paper outlines a simple electrochemical technique adapted towards point-of-care cancer detection and treatment monitoring from biofluids using a label-free detection strategy. The mutations used for analysis were the KRAS G12D and G13D mutations, which are both important in the initiation, progression and drug resistance of many human cancers, leading to a high mortality rate. A low-cost DNA sensor was developed to specifically investigate these common circulating tumour markers. Initially, we report on some developments made in carbon surface pre-treatment and the electrochemical detection scheme which ensure the most sensitive measurement technique is employed. Following pre-treatment of the sensor to ensure homogeneity, DNA probes developed specifically for detection of the KRAS G12D and G13D mutations were immobilized onto low-cost screen printed carbon electrodes using diazonium chemistry and 1-ethyl-3-(3-dimethylaminopropyl) carbodiimide hydrochloride/N-hydroxysuccinimide coupling. Prior to electrochemical detection, the sensor was functionalised with target DNA amplified by standard and specialist PCR methodologies (6.3% increase). Assay development steps and DNA detection experiments were performed using standard voltammetry techniques. Sensitivity (as low as 0.58 ng/μL) and specificity (>300%) was achieved by detecting mutant KRAS G13D PCR amplicons against a background of wild-type KRAS DNA from the representative cancer sample and our findings give rise to the basis of a simple and very low-cost system for measuring ctDNA biomarkers in patient samples. The current time to receive results from the system was 3.5 h with appreciable scope for optimisation, thus far comparing favourably to the UK National Health Service biopsy service where patients can wait for weeks for biopsy results.

Keywords: electrochemical; DNA biosensors; KRAS; liquid biopsy; cancer point-of-care diagnostic tests

Citation: Attoye, B.; Baker, M.J.; Thomson, F.; Pou, C.; Corrigan, D.K. Optimisation of an Electrochemical DNA Sensor for Measuring KRAS G12D and G13D Point Mutations in Different Tumour Types. *Biosensors* **2021**, *11*, 42. https://doi.org/10.3390/bios11020042

Received: 19 December 2020
Accepted: 1 February 2021
Published: 5 February 2021

Publisher's Note: MDPI stays neutral with regard to jurisdictional claims in published maps and institutional affiliations.

1. Introduction

In recent years, analytical electrochemistry has emerged as a powerful tool for the rapid in-vitro analysis of biological analytes for the early detection of certain diseases, such as cancer [1]. Cancer is a genetic disease by nature, caused by mutations in certain genes thereby resulting in cellular malfunction [2]. Imaging tests can sometimes be inconclusive

Biosensors **2021**, *11*, 42. https://doi.org/10.3390/bios11020042   https://www.mdpi.com/journal/biosensors

and broadly do not provide information on the stage or type of cancer, so further biopsy is needed [3]. Serious medical risks and related metastasis may ensue from gathering multiple biopsies from different regions of a primary tumour [4].

The period at which a tumour shows clinical symptoms usually corresponds with the later stages of progression (e.g., Phases III and IV), when the cancer is metastatic or unresectable, causing surgery and therapy to be less effective. In addition, surgical biopsy procedures are not possible or recommended for some patients; therefore, liquid biopsies that are able to detect the presence of tumour DNA hold promise as a non-invasive alternative.

Most body fluids, including blood, contain tumour biomarkers and short fragments of cell-free DNA (cfDNA) that can be detectable as shown in Figure 1A below. In cancer patients, a fraction of cfDNA called circulating tumour DNA (ctDNA) can be found, which emerges from tumours and may feature the same mutations and genetic modifications as those present in the primary tumour [5]. While circulating tumour cells (CTCs) that have been shed into the vasculature of a primary tumour are also transported around the body in circulation, they are present at quantities of around 10 cells/mL of blood, suggesting that only very low concentrations are present in clinical samples. In contrast to ctDNA, CTCs are rare in peripheral blood and are difficult to separate from other cells, increasing the credibility for the use of ctDNA in liquid biopsy applications. The mechanism of ctDNA release from tumour cells is poorly understood [6], however it is thought to be released in small quantities following apoptosis or necrosis. ctDNA typically comprises 0.01–1% of the circulating free DNA in blood [7] and it is important to note that this can be shed as both double stranded and single stranded DNA [8]. At present, ctDNA can be detected in blood and other body fluids like lymph, urine and stool [9]. Due to the small fraction of ctDNA concealed by large background levels of wild-type cell-free DNA, sensitive amplification reactions such as polymerase chain reaction (PCR) will need to be implemented to achieve detection and discrimination above wild type signals. A point of care (PoC) measurement of circulating tumour DNA (ctDNA) may offer a non-invasive strategy for evaluating response to treatment, monitoring disease recurrence, capturing tumour heterogeneity and gaining insights into a tumour's mutational profile [9,10].

Single nucleotide variations (SNV) in the Kirsten rat sarcoma viral oncogene homolog, commonly abbreviated 'KRAS' are present across many human tumour types with KRAS G12D and G13D being specific variants observed. KRAS is a member of the RAS family of proteins which are a part of at least six signalling pathways in a healthy human cell and is the most commonly mutated protein across many human tumour types [11]. KRAS mutations take place in approximately 90% of pancreatic cancers [12], 30% of lung cancers [13], 60% of thyroid cancers and 43% of colorectal cancers [14]. KRAS activated mutations drive cancer initiation, progression and drug resistance, directly leading to nearly a million deaths per year. SNVs have been used as biomarkers for predicting disease risk [15,16], and its combination with liquid biopsies will create innovations in biomarker detection that will enhance clinical outcomes for patients at all cancer stages [17].

From a PoC viewpoint, Electrochemical DNA biosensors represent an exciting approach in the detection of clinically important biomarkers due to their rapidity and simplicity [18–20]. Electrochemical biosensors are used to directly convert a biological binding event to an electronic signal [21]. A range of electrode materials and electrochemical measurement approaches have been employed for sensitive measurements [8,22–24]. The possibilities of electrochemical biosensors, once matured as a technology to provide efficient clinical workflows, is vast. In electrochemical DNA biosensing, a change in signal is obtained when recognition and hybridisation of two opposing strands of DNA occur as a result of their base-pair complementarity. A double stranded DNA sequence with tumour-specific mutations can indicate the diagnosis of a specific cancer [24]. As the concentration of ctDNA is directly proportional to the tumour grade, attaining high sensitivity for the DNA sensor is key for the early detection of disease, developing tailored therapies and monitoring therapy efficiency.

Given the continued need for the miniaturization of advanced electronics, the area of screen printing techniques has been adapted for electronic circuit fabrication. Screen printed electrodes (SPEs) are evolving as they are easy to use and can be produced on a large scale. SPEs are also very practical as they are disposable and low cost when manufactured in large volumes. SPEs are usually composed of working electrodes made of conductive inks like carbon, platinum, gold or silver. Although, carbon with organic solvents, binding pastes and some additives that provide functional characteristics are contained in conductive inks found in Screen Printed Carbon electrodes (SPCEs), they can be modified in order for their electrochemical properties to be improved [24]. Carbon electrodes are also chemically inert, specifically at negative potential ranges in all media, making them particularly suitable electrode sensors for electroanalytical chemistry, providing an advantage over metal electrodes [25]. SPCEs are simple, sensitive, cost-effective (~£2 each) and disposable, making them preferable for rapid electrochemical analyses and suitable as electrodes for characterizing the processes implemented herein, specifically for the detection of ctDNA.

The SPCE sensor shown in this study was developed by characterising the surface of the electrode chip to determine the treatments and buffers with optimal sensitivity. In order to make the surface of the SPCE as homogenous as possible, it is important that they are pre-treated. These pre-treatments remove any binder residues left on the carbon surface after the curing process [26–29], with well-established electrochemical oxidative pre-treatments not only showing removal of binder residues left on the surface of SPCE after curing but also improvement of carbon surface sensitivity [30]. In this study, two common buffers are compared for pre-treatment: NaOH and NaCl. Up until now, few studies have been done on electrochemically pre-treating and characterising the surface of activated screen printed carbon electrodes [28–30]. We make the choice of a characterisation redox buffer after observing the effect of surface chemistry in relation to electron transfer rates using an inner-sphere redox mediator (Ferri-ferrocyanide) and an outer-sphere redox mediator (ruthenium hexaminechloride). Further voltammetric characterisation is performed to reveal DNA hybridisation effects and thus mutation detection in both potassium ferri-ferrocyanide (1 mM $Fe(CN)_6^{3-/4-}$ in 0.1× PBS) and ruthenium hexaminechloride (1 mM $Ru(NH_3)_6Cl_3$ in 0.1× PBS) solutions.

This work presents a KRAS G12D and G13D DNA oligonucleotide probe modified sensor array that can accurately detect mutant KRAS amplicons and therefore forms the basis of a system for the accurate detection of ctDNA in patient samples and monitoring of response during treatment. This was achieved by amplifying mutant DNA isolated from a human cancer cell line recovered from clinical samples, using electrochemical techniques and SPCEs to detect a clinically relevant mutation, comparing the signal change from DNA hybridisation experiments involving amplified KRAS mutant samples and amplified wild-type KRAS samples, varying concentration of amplified products to determine concentration effects and establishing a limit of detection for the DNA amplification reaction. Cyclic Voltammetry (CV), Square Wave Voltammetry (SWV) and Differential Pulse Voltammetry (DPV) are routinely used electrochemical measurement techniques that supply information on electron transfer reaction kinetics of any combined chemical reaction [31]. In these techniques, a potential waveform is applied to the working electrode (WE). The peak current obtained is directly influenced by hybridisation between target and immobilised probe DNA strands [32]. In this study, DPV, SWV and CV were used depending on whether electrodes needed to be cleaned, electrografted, or characterised during sensor measurement. Considering the choice of steps and ease of use of the assay being developed, the system can be very easily automated and integrated into a final device capable of fast and seamless clinical measurements. The presented work builds on a recent publication [1] showing the possible detection of KRAS G12D mutations, by developing understanding or surface pre-treatment steps (essential to realising a reproducible analytical technique) and by introducing the detection of the KRAS G13D mutation which expands the assay towards a multi-marker assay and permits the analysis of more tumour types.

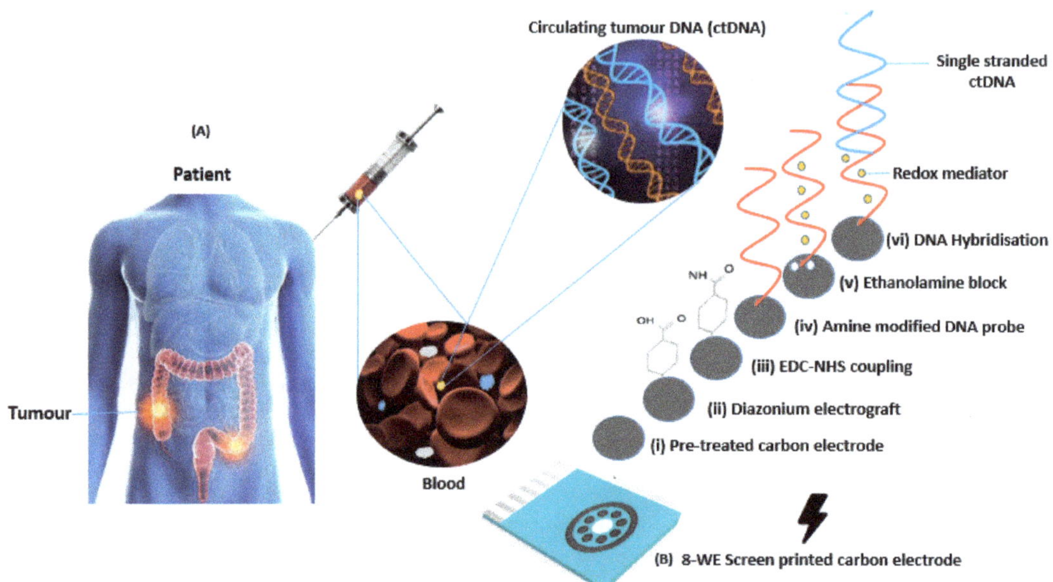

Figure 1. (**A**) Schematic showing circulating tumour DNA (ctDNA) retrieval and analysis [33–35] (**B**) Image of a screen-printed electrode array employing eight working electrodes with a common Ag reference and carbon counter electrodes along with a schematic showing modification steps and DNA functionalisation.

2. Materials and Methods

2.1. Reagents

Supermix for probes, Droplet digital PCR (ddPCR) assays, DG8TM cartridges and Droplet Generation Oil were obtained from Bio-Rad Laboratories Ltd., Hertfordshire, UK. Deionized water, sodium chloride, sodium hydroxide, phosphate-buffered saline (PBS), sodium nitrate, 4-aminobenzoic acid, hydrochloric acid, ethanolamine, 2-(N-morpholino) ethanesulfonic acid (MES), 1-ethyl-3-(3-dimethylaminopropyl) carbodiimide hydrochloride (EDC), N-hydroxysuccinimide (NHS), hexammineruthenium (III) chloride, potassium ferricyanide, potassium chloride and potassium ferrocyanide were all purchased from Sigma–Aldrich, (Dorset, UK). Two hundred and fifty units of HotStarTaq Plus and dNTP Mix, PCR Grade (200 μL), were purchased from Qiagen, (Manchester, UK). Phusion Direct PCR kit was purchased from thermo fisher scientific (Renfrew, UK).

2.2. Sensor Development and Set-Up

A multiplex system comprising screen-printed multi-carbon electrodes (DRP 8W110), a potentiostat, a multiplexer and a connector were set up for electrochemical measurements. The chip containing eight carbon working electrodes with diameters of 2.95 mm each, a carbon counter and silver reference electrode as shown in Figure 1B above were obtained from DropSens (Oviedo, Spain) with chip dimensions of 50 × 27 × 1 mm (L × W × D). The screen-printed fabrication process was specified by the manufacturers.

2.3. Electrode Preparation and Surface Functionalisation

All electrochemical measurements were recorded using PS-Trace software. DNA hybridisation experiments were performed using a covalently attached layer of single-stranded DNA probes. The surface functionalisation protocol is illustrated in Figure 1B. To prepare the surface of the carbon electrodes for DNA probe attachment, it was necessary to first use a surface pre-treatment method by applying 1.4 V for 1 min in 0.5 M acetate buffer solution (ABS) containing 20 mM NaCl 8 (pH 4.8) via CV. An alternative pre-treatment technique explored for optimisation comparison required soaking the SPCEs in 3 M NaOH for 1 h as an initial step and anodizing at 1.2 V using a scan rate of 0.5 V/s via CV. Next, 2 mM NaNO$_2$ solution with 2 mM 4-aminobenzoic acid was prepared in 0.5 M HCl and stirred for approximately 5 min at room temperature to produce a diazonium compound. The activated diazonium solution was then scanned using CV from +0.4 to −0.6 V at a scan rate of 100 mV/s followed by a wash with deionised (DI) water. The resulting 4-carboxyphenyl (AP) film was activated on the electrode's surface with 100 mM EDC and 20 mM NHS in 100 mM MES buffer (pH 5.0) for 60 min to form an ester that allowed for efficient conjugation to the amine-modified ssDNA probe. A 1 mM ruthenium and 1 mM potassium ferri-ferrocynide buffer were compared to analyse electron transfer rates. Ferricyanide buffer (5 mM) was used to characterise the sensor surface for DNA detection. All the reported steps and measurements were carried out at room temperature, unless otherwise stated.

2.4. Genomic DNA Sample Preparation, DNA Probe Design, and Sample Amplification

Copies of the KRAS pG12D and pG13D mutant and wild-type DNA were amplified from genomic DNA (gDNA) isolated from SK-UT-1 cells (pG12D) and HCT116 cells (pG13D). Levels of both mutant and wild-type DNA were determined using ddPCR assays in combination with a QX200TM Droplet DigitalTM PCR system (Bio-Rad Laboratories Ltd., Hertfordshire, UK) following the manufacturer's instructions. Briefly, 5–10 ng of gDNA isolated from SK-UT-1 cells was combined with ddPCR Supermix for probes (No dUTP) and fluorescein amidite (FAM)-labelled KRAS p.G12D (KRAS p.G12D c.35G>A, Human (dHsaMDV2510596)) or KRAS p.G13D (KRAS p.G13D c.35G>A, Human (dHsaMDV2510598)) primers/probe and hexachloro-fluorescein 9 (HEX)-labelled KRAS WT primers/probes (KRAS WT for p.G12D c.35G>A and KRAS WT for p.G13D c.38G>A) in the presence of HaeIII restriction enzyme and in a volume of 20 µL. Reaction samples were loaded onto a DG8TM cartridge with 70 µL of droplet generation oil for Probes according to the Droplet Generator Instruction Manual (Bio-Rad Laboratories Ltd., Hertfordshire, UK). The PCR cycling conditions for the generated droplets were as follows: initial enzyme activation at 95 °C for 10 min, followed by 40 cycles of denaturation at 94 °C for 30 s, and annealing/extension at 55 °C for 1 min, after which it ended with a final enzyme deactivation at 98 °C for 10 min. Data acquisition after thermal cycling was performed using the QX200 Droplet Reader and the QuantaSoft Software (Bio-Rad Laboratories Ltd., Hertfordshire, UK).

The PCR primers and probes designed in this study were based on the published sequence of KRAS pG12D and pG13D under accession number NC_000012.12 [36]. Amine-modified synthetic oligonucleotides (KRAS G12D and KRAS G13D) were designed for use as probes, as shown in Table 1 below, with a concentration of 200 µM obtained from Sigma–Aldrich, UK, and stored at −80 °C prior to aliquoting for use as probes. A wild-type probe (without the single base mutation) was also designed for use as a negative control. The DNA probe stocks were diluted to a concentration of 2 µM in 0.1× PBS prior to immobilisation. Primer-BLAST software was used to design the PCR primers used in this study. The forward primers and reverse primers had an estimated GC content of 40–55%, an estimated product length of 88 with low self-complementarity.

Further amplification of extracted wild-type, KRAS G12D and G13D-mutated DNA samples was carried out using the Phusion Direct PCR kit following the protocol and reaction set-up guide outlined by Thermo Scientific, UK. Phusion Blood II DNA polymerase (1 µL), 2× PCR Buffer (25 µL), 50 mM EDTA (2.5 µL), 50 mM MgCl$_2$ solution (1.5 µL) and 100% DMSO (2.5 µL) were all included in the reaction mix and dispensed into appropriate PCR tubes. 2.5 µL of Template DNA containing 18.4 ng/µL double stranded DNA, 5 µL forward and reverse primers and 10 µL ultrapure water were added to the master PCR tube containing the reaction mix, and the thermal cycler was programmed to start with an initial heat-activation step at 98 °C for 300 s. Temperature specifications for denaturing, annealing and extending were set at 90 s for 94 °C, 65 °C and 72 °C, respectively. A final extension for 60 s at 72 °C was set, and the PCR conditions were set for 37 cycles. The PCR amplification of wild-type and KRAS G13D samples was performed using the miniPCR thermal cycler [37], and amplicon yields of 117 ng/μL were confirmed using the Qubit 4 fluorometer and dsDNA broad range quantification assay [38].

Table 1. List of DNA sequences employed in this study.

KRAS G13D Probe and Primer Sequences	
23 Bases Wild-Type Hybridisation Probe	TGGAGCTGGTGGCGTAGGCAAGA
23 Bases Mutant Hybridisation Probe	TGGAGCTGGTGACGTAGGCAAGA
Forward Primer (Wild-Type)	TGTGGTAGTTGGAGCTGGTG
Forward Primer (Mutant)	TGTGGTAGTTGGAGCTGATG
PCR Probe (Mutant)	TCTTGCCTACGCCACCAGCTCCA
Reverse Primer	TTGTGGACGAATATGATCCAACA
KRAS G12D Probe and Primer Sequences	
23 bases Wild-type Hybridisation Probe	AGTTGGAGCTGGTGGCGTAGGCA
23 bases Mutant Hybridisation Probe	AGTTGGAGCTGATGGCGTAGGCA
Forward Primer (Wild-type)	TGTGGTAGTTGGAGCTGGTG
Forward Primer (Mutant)	TGTGGTAGTTGGAGCTGATG
Reverse Primer	TTGTGGACGAATATGATCCAACA

3. Results and Discussion

3.1. Assay Workflow and Development

The use of an SPCE with multiple working electrodes allows each electrode to be individually modified and rapidly carry out simultaneous measurements of peak currents. Electrografting using in situ generated diazonium cations is important for modifying the surface of the SPCE by allowing the formation of covalent bonds between the carbon surface and organic films [39–41]. The EDC molecule is an established zero-length cross-linking agent that has been employed in coupling carboxyl groups to primary amines in various applications [41]. One of the main benefits of EDC coupling is its water solubility that allows direct bioconjugation without prior organic solvent dissolution. To improve the stability of the active ester, NHS was introduced to modify the amine-reactive chemical substance by converting it to an active NHS ester, thus maximising the efficiency of the EDC-mediated coupling reactions. The reproducibility of this hybridisation sensor was explored by simultaneously analysing all eight WEs from the multi-electrode chip after pre-treatment and electrochemical signal changes of similar amplitude, direction and magnitude were observed. Figure 2A below shows the effect of each modification step on DPV peak current using ferri-ferrocyanide. The low peak current observed after diazonium reduction can be attributed to the thickness of the film resulting from the covalent bonds created on the surface of the electrode. In Figure 2B, this effect is reversed and a higher peak current with two peaks are noted, suggesting a high sensitivity to the organic films. From the ferri-ferrocyanide characterisation, the NHS-EDC peak reflects both coupling initiation and activation on the surface of the electrode which results in an enhanced oxidation due to a neutrally charged NHS ester, leading to a negative potential shift and an increase in peak current. DNA is negatively charged and thus resulted in a decrease in peak current when immobilised on the sensor surface. There is also an electrostatic repulsion between negative ferri-ferrocyanide ions and the negative phosphate group in the DNA structure. The double peaks from the ruthenium hexaminechloride characterisation make it more difficult to identify the correct peak current. In this case, we attribute the right hand peak to the ruthenium hexamine chloride redox reaction from free solution and the smaller smeared out left hand peak to the ruthenium redox reactions taking place at higher potentials because the redox reporter is trapped in organic layers and electrostatically associating with the DNA strands on the probe modified electrode surface. The multiplexed analysis we used greatly reduced the analysis time because of the high throughput of samples and minimised reagent consumption. After introducing the probe solution to the surface of the modified electrodes, the remaining active groups on the electrode were blocked using ethanolamine to produce a consistent sensing layer in order to facilitate DNA specificity and stability in terms of the DNA binding response. Blocking the free surface on the electrode resulted in an increase in peak current in ferri-ferrocyanide characterisation. A single but decreased current peak is shown in Figure 2B after blocking and this can be attributed to the sensitivity of ruthenium hexaminechloride to the consistent layer on the outer surface of the electrode. The findings from these modification characterisations are in line with previous studies [42–45] and noting the observations on SPCEs, our optimal characterisation buffer for these studies is ferri-ferrocyanide.

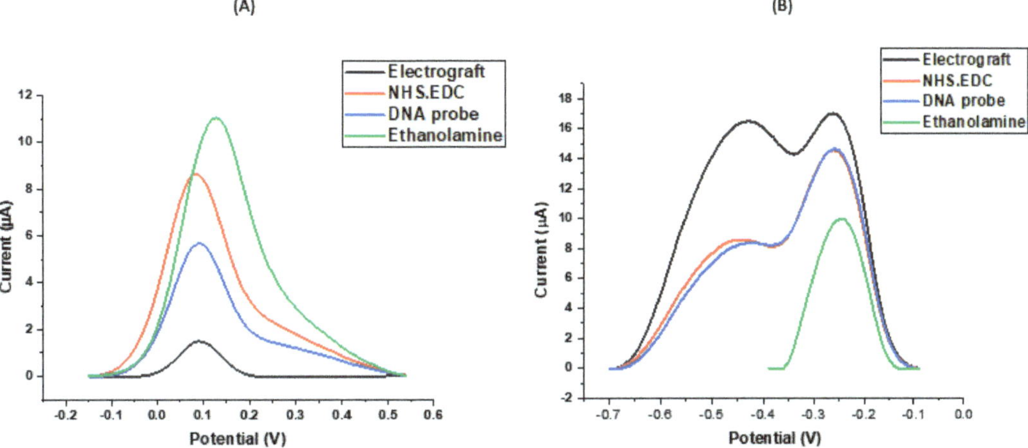

Figure 2. Examines the effect of two different redox agents on the DPV peak current after each functionalisation step on SPCE sensor response characterised using (**A**) 1 mM ferri-ferrocyanide buffer in $0.1\times$ PBS and (**B**) 1 mM ruthenium hexaminechloride in $0.1\times$ PBS.

A growing demand for reliable detection devices motivates much biosensor development [44], so it is therefore necessary to improve assay reproducibility and one crucial aspect of this is surface pre-treatment. A high level of consistency in the peak current, potential and width was observed after repeated cycling of each bare electrode, however optimisation work was carried out to ensure that our assay was as sensitive as possible. In Figure 3, it was noted that although the bare SPCE exhibited an admirable sensitivity with the highest peak current of those presented (Figure 3A), after all surface modification steps and DNA hybridisation was carried out, the SPCE pre-treated using acetate buffer containing NaCl showed the most suitable response. As previous studies have shown an increase in surface roughness after pre-treating screen printed electrodes [29,30], we can infer that the improved electrochemical performance after DNA hybridisation on pre-treated electrodes resulted from the ability of the target DNA strands to bind readily to the surface of the probe modified electrode. A high percentage signal change directly implies that a significant reduction in peak current upon target hybridisation has occurred. For a ferri-ferrocyanide characterisation buffer, this can be attributed to the repulsion between negative charges of target DNA, probe DNA and negative ferri-ferrocyanide ions. A wider peak separation (ΔE_p) than that predicted by the Nernst equation is also observed in SPCEs characterised by ferri-ferrocyanide (due to surface effects). Both redox buffers exhibited acceptable reversibility on pre-modified SPCEs. From Figure 3B, the changes in buffer characterisation observed on the DNA modified SPCEs confirms ruthenium hexaminechloride is an outer-sphere electron transfer mediator that is only affected by changes in the electroactive area, while the redox couple ferri-ferrocyanide is useful for determining the existence of functional groups due to its inner sphere sensitivity. The ferri-ferrocyanide buffer gave the most consistent signal change upon DNA hybridisation on the modified SPCEs, particularly when pre-treated with NaCl, therefore giving us a system which could be taken through to a full assay development stage. Upon hybridisation, NaOH pre-treatment indicated a lower sensitivity and a higher variation of sensor surface, thus raising doubts on the suitability of NaOH treatment for SPCEs in DNA hybridisation work.

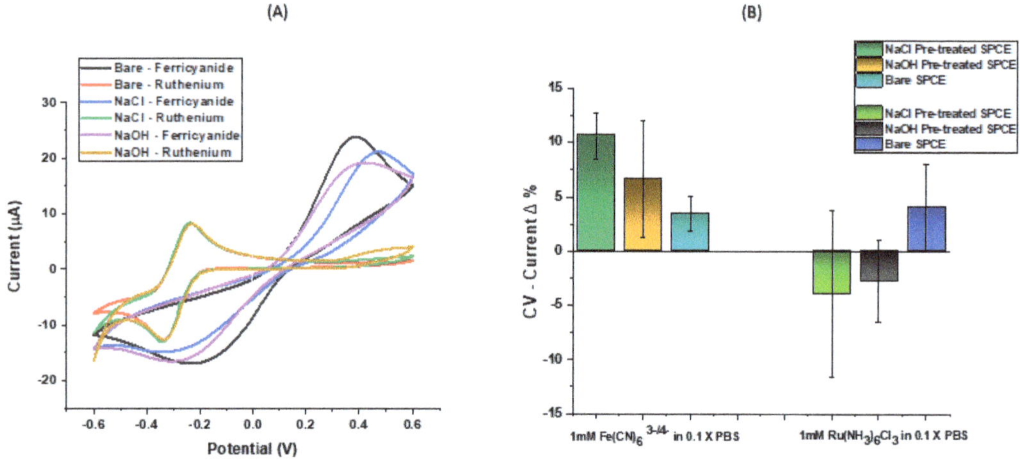

Figure 3. CV measurements showing the effect of an inner sphere and outer sphere redox mediator on electron transfer for different SPCE surfaces on (**A**) non-modified surface and (**B**) modified and DNA functionalised surfaces.

3.2. DNA Sensor Hybridisation Specificity

After observing consistent behaviour of modified electrodes on the same chip, the next step was to test the assay's response to incubation in a representative KRAS sample using the designed probes. We explored the ability of the probe-modified electrodes to discriminate between G13D mutant and wild-type KRAS sequences in representative samples. To investigate specificity levels and gain an initial impression of assay sensitivity, a series of electrodes were functionalised with KRAS G13D mutant and wild-type probe sequences. The results of these experiments are summarised in Figure 4A, which shows the percentage change in the CV peak current following target hybridisation. For macroscale electrodes functionalised with biological molecules, such as DNA or antibodies, the expectation is that differential pulse voltametric peak currents will reduce upon target hybridisation. It has been observed that these effects can be reversed when micro- or nanoscale electrodes are employed [20,46], but for this study, the electrodes used were comfortably on the macro scale (diameter = 2.95 mm). For nanomolar (>10 nM) and micromolar concentrations, an increase in the peak current following hybridisation was consistently observed (and has also been observed in other data from our lab involving SPCEs) [1] for carbon electrodes which is likely explained by the high surface density of hybridised DNA amplicons changing the interfacial properties of the electrode and, therefore, altering the electrochemical response. The underlying physical mechanism of this effect is actively under investigation. Figure 4A shows that when mutant and wild-type oligonucleotide probe sequences functionalised SPCEs were incubated in a representative sample containing the G13D mutation, there was hybridisation in both cases; the signal change was greater for the wild-type probe because of the high background of wild-type DNA and the comparatively low fraction of mutated KRAS G13D present in the representative sample. Similar behaviour was observed for KRAS G12D probe functionalised electrodes for a representative sample for that particular mutation, showing the wild-type KRAS DNA hybridised strongly to the nucleic acid modified carbon surfaces. As a result of these findings and the inability to electrochemically discriminate between positive and negative samples owing to the strong influence of background DNA in the sample, DNA amplification strategies were developed and tested in order to ensure the production of unequivocal detection of ctDNA mutations.

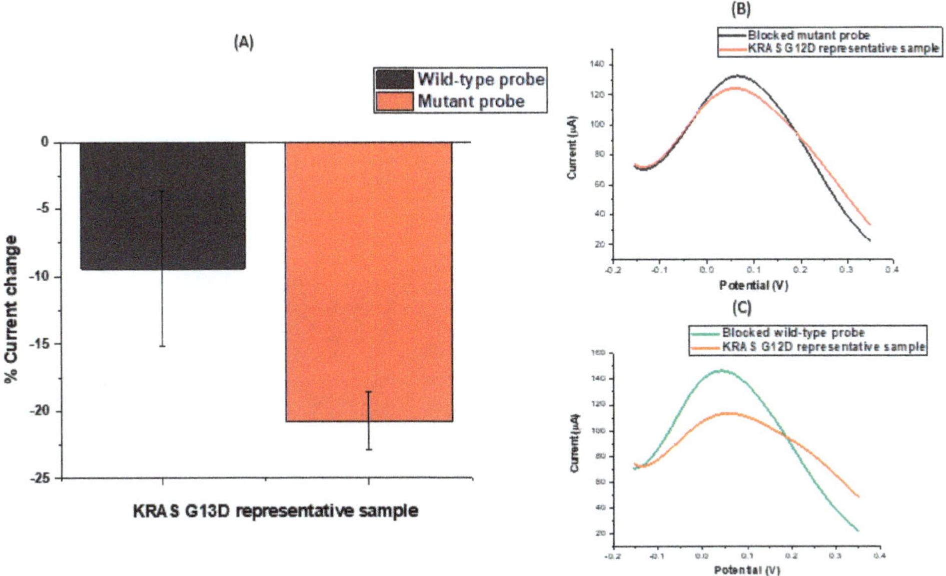

Figure 4. (**A**) CV percentage signal change in response to mutant and wild-type probes hybridized with genomic KRAS G13D ssDNA. (**B,C**) DPV signal changes in response to incubated KRAS G12D ssDNA hybridised with mutant and wild-type (WT) oligonucleotides, respectively.

3.3. KRAS G13D Amplification and Negative Control

In order to selectively amplify the mutant target from a pool of mutant and wild-type sequences in a sample, primers used for the PCR amplification were tailored by varying the single nucleotide responsible for the mutation in the primer sequence. Adopting this approach allowed us to effectively enrich the number of mutated DNA sequences in the sample without amplifying the wild-type in order to produce a signal change above the background signal generated by the KRAS wild-type DNA non-specifically associating with the oligonucleotide probe sequences for KRAS G13D. In selecting the approach reported here, ctDNA detection could potentially be coupled to a DNA amplification reaction, because it allows the possibility of developing a multiplexed panel of DNA sequences on a single chip, meaning that commonly mutated genes could all be identified in parallel (e.g., KRAS, TP53, BRCA1/2, IDH-1). This concept of developing biomarker "panels" is thought to be one of the key advantages of this approach [47]. From Figure 5A, when the wild-type probe-sequence modified electrodes were hybridised with KRAS G13D amplicons, alterations in the peak current were not observed, indicating no significant hybridisation. The mutant amplicons when incubated with mutant probe modified electrodes gave rise to a very significant signal change (~350%), indicating hybridisation with ultra-concentrated DNA samples (nano–micromolar concentration ranges) because of the strong positive signal change. An opposite response is noted in Figure 5B where wild-type amplicons resulting from DNA amplification using wild type primers are hybridised using wild-type probe modified electrodes. In this case, the wild-type hybridisation exhibited a significant signal change while mutant probe modified electrodes showed no significant hybridisation, representing an additional control. These findings were thoroughly satisfying, i.e., that the surface-tethered KRAS G13D mutant probe sequence could, in fact, discriminate between the mutant and wild-type samples based on the presence or absence of PCR amplicons for KRAS G13D with high sensitivity. This in fact represented a type of double specificity for the PCR-based assay, because the primer design had already been shown to specifically amplify the mutated sequence so coupling in the specificity of the electrochemical probe sequence meant that the assay would be able to successfully discriminate mutant amplicons from the sample. Having established the specificity of the assay and the nature of the electrochemical change, the next step involved verifying the sensitivity of the assay and dose–response effects for the KRAS G13D mutant PCR product.

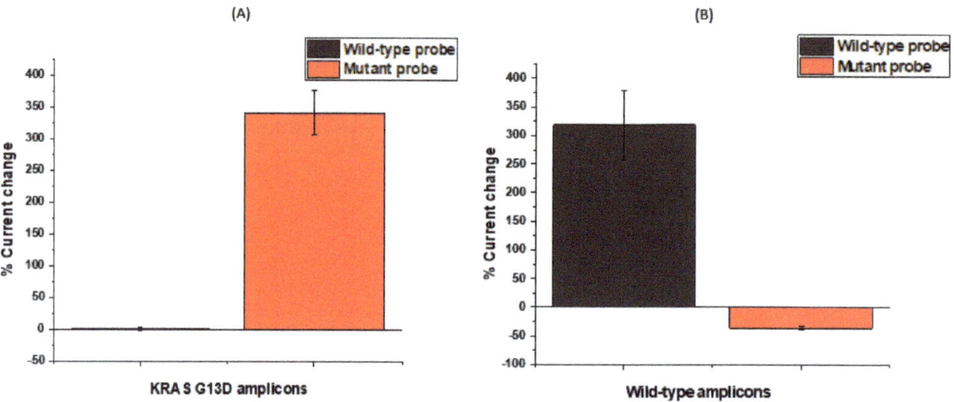

Figure 5. Successful amplification of (**A**) KRAS G13D mutants further confirmed by a large percentage signal change ratio between mutant probe and G13D mutant amplicons (**B**) KRAS wild-type further confirmed by a large percentage signal ratio between wild-type probe and wild-type DNA.

3.4. Concentration Dose Response

After establishing PCR primer specificity, ssDNA probe specificity and electrochemical signal changes in the correct direction and magnitude, it was important to investigate dose–response effects. In these experiments, non-amplified and amplified samples were diluted and a dose-response curve was constructed (see Figure 6A,B). We expected a reduction in the peak current when specific DNA hybridisation had taken place, and this was found to be the case for lower concentrations of DNA (pico-to-low nanomolar concentrations). For the unamplified sample (Figure 6A), the lowest concentration (0.85 ng/µL) demonstrated the lowest reduction in oxidative peak current post-hybridisation with signal change increasing as sample concentration increased. The problem here, however, was the specificity of the probe–target interaction (as shown earlier) and the relatively small signal change brought about by incubation with unamplified samples. The signal changes were negative, due to the fact that these were relatively low concentrations of DNA, leading to limited hybridisation. On the other hand, the amplified sample produced a dose–response curve with higher signal changes which were positive in direction due to the specific enrichment of the mutant sequence concentration with smaller standard deviations because of the hybridisation of strands with high complementarity (the unamplified samples contained fewer point mutations) and, in effect, the full fraction of cfDNA from the sample.

Achieving good sensitivity is very important as the concentration of circulating free DNA released by tumour cells is usually in proportion to the stage of cancer [48]. We were able to detect as low as 4.4 mutated copies per ng of DNA against a genomic DNA background also containing the wild type at levels of 565 copies/ng of DNA. We saw assay signal increases of as much as 300% as shown in Figure 5 with these quantities of mutated and wild type DNA. The specificity and sensitivity results are not complete and cannot be fully stated but the data presented and discussed here show that we can get appreciable changes in the electrochemical signal from relatively low copy numbers of the mutated gene compared to the highly abundant wild type gene also present in the sample. Further work will involve fully defining the assay sensitivity and specificity. As circulating nucleic acids are present in blood at ng/mL levels, which based on the fragment length is analogous to a picomolar concentration, a minimum of femtomolar sensitivity will be beneficial for detection of tumour-specific sequences [17]. Many published biosensor studies realised such sensitivity levels through the use of exotic electrode modifications, typically involving the fabrication of electrodes modified with graphene, nanoparticles, carbon nanotubes, etc. In our case, we opted to keep the electrode substrate low cost, easy to produce and coupled to a PCR reaction to achieve the desired sensitivity and specificity. Whilst our approach leads to a trade-off in terms of time to result, it establishes specific amplification and sensitive and specific hybridisation signals, giving confidence in the result whilst achieving an overall time to result which is a significant improvement over the current clinical practice. The ctDNA concentration response shown in Figure 6 shows a clear dose–response effect which predicts that an increase in ctDNA, per unit concentration, will result in a larger electrochemical signal response in the positive direction (i.e., increasing DPV peak current). Since levels of ctDNA are strongly correlated with tumour stage and response to therapy [49], there is a clear potential for this system to be applied in measuring how a patient's cancer treatment is progressing.

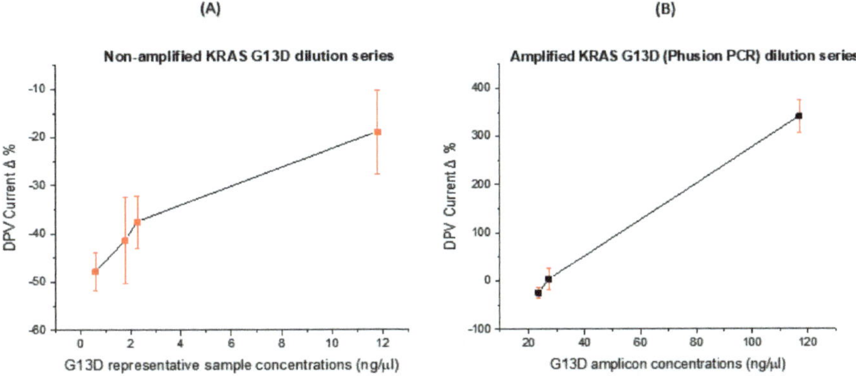

Figure 6. ctDNA response at different concentrations in a dilution series (**A**) DPV Peak currents percentage change for genomic KRAS G13D representative clinical sample at different concentrations (**B**) DPV Peak currents percentage change for KRAS G13D amplicons at different concentrations.

The findings of this study on ctDNA amplification are in agreement with several previous studies that were also able to successfully detect ctDNA KRAS mutations in patient samples using the ddPCR technique [50,51]. Electrochemical detection will quickly and accurately screen for cancer so treatment can be initiated as quickly as possible. Compared to other low-cost mutation detection technique like StripAssay, our sensor device is more reproducible, sensitive, and easier to manufacture and operate, especially from a clinical point of view. In addition, this study shows that electrochemical sensors can be directly coupled to a PCR reaction that uses standard primers and reagents and does not require optimisation, meaning that amplification reactions for other ctDNA markers can be developed off-chip and transferred directly into the assay to produce a ctDNA panel.

The current time to result for cancer detection in a clinical setting is two–three days (including sample transportation) for a non-complicated biopsy analysis and 7–10 days for a complicated biopsy analysis [52]. In the UK, the National Health Service mutation typing following biopsy can take up to nine weeks [52]. In summary, the DNA isolation from blood, clean up and PCR amplification took around 150 min. The ctDNA target incubation took approximately 60 min, while the CV and DPV pre- and post-hybridisation measurements for each electrode took less than 10 min. This gives a sum total of 3.5 h. However, through optimisation and device integration, we believe there is considerable room for optimisation in terms of time to result. The current analysis time of 3.5 h is a big stride towards PoC provision for ctDNA profiling in a healthcare setting. Further optimisation can be made using isothermal amplification which can cut down the number of thermal cycles and, in turn, decrease the overall amplification time from 1 h to 30 min [53]. As our previously published work shows that we were able to detect KRAS amplicons in plasma to mimic a 'clinical sample' [54], near future work will explore the detection of non-specified clinical samples containing different KRAS mutations and mutations in other genes involved in cancer, e.g., P53 and BRCA1. Analysing multiple mutations simultaneously in a given sample without prior knowledge of the alterations using multiplex techniques and direct detection of ctDNA from cancer patient samples will support the future direction of PoC clinical testing.

4. Conclusions

We were able to successfully produce a simple DNA sensor requiring no labelling processes or external indicators using a multi-carbon electrode. An electrochemical detection scheme involving a DNA hybridisation technique and screen-printed carbon electrodes were developed and shown through a series of comparative measurements to be sensitive and specific for the KRAS G12D and G13D mutations. The DNA modified sensors demonstrated superior performance to electrochemical pre-treatment with acetate buffer containing NaCl and characterisation using ferri-ferrocyanide buffer. Improved sensor sensitivity was achieved by designing a PCR reaction capable of amplifying either mutant KRAS G13D or wild-type KRAS through primer choice from representative patient samples. Cyclic voltammetry and Square Wave measurements were very sensitive for charactering the surface of the modified SPCEs and Differential Pulse voltammetry measurements provided the desired response and indicated detection was possible from samples containing as few as 0.58 ng/µL concentrations of amplicons. In addition, the response was found to be consistent with previously observed results, i.e., large signal decreases being evident upon amplification of the mutant allele, offering the promise of quantitation of mutant sequences from clinical samples. Both non-complementary DNA probes and wild-type DNA amplification reaction was successfully used as control. These results increase the prospect of simple, rapid and cost-effective measurement of nucleic acid tumour markers from blood and other body fluids. The current time to result of the electrochemical sensor was 3.5 h, providing notable scope for optimisation. It is essential to note that the sensor being developed can be potentially used for both early detection of cancer and monitoring the response to cancer treatment.

Author Contributions: Conceptualization, D.K.C.; methodology, B.A.; validation, D.K.C.; formal analysis, B.A.; investigation, B.A.; resources, F.T. and C.P.; data curation, B.A.; writing—original draft preparation, B.A.; writing—review and editing, B.A., D.K.C., C.P. and F.T.; supervision, D.K.C., and M.J.B.; project administration, B.A. and D.K.C. All authors have read and agreed to the published version of the manuscript.

Funding: The lead author was supported by EPSRC (grant reference number EP/L015595/1). F.T. and C.P. were supported by CRUK and ECMC core funding.

Institutional Review Board Statement: The study was approved by the Ethics Committee of the University of Strathclyde (protocol code UEC17/83—17/04/2018).

Informed Consent Statement: Not applicable. The HCT116 and SK-UT-1 cell lines were obtained from American Type Culture Collection organization (ATCC, catalogue #CCL-247 TM and HTB-114TM).

Data Availability Statement: Data is contained within this article and additional data can be found in preceding article https://doi.org/10.3390/IECB2020-07067.

Acknowledgments: Special gratitude goes to CDT Medical Devices at the University of Strathclyde through which the grant was received.

Conflicts of Interest: The authors declare no conflict of interest. The funders had no role in the design of the study; in the collection, analyses, or interpretation of data; in the writing of the manuscript, or in the decision to publish the results.

References

1. Attoye, B.; Baker, M.; Pou, C.; Thomson, F.; Corrigan, D.K. Electrochemical DNA Detection Methods to Measure Circulating Tumour DNA for Enhanced Diagnosis and Monitoring of Cancer. *Proceedings* **2020**, *60*, 15.
2. National Cancer Institute. Genetic Changes and Cancer. 2019. Available online: https://www.cancer.gov/about-cancer/causes-prevention/genetics (accessed on 12 March 2020).
3. Othman, E.; Wang, J.; Sprague, B.L.; Rounds, T.; Ji, Y.L.; Herschorn, S.D.; Wood, M.E. Comparison of false positive rates for screening breast magnetic resonance imaging (MRI) in high risk women performed on stacked versus alternating schedules. *Springerplus* **2015**, *4*, 4–9. [CrossRef] [PubMed]
4. Shyamala, K.; Girish, H.C.; Murgod, S. Risk of tumor cell seeding through biopsy and aspiration cytology. *J. Int. Soc. Prev. Community Dent.* **2014**, *4*, 5–11. [CrossRef] [PubMed]
5. Vendrell, J.A.; Taviaux, S.; Béganton, B.; Godreuil, S.; Audran, P.; Grand, D.; Clermont, E.; Serre, I.; Szablewski, V.; Coopman, P.; et al. Detection of known and novel ALK fusion transcripts in lung cancer patients using next-generation sequencing approaches. *Sci. Rep.* **2017**, *7*, 12510. [CrossRef]
6. Gorgannezhad, L.; Umer, M.; Islam, N.; Nguyen, N.; Shiddiky, M.J.A. Lab on a Chip Circulating tumor DNA and liquid biopsy: Opportunities, challenges, and recent advances in detection technologies. *Lab Chip* **2018**, 1174–1196. [CrossRef]
7. Chu, Y.; Cai, B.; Ma, Y.; Zhao, M.; Ye, Z.; Huang, J. Highly sensitive electrochemical detection of circulating tumor DNA based on thin-layer MoS_2/graphene composites. *RSC Adv.* **2016**, *6*, 22673–22678. [CrossRef]
8. Schwarzenbach, H.; Hoon, D.S.B.; Pantel, K. Cell-free nucleic acids as biomarkers in cancer patients. *Nat. Rev. Cancer* **2011**, *11*, 426–437. [CrossRef]
9. Taly, V.; Pekin, D.; Benhaim, L.; Kotsopoulos, S.K.; Corre, D.L.; Li, X.; Atochin, I.; Link, D.R.; Griffiths, A.D.; Pallier, K.; et al. Multiplex picodroplet digital PCR to detect KRAS mutations in circulating DNA from the plasma of colorectal cancer patients. *Clin. Chem.* **2013**, *59*, 1722–1731. [CrossRef]
10. Wan, J.C.M.; Massie, C.; Garcia-Corbacho, J.; Mouliere, F.; Brenton, J.D.; Caldas, C.; Pacey, S.; Baird, R.; Rosenfeld, N. Liquid biopsies come of age: Towards implementation of circulating tumour DNA. *Nat. Rev. Cancer* **2017**, *17*, 223. [CrossRef]
11. AACR Project Genie Consortium. AACR Project GENIE: Powering precision medicine through an international consortium. *Cancer Discov.* **2017**, *7*, 818–831. [CrossRef]
12. Kulemann, B.; Rösch, S.; Seifert, S.; Timme, S.; Bronsert, P.; Seifert, G.; Martini, V.; Kuvendjiska, J.; Glatz, T.; Hussung, S.; et al. Pancreatic cancer: Circulating Tumor Cells and Primary Tumors show Heterogeneous KRAS Mutations. *Sci. Rep.* **2017**, *7*, 4510. [CrossRef]
13. Shackelford, R.E.; Whitling, N.A.; McNab, P.; Japa, S.; Coppola, D. KRAS Testing: A Tool for the Implementation of Personalized Medicine. *Genes Cancer* **2012**, *3*, 459–466. [CrossRef]
14. Bettegowda, C.; Sausen, M.; Leary, R.J.; Kinde, I.; Wang, Y.; Agrawal, N.; Bartlett, B.R.; Wang, H.; Luber, B.; Alani, R.M.; et al. Detection of circulating tumor DNA in early- and late-stage human malignancies. *Sci. Transl. Med.* **2014**, *6*. [CrossRef]
15. Strausberg, R.L.; Buetow, K.H.; Emmert-Buck, M.R.; Klausner, R.D. The Cancer Genome Anatomy Project: Building an annotated gene index. *Trends Genet.* **2000**, *16*, 103–106. [CrossRef]
16. Imyanitov, E.N. Gene polymorphisms, apoptotic capacity and cancer risk. *Hum. Genet.* **2009**, *125*, 239–246. [CrossRef]
17. Kelley, S.O. What Are Clinically Relevant Levels of Cellular and Biomolecular Analytes? *ACS Sens.* **2017**, *2*, 193–197. [CrossRef]
18. Crossley, L.; Attoye, B.; Vezza, V.; Blair, E.; Corrigan, D.K.; Hannah, S. Establishing a Field-Effect Transistor Sensor for the Detection of Mutations in the Tumour Protein 53 Gene (TP53)—An Electrochemical Optimisation Approach. *Biosensors* **2019**, *9*, 141. [CrossRef]
19. Russell, C.; Ward, A.C.; Vezza, V.; Hoskisson, P.; Alcorn, D. Biosensors and Bioelectronics Development of a needle shaped microelectrode for electrochemical detection of the sepsis biomarker interleukin-6 (IL-6) in real time. *Biosens. Bioelectron.* **2019**, *126*, 806–814. [CrossRef] [PubMed]
20. Butterworth, A.; Ward, A.C. Analytical Methods Electrochemical detection of oxacillin resistance with SimpleStat: A low cost integrated potentiostat and sensor platform. *Anal. Methods* **2019**, *11*, 1958–1965. [CrossRef]
21. Dorothee, G.; Robert, M.; Janos Voros, E.R. Electrochemical Biosensors—Sensor Principles and Architectures. *Sensors* **2008**, *8*, 1400–1458.
22. Shoaie, N.; Forouzandeh, M.; Omidfar, K. Voltammetric determination of the Escherichia coli DNA using a screen-printed carbon electrode modified with polyaniline and gold nanoparticles. *Microchim. Acta* **2018**, *185*, 1–9. [CrossRef]

23. Li, P.Q.; Piper, A.; Schmueser, I.; Mount, A.R.; Corrigan, D.K. Correction: Impedimetric measurement of DNA–DNA hybridisation using microelectrodes with different radii for detection of methicillin resistant Staphylococcus aureus (MRSA). *Analyst* **2017**, *142*, 1946–1952. [CrossRef]
24. Lee, J.; Arrigan, D.W.M.; Silvester, D.S. Sensing and Bio-Sensing Research Mechanical polishing as an improved surface treatment for platinum screen-printed electrodes. *Sens. BioSens. Res.* **2016**, *9*, 38–44. [CrossRef]
25. Yang, N.; Waldvogel, S.R.; Jiang, X. Electrochemistry of Carbon Dioxide on Carbon Electrodes. *ACS Appl. Mater. Interfaces* **2016**. [CrossRef]
26. Iniesta, J.; Thakur, B.; Carmo, R.; Banks, C.E. Can the mechanical activation (polishing) of screen-printed electrodes enhance their electroanalytical response? *Analyst* **2016**, *141*, 2791–2799. [CrossRef]
27. Mahmoodi, P.; Rezayi, M.; Rasouli, E.; Avan, A.; Gholami, M. Early-stage cervical cancer diagnosis based on an ultra-sensitive electrochemical DNA nanobiosensor for HPV-18 detection in real samples. *J. Nanobiotechnol.* **2020**, 1–12. [CrossRef]
28. González-sánchez, M.I.; Gómez-monedero, B.; Agrisuelas, J.; Iniesta, J.; Valero, E. Electrochemistry Communications Highly activated screen-printed carbon electrodes by electrochemical treatment with hydrogen peroxide. *Electrochem. Commun.* **2018**, *91*, 36–40. [CrossRef]
29. Wei, H.; Sun, J.J.; Xie, Y.; Lin, C.G.; Wang, Y.M.; Yin, W.H.; Chen, G.N. Enhanced electrochemical performance at screen-printed carbon electrodes by a new pretreating procedure. *Anal. Chim. Acta* **2007**, *588*, 297–303. [CrossRef] [PubMed]
30. Du, C.X.; Han, L.; Dong, S.L.; Li, L.H.; Wei, Y. A novel procedure for fabricating flexible screen-printed electrodes with improved electrochemical performance. *IOP Conf. Ser. Mater. Sci. Eng.* **2016**, *137*, 1–8. [CrossRef]
31. Elgrishi, N.; Hammon, K.; McCarthy, B.; Eisenhart, T.; Dempsey, J. A Practical Beginner's Guide to Cyclic Voltammetry. *J. Chem. Educ.* **2017**, 1–10. [CrossRef]
32. Rashid, J.I.A.; Yusof, N.A. The strategies of DNA immobilization and hybridization detection mechanism in the construction of electrochemical DNA sensor: A review. *Sens. BioSens. Res.* **2017**, *16*, 19–31. [CrossRef]
33. Focused Collection. Illustration of Transparent Blue Silhouette of Male Body with Colored Colon Cancer. 2020. Available online: https://focusedcollection.com/236856886/stock-photo-illustration-transparent-blue-silhouette-male.html (accessed on 16 December 2020).
34. Dana-Farber Cancer Institute. What Is the Relationship Between Anemia and Cancer? 2020. Available online: https://blog.dana-farber.org/insight/2018/09/relationship-anemia-cancer/.ShutterstockctDNAimages (accessed on 27 November 2020).
35. Shutterstock. ctDNA Images. 2020. Available online: https://www.shutterstock.com/search/ctdna (accessed on 27 November 2020).
36. NCBI. 2020. Available online: https://www.ncbi.nlm.nih.gov/gene/3845 (accessed on 20 January 2020).
37. Amplyus. Bluegel-Electrophoresis. 2017. Available online: https://www.minipcr.com/ (accessed on 7 October 2019).
38. Thermofisher Scientific. Qubit 4 Fluorometer. 2020. Available online: https://www.thermofisher.com/uk/en/home/industrial/spectroscopy-elemental-isotope-analysis/molecular-spectroscopy/fluorometers/qubit/qubit-fluorometer.html (accessed on 14 October 2020).
39. Hetemi, D.; Noël, V.; Pinson, J. Grafting of diazonium salts on surfaces: Application to biosensors. *Biosensors* **2020**, *10*, 4. [CrossRef]
40. Li, N.; Chow, A.M.; Ganesh, H.V.S.; Ratnam, M.; Brown, I.R.; Kerman, K. Diazonium-modified screen-printed electrodes for immunosensing growth hormone in blood samples. *Biosensors* **2019**, *9*, 88. [CrossRef]
41. Obaje, E.A.; Cummins, G.; Schulze, H.; Mahmood, S.; Desmulliez, M.P.Y.; Bachmann, T.T. Carbon screen-printed electrodes on ceramic substrates for label-free molecular detection of antibiotic resistance. *J. Interdisc. Nanomed.* **2016**, *1*, 93–109. [CrossRef]
42. Ocaña, C.; Hayat, A.; Mishra, R.K.; Vasilescu, A.; del Valle, M.; Marty, J.L. Label free aptasensor for Lysozyme detection: A comparison of the analytical performance of two aptamers. *Bioelectrochemistry* **2015**, *105*, 72–77. [CrossRef] [PubMed]
43. Chalupczok, S.; Kurzweil, P.; Hartmann, H.; Schell, C. The Redox Chemistry of Ruthenium Dioxide: A Cyclic Voltammetry Study—Review and Revision. *Int. J. Electrochem.* **2018**, *2018*, 1273768. [CrossRef]
44. Loew, N.; Fitriana, M.; Hiraka, K.; Sode, K.; Tsugawa, W. Characterization of electron mediator preference of aerococcus viridans-derived lactate oxidase for use in disposable enzyme sensor strips. *Sens. Mater.* **2017**, *29*, 1703–1711. [CrossRef]
45. Su, W.Y.; Wang, S.M.; Cheng, S.H. Electrochemically pretreated screen-printed carbon electrodes for the simultaneous determination of aminophenol isomers. *J. Electroanal. Chem.* **2011**, *651*, 166–172. [CrossRef]
46. Manzanares-Palenzuela, C.L.; Fernandes, E.G.R.; Lobo-Castañón, M.J.; López-Ruiz, B.; Zucolotto, V. Impedance sensing of DNA hybridization onto nanostructured phthalocyanine-modified electrodes. *Electrochim. Acta* **2016**, *221*, 86–95. [CrossRef]
47. Perakis, S.; Speicher, M.R. Emerging concepts in liquid biopsies. *BMC Med.* **2017**, 1–12. [CrossRef]
48. Wang, X.; Wang, L.; Su, Y.; Yue, Z.; Xing, T. Plasma cell-free DNA quantification is highly correlated to tumor burden in children with neuroblastoma. *Cancer Med.* **2018**, 3022–3030. [CrossRef] [PubMed]
49. Reece, M.; Saluja, H.; Hollington, P.; Karapetis, C.S.; Vatandoust, S.; Young, G.P.; Symonds, E.L. The Use of Circulating Tumor DNA to Monitor and Predict Response to Treatment in Colorectal Cancer. *Front. Genet.* **2019**, *10*. [CrossRef] [PubMed]
50. Demuth, C.; Spindler, K.L.G.; Johansen, J.S.; Pallisgaard, N.; Nielsen, D.; Hogdall, E.; Vittrup, B.; Sorensen, B.S. Measuring KRAS Mutations in Circulating Tumor DNA by Droplet Digital PCR and Next-Generation Sequencing. *Transl. Oncol.* **2018**, *11*, 1220–1224. [CrossRef] [PubMed]

51. Coutinho, F.; Pancas, R.; Magalha, E.; Bernardo, E.; Antunes, M.J. Diagnostic value of surgical lung biopsy: Comparison with clinical and radiological diagnosis. *Eur. J. Cardio Thorac. Surg.* **2008**, *33*. [CrossRef] [PubMed]
52. NHS. Biopsy. 2018. Available online: https://www.nhs.uk/conditions/biopsy/ (accessed on 18 July 2019).
53. Chung, C.H.; Kim, J.H. One-step isothermal detection of multiple KRAS mutations by forming SNP specific hairpins on a gold nanoshell. *Analyst* **2018**, *143*, 3544–3548. [CrossRef]
54. 54. Attoye, B.; Pou, C.; Blair, E.; Rinaldi, C.; Thomson, F.; Baker, M.J.; Corrigan, D.K. Developing a Low-Cost, Simple-to-Use Electrochemical Sensor for the Detection of Circulating Tumour DNA in Human Fluids. *Biosensors* **2020**, *10*. [CrossRef]

 **biosensors**

Article

Electrical Characterization of Cellulose-Based Membranes towards Pathogen Detection in Water [†]

Grégoire Le Brun [1,*], Margo Hauwaert [1], Audrey Leprince [2], Karine Glinel [3], Jacques Mahillon [2] and Jean-Pierre Raskin [1]

[1] Institute of Information and Communication Technologies, Electronics and Applied Mathematics, UCLouvain, 1348 Louvain-la-Neuve, Belgium; margo.hauwaert@student.uclouvain.be (M.H.); jean-pierre.raskin@uclouvain.be (J.-P.R.)

[2] Laboratory of Food and Environmental Microbiology, Earth and Life Institute, UCLouvain, 1348 Louvain-la-Neuve, Belgium; audrey.leprince@uclouvain.be (A.L.); jacques.mahillon@uclouvain.be (J.M.)

[3] Institute of Condensed Matter and Nanosciences (Bio and Soft Matter), UCLouvain, 1348 Louvain-La-Neuve, Belgium; karine.glinel@uclouvain.be

* Correspondence: gregoire.lebrun@uclouvain.be

[†] This paper is an extended version of our paper published in: Le Brun, G.; Hauwaert, M.; Leprince, A.; Glinel, K.; Mahillon, J.; Raskin, J.-P. Electrochemical Characterization of Nitrocellulose Membranes towards Bacterial Detection in Water. In Proceedings of the 1st International Electronic Conference on Biosensors, 2–17 November 2020.

Citation: Le Brun, G.; Hauwaert, M.; Leprince, A.; Glinel, K.; Mahillon, J.; Raskin, J.-P. Electrical Characterization of Cellulose-Based Membranes towards Pathogen Detection in Water . *Biosensors* **2021**, *11*, 57. https://doi.org/10.3390/bios11020057

Received: 31 January 2021
Accepted: 19 February 2021
Published: 21 February 2021

Publisher's Note: MDPI stays neutral with regard to jurisdictional claims in published maps and institutional affiliations.

Abstract: Paper substrates are promising for development of cost-effective and efficient point-of-care biosensors, essential for public healthcare and environmental diagnostics in emergency situations. Most paper-based biosensors rely on the natural capillarity of paper to perform qualitative or semi-quantitative colorimetric detections. To achieve quantification and better sensitivity, technologies combining paper-based substrates and electrical detection are being developed. In this work, we demonstrate the potential of electrical measurements by means of a simple, parallel-plate electrode setup towards the detection of whole-cell bacteria captured in nitrocellulose (NC) membranes. Unlike current electrical sensors, which are mostly integrated, this plug and play system has reusable electrodes and enables simple and fast bacterial detection through impedance measurements. The characterized NC membrane was subjected to (i) a biofunctionalization, (ii) different saline solutions modelling real water samples, and (iii) bacterial suspensions of different concentrations. Bacterial detection was achieved in low conductivity buffers through both resistive and capacitive changes in the sensed medium. To capture *Bacillus thuringiensis*, the model microorganism used in this work, the endolysin cell-wall binding domain (CBD) of Deep-Blue, a bacteriophage targeting this bacterium, was integrated into the membranes as a recognition bio-interface. This experimental proof-of-concept illustrates the electrical detection of 10^7 colony-forming units (CFU) mL^{-1} bacteria in low-salinity buffers within 5 min, using a very simple setup. This offers perspectives for affordable pathogen sensors that can easily be reconfigured for different bacteria. Water quality testing is a particularly interesting application since it requires frequent testing, especially in emergency situations.

Keywords: paper-based sensors; nitrocellulose; impedance measurements; dielectric properties; parallel-plate electrodes; interdigital electrodes; endolysins; *Bacillus thuringiensis*

1. Introduction

Access to safe and sufficient water is a prerequisite for to the development of human communities and the enhancement of economic activities [1]. Identification, control and prevention of ground and surface water pollution require both frequent water quality testing and diligent water management. The detection of pathogenic bacteria such as *Escherichia coli*—used as an indicator for fecal water contamination—usually deploys techniques that include colony count, DNA analysis after polymerase chain reaction (PCR) or enzyme-linked immunosorbent assay (ELISA) [2,3]. Despite the favorable performance

123

(detection limits of 1 CFU mL^{-1}) of these techniques, their high cost, their requirement of well-equipped laboratory facilities, and the time constraint (at least several hours per test) have prompted the development of portable, simple and low-cost tools suitable for a rapid (less than 1 h) and precise detection methods of pathogens. Furthermore, the sanitary pandemic caused by the infectious severe acute respiratory syndrome coronavirus 2 (SARS-CoV-2) highlighted the field of action of point-of-care (PoC) devices because it can meet the need of mass screening, in particular to screen the virus in water environments. Wastewater-based epidemiology (WBE) is a promising approach to predict the potential spread of the infection by testing for infectious agents in wastewater, and has been approved as an effective way to obtain information on diseases and pathogens [4,5].

Microfluidic paper analytical devices (μPADs), on the one hand, are the dominant PoC biosensors for the rapid detection of pathogens in both healthcare and environmental monitoring, especially in situations with scarce resources [6,7]. Given the accessibility of this technology, it is appropriate for either water scientists or citizen groups, without requiring specific training [8]. Paper is a valuable platform for biodetection as it presents several assets. First, it has beneficial spontaneous microfluidic properties through capillarity. Second, it facilitates the attachment of bioreceptors which are often proteins such as antibodies, that contribute to specificity towards pathogens. Third, it is low-priced and it allows for straightforward manufacturing and disposability. These benefits are already used in lateral flow assays (LFA) such as pregnancy test strips, in which analytes of interest are passively drained through a nitrocellulose (NC) membrane, a paper derivative, towards the detection zone where they are immobilized by specific bioreceptors. Current paper-based sensors mostly require the use of labels (such as gold nanoparticles, which have an intense red color), conjugated with antibodies to achieve specific optical detection of the immobilized analytes [9]. However, they have two main disadvantages that inhibit their use in the field of water potability, which has demanding limits of detection for pathogens and pollutants. Optical LFA have a reduced sensitivity since only the top 10 μm depth of the paper contributes to the colorimetric signal due to the opaqueness of the membrane [10]. Additionally, the measurement result is mostly qualitative or semi-quantitative.

Electrical biosensors, on the other hand, rely on the monitoring of changes in material electrical properties when bacteria bind in close proximity from the surface of, e.g., interdigital electrodes (IDE) designed on a solid substrate [11,12]. The signal response, often proportional to the number of bacteria, is used as an electrical fingerprint of the sample to provide fast, precise and quantified information about the bacteria presence in water. However, the particular mechanisms of electron transfer between electrodes and specific bacterial cells, as well as within the cells, are still under fundamental studies in bioelectrochemistry [13]. Furthermore, the production and usage of electrical biosensors face significant obstacles. Grafting a biorecognition layer, e.g., bacteriophages (phages) or antibodies, on the surface of conventional surface-based electrical biosensors faces problems such as reproducibility, uniformity and stability over time [14]. The functionalization protocol needs to be adapted to every surface material and grafting molecule. In addition, the capture percentage of bacteria by the biorecognition layer is relatively low since only bacteria in close vicinity to the surface bind to the specific receptors. Many of the target pathogens thus flow over the electrode without binding, decreasing the sensor sensitivity. Finally, conventional surface-based electrical biosensors utilize gold electrodes functionalized with bioreceptors/antibodies using classical thiol chemistry [15,16]. However, insufficient chemical stability of thiolates is one of the most serious problems for their applications in ambient and aqueous environments [17].

Despite the aforementioned drawbacks of both individual sensing technologies, studies have shown that the integration of highly sensitive electronic detection methods with LFA is an attractive approach to circumvent these and capitalize on the advantages of paper substrates and electrical biosensors [18]. Given the favorable electrostatic properties of nitrocellulose, bioreceptors are readily immobilized through the whole pore volume of the paper membrane, thus drastically increasing the number of interactions with targeted pathogens. Unlike surface-based methods, electrical measurements taking advantage of paper porosity thus allow to quantify the number of bioreceptor-bacteria conjugates in the whole tested sample volume. However, the development of such sensors remains very challenging for three main reasons.

First, one of the key factors affecting the analytical performance of μPADs is the bioreceptor used to capture the bacteria in the test zone. Antibodies, commonly used as a biointerface in μPADs, are rather expensive. As a result, there is a growing interest in developing proteins as alternative receptors for LFA. Particularly promising are bacteriophages, viruses that specifically infect bacteria and produce lytic enzymes called endolysins that show strong affinity and high specificity towards target bacteria [19,20].

Second, due to significantly different properties of paper-based and more conventional substrates for electronic circuits, innovative design methods are needed. Alternative manufacturing techniques, e.g., printing sensors such as IDE, are being explored as integrated sensing devices [21–23]. However, integrating electronics on paper substrates is difficult to implement because of the inhomogeneous nature of the paper, resulting in low electrode resolutions (~hundreds of μm) and high electrical resistances [24] with respect to the classical design of microelectrodes suitable for bacteria sensing (a few μm finger gaps) [25].

Third, prior works that attempted to accommodate electrical bacteria detection on common μPADs were mostly based on direct charge transfer measurements [26,27]. However, these require cumbersome redox probes. Furthermore, electrical measurements are influenced by the ionic strength of the analyzed solution. Direct current measurements are particularly affected by this ionic background noise, since they only measure the solution resistance, inversely proportional to the number of ions. When dealing with aqueous samples presenting varying electrical conductivities, therefore, it is challenging to calibrate the sensor and to differentiate the electrical response of target compounds from the electrical background signal.

In this paper we address the aforementioned challenges by emphasizing three contributions towards paper-based electrical biosensing for simple, rapid and affordable bacteria detection in water.

First, we capitalize on the natural capillarity of the NC membrane to wick bacterial suspensions to the testing zone, where the membrane is functionalized with the recently characterized cell-wall binding domain (CBD) derived from the PlyB221 endolysin encoded by phage Deep-Blue targeting *Bacillus thuringiensis* [28], used as model microorganism in this work. This ensures high binding and immobilizing capacity towards this specific bacteria.

Second, we demonstrate the relevance of the simple parallel-plate setup presented in [29]. This common material dielectric measurement system can be judiciously used as sensor to perform electrical measurements on NC membranes inserted in between the electrodes. This plug-and-play setup eliminates the unpractical need to integrate electronics components directly onto the paper matrix.

Third, impedance sensing is proposed to monitor the porous NC membrane permittivity changes caused by various electrolyte solutions. Since no electron transfer occurs at the electrode surface, changes in electrical properties are mainly observed through volumic impedance properties. By measuring the later with an AC signal, both resistive and capacitive properties are estimated at high frequencies, which enables the label-free and non-intrusive detection of bacteria immobilized by specific bioreceptors. Indeed, their presence in the membrane affects both the global conductivity and permittivity [25,30].

The remainder of this paper is organized as follows. We began by validating the CBD biointerface, both electrically and optically. In order to characterize the system behavior and ionic noise for aqueous solutions with different ionic strengths, we then analyzed the sensor response and resolution to saline solutions at different salt concentrations, used as models of real aqueous samples. An equivalent electrical model of the sensing system was developed to quantify the impact of ionic concentration on the total measured impedance.

Then, the sensing principle was validated in the presence of bacterial cells. A proof of concept of the simple and rapid (<5 min) parallel-plate biosensor was demonstrated by detecting *B. thuringiensis* cells in low-conductive buffers. Finally, the bacterial detection results with the plug-and-play parallel-plate setup were compared with a planar fringing field electrodes system, composed of IDE directly applied on a single side of the NC membrane. The sensing principles of both parallel-plate and IDE devices were modeled and analyzed using small-signal electrical equivalent circuits, highlighting the contribution of ions in both bacterial detection mechanism. Their potentials, advantages and limitations are also discussed.

2. Materials and Methods

2.1. Materials

Nitrocellulose membranes on a polyester backing (UniSart Lateral Flow CN95 Backed, 20 μm nominal pore size) were purchased from Sartorius (Göttingen, Germany). Phosphate-buffered saline (PBS) and sodium chloride solution (NaCl, 1 M) were purchased from Sigma-Aldrich (St. Louis, MO, USA). Deionized (DI) water was produced in our facilities (conductivity $\sigma = 6.6 \times 10^{-6}$ S/m).

2.2. Biological Procedures

2.2.1. Bacterial Strains and Growth Conditions

B. thuringiensis GBJ002 expressing the cyan fluorescent protein (CFP) was used as proof-of-concept for this study. Bacteria were plated on lysogeny broth (LB)-agar plate containing kanamycin (50 μg/mL) and acid nalidixic (25 μg/mL) and grown overnight (O/N) at 30 °C. Next, an individual colony was used to inoculate 5 mL of LB supplemented with antibiotics and the suspension was incubated O/N at 30 °C with agitation (120 rpm). Then, the culture was washed twice with an appropriate buffer (water or diluted PBS) to remove residual LB (centrifugation at 10,000× *g* for 5 min at room temperature (RT)). The pellet was finally re-suspended in the appropriate buffer yielding a bacterial suspension of ca. 10^8 CFU mL^{-1}.

2.2.2. Phage Endolysins Expression and Purification

A detailed description of the expression and purification of the CBD of PlyB221 endolysin, encoded by phage Deep-Blue can be found in [28]. The CBD was fused to a green fluorescent protein (GFP) for fluorescence assays. The protein concentration was adjusted to 1 mg/mL.

2.2.3. Preparation and Characterization of the Cell-Wall Binding Domain (CBD) Biointerface for Specific Bacteria Capture

A protocol for optimal biofunctionalization of the NC membranes with endolysin CBD based on simple physisorption was developed (Figure 1). Membranes were prepared by individually depositing and shaking 50 μL of 1 mg/mL solution of purified CBD on the NC. Membranes were then dried in an oven for 60 min at 37 °C followed by desiccation for 30 min at RT to fix the proteins to the membrane. Following desiccation, membranes were washed twice during 2 min with deionized (DI) water to remove proteins in excess. Membranes were wiped, dried, and finally stored in a desiccation chamber at RT. Effective binding of specific bacteria to the biointerface was assessed by depositing a droplet of *B. thuringiensis* suspension on the biofunctionalized membranes followed by a washing step (Figure 1). Confocal laser scanning microscopy (Zeiss LSM 710) experiments were performed to observe bacteria immobilization in the membrane depth.

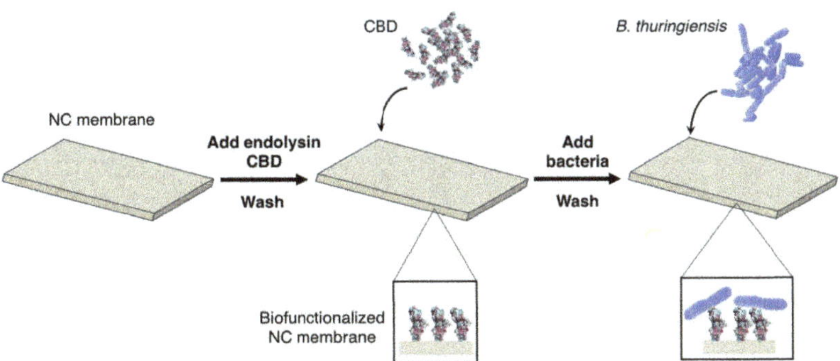

Figure 1. Protocol of the nitrocellulose (NC) membrane biofunctionalization with endolysin cell-wall binding domain (CBD) and validation of the biointerface through capture of *B. thuringiensis* cells. (1) The deposition of the phage endolysin CBD in the NC membrane is followed by drying, washing and desiccator steps. (2) The bacteria are then deposited on the membrane and specifically captured by the CBD.

2.3. Parallel-Plate Setup

2.3.1. Impedance Sensing

Figure 2A shows a schematic of the experimental system for the simple parallel-plate measurements. The nitrocellulose membranes were held between the two parallel-plates of the dielectric test fixture (16451B, Agilent, Santa Clara, CA, USA), connected to an impedance analyzer (LCR 4284A, Agilent, Santa Clara, CA, USA). The impedance spectroscopy measurements were carried out with the LCR, remotely controlled by a computer through the Labview software (Labview National Instrument, Austin, TX, USA) to perform an automatic sweep from 1 kHz to 1 MHz, at voltage amplitude of 200 mV. Before impedance measurement, an open-circuit calibration was performed. The impedance data were extracted in a magnitude-phase data-structure. The ZVIEW software (Scribner Associates Inc., Southern Pines, NC, USA) is used to fit the electrical equivalent model of the parallel-plate sensing system, as presented in Figure 2A, and extract the model parameter values from the measurements.

The material real and imaginary dielectric constant are derived for a given membrane thickness by measuring its capacitance and dissipation factor. An electrical model is used to differentiate the actual dielectric properties of the NC from those of the polyester backing (Figure 2A).

2.3.2. Sensor Modelling

An analytical model of the NC membranes soaked with biological solutions and inserted in between the two parallel-plate electrodes is established to provide a physical understanding of the system. In the presence of an electrolyte, the equivalent electrical model considers a double layer capacitance (C_{dl}). The equivalent circuit also considers the NC membrane, modelled through the parallel association of its capacitive (C_{NC}) and conductive (R_{NC}) properties, as well as the polyester backing, modelled using its capacitive properties only ($C_{Backing}$), since its dielectric losses are considered to be negligible in the studied frequency range. The value of the polymer backing capacitance (~100 pF) is determined experimentally through measurement of its permittivity by means of dielectric measurements (see Section 2.3.1). C_{NC} and R_{NC}, representing the electrical properties of the NC membranes saturated with saline solutions or bacterial suspensions, strongly depend on the ionic concentration in the solution. The double layer capacitance, representing the interfacial properties is given by [19]:

$$C_{dl} = \frac{\varepsilon_0 \varepsilon_{r,sol}}{\sqrt{\dfrac{\varepsilon_0 \varepsilon_{r,sol} k_B T}{2\, q^2 N_{av} c_{ions} 10^3}}} A_e \tag{1}$$

with C_{dl} the double layer capacitance, k_B the Boltzmann constant, c_{ions} molar ionic concentration of the solution in which the double layer occurs, $\varepsilon_{r,sol}$ the relative permittivity of this solution and A_e the surface of the parallel-plate electrode. The values for C_{dl} at different ionic concentrations were simulated using COMSOL Multiphysics v.5.4 (COMSOL AB, Stockholm, Sweden) based on Equation (1), and lie in the 10–100 µF range. Regarding the values of the other parameters, their influence can therefore be neglected at the considered frequencies between 1 kHz and 1 MHz.

The equivalent model has three cut-off frequencies:

$$f_1 = \frac{C_{dl} + C_{Backing}}{2\pi \cdot R_{NC} \cdot \left[C_{dl} C_{Backing} + C_{NC} \left(C_{dl} + C_{Backing} \right) \right]} \tag{2}$$

$$f_2 = \frac{C_{dl} + C_{Backing}}{2\pi \cdot C_{dl} C_{Backing}} \tag{3}$$

$$f_3 = \frac{1}{2\pi \cdot R_{NC} C_{NC}} \tag{4}$$

2.4. Interdigital Electrodes (IDE) Setup

2.4.1. Interdigital Electrode Design and Fabrication on Nitrocellulose (NC) Membranes

The deposition of IDE on nitrocellulose substrate was conducted using a physical vapor deposition (PVD) e-gun evaporation technique. Gold IDE were deposited by applying patterned nickel masks on top of the nitrocellulose membrane. This deposition technique is more cumbersome than screen-printing and inkjet printing techniques, which are usually used for deposition of electrodes on paper [31]. However, these have the disadvantage of not allowing precise electrode deposition on chemically untreated nitrocellulose. Hence, PVD is chosen because it allows for precise deposition without inducing variability through additional treatment. The IDE finger width and interdigit gap are 200 µm, which enables the detection of dielectric properties over the whole NC membrane depth (~140 µm) [32].

2.4.2. IDE Impedance Sensing

Figure 2B shows a schematic of the experimental system for IDE measurements. The IDE were connected through toothless crocodile clips to an impedance analyzer (LCR 4284A, Agilent, Santa Clara, CA, USA) through BNC connectors. The impedance spectroscopy measurements were carried out with the LCR, remotely controlled by a computer through the Labview software (Labview National Instrument, Austin, TX, USA) to perform an automatic sweep from 1 kHz to 1 MHz, at voltage amplitude of 20 mV. Before impedance measurement, an open-circuit calibration was performed without any electrical contacts between the crocodile clips. The impedance data were extracted in a magnitude-phase data-structure.

2.4.3. IDE Sensor Modelling

The equivalent circuit of the IDE in Figure 2B incorporates the surficial phenomenon of double layer capacitance through C_{dl} and the volumic phenomena through C_{air}, R_{air}, C_{NC}, and R_{NC}, corresponding to the upper air layer and lower nitrocellulose layer, respectively. Given the width of IDE fingers, the backing is not taken into account (Section 2.4.1). The double layer capacitance for IDE electrodes in contact with a given solution is extended from (1) to:

$$C_{dl} = \frac{\varepsilon_0 \varepsilon_{r,sol}}{\sqrt{\dfrac{\varepsilon_0 \varepsilon_{r,sol} k_B T}{2\, q^2 N_{av} c_{ions} 10^3}}} A_e (N-1) \tag{5}$$

with A_e the surface per electrode finger and N the number of fingers. The equivalent resistance and capacitance of the nitrocellulose and air volume are given by

$$\begin{aligned} R_{NC} &= K_{cell}\sigma_{NC}^{-1} & R_{air} &= K_{cell}\sigma_{air}^{-1} \\ C_{NC} &= K_{cell}^{-1}\varepsilon_0\varepsilon_{r,NC} & C_{air} &= K_{cell}^{-1}\varepsilon_0\varepsilon_{r,air} \end{aligned} \tag{6}$$

with K_{cell} the cell constant, ε_r and σ the relative permittivity and conductivity of the sensed nitrocellulose and air volumes, respectively. The cell constant is determined experimentally and incorporates the geometric properties of the IDE [33]. Hence, it does not vary with frequency nor with the electrical properties of the material.

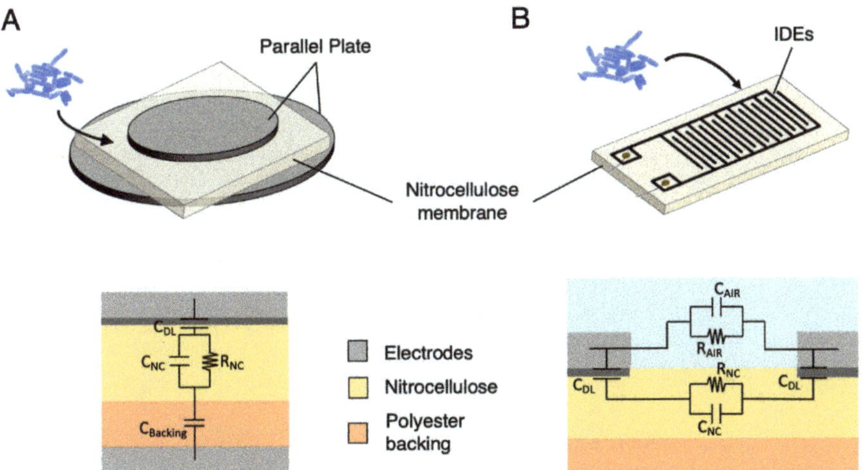

Figure 2. Experimental setups and corresponding electrical models investigated towards bacteria detection in this work. (**A**) Parallel-plate probes are a common material dielectric measurement system. The bacterial sample is deposited on the NC membrane and conducted to the test zone by capillarity. A simple electrical model is proposed to consider the dielectric effect of the polyester backing supporting the NC and the electrical double layer that arises from charge redistribution at the interface between the electrolyte and the probe. (**B**) Interdigital electrodes are generally used as sensors to monitor impedance changes at the proximity of the metallic fingers, here deposited on NC membrane. The bacterial samples are deposited on top of the interdigital electrode (IDE)-NC sensor. The model does not include the polyester backing as its impact on the impedance seen by the IDE is negligible due to its depth.

2.5. Sensing of Saline Solutions as Models for Real Water Samples

Prior to electrical measurements, sodium chloride solutions of various concentrations (c_{NaCl}), 10^{-5}, 10^{-4}, 5×10^{-4}, 10^{-3}, 5×10^{-3}, 10^{-2}, 10^{-1} mol/L (M), were prepared by dilutions of a 1 M NaCl solution in DI water, in order to model the electrical properties of different types of real water sample and biological buffer (Table 1). Regarding the impedance measurements considered in this study, the parameters of interest to be modelled are the mean ionic strength of the liquid (through adjustment of c_{NaCl}), and hence its dielectric properties (electrical conductivity and permittivity).

Then, 50 µL of the diluted sodium chloride solutions were then deposited on NC membranes, placed in between the parallel-plate electrodes, and reference dielectric or impedance measurements were performed within 5 min. The parallel-plates were wiped between each measurement to remove remaining water and salt on the electrodes.

Table 1. Modelling of the electrical properties at 20 °C of real aqueous samples and biological buffers using saline solutions at different concentrations. Mean concentrations and, therefore, dielectric properties are considered as the physico-chemical content of the water samples can vary from place to place [34,35].

Modelled Solutions	c_{NaCl} [M]	$\varepsilon_{r,sol}$ [/]	σ_{sol} [S/m]
PBS:1000	1.6×10^{-4}	~80	1.8×10^{-3}
Drinking/surface water	~10^{-3}	~80	10^{-1}–10^{-2}
PBS/highly saline water	1–5×10^{-1}	~70	1–5

2.6. Bacteria Detection in Physiological Buffers

Before depositing bacterial suspensions, 50 µL of PBS buffer diluted 1000× (PBS:1000) in DI water was deposited on a previously biofunctionalized membrane, and reference dielectric or impedance measurements were performed within 5 min with parallel-plates or IDE, respectively. PBS:1000 was chosen as biological buffer because the largest detection sensitivities were shown to be achieved with low-salt buffer solutions [25], and such low-salt buffers have electrical properties similar to real water samples. Five min is a time limit before which it is assumed that the wet impedance has not changed by more than 5% due to drying. Then, suspensions of 10^8, 10^7 and 10^6 CFU mL^{-1} of stationary-state *B. thuringiensis* resuspended in PBS 1:1000 were deposited (50 µL) on top of the membrane sample with a micropipette (Figure 2A,B) and spread within the membrane due to capillarity, and impedance or dielectric measurements were performed. In order to observe the global sensitivity of the setup, including both sensitivities of the impedance modulus and phase to bacteria presence, the sensitivities (S) in the explored frequency range were computed as the amplitude of the differences in complex impedance measured with and without bacterial cells, in percent:

$$S = \left| \frac{Z_{Bact} - Z_{PBS}}{Z_{PBS}} \right| \tag{7}$$

3. Results

3.1. Characterization of the CBD-Biofonctionalized Nitrocellulose Membrane

3.1.1. Optical Characterization of the CBD Biointerface

In this section, we validate the biofunctionalization of the NC membrane with a CBD-biointerface aimed at capturing *B. thuringiensis* whole cells for subsequent electrical detection. The binding of CBD to bacterial cells was evaluated in a cell wall decoration assay as described by [28], relying on the homogeneous adsorption of GFP-CBD to *B. thuringiensis* cells observed by fluorescence microscopy. This confirms their potential as specific immobilization probes for LFA biosensor schemes. In [36], we observed that deposited specific proteins (antibodies) were completely and uniformly distributed throughout the thickness of the NC membrane, promising the capture of bacterial cells throughout the volume and thus enabling electrical detection over the whole volume. In this paper we demonstrate that, by applying the developed protocol, GFP-CBD has been successfully deposited over the whole NC volume (Figure 3). Confocal microscopy images captured after deposition of the bacterial suspension and subsequent washings of the membrane also demonstrates strong affinity and robust capture of *B. thuringiensis* by the CBD within the NC, confirming the potential of the CBD as an immobilized probe in paper detection schemes. Furthermore, the porous structure of the nitrocellulose membrane is highlighted by the biofunctionalization of the substrate with fluorescent bioreceptors which experimentally confirms the mean pore size of the membrane (around 20 µm). The size of *B. thuringiensis* cells, about 0.5–1.0 µm × 2–5 µm [37], and their tendency to form aggregates [38], justifies this pore diameter since clogging and retention to the membrane should be avoided.

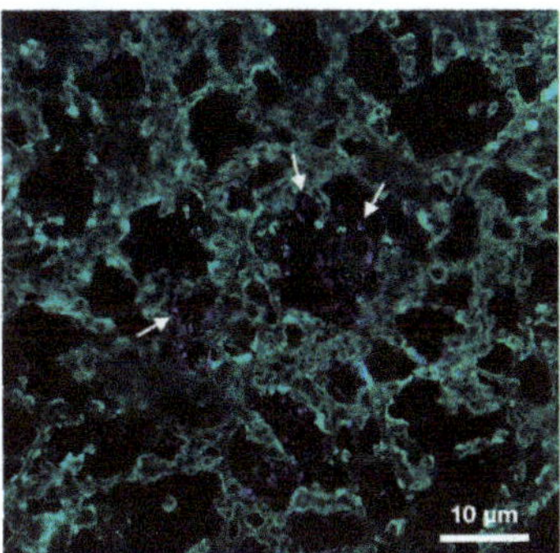

Figure 3. Confocal fluorescence microscopy image of immobilized *B. thuringiensis* cells (blue) in the pores of a CBD-biofunctionalized nitrocellulose membrane (turquoise). The white arrows indicate the presence of captured bacteria on the surface of the membrane pores.

3.1.2. Electrical Characterization of Dry and Biofunctionalized Nitrocellulose Membranes

In order to characterize the impact of the biofunctionalization on the membrane electrical properties, we performed dielectric measurements with the parallel-plate electrodes on raw and CBD-biofunctionalized NC membranes, under dry conditions. The permittivity of a raw nitrocellulose membrane drops from 1.55 to 1.45 between 1 kHz and 1 MHz (Figure 4). The biofunctionalization causes a permittivity reduction of approximatively 1–3%.

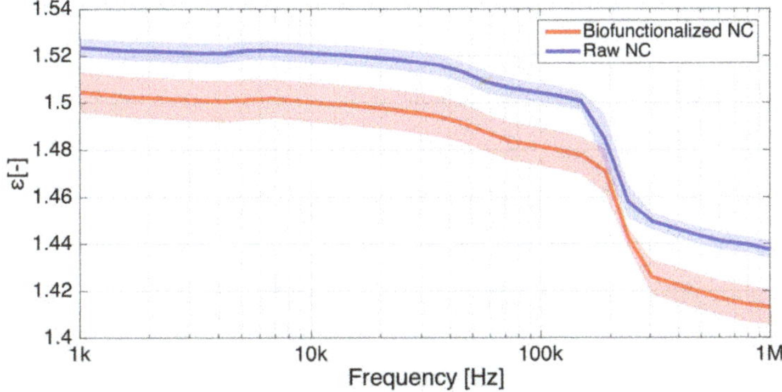

Figure 4. Relative permittivity of raw and CBD-biofunctionalized nitrocellulose membranes over 1 kHz–1 MHz. The biofunctionalization process causes a small decrease in the permittivity. Total number of samples: 7. Shaded area surrounding the measurement curves: standard error (σ).

The shaded area around the measurement curves in Figure 4 express the standard errors bars over the frequency range of interest, and indicate that results show variability to a certain extent, which can be explained by two parameters. First, the NC membranes present a foam-like structure (Figure 3) resulting in structural anisotropy and surface inhomogeneity, which renders absolute measures of the permittivity difficult as the solver algorithm assumes that the material under test is homogenous [39]. Second, the biofunctionalization protocol can substantially modify the permittivity of the NC sheets by altering its pore surface properties. Indeed, under conditions of low humidity reached after the desiccator step in the protocol, nitrocellulose membranes accumulate a significant static charge [9], affecting the dielectric measurements.

3.2. Impact of the Electrolyte Conductivity on the Parallel-Plate Sensor Response

Table 2 summarizes the observed changes in the equivalent circuit elements of Figure 2A for different NaCl concentrations. Between 1 kHz and 1 MHz, we observe that the impedance measurements are particularly sensitive to the electrical conductivity (σ_{sol}) of the electrolyte, as it directly influences R_{NC}. Thus, to understand the response of the parallel-plate sensing system and discriminate the bacteria electrical contribution, it is of utmost importance to characterize the effect of the background ionic noise resulting from remaining dissolved salt in the solution. Not monitoring or controlling the ionic concentration of the electrolyte where bacteria are suspended could lead to misinterpretation of the electrical results, as the supposed detection of elements could be caused by changes in the background ionic strength.

The concentrations of the saline solutions were chosen to represent background ionic noise for different water sources of interest (Table 1). Impedance measurements were first carried out to investigate the impedance modulus and phase dependence upon the salt concentration between 1 kHz and 1 MHz. Figure 5A,B show that the sensor discriminates the different saline concentrations through shifts of both impedance magnitude and phase. A global decrease of the impedance magnitude is observed with increasing NaCl concentration, since increased salt concentration increases the conductivity σ_{sol}. This magnitude decrease is higher between 10 kHz and 1 MHz, where the impedance phase is mostly resistive: this is where the highly varying R_{NC} has the most effect on the total complex impedance. Even if these peaks tend towards resistive impedance angles, they stay lower than $-45°$ given that no direct conduction path exists between the two parallel electrodes due to the isolating backing (Figure 2A). At both sides of these peaks, the phase decrease indicates a transition from a mixed to an exclusively capacitive behavior, led by the volume capacity C_{NC} at higher frequencies. We observe a frequency shift of the peak, shifting towards higher frequencies when the salt concentration increases. This can be explained by Equations (2) and (4): f_1 and f_3 correspond to the middle of the upwards and downwards flank of the peak. These cut-off frequencies increase when the salt concentration increases, given the variation of R_{NC} and C_{NC} in Table 2.

The working frequencies of the sensor towards detection of saline solutions corresponding to ionic noise of interest (Table 1) lies in the range 10–200 kHz. The limit of detection of the system (LOD) lies between 10^{-5} M and 10^{-4} M since the sensing device was not able to differentiate significantly ($<3\sigma$) impedance modulus and phases. This LOD corresponds to very low salinity electrolytes, less conductive than most of the buffers considered in biological detection schemes, which is beyond the scope of interest for the sensor-applications.

In order to quantify the impact of changes in salt concentration on the global system impedance, we extracted the values of R_{NC} and C_{NC} for the different salt concentrations based on the simple electrical equivalent model of the parallel-plate setup (see Table S1 for the data). The system is sensitive to both resistive and capacitive effects, even though, comparatively, the increasing ionic strength of the solution is more sensed through R_{NC} than C_{NC} (Table 2). In addition, the quantitative evaluation of R_{NC} and C_{NC} also supports the LOD of 10^{-4} M, as the difference in impedance, resistance and capacitance with the 10^{-5} M solution is less than 3 times the standard deviation (σ).

Dielectric measurements were carried out to corroborate the impedance measurements. Relative permittivity of NC membranes soaked with 10^{-4} M and 10^{-3} M NaCl solutions, modeling respectively highly diluted PBS (PBS:1000) and slightly saline solutions, was extracted over the frequency range of interest (Figure 5C). The permittivity of the saturated membrane with backing shows an increase of the system permittivity with the salt concentration. This is reflected by an increasing value of C_{NC} extracted from the impedance measurements (Table 2), even if the proportions are different since this system permittivity considers also the double layer and backing permittivity while C_{NC} only incorporates the volume of the nitrocellulose.

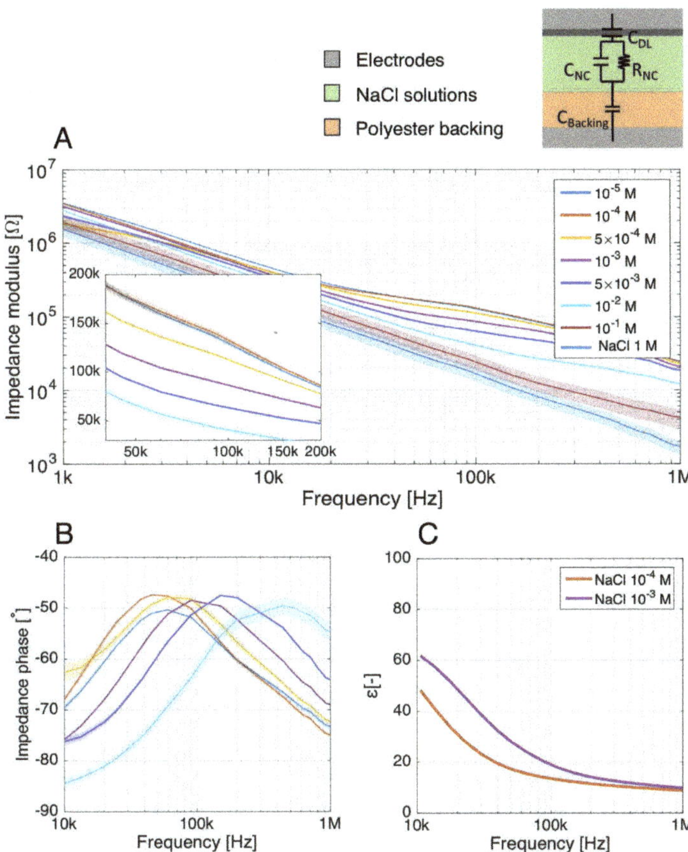

Figure 5. (**A**) Impedance measurements of the nitrocellulose (NC) membrane with isolating backing as seen by the parallel-plate setup. The NC membrane is saturated with saline solutions of different molar concentrations, modelling the dielectric properties of real water samples. A significant decrease in the impedance modulus results from an increase of the ionic strength, expressed as a drop in the membrane resistance R_{NC}. (**B**) The impedance phase evolution over 10 kHz–1 MHz highlights the contribution of both R_{NC} and C_{NC} to the impedance of the system. As the resistive peak in the phase shifts with the ionic strength, the system is, therefore, sensitive to saline electrolyte through both R_{NC} and C_{NC} changes. (**C**) Dielectric relative permittivity of the system as seen by the parallel-plate setup, with different diluted salt solutions in the nitrocellulose membrane. Number of samples: 12 from two independent experiments. Shaded area surrounding the measurements curves: standard deviation (σ). σ smaller than the measurement line thickness if not visible on the graph.

Table 2. Monitoring of the nitrocellulose membrane resistance R_{NC} and capacitance C_{NC}, respectively showing consecutive relative decreases and increments with the concentration of the NaCl solutions used to model different types of water samples through their conductivities. The parallel-plate setup is more sensitive to R_{NC}, but is still responsive to capacitance changes of the membrane.

	10^{-4} M	5×10^{-4} M	10^{-3} M	5×10^{-3} M	10^{-2} M	10^{-1} M
ΔR_{NC}	1.9×10^{4} Ω	-15%	-35%	-47%	-57%	-90%
ΔC_{NC}	8.34 pF	$+5\%$	$+17\%$	$+15\%$	$+35\%$	$+22\%$

3.3. Detection of B. thuringiensis Cells with the Parallel-Plate Setup

After the characterization of the sensor response to ionic background noise in aqueous solutions, impedance measurements of *B. thuringiensis* resuspended in the low-salt buffer PBS:1000 was investigated to extend the electrical model assessed in Section 2.3 to the detection of label-free, whole bacterial cells. As the impedance measurements are extremely sensitive to the electrical conductivity σ_{sol} of the electrolyte, it is almost impossible without labels to directly predict the bacterial concentration from a single measure since two samples with identical bacterial loads but different conductivities would result in different signals. Effective discrimination of the electrical footprint of the bacterial cells from the ionic background noise thus requires comparing the sensor signal to an appropriate control value.

Therefore, we performed differential measurements, comparing the signal obtained with samples of 10^8, 10^7 and 10^6 CFU mL^{-1} of *B. thuringiensis* to blank PBS:1000 measurements (Figure 6A). Significant differences with and without bacterial cells were observed between 10 kHz and 1 MHz for both impedance modulus and phase, where the membranes soaked with bacterial solutions show a lower value of impedance modulus than the blank reference PBS:1000 buffer. The measurements for bacterial cells and PBS:1000 follow the same typical decrease of impedance modulus than observed for low salinity solutions (Figure 5A). The phase peak of the membrane filled with 10^8 CFU mL^{-1} bacterial suspension (Figure 6B) is also subjected to a shift towards higher frequencies. The bacteria and PBS:1000 impedance modulus and phase curves show good fitting with the impedance modulus and phase of the NaCl solutions. In particular, PBS:1000 impedance curves lie in between the impedance curves of 2×10^{-4} M and 5×10^{-4} M NaCl solutions, while the impedance of the solution containing the 10^8 CFU mL^{-1} bacterial suspension lies between the 5×10^{-4} M and 10^{-3} M curves. Regarding the PBS:1000 buffer, the approximate correspondence of its dielectric properties to the range of 2–5×10^{-4} M NaCl solution is confirmed as the salt concentration used to model this buffer is precisely 1.6×10^{-4} M (Table 1). To deepen the comparison between the bacterial and saline solutions, we have extracted the membrane resistance R_{NC} and capacitance C_{NC} of the electrical model under the PBS:1000 and 10^8 CFU mL^{-1} bacteria conditions (see Table S2 for the data). R_{NC} and C_{NC} extracted for the PBS:1000 condition diverges of around 5% from the 5×10^{-4} M NaCl condition, against about 10% between the solution containing 10^8 CFU mL^{-1} bacteria and the 10^{-3} M NaCl model solution. For their part, the 10^7 and 10^6 CFU mL^{-1} bacterial suspensions show a significant drop in impedance modulus (Figure 6A, insert) relative to the PBS:1000 buffer. However, unlike the 10^8 CFU mL^{-1} suspensions, the shift in the phase peak is not significant for the 10^7 and 10^6 CFU mL^{-1} bacteria solutions (Figure 6B) and. therefore, cannot be used to assess the bacteria presence in the membrane. In addition, the impedance modulus and phases of 10^7 and 10^6 CFU mL^{-1} suspensions are overlapping over the whole spectrum. This suggests an intrinsic limit of detection of the parallel plate towards bacteria detection of about 10^7 CFU mL^{-1}.

Dielectric measurements were carried out to substantiate the bacterial detection results through impedance measurements (Figure 6C). Here again, the relative shift between PBS:1000 with and without bacteria follows a similar tendency to the model saline solutions. This tends to confirm the similarities in dielectric property variations between solutions with and without bacteria, and the differences in salt concentration. It is thus consistent to pose the hypothesis that bacteria are sensed through increase in ion concentration.

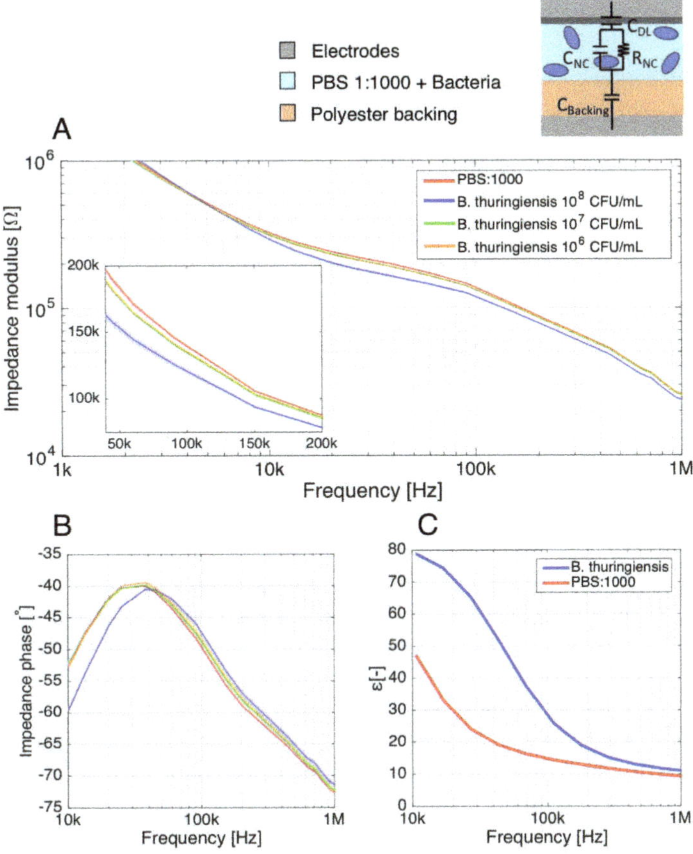

Figure 6. (**A**) Impedance measurements performed on a nitrocellulose (NC) membrane as seen by the parallel-plate setup. The NC membrane is saturated with phosphate-buffered saline (PBS):1000 (reference buffer) or with PBS:1000 containing 10^8, 10^7 and 10^6 CFU mL^{-1} *B. thuringiensis* cells. The global impedance is experimentally shown to decrease, and the phase is subjected to a shift in presence of bacterial cells in the buffer, showing high similarities with the response of slightly saline solutions. (**B**) The impedance phase evolution over 10 kHz–1 MHz, presenting a phase peak, indicates an interplay of R_{NC} and C_{NC} in the impedance of the system when subjected to bacteria. (**C**) Dielectric permittivity of the system as seen by the parallel-plate setup, when subjected to low-salt buffer with and without bacterial cells. Number of samples: 12 from two independent experiments. Shaded area surrounding the measurements curves: standard deviation (σ).

3.4. Comparison with Another System: B. thuringiensis Detection with the IDE Setup

In order to assess the detection results obtained with the plug-and-play parallel-plate system where the electrodes are deported, we considered to deposit metallic IDE directly on top of the nitrocellulose membrane to reach an integrated sensing device.

3.4.1. Gold IDE Deposited on Nitrocellulose Membranes

Au-IDE were successfully deposited on top of the NC membranes, showing good adherence with the support (Figure 7). The Au-deposited thin-film follows the porous microstructure of the membrane, and shows good conductivity.

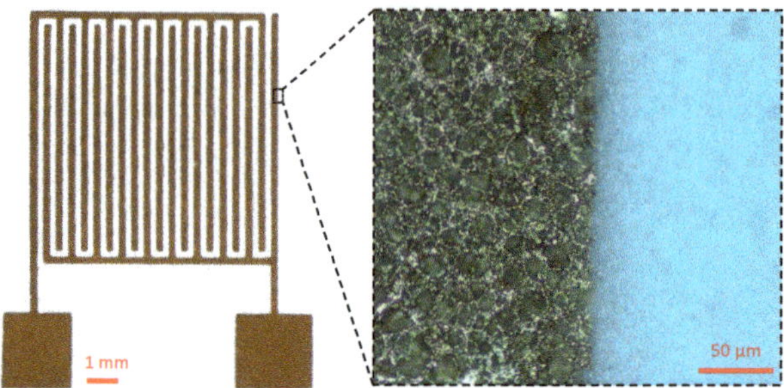

Figure 7. Optical microscopy images of Au-IDE (200 μm of interdigit gap) deposited on a nitrocellulose membrane. The inset is a zoom in image (magnification 20×) of the electrode showing the Au deposition on the membrane surface as well as in the first microns of the membrane thickness due to NC porosity.

3.4.2. Detection of *B. thuringiensis* with the IDE Setup

Interdigital electrodes, which are among the most commonly used periodic electrode structures for fringing field detection [32], were used for impedance measurements in order to substantiate the parallel-plate measurement results. An important advantage of this electrode design is that only a single-side access to the test material is required.

Figure 8A shows a lower impedance modulus due to bacterial presence, with a resistive angle. This expresses the fact that bacterial solutions are sensed through a decrease in the solution resistance resulting from an increase in conductivity. Given the single-side access to the material, the conductive phenomena of the nitrocellulose soaked with PBS:1000 are not shielded by the isolating backing layer, resulting in highly resistive impedance phase over the 1kHz–1 MHz range.

The sensitivity towards 10^8 CFU mL^{-1} *B. thuringiensis* cells was evaluated over the whole spectrum, and presents a plateau of >18% at 10 kHz–0.3 MHz, resulting essentially from the almost stable difference of the impedance modulus in this frequency range. Due to the very resistive nature of the phase over this range, the modulus remains quasi-independent of the frequency.

3.4.3. Sensitivities towards *B. thuringiensis* Cells

The sensitivities towards 10^8 CFU mL^{-1} *B. thuringiensis* cells in low salinity buffers were evaluated over the whole spectrum for both parallel-plate and IDE setups. The spectral sensitivities of the systems were computed from the complex impedance of the blank PBS:1000 buffer and the bacterial suspension (Figure 9).

Regarding the parallel-plate (Figure 9A), a maximal sensitivity of about 21% is observed around 30 kHz for the parallel-plate setup. The sensitivity computed from the complex impedance is superior to the modulus sensitivity (around 17% at 40 kHz) as it includes both contribution from the impedance modulus change and phase shift around this frequency (depending on the interplay between R_{NC} and C_{NC}). In this system, the sensitivity does not remain constant over the spectrum as C_{NC} has a non-negligible contribution to the detection.

Contrary to the parallel-plate, the sensitivity of the IDE setup presents a plateau of >18% at 10 kHz–0.3 MHz (Figure 9B), resulting essentially from the almost stable difference of the impedance modulus in this frequency range. Due to the very resistive nature of the phase over this range, the modulus remains quasi-independent of the frequency.

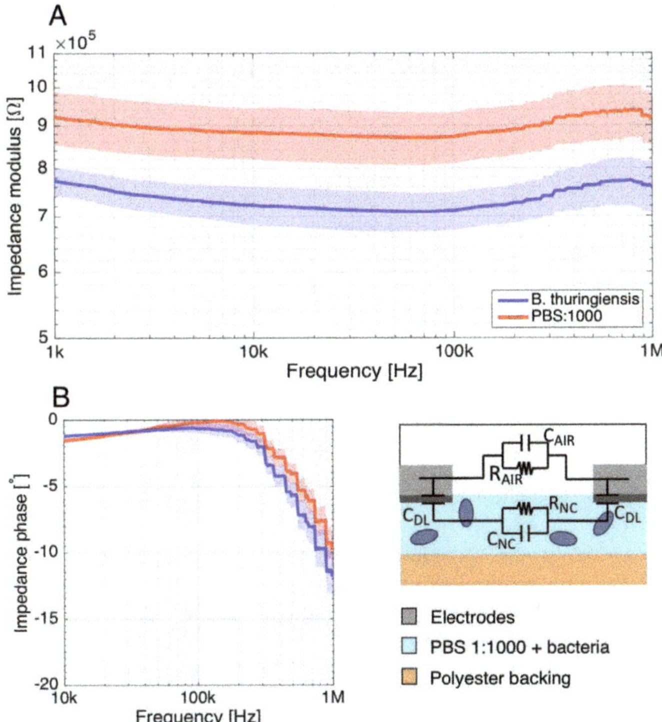

Figure 8. (**A**) Impedance measurements performed on nitrocellulose (NC) membrane as seen by the IDE, with and without bacterial cells in PBS:1000 buffers. The impedance modulus presents a constant impedance shift representing a decrease in the solution resistance. (**B**) Since a direct conduction path exists between their electrodes, the impedance seen by the IDE has a character that is rather resistive as expressed by the highly resistive phase over a large spectrum. Number of samples: 12 from two independent experiments. Shaded area surrounding the measurements curves: standard deviation (σ).

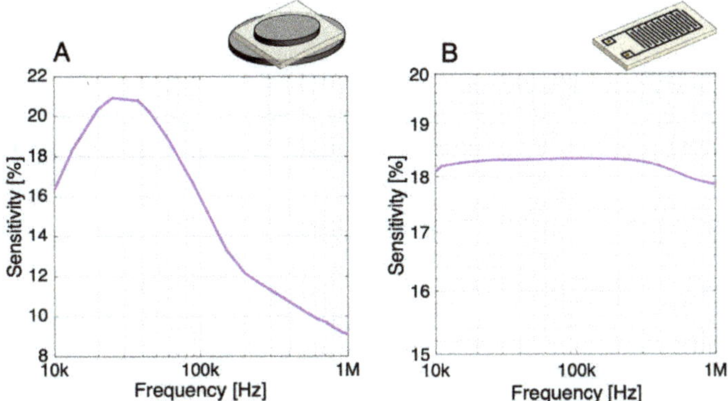

Figure 9. Spectral sensitivities of the (**A**) parallel-plate and (**B**) IDE setup towards 10^8 CFU mL^{-1} *B. thuringiensis* cells in low salinity buffers. The sensitivity includes both contribution from the modulus and the phase shift. The IDE sensitivity expresses their high response towards resistance changes through a plateau between 10 kHz and 0.3 MHz, while the parallel-plate sensitivity presents a peak because of the interplay between R_{NC} and C_{NC} in the sensing mechanism.

4. Discussion

The main objective of this work was to accommodate electrical sensing on paper substrates towards simple, rapid, quantitative bacteria detection in aqueous solutions.

To this end, we first developed a biointerface by functionalizing the NC membrane with endolysin cell-wall binding domain (CBD) aimed at capturing bacterial cells through the whole membrane pore surface. The efficiency of CBD specific binding to *B. thuringiensis* was demonstrated in [28]. In this paper, we validated their potential as immobilized bioreceptors on porous NC membranes, taking advantage of their highly selective binding capacity to capture *B. thuringiensis* bacterial cells present in low-salinity buffers. As the activity of the endolysins was shown to decrease at salt concentrations higher than 200 mM [22], it is necessary to assess the binding capacity of the CBD biointerface towards bacteria under different NaCl concentrations. This assessment is required to determine the range of aqueous solutions that can be considered for bacteria detection without loss of the sensor specificity. Also, the specificity of the detection systems towards *B. thuringiensis* is usually assessed with the absence of electrical signals in the absence of bioreceptors or in the presence of only non-target pathogens. Therefore, the detection specificity remains to be studied, and will be the topic of future works. When considering specificity of the detection, it will be of upmost importance to use a complete lateral flow assay with controlled flow as support to perform the electrical detection and a related robust detection protocol integrating the necessary washing steps.

Second, we took advantage of the NC support to develop an innovative volume-based electrical detection setup by applying the membrane between the electrodes of a parallel-plate prober for dielectric measurements. This setup forms a simple plug-and-play sensing device which is responsive to the electrical properties of the NC membrane. A nitrocellulose membrane functionalized with a specific biointerface is indeed an adequate substrate to support the volume-based electrical detection of bacterial suspensions. Such a detection scheme is attractive as the whole sample volume contributes to the sensing, increasing the contribution from targeted bacteria with respect to surface-based methods. The parallel-plate detection scheme based on impedance spectroscopy proposed in this work was much more straightforward both in fabrication and handling than most of the paper-based electrical biosensors encountered in the literature. They often rely on direct current monitoring requiring to apply more complex electrode systems to the paper (usually three electrodes, for reference, working and counter electrode) and involving electrochemical reactions which are relatively complex to interpret in ionic medium [40,41]. These paper-based sensors usually rely either on the application of conducting pastes showing high electrode and contact resistance (generally requiring custom, hand-made electrical connections) [41,42], or more complex and expansive fabrication processes from the microelectronics industry.

Third, we quantified the presence of 10^7 to 10^8 CFU mL^{-1} *B. thuringiensis* in diluted physiological buffer (PBS diluted 1000×). An important feature is the rapidity of the sensing mechanism: the whole detection protocol lasts less than 5 min. Using a simple electrical model including the electrical properties of the NC membrane and the capacitive contribution of the polyester backing, the interplay of both capacitive and resistive properties of the electrolyte are observed. The detection mechanism, based on ion concentration, was determined by the parallel-plate characterization of NC membrane filled with different saline solutions, and helped us understanding the sensor differential response for various electrolyte solutions. Similarities can be drawn in the complex impedance difference between the saline solutions and for biological buffer with and without bacterial cells. In particular, close correspondence is observed between electrical properties of PBS:1000 and ~2×10^{-4} M NaCl solution, while signals from the 10^8 CFU mL^{-1} bacterial suspensions are representative of those of the ~10^{-4} M condition. These results, supported by dielectric measurements, suggest that the electrical model proposed for the sensing of NC membranes soaked with ionic solutions can be extended to bacterial solutions. Furthermore, it also endorses the concept of bacteria detection through its surroundings ions, which has already been discussed in [25]: the differences in sensor response are attributed to the slight difference in ionic content, i.e., electrical conductivity, between sterile PBS:1000 and bacterial resuspension in PBS:1000. Indeed, centrifugation steps lead bacterial cells to release ions due to osmotic pressure and damaged cell walls [43].

Both for saline solutions and bacterial suspensions, an interesting phase shift occurs in the phase peaks between 10 and 1000 kHz, driven by changes in f_3 due to both C_{NC} and R_{NC} variations. Although the resistive character of the sensor is predominant (changes in R_{NC} affect the total module more than C_{NC}), the possibility to monitor NC membrane property changes through variation of C_{NC} renders this setup versatile in use and potentially more robust against ionic noise, strongly affecting R_{NC}. To assess the relative advantage of the parallel-plate setup over traditional paper-based sensors, we also applied gold microelectrodes (IDE) to the NC membrane using a standard microfabrication process. The IDE sensor is only reactive towards R_{NC}, and shows a high sensitivity to bacteria over a larger frequency range than the parallel-plate. Indeed, the electrode disposition does not prevent a direct conduction path between the two electrodes.

Despite its dependence on bacterial concentration, two reasons make the detection of bacteria through higher ionic content unsuitable for bacterial sensing. First, the resulting signal is strongly affected by experimental procedures such as manipulation, contamination and experimental conditions, all affecting the baseline sample conductivity. Second, such a sensing principle is useless for real applications dealing with detection in highly saline solutions, whose high electrical conductivity is hardly impacted by bacterial ion release. Thus, the detection of bacteria through ionic contribution has the main disadvantage that it lacks robustness in a complex environment with various living organisms that are potential ion-sources, resulting in a low signal-to-noise ratio (SNR). The specificity and sensitivity of the electrical detection can be improved by nanoparticles (NP) to specifically label whole bacterial cells. In [44–46], *E. coli* O157:H7 were detected through amplified conductance and permittivity changes by means of the conjugation of specific graphene or gold NPs to the bacteria. In other works, the highly intensive response of Si-NPs and Au-NPs conjugated to bioanalytes at radio frequencies (RF) was used to amplify the dielectric contrast of bioanalytes in solutions [47,48] or to act as microantenna in NC membranes [49].

The ability of the parallel-plate setup to monitor changes of both conductive and dielectric properties in the NC membrane makes it possible to select the type of NP (conductive or dielectric) that increases the SNR the most. Conjugating diverse types of NP with bacterial cells offers promising perspectives for highly specific electrical bacterial detection on lateral flow assays, and will be the focus of upcoming works.

5. Conclusions

In this paper, an innovative method for the electrical characterization of cellulose-based membranes was proposed towards the development of low-cost, quantitative paper-based biosensors. It consists in the use of a simple parallel-plate setup as sensor to perform impedance measurements on the membrane inserted within the electrodes. The sensing principle was studied and validated by detecting saline solutions of different molar concentrations spread within nitrocellulose membranes. We then demonstrated the proof-of-concept of the detection of bacteria: impedance-based detection of 10^8 CFU mL^{-1} of *B. thuringiensis* cells was presented without labeling nor signal enhancement strategies. The bacteria were detected through an overall increase in ions in the membrane caused by their presence. Impedance measurements were also performed with interdigital electrodes integrated on the membrane and confirmed the parallel-plates results. Finally, newly discovered endolysin CBD were introduced as specific bioreceptors and deposited inside the nitrocellulose membrane, enabling successful bacterial cell capture over the whole membrane volume.

In conclusion, by combining the benefits of a cellulose-based membrane, novel protein bioreceptors and precise impedance measurements with a reusable plug-and-play setup, we obtained promising results towards the development of an affordable and sensitive biosensor with a speed of response under 5 min. To overcome the limitations presented by our system and reach the very low limit of detection required for example for drinking water assessment, it is necessary to integrate signal enhancement strategies. Future research works will explore the use of nanoparticles as labels to increase the analytical performances of the device. The development of such a versatile tool creates novel opportunities in situations that require rapid and frequent pathogen detection, such as the detection of *E. coli* in drinking water. Sensing applications could be extended to the detection of various pathogens and viruses as well, through appropriate and direct modification of the biointerface. This may prove particularly useful in emergency situations in light of the recent coronavirus disease 2019 (COVID-19) pandemic.

Supplementary Materials: The following are available online at https://www.mdpi.com/article/10.3390/bios11020057/s1, Table S1: Results from the fitting of impedance data (Figure 5) to the simple electrical equivalent model used for the parallel-plate setup. Table S2: Results from the fitting of impedance data (Figure 6) to the simple electrical equivalent model used for the parallel-plate setup.

Author Contributions: Conceptualization, G.L.B., M.H. and J.-P.R.; methodology, G.L.B., M.H. and A.L.; software, G.L.B. and M.H.; validation, G.L.B. and M.H.; formal analysis, G.L.B. and M.H.; investigation, G.L.B. and M.H.; resources, J.-P.R., K.G. and J.M.; data curation, G.L.B. and M.H.; writing—original draft preparation, G.L.B. and M.H.; writing—review and editing, J.-P.R., K.G., A.L., J.M.; visualization, G.L.B.; supervision, J.-P.R., K.G., J.M.; project administration, J.-P.R., K.G., J.M.; funding acquisition, G.L.B., J.-P.R., K.G., J.M. All authors have read and agreed to the published version of the manuscript.

Funding: This research was funded by National Fund for Scientific Research (FRIA grant to G.L.B. and FNRS grant to J.M.) and the Research Department of the Communauté française de Belgique (Concerted Research Action, ARC no.17/22-084, grant to A.L.).

Institutional Review Board Statement: Not applicable.

Informed Consent Statement: Not applicable.

Data Availability Statement: The data presented in this study are available on request from the corresponding author.

Acknowledgments: We thank all the engineers and technicians of the UCL-WINFAB cleanrooms and WELCOME platform for IDE depositions and help in electrical measurements, respectively; W. Xu (IMCN, UCLouvain) for her help in biofunctionalization protocol development; M.-C. Eloy (LIBST, UCLouvain) for confocal experiments; and researchers from the MIAE lab for protein and bacteria productions. K.G. is a Senior Research Associate of F.R.S.-FNRS.

Conflicts of Interest: The authors declare no conflict of interest. The funders had no role in the design of the study; in the collection, analyses, or interpretation of data; in the writing of the manuscript; or in the decision to publish the results.

References

1. WHO Report. 2019. Available online: https://www.who.int/news-room/fact-sheets/detail/drinking-water (accessed on 25 September 2020).
2. Váradi, L. Methods for the detection and identification of pathogenic bacteria: Past, present, and future. *Chem. Soc. Rev.* **2017**, *46*, 4818–4832. [CrossRef] [PubMed]
3. Lazcka, O.; Del Campo, F.J.; Munoz, F.X. Pathogen detection: A perspective of traditional methods and biosensors. *Biosens. Bioelectron.* **2007**, *22*, 1205–1217. [CrossRef] [PubMed]
4. Yang, Z.; Mao, K.; Zhang, H. Can a paper-based device trace covid-19 sources with wastewater-based epidemiology. *Environ. Sci. Technol.* **2020**, *54*, 3733–3735.
5. Larsen, D.A.; Wigginton, K.R. Tracking COVID-19 with wastewater. *Nat. Biotechnol.* **2020**, *38*, 1151–1153. [CrossRef]
6. Parolo, C.; Merkoçi, A. Paper-based nanobiosensors for diagnostics. *Chem. Soc. Rev.* **2013**, *42*, 450. [CrossRef] [PubMed]
7. Martinez, A.W.; Phillips, S.T.; Carrilho, E.; Whitesides, G.M. Diagnostics for the Developing World: Microfluidic Paper-Based Analytical Devices. *Anal. Chem.* **2010**, *82*, 3–10. [CrossRef]
8. Busa, L.S.A.; Mohammadi, S.; Maeki, M.; Ishida, A.; Tani, H.; Tokeshi, M. Advances in Microfluidic Paper-Based Analytical Devices for Food and Water Analysis. *Micromachines* **2016**, *7*, 86. [CrossRef]
9. Merkoçi, A. Paper Based Sensors. *Compr. Anal. Chem.* **2020**, *89*, 2–464.

10. EMD Millipore. *Rapid Lateral Flow Test Strips-Considerations for Product Development*; EMD Millipore Corporation: Billerica, MA, USA, 2013; Volume 29, pp. 702–707.

11. Rajapaksha, R.D.A.A.; Afnan Uda, M.N.; Hashim, U.; Gopinath, S.C.B.; Fernando, C.A.N. Impedance based Aluminium Interdigitated Electrode (Al-IDE) Biosensor on Silicon Substrate for Salmonella Detection. In Proceedings of the 2018 IEEE International Conference on Semiconductor Electronics (ICSE), Kuala Lumpur, Malaysia, 15–17 August 2018.

12. Thivina, V.; Hashim, U.; Arshad, M.K.M.; Ruslinda, A.R.; Ayoib, A.; Nordin, N.K.S. Design and fabrication of Interdigitated Electrode (IDE) for detection of Ganoderma boninense. In Proceedings of the 2016 IEEE International Conference on Semiconductor Electronics (ICSE), Kuala Lumpur, Malaysia, 17–19 August 2016.

13. Bollella, P.; Katz, E. Enzyme-Based Biosensors: Tackling Electron Transfer Issues. *Sensors* **2020**, *20*, 3517. [CrossRef] [PubMed]

14. Van Overstraeten-Schlögel, N.; Lefèvre, O.; Couniot, N.; Flandre, D. Assessment of different functionalization methods for grafting a protein to an alumina-covered biosensor. *Biofabrication* **2014**, *6*, 3. [CrossRef] [PubMed]

15. Amini, K.; Ebralidze, I.I.; Chanb, N.W.C.; Kraatz, H.-B. Characterization of TLR4/MD-2-modified Au sensor surfaces towards the detection of molecular signatures of bacteria. *Anal. Methods* **2016**, *8*, 7623–7631. [CrossRef]

16. Cimafonte, M.; Fulgione, A.; Gaglione, R.; Papaianni, M.; Capparelli, R.; Arciello, A.; Bolletti Censi, S.; Borriello, G.; Velotta, R.; Della Ventura, B. Screen Printed Based Impedimetric Immunosensor for Rapid Detection of Escherichia coli in Drinking Water. *Sensors* **2020**, *20*, 274. [CrossRef] [PubMed]

17. Vericat, C.; Vela, M.E.; Benitez, G.; Carrob, P.; Salvarezza, R.C. Self-assembled monolayers of thiols and dithiols on gold: New challenges for a well-known system. *Chem. Soc. Rev.* **2010**, *39*, 1805–1834. [CrossRef]

18. Nie, Z.; Nijhuis, C.A.; Gong, J.; Chen, X.; Kumachev, A.; Martinez, A.W.; Narovlyansky, M.N.; Whitesides, G.M. Electrochemical sensing in paper-based microfluidic devices. *Lab Chip* **2010**, *10*, 477–483. [CrossRef] [PubMed]

19. Kong, M.; Sim, J.; Kang, T.; Nguyen, H.H.; Park, H.K.; Chung, B.H.; Ryu, S. A novel and highly specific phage endolysin cell wall binding domain for detection of *Bacillus cereus*. *Eur. Biophys. J.* **2015**, *44*, 437–446. [CrossRef] [PubMed]

20. Bai, J.; Kim, Y.-T.; Ryu, S.; Lee, J.-H. Biocontrol and Rapid Detection of Food-Borne Pathogens Using Bacteriophages and Endolysins. *Front. Microbiol.* **2016**, *7*, 474. [CrossRef]

21. Jenkins, G.; Wang, Y.; Xie, Y.L.; Wu, Q.; Huang, W.; Wang, L.; Yang, X. Printed electronics integrated with paper-based microfluidics: New methodologies for next-generation health care. *Microfluid. Nanofluid* **2015**, *19*, 251–261. [CrossRef]

22. Tobjörk, D.; Österbacka, R. Paper Electronics. *Adv. Mater.* **2011**, *23*, 1935–1961. [CrossRef] [PubMed]

23. Smith, S.; Land, K.; Joubert, T.-H. Printed Functionality for Point-of-Need Diagnostics in Resource-Limited Settings. In Proceedings of the 20th International Conference on Nanotechnology (IEEE-NANO), Montreal, QC, Canada, 29–31 July 2020. Virtual Conference.

24. Joubert, T.-H.; Bezuidenhout, P.H.; Chen, H.; Smith, S.; Land, K.J. Inkjet-printed Silver Tracks on Different Paper Substrates. *Mater. Today Proc.* **2015**, *2*, 3891–3900. [CrossRef]

25. Couniot, N.; Vanzieleghem, T.; Rasson, J.; Van Oversrtraeten-Schlögel, N.; Poncelet, O.; Mahillon, J.; Francis, L.A.; Flandre, D. Lytic enzymes as selectivity means for label-free, microfluidic and impedimetric detection of whole-cell bacteria using ALD-Al2O3 passivated microelectrodes. *Biosens. Bioelec.* **2015**, *67*, 154–161. [CrossRef]

26. Pal, S.; Alocilja, E.C.; Downes, F.P. Nanowire labeled direct-charge transfer biosensor for detecting Bacillus species. *Biosens. Bioelec.* **2007**, *22*, 2329–2336. [CrossRef]

27. Luo, Y.; Nartker, S.; Wiederoder, M.; Miller, H.; Hochhalter, D.; Drzal, L.T.; Alocilja, E.C. Novel Biosensor Based on Electrospun Nanofiber and Magnetic Nanoparticles for the Detection of E. coli O157:H7. *IEEE Trans. Nanotechnol.* **2012**, *11*, 676–681. [CrossRef]

28. Leprince, A.; Nuytten, M.; Gillis, A.; Mahillon, J. Characterization of PlyB221 and PlyP32, Two Novel Endolysins Encoded by Phages Preying on the Bacillus cereus Group. *Viruses* **2020**, *12*, 1052. [CrossRef] [PubMed]

29. Le Brun, G.; Hauwaert, M.; Leprince, A.; Glinel, K.; Mahillon, J.; Raskin, J.-P. Electrochemical Characterization of Nitrocellulose Membranes towards Bacterial Detection in Water. *Proceedings* **2020**, *60*, 61.

30. Chuang, C.H.; Shaikh, M. Label-free impedance biosensors for Point-of-Care diagnostics. *Point Care Diagn. New Prog. Perspect.* **2017**, *3*, 171–201.

31. Liang, T.; Zou, X.; Mazzeo, A.D. A Flexible Future for Paper-based Electronics. *Proc. SPIE* **2016**, *9836*, 98361D.

32. Mamishev, A.V.; Sundara-Rajan, K.; Yang, F.; Du, Y.; Zahn, M. Interdigital Sensors and Transducers. *Proc. IEEE* **2004**, *92*, 808–845. [CrossRef]

33. Igreja, R.; Dias, C.J. Analytical evaluation of the interdigital electrodes capacitance for a multi-layered structure. *Sens. Actuators A* **2004**, *112*, 291–301. [CrossRef]

34. ITU-R. *Electrical Characteristics of the Surface of the Earth*; Recommendation ITU-R: Geneva, Switzerland, 2017.

35. Vlaamse Milieumaatschappij. *Kwaliteit Van Het Drinkwater*; Vlaamse Milieumaatschappij: Aalst, Belgium, 2015.

36. Le Brun, G.; Raskin, J.-P. Material and manufacturing process selection for electronics eco-design: Case study on paper-based water quality sensors. *Procedia CIRP* **2020**, *90*, 344–349. [CrossRef]

37. Lagadic, L.; Caquet, T. Bacillus thuringiensis. In *Encyclopedia of Toxicology*, 3rd ed.; Academic Press: Cambridge, MA, USA, 2014; pp. 355–359. ISBN 9780123864550.

38. Dorken, G.; Ferguson, G.P.; French, C.E.; Poon, W.C.K. Aggregation by depletion attraction in cultures of bacteria producing exopolysaccharide. *J. R. Soc. Interface* **2012**, *9*, 3490–3502. [CrossRef] [PubMed]

39. Agilent Technologies. *Agilent E4991A RF Impedance/Material Analyzer: Installation and Quick Start Guide*, 10th ed.; Agilent Technologies: Santa Clara, CA, USA, 2012.
40. Kumar, S.; Nehra, M.; Mehta, J.; Dilbaghi, N.; Marrazza, G.; Kaushik, A. Point-of-Care Strategies for Detection of Waterborne Pathogens. *Sensors* **2019**, *19*, 4476. [CrossRef] [PubMed]
41. da Costa, T.H.; Song, E.; Tortorich, R.P.; Choi, J.-W. A Paper-Based Electrochemical Sensor Using Inkjet-Printed Carbon Nanotube Electrodes. *ECS J. Solid State Sci. Technol.* **2015**, *4*, 3044–3047. [CrossRef]
42. Lombardi, J.; Poliks, M.D.; Zhao, W.; Yan, S.; Kang, N.; Li, J.; Luo, J.; Zhong, C.-J.; Pan, Z.; Almihdhar, M.; et al. Nanoparticle Based Printed Sensors on Paper for Detecting Chemical Species. In Proceedings of the IEEE 67th Electronic Components and Technology Conference, Orlando, FL, USA, 30 May–2 June 2017.
43. Peterson, B.W.; Sharma, P.K.; van der Mei, H.C.; Bussche, H.J. Bacterial Cell Surface Damage Due to Centrifugal Compaction. *Appl. Environ. Microbiol.* **2012**, *78*, 120–125. [CrossRef] [PubMed]
44. Yao, L.; Wang, L.; Huang, F.; Cai, G.; Xi, X.; Lin, J. A microfluidic impedance biosensor based on immunomagnetic separation and urease catalysis for continuous-flow detection of *E. coli* O157:H7. *Sens. Actuators B* **2018**, *259*, 1013–1021. [CrossRef]
45. Hassan, A.-R.H.A.-A.; de la Escosura-Muñiz, A.; Merkoçi, A. Highly sensitive and rapid determination of Escherichia coli O157:H7 in minced beef and water using electrocatalytic gold nanoparticle tags. *Biosens. Bioelectron.* **2015**, *67*, 511–515. [CrossRef]
46. Pandey, A.; Gurbuz, Y.; Ozguz, V.; Niazi, J.H.; Qureshi, A. Graphene-interfaced electrical biosensor for label-free and sensitive detection of foodborne pathogenic *E. coli* O157:H7. *Biosens. Bioelectron.* **2017**, *91*, 225–231. [CrossRef]
47. Moreno-Hagelsieb, L.; Foultier, B.; Laurent, G.; Pampin, R.; Remacle, J.; Raskin, J.-P.; Flandre, D. Electrical detection of DNA hybridization: Three extraction techniques based on interdigitated Al/Al2O3 capacitors. *Biosens. Bioelectron.* **2007**, *22*, 2199–2207. [CrossRef] [PubMed]
48. Chien, J.H.; Kuo, L.S. Protein detection using a radio frequency biosensor with amplified gold nanoparticles. *Appl. Phy. Lett.* **2007**, *91*, 143901. [CrossRef]
49. Yuan, M.; Alocilja, E.C. A Novel Biosensor Based on Silver-Enhanced Self-Assembled Radio-Frequency Antennas. *IEEE Sens. J.* **2014**, *14*, 4. [CrossRef]

Article

Non-Invasive Determination of Glucose Concentration Using a Near-Field Sensor

Aleksandr Gorst, Kseniya Zavyalova * and Aleksandr Mironchev

Radiophysics Faculty, Tomsk State University, 634050 Tomsk, Russia; gorst93@gmail.com (A.G.);
mironchev42@mail.ru (A.M.)
* Correspondence: ksu.b@mail.ru

Abstract: The article presents a model of a near-field sensor for non-invasive glucose monitoring. The sensor has a specific design and forms a rather extended near-field. Due to this, the high penetration of electromagnetic waves into highly absorbing media (biologic media) is achieved. It represents a combined slot antenna based on a flexible RO3003 substrate. Moreover, it is small and rather flat (25 mm in diameter, 0.76 mm thick). These circumstances are the distinguishing features of this sensor in comparison with microwave sensors of other designs. The article presents the results of numerical modeling and experimental verification of a near-field sensor. Furthermore, a phantom of human biological media (human hand) was created for experimentation. In the case of numerical modeling, the sensor is located close to the hand model. In a full-scale experiment, it is located close to the phantom of the human hand for the maximum interaction of the near-field with biological materials. As a result of a series of measurements for this sensor, the reflection coefficient is measured, and the dependences of the reflected signal on the frequency are plotted. According to the results of the experiments carried out, there is a clear difference in glucose concentrations. At the same time, the accuracy of determining the difference in glucose concentrations is high. The values of the amplitude of the reflected signal with a change in concentration differ by 0.5–0.8 dB. This sensor can be used for developing a non-invasive blood glucose measurement procedure.

Keywords: sensor; combined slot antenna; diabetes; dielectric permeability; electromagnetic fields; glucose concentration; near-field sensor; non-invasive measurements; microwave sounding

Citation: Gorst, A.; Zavyalova, K.; Mironchev, A. Non-Invasive Determination of Glucose Concentration Using a Near-Field Sensor. *Biosensors* **2021**, *11*, 62. https://doi.org/10.3390/bios11030062

Received: 31 January 2021
Accepted: 23 February 2021
Published: 26 February 2021

Publisher's Note: MDPI stays neutral with regard to jurisdictional claims in published maps and institutional affiliations.

1. Introduction

The number of people with diabetes is growing every year. Back in 2002, Jones M. and Harrison J. M. described in their article [1] that the World Health Organization (WHO) expects an increase in the number of people with diabetes to 300 million by 2025. According to a survey conducted by the International Diabetes Federation (IDF) already in 2013, three-hundred eighty-two million patients worldwide had diabetes. Seven years later, in 2020, the number of people living with diabetes was 463 million. Diabetes mellitus is a serious disease that can lead to many complications. There are several types of the disease. The main types are type 1 and type 2 diabetes. They are the most prevalent. In both of these cases, a person with diabetes needs to control blood sugar levels to avoid the complications of the disease. Self-testing devices require small blood sampling using different needles, which cause pain and discomfort to the user of the device. Today, a large number of methods and devices are being developed for determining glucose levels using non-invasive and continuous methods. This approach will allow people with diabetes to avoid the discomfort of measuring blood sugar and monitor it throughout the day. The most well-known non-invasive methods for measuring blood glucose are Raman spectroscopy, impedance spectroscopy, near-infrared spectroscopy, photoacoustic spectroscopy, and others.

Raman spectroscopy [2,3] is based on the measurement of scattered light. The disadvantages of this method are the instability of the intensity and wavelength of laser beams during probing, a long time for obtaining data, as well as errors associated with chemical substances in the tissues. Impedance spectroscopy is based on the measurement of resistance when the radiation frequency changes. To measure the glucose level, several sensors located in the area of the veins in human hands are needed [4]. Near-infrared spectroscopy is based on the transmission of near-infrared radiation through a vascular region of the body (finger, earlobe, etc.). In this case, the glucose concentration is calculated on the basis of the received spectral information [5,6]. All measurements in near-infrared spectroscopy are based on the transmission of light through or into the sample and the measurement of the intensity (transmitted or reflected) of the beam. Spectrometers for measurements in near-infrared spectroscopy have a suitable light source (such as a highly stable quartz tungsten lamp), a monochromator or interferometer, and a detector. Conventional monochromators are acousto-optic tunable filters, diffraction gratings, or prisms. Mid-infrared spectroscopy is based on the absorption of light by glucose molecules [7–9]. This method uses a beam of light to travel through a crystal in contact with the skin. Thus, the electromagnetic field generated by the reflected light reaches the dermis (the layer of skin that contains the most glucose). The disadvantage of this method is the dependence of the obtained data on the water content in the dermis. Therefore, the reliability of the results largely depends on the degree of hydration. There is a technology based on ultrasonic sensing: photoacoustic spectroscopy [10,11]. This method is based on the acoustic response of a liquid using laser light. This method is similar to mid-infrared spectroscopy. There are also other less known methods for determining the level of glucose in the blood [12,13].

With the existing variety of physical methods used for the non-contact determination of blood glucose concentration, the problem of creating non-invasive glucometers has not yet been solved. At the same time, conventional invasive glucometers have been actively improving over the past twenty years: chemical analysis has been replaced by electrochemistry; the devices are equipped with internal memory and other conveniences; their use has become easier, but still painful. It should be noted that invasive analysis is a direct method; in this case, a sample (blood drop) containing glucose is directly examined. While analyses of non-invasive glucometers are indirect methods, they are based on data obtained, as a rule, by spectral means. In some cases (IR spectroscopy), an attempt is made to quantitatively analyze blood glucose without removing a sample from the body; in others (for example, electrical and thermal characteristics), the study of factors associated with glucose levels is carried out in a very complex and ambiguous way. In any situation, the influence of surface tissue structures, the individual characteristics of the skin and the composition of the intercellular fluid, the functional complexity of blood components, and many difficult parameters of the external and internal environment are very great. The glucose percentage is a continuously changing parameter.

At the same time, in recent years, many laboratories and companies have been developing non-invasive glucometers, and some of the ready-made solutions have even received certification. In 2014, French researchers Chretiennot T., Dubuc D., and Grenier K.proposed a resonant microwave biosensor, which, according to the authors, achieved the accuracy of chemical glucometry methods. They published the paper [14]. However, this method cannot be rightfully considered non-invasive: it still requires the extraction of physiological fluids for analysis. A glucometer in the form of a patch (sugarBEAT) was created by the British company Nemaura Medical and certified in Europe. The sugarBEAT patch is only 1mm thick. It measures the glucose level in the tissue fluid of the upper layer of the skin (in sweat) and transmits the measurement results every 5 min via a digital interface. It is worth noting that sugarBEAT does not relieve the traditional finger puncture procedure, but it only needs to be done once to calibrate the meter before attaching it. The term of its operation can be up to two years. The American company M10, together with Seoul National University, created a prototype of a wearable device. It can not only measure the level of glucose in the blood, but also administer the required dose of insulin. The small patch contains sensors for glucose, temperature, humidity, and pH (sweat particles are used for measuring), as well as microneedles for injecting the drug and a heater (with its help, the needles are activated).

The SugarSenz Velcro glucometer of the American company Glucovation is attached to the stomach and constantly measures the glucose level with the possibility of digital data transmission. However, it also pierces the skin, not like conventional glucometers, but to the subcutaneous layer, in which the glucose level can also be measured. Google X lab worked on the creation of contact lenses, which, using a special sensor, could measure glucose levels not in blood, but in tears. However, glucose in the extracellular fluid is different from blood glucose. On average, the physicochemical reaction between glucose and the electrode surface takes three to 20 min. During this time, glucose passes from the blood into the intercellular fluid, and only then, the received signal undergoes algorithmic processing. Therefore, such devices are suitable for detecting trends in sugar levels, but are not suitable for the instant and accurate analysis required in real conditions. GlucoTrack DF-F is a non-invasive blood glucose meter of the Israeli company Integrity Applications (received a certificate from the European Commission). It uses three ways to measure blood sugar: ultrasonic, electromagnetic, and thermal. All of these measurements are taken using a miniature clip-on transducer that clings to the ear. With three measurements, the meter can give a more accurate reading. The disadvantages are the need for regular device calibration and the lack of accuracy.

Scientists at the Swiss Institute in Lausanne (EPFL) have created a miniature implant that can measure various parameters of a person's blood, such as glucose or cholesterol levels. The sensor is installed under the skin, and above it, a small unit with a battery and a wireless transmitter is attached to the skin. C8 MediSensors (a non-invasive blood glucose meter of the American company C8 MediSensors (European certificate)) is based on the optical principle and does not require a blood sample. The meter sensor is attached to the skin in the abdomen and transmits measurements digitally. The measurement uses Raman spectroscopy technology.

GlucoWatch is a glucometric watch developed by Cygnus Inc (Petoskey, MI, USA). The sensing element is a sensor in contact with the skin. Using a weak electrical current, the sensors pull glucose out of the skin cells. The glucose entered into the sensor is measured and converted into blood glucose, and the result is displayed on the screen and, together with the date and time of the analysis, is recorded in the device's memory. Measurements are taken every 10 min for 13 h. The tests revealed the following: (1) the device is inconvenient in operation (the manual is a book of 112 pages) and expensive to maintain; its use requires a complex individual calibration; it is not suitable for some types of skin; the analysis area on the hand must change every time; (2) the data obtained with its help are less accurate than measurements on a conventional glucometer; in some cases, the error can reach 25–30%; (3) the change in the level of glucose in the cell fluid occurs with a delay in relation to the change in blood glucose, and as a result, GlucoWatch lags behind the true state of affairs by ten or more minutes.

Omelon V-2 is a technology developed in Russia. The principle of its action is based on the fact that muscle and vascular tone are dependent on glucose levels. The device measures the pulse wave, vascular tone, and blood pressure several times, based on which it calculates the sugar level. The high percentage of coincidences of the calculated indicators with laboratory data allowed this blood glucose monitor to be launched into mass production. However, the device has dimensions of $155 \times 100 \times 45$ cm, which does not allow carrying it in a pocket, and the correctness of the readings depends on the observance of the rules for measuring pressure: the correspondence of the cuff to the girth of the arm, the patient's calmness, and the absence of movement during the operation of the device.

The Israeli firm Integrity Applications solved the problem of painless, fast, and accurate blood sugar measurement by combining ultrasonic, thermal, and electromagnetic technologies in the GlucoTrack DF-F glucometer model. The GlucoTrack Model DF-F is intended for use by adults (over 18 years of age) with type 2 diabetes. The GlucoTrack Model DF-F is an expensive monitoring device and cannot be used for diagnostics.

The Symphony tCGM System, developed by Echo Therapeutics (Franklin, MA, USA), measures sugar levels transdermally. However, for the correct installation of the sensor and its accurate operation, it is necessary to pre-treat the skin with a special device, the Prelude SkinPrep System; it performs a superficial peeling of the skin area on which the study will be carried out by improving the electrical conductivity of the skin. After preparation, a sensor is attached to the skin area, and after a while, the device displays data, including indicators of glycemia and the percentage of body fat. This information can also be transferred to a smartphone. The accuracy of the device reaches 95%, which is slightly lower than that of standard invasive glucometers. This technology is under development.

Japanese startup Quantum Operation unveiled at CES 2021 a smartwatch-style device that is supposedly capable of accurately measuring blood glucose without puncturing the skin. The principle of action is based on spectroscopy (Raman spectroscopy). Judging by the information from the sources, the watch does not constantly monitor the glucose level, but does it after activating the corresponding function, and one must wait for about 20 s for the result. This is less effective than real-time monitoring, which is provided, for example, by the well-known partially invasive FreeStyle Libre device.

The large number of technologies presented to date for non-invasive glucometers shows the extreme popularity of the research area (only a few of them are noted above). However, despite the variety of methods being developed, a device that meets all the necessary requirements such as complete non-invasiveness, the accuracy of readings within the permissible error (laboratory tests are taken as basic ones, and it is believed that the readings of the glucometer should not differ from them by more than 10–15%), ease of use, lack of consumables, price, etc., has not yet been created. Thus, the development of a non-invasive method for the determination of glucose remains an urgent task today. This article is devoted to the development of a sensor based on near-field microscopy. The near-field method provides deep penetration of electromagnetic waves even in a highly absorbing environment, which, in particular, is human biological tissue.

The work is organized as follows. At the beginning of the work, a model of the sample of the biological medium in the form of a human hand is calculated, and the dependences of the real part of the dielectric permittivity of various concentrations of glucose in physiological saline on the frequency are plotted based on the experiment. This is done to bring numerical simulations closer to real conditions. Next, a numerical simulation of the sensor is carried out, based on which a real model of the sensor is created. Furthermore, a flat-layered phantom of a human hand is created for the experiments. Then, the analysis and results of numerical and field experiments of the proposed sensor design are presented.

2. Materials and Methods

2.1. Hand Model

To develop a new sensor for non-invasive diagnostics of glucose levels, it is necessary to take into account several factors that can influence the measured value of the reflected electromagnetic signal. It should also be borne in mind that the concentration of glucose in different tissues of a human is distributed differently; the thickness of tissue layers of different people is different; and these are previously unknown values for non-invasive measurements. When developing non-invasive methods for measuring glucose by electromagnetic methods, first of all, it is necessary to consider the effect of skin layers.

This is important because skin, like blood, has a high dielectric permittivity. Therefore, the electromagnetic wave attenuates in it more strongly than in other layers. Each layer of the skin has its dielectric characteristics, which may differ from human to human due to differences in the morphology and thickness of skin layers, the concentration of tissue/blood components (such as glucose), cutaneous blood perfusion, etc.

The skin has a great influence on glucose measurement; it can vary in the range of 2–10 mm when looking at the whole human body. The skin has a multilayer structure and includes stratum corneum; its thickness ranges from 15 to 150 microns in various parts of the body; the epidermis, the upper outer layer of human skin, which includes about 15–35% interstitial fluid, without blood vessels; the dermis, the skin itself, is a connective tissue and contains arterioles, venules, capillaries, and about 40% interstitial fluid. Furthermore, the dermis includes subcutaneous fatty tissue and loose fibrous connective tissue. The distribution and thickness of the skin depend on heredity, sex hormones, and human living conditions. On average, the thicknesses vary in the following ranges: the epidermis thickness is 0.068–0.146 mm; the stratum corneum of the epidermis is 0.021–0.049 mm; dermis is 1.89–3.04 mm; the subcutaneous fat is 0.03–1.41 mm. The next layer is subcutaneous fat, which has a wide range of thicknesses. The thickness of the fat layer on the forearm is the smallest compared to other parts of the body. Before considering muscle fibers, it is required to consider the saphenous veins passing through the human forearm.

To measure glucose with a standard glucometer, blood is drawn from the subcutaneous capillary vessels. In a medical examination, blood is drawn from a vein. The fundamental difference between these two methods is measurement accuracy. It should be understood that in medicine, the level of glucose (glycated hemoglobin) is determined in practice by venous blood. The content of these substances in venous and arterial blood is somewhat different. For a more accurate calculation of the concentration of glucose, the venous blood was considered in our study. A vein has a multi-layered structure. Depending on the type of vessels, they have different thicknesses, densities, and permeabilities. Large vessels additionally contain small blood and lymphatic capillaries.

Due to different pressures during the entire period of a person's life, the blood supply (blood volume in the measurement area) changes; along with this, the diameter of the vessels also changes. People with symptoms of tachycardia (increased heart rate) have smaller vessels because a rapid heartbeat decreases the efficiency of the heart since the ventricles do not have time to fill with blood. As a result, blood pressure decreases, and blood flow to organs decreases; hence, the area for the microwave signal response is smaller.

Another factor that can affect the dielectric properties of blood is blood hematocrit. Blood hematocrit refers to the percentage of red blood cells in the blood. Differences in the size, morphology, and distribution of red blood cells in human blood lead to changes in its dielectric properties, regardless of glucose concentration, and thereby affect the accuracy of glucose determination using measurement methods based on dielectric properties. Normally, this indicator is 40–48% for men and 36–42% for women. At low frequencies (about 3 MHz), the spread in the dielectric permittivity is quite high in the region of 30% with a change in hematocrit of 5% [15]. This scatter is associated with a rapid change in blood dielectric permittivity from frequency, which is clearly shown in Figure 1. With increasing frequency, this effect disappears due to the linear behavior of the dependence of the real part of the dielectric permittivity. In this case, the change will vary by 0.01–0.02%.

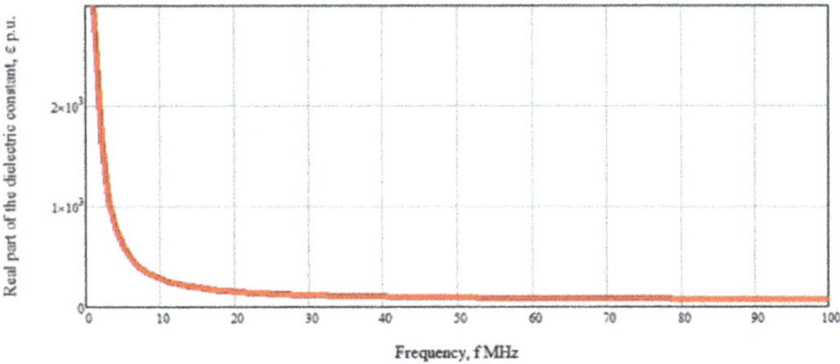

Figure 1. Dependence of the real part of the dielectric permittivity on frequency.

Based on the results of the literature review and the analysis of the studied information, the summary Table 1 was formed. It shows the thicknesses of all materials used for modeling.

Table 1. Thickness of the materials used in modeling.

Name of Materials	Thickness, mm
Stratum corneum of the epidermis	0.02
Epidermis	0.04
Dermis	1.83
Subcutaneous adipose tissue	1
Hand vein	4
Hand vein wall	0.5
Fat	6

The values for the thicknesses of the muscles and bones of the human forearm are not given in the table, since when the near-field penetrates these tissues, a strong signal attenuation occurs, and the response from them will be at the noise level. Therefore, these layers in the calculation of the thicknesses of the materials used in the simulation can be neglected. Based on the given data, we created the model of a biological medium sample in the form of a human hand (Figure 2).

The model has a flat-layered structure. All previously described materials were modeled, and a 13 mm thick muscle layer was added. We used a flat structure because the used antenna substrate is flexible. In the future, when creating a real prototype of the sensor, this will make it possible to attach it directly to the area of human skin.

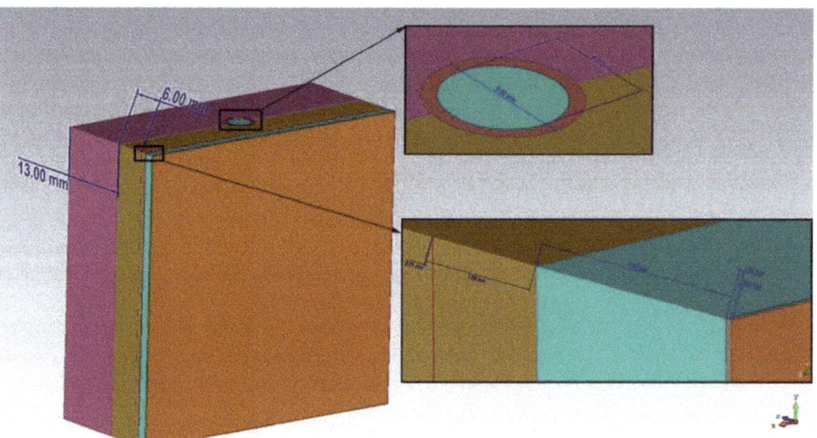

Figure 2. Model of the sample of the biological medium in the form of a human hand.

In the hand blood model, we used the data of the dielectric permittivity of saline with the following glucose concentrations: 0, 1, 3, 4, 5, 7, 9, and 10 mmol/L (Figure 3).

The data were analyzed using a PNA-L Network Analyzer (N5230C) and Dielectric Probe Kit-85070E Slim Form from Agilent Technologies. In subsequent simulations for blood with varying glucose concentrations, we used data obtained experimentally for physical solutions (saline) with different dextrose concentrations. Thus, it was possible to bring numerical modeling closer to real conditions.

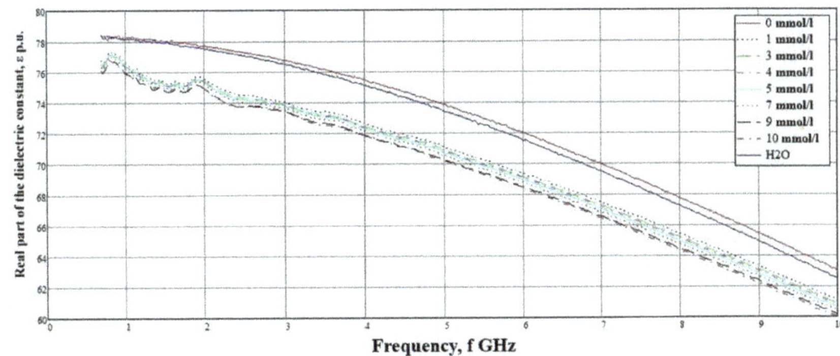

Figure 3. Dependence of the real part of the dielectric permittivity of solutions on frequency.

2.2. Sensor Design

The development of the near-field sensor was based on a combined slot antenna. The brown color in Figure 4 indicates the ideal conductor and green the flexible dielectric RO3003 with relative permittivity $\varepsilon = 3$. The shape of the sensor resembles a coin in the center of which there is a metalized circle with a slot in it. Then, using two rectangular metalized sections, the circle is connected to a conductive frame (Figure 4a). The backside of the sensor is a dielectric layer with a microstrip line close to a central circle and a small cut in the dielectric up to the conductive frame for connecting to the supply element (Figure 4b). The probing near-field is formed on the side with a slot (Figure 4a). Due to the presence of several radiating elements (slit and frame), the formation of the reactive parts of a specific interference energy flow takes place. These features can increase the sensor's sensitivity for near-field diagnostics of biological media and objects.

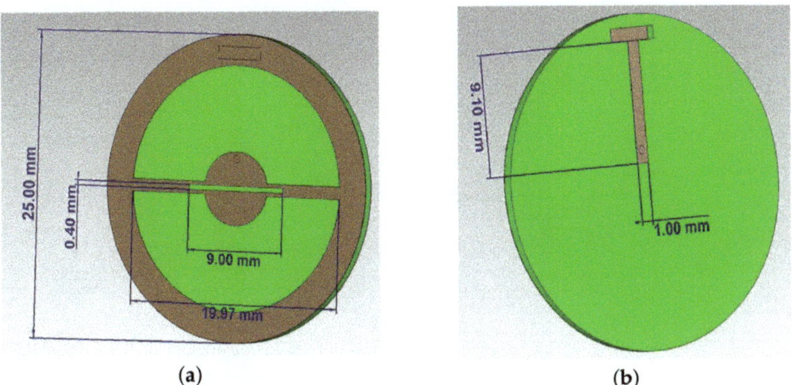

(a)   (b)

Figure 4. Near-field sensor model. Front (**a**) and back (**b**) views.

The VSWR of the sensor is also considered in terms of coherence with a sample of a biological medium in the form of a model of a human hand (Figure 5). The graph shows two sections of the curve with the highest matching of the sensor and the sample under study: the first at a frequency of 1 GHz (VSWR = 1.4) and the second in the range of 2.1–5 GHz (VSWR = 1.4). Note that almost the entire VSWR graph is below Level 3.

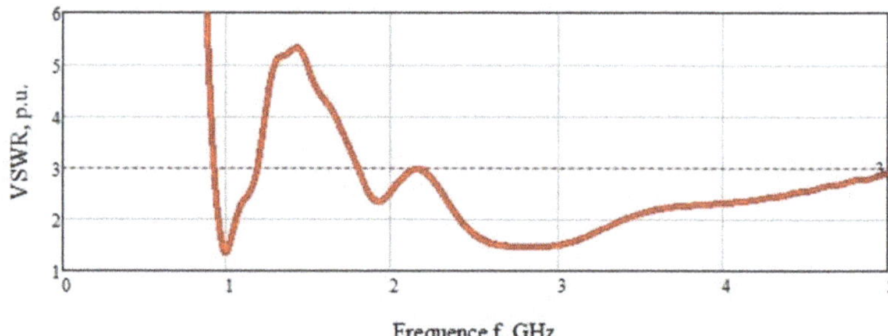

Figure 5. VSWRsensor coherence with the sample of the biological medium.

In the numerical simulation of the experiment to determine the concentration of glucose, the sensor was closely applied to a sample of a biological medium in the form of a human hand. The measurements were carried out in the frequency range of 0.5–5 GHz. We used this range to cover the entire frequency range in which the VSWR of the sensor is less than Level 3.

3. Experiments

3.1. Creation of a Sensor and a Phantom of Human Biological Tissues

The sensor production was based on a calculated model. We used the same materials as in the simulation: a Rogers3003 substrate was use;, the sensor was made with dimensions of 25 mm in diameter; and a 50 Ohm port was soldered to it. A photograph of the sensor is shown in Figure 6.

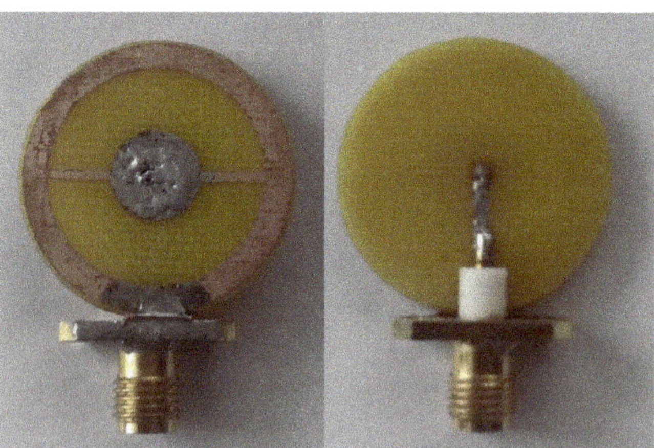

Figure 6. Photo of a near-field sensor.

The production of the biological phantom was based on graphite, polyurethane, and acetone [16]. Such a structure is strong enough for creating thin materials such as stratum corneum, epidermis, and the capillary layer. The listed layers are the thinnest. The calculated data on the components for the phantom layers are presented in Table 2.

Table 2. Calculated data for creating biological tissues.

Name of Materials	Polyuret. HP40 %	Two-Comp. Polyuret.%	Graphite, %	Acetone, mL/100 g
Muscle	30	30	32	6.8
Fat	40	40	20	0.00
Capillaries	30	30	33.8	6.2
Dermis	30	30	32.3	7.7
Epidermis	32	32	30	6.0
Stratum corneum of the epidermis	32	32	29.4	6.6

The phantoms were made in the following way. Polyurethane HP40 and two-component polyurethane for forms were used in equal proportions. They were mixed according to the manufacturer's instructions. Immediately thereafter, powder graphite and acetone were slowly added and mixed with the polyurethane base. The curing process took 12–16 h, but it was important to add the powders when the mixture was most flexible. The higher the mass percentage of the powder, the higher the dielectric properties of the sample were. However, the resulting mixture became lumpy when a large percentage of graphite was added. A small amount of acetone was added to ensure uniform mixing of each sample, as well as to create samples with a higher relative permittivity. The values of acetone affected the dielectric properties of the phantoms, so its amount was precisely adjusted to achieve the required values of the real part of the dielectric permittivity. The obtained flat materials were measured using a setup for measuring electrophysical parameters [17], as shown in Figure 7.

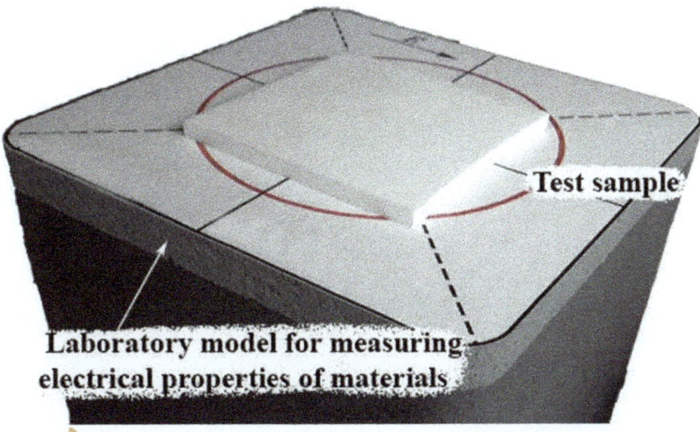

Figure 7. Installation for measuring the electrophysical parameters of the material [17].

Based on the measurement results, graphs of the real parts of the dielectric permittivity versus frequency were built for each of the materials (Figures 8 and 9). As can be seen from the graphs, the values of the real part of the dielectric permittivity are similar to the materials presented in the simulation.

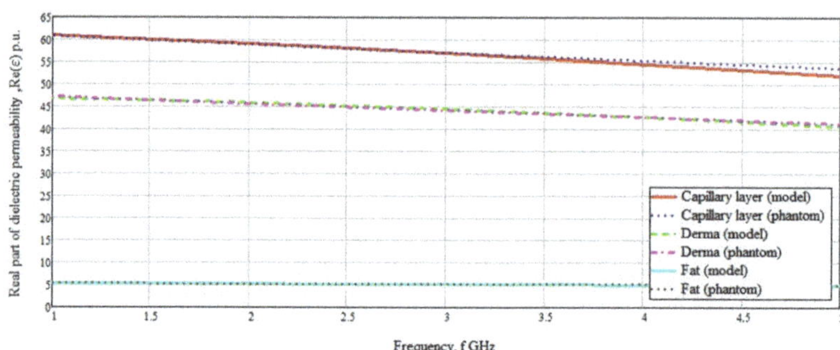

Figure 8. The real part of the dielectric permittivity of the model materials and the created phantoms.

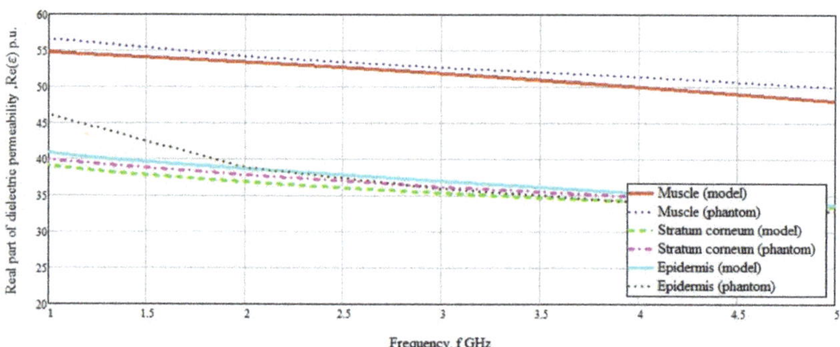

Figure 9. The real part of the dielectric permittivity of the model materials and the created phantoms.

It can be observed that in Figure 8, the values for both the calculated and created materials match with high accuracy. A difference is seen for the capillary layer in the range from 4 to 5 GHz, but this range did not have much influence on the sensor measurements. The range we chose was 1–3 GHz. In this range, the difference in the values of the real part of the dielectric permittivity was minimal. The greatest difference in the values of the real part of the dielectric permittivity was observed in the muscles, stratum corneum, and epidermis.

The difference in the values of the real part of the dielectric permittivity averaged 1–2 rel.units. An abnormal case is presented by epidermal values. This deviation was in the frequency range we were measuring. Since the thickness of the epidermis does not exceed 0.1 mm, this value can be neglected because the reflected signal will undergo minimal changes. The graphs of stratum corneum and epidermis overlap because their values of the real part of the dielectric permittivity are similar in structure.

The used data made it possible to create a phantom of human biological tissues. The created area belongs to the forearm of the hand, because in this place lies a large in diameter vein. Each of the layers was made separately in different thickness molds. Subsequently, layers were cut into identical pieces of 150 × 150 mm in size (Figure 10). The size was selected based on the size of the antennas for measuring blood glucose concentration. The layers were layered sequentially. Due to the presence of acetone and graphite, each of the layers tightly adhered and glued to the other.

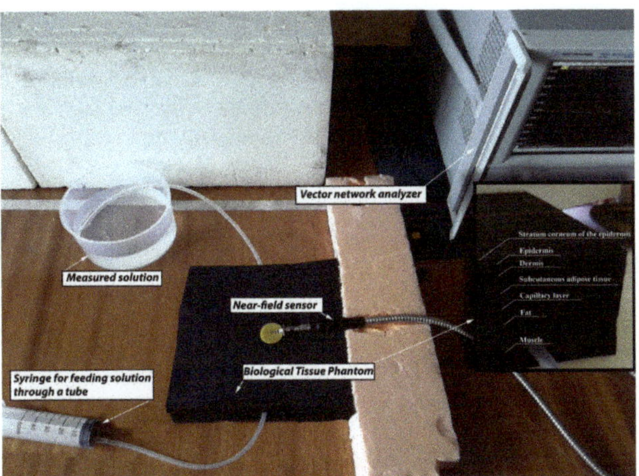

Figure 10. Manufactured graphite based phantom and the measuring setup for determining the level of glucose concentration in saline solution.

The measuring setup (Figure 10) was based on PNA-L Network Analyzer (N5230C) from Agilent Technologies. A silicone tube with an inner diameter of 5 mm was used as a venous vessel. A syringe was used to place saline of the required concentration of glucose inside the tube, and when the liquid was squeezed out, it passed through the tube, thereby simulating the flow of blood through the veins of a human. Such an approach allowed us to change the glucose concentration without physically affecting the antenna (transferring the antenna from one phantom to another), thereby making more accurate measurements.

4. Results and Discussion

4.1. Simulation Results

The used sensor was located close to the hand model for the maximum interaction of the near-field with biological materials. The dependences of the reflected signal on the frequency are plotted (Figure 11a) as a result of a series of measurements for this sensor model. You may notice that the graphs are poorly distinguishable. This is due to small changes in the dielectric permittivity. When zooming into a small area at 1 GHz, it can be seen that the most distinguishable reflected signal is observed for the hand model at 0 mmol/L blood glucose. The rest of the values differ only by a thousandth. Thus, the values for the concentration of 1 and 3 mmol/L are −16.672 and −16.671 dB, respectively (Figure 11b). In this graph, it is not possible to visually determine the various concentrations.

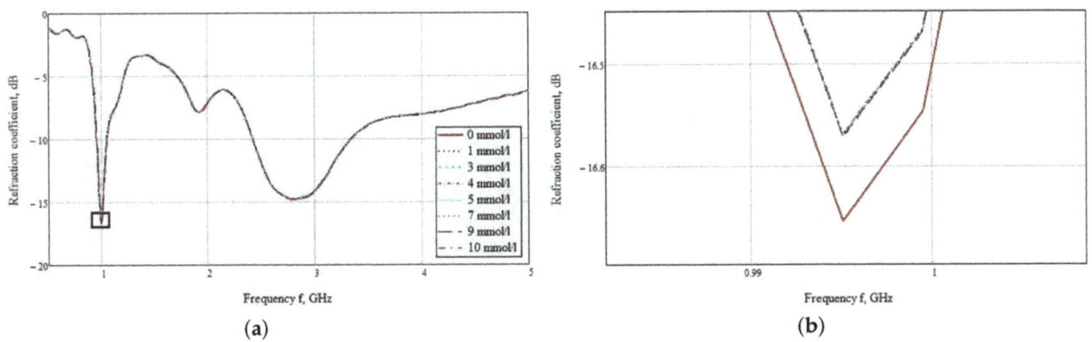

Figure 11. Frequency dependence of the reflected signal for a hand model with different glucose concentrations. Frequency range 0.5–5 GHz (**a**) and 0.5–1.5 GHz (**b**).

The subtraction of zero concentration was carried out, according to the result of which the graphs presented in Figure 12 were obtained. Differences are visible at the amplitude peaks of 1 GHz and the range of 1.5–1.8 GHz. Considering these graphs in more detail, we see a clear difference in concentrations at the frequencies presented above. To build a table with amplitude values, we used the maximum at 1.07 GHz.

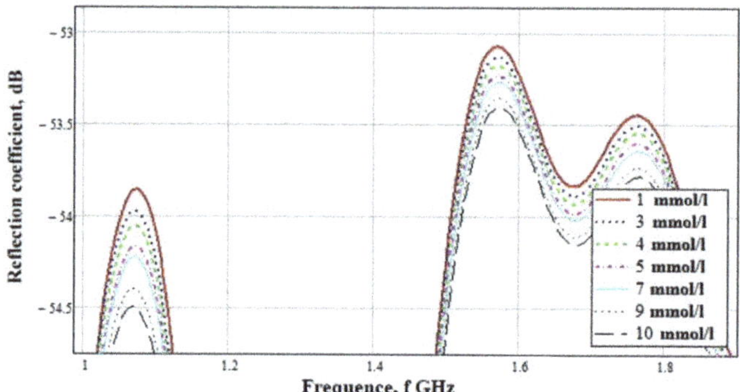

Figure 12. Parameter S11 for a model of a hand minus blood with a concentration of 0 mmol/L.

Table 3 shows the numerical simulation data for a near-field sensor at a frequency of 1.07 GHz. This demonstrates its advantage in the accuracy of determining the concentration difference from the mean values. The difference of 1 mmol/L is 0.1 dB.

Table 3. The results of modeling the amplitude of the reflected signal by the proposed new sensor for different glucose concentrations.

Glucose Concentration, mmol/L	Frequency, GHz	Amplitude, dB
1	1.07	−53.85
3	1.07	−53.97
4	1.07	−54.05
5	1.07	−54.16
7	1.07	−54.22
9	1.07	−54.39
10	1.07	−54.49

As a result of theoretical studies and numerical modeling, the design of a new combined sensor based on a resonant antenna and a near-field effect was proposed to determine the concentration of glucose in a biological medium in the form of a human hand.

4.2. Experimental Results

The simulation results showed that the developed sensor is able to distinguish between different levels of blood glucose concentration. The sensor was tightly attached to the test sample, and the simulation showed the best result at a frequency close to 1 GHz.

An experimental study showed that the maximum difference in amplitudes is in the frequency range 1.45–1.55 GHz (Figure 13). This bias is due to the inaccuracy of the sensor manufacturing compared to simulation (1.07 GHz). Considering this range, it can be seen that the amplitudes at different concentrations vary.

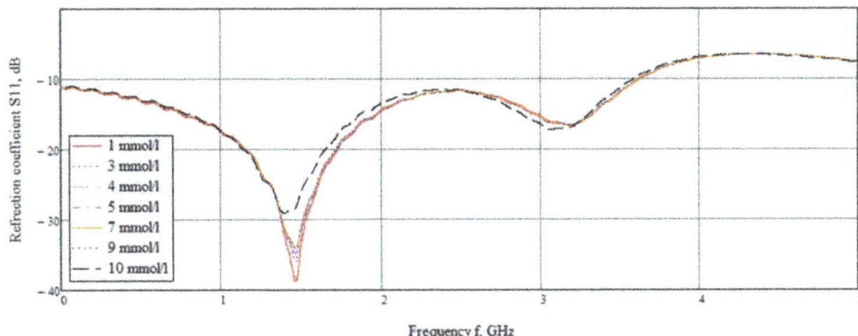

Figure 13. Dependence of the reflected signal on frequency for a near-field sensor in the frequency range 0–5 GHz.

A detailed examination (Figure 14) showed that the reflected signals are lined up in the correct sequence, but the difference between the values of 7 and 9 mmol/L is extremely small. This result is associated with the non-linear behavior of the dielectric permittivity in the frequency range of 1–2 GHz (Figure 3).

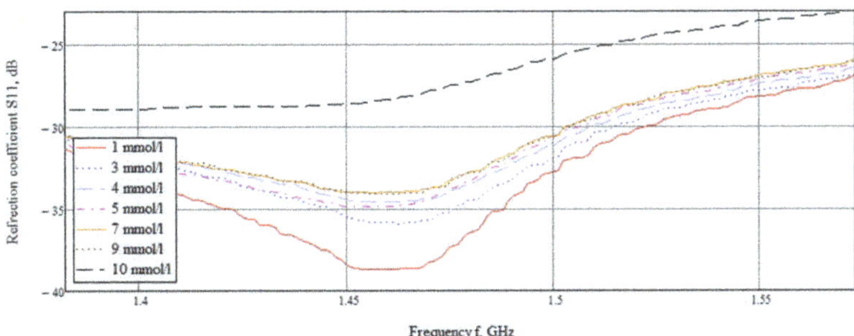

Figure 14. Dependence of the reflected signal on frequency for a near-field sensor in the frequency range 1.3–1.57 GHz.

The rest of the concentrations are arranged in order, which indicates the correctness of this approach to the study of the concentration of glucose in the blood. Table 4 shows the average values of the amplitude of the reflected signal from the sensor for different glucose concentrations at a frequency of 1.53 GHz. Based on the results of the 10 conducted experiments, a confidence interval was constructed with errors for each individual glucose concentration value.

Table 4. The results of measuring the amplitude of the reflected signal by the proposed new sensor for different glucose concentrations.

Glucose Concentration, mmol/L	Frequency, GHz	Amplitude, dB
1	1.53	-29.179 ± 0.07
3	1.53	-28.676 ± 0.09
4	1.53	-28.434 ± 0.06
5	1.53	-27.979 ± 0.11
7	1.53	-27.814 ± 0.15
9	1.53	-27.743 ± 0.05
10	1.53	-24.239 ± 0.18

As can be seen from the table, the values of the amplitude of the reflected signal with a change in concentration differ by 0.15–0.4 dB; on average, the amplitude is 0.27 dB. The obtained sensitivity exceeded the simulation results (0.1 dB).

Based on the obtained results, the dependence of the reflected signal (reflection coefficient S11) on various glucose concentrations in saline solution is plotted (Figure 15). The crosses mark the data obtained experimentally at a frequency of 1.53 GHz. A regression line is drawn along these experimental data, and an error interval was built for them taking into account the data presented in Table 4. It can be seen that at the minimum glucose concentration, the obtained values are aligned according to the regression curve. Furthermore, the concentration of 10 mmol/L matches the regression curve. In the case of concentrations of 7 and 9 mmol/L, the values differ from the expected ones, so the maximum difference is seen at a sugar level of 9 mmol/L. This deviation is associated with the non-linear behavior of the dielectric constant value at low frequencies (Figure 3).

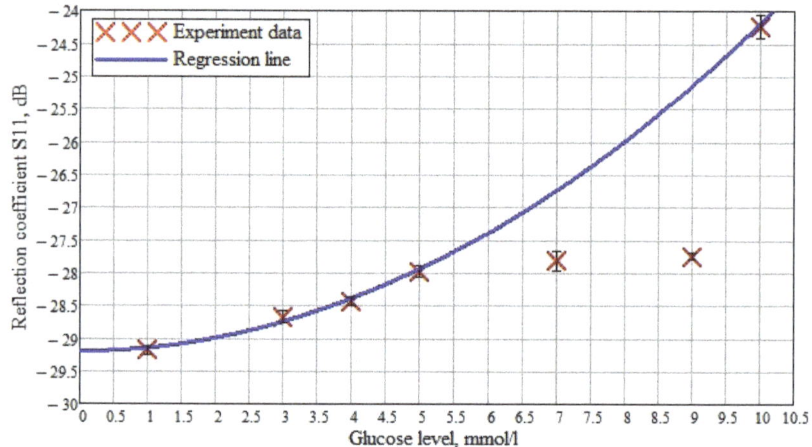

Figure 15. The dependence of the reflected signal on the concentration of glucose in saline at a frequency of 1.53 GHz.

Below is a comparison of the data obtained using the developed sensor with other developments (Table 5). To compare the results of this study with others, we selected the results where both the reflected signal and the transmitted signal were used as the measured parameters.

Table 5. Comparison of the developed sensor with other modern sensors for determining glucose levels.

Reference	Concentration, mg/mL	Frequency, GHz	Sensitivity Parameter dB per mg/mL	S
[18]	0.78–50	1.4–1.9	S_{11}	0.18
[19]	0–300	2–2.5	S_{11}	0.003
[20]	0–3	60–80	S_{12}	0.23
[21]	0.7–1.2	50–70	S_{12}	0.8–1
[22]	40–200	2.5–6	S_{12}	0.01
This work	0–1.81	1.45–1.55	S_{11}	1.7–3.4

In [18], a microstrip glucose sensor measured the reflected signal and the input impedance. The authors managed to achieve a sensitivity of 0.18 dB/(mg/dL) at a frequency of 1.4–1.9 GHz. In [19], a method based on the change in the reflection coefficient between the dielectric waveguide resonator and the measured liquid was used. The proposed sensor allows detecting changes of 0.5 mg/mL at a sensor sensitivity of 0.003 dB/(mg/mL).

Among the presented sensors from Table 5, the most sensitive is the sensor of [21]. It is based on the measurements of the transmitted signal with resonant frequencies in the range of 50–70 GHz. This method is based on the shift of the resonant frequency. The sensitivity of the sensor was 0.8–1 dB/(mg/mL) in an experimental study and 7.7 dB/(mg/mL) in simulation. Comparing the obtained data with the literature data, we can talk about the high sensitivity of the presented sensor in comparison with others. The proposed near-field sensor for non-invasive glucose monitoring surpassed all sensors in Table 5 at its sensitivity, which was 1.7–3.4 dB/(mg/mL).

5. Conclusions

In this work, a sensor based on a combined slot antenna is demonstrated. This sensor has an extended near-field with a high penetration depth of the electromagnetic field. A study of the biological structure of the human hand in the elbow bend is carried out, and subsequently, a simplified numerical model is created. We also measure the reflected signal from the hand model with a shallow vein. The sensor gives a high echo response with a small change in the glucose level in saline. The obtained data indicate the possibility of determining the concentration of glucose in the blood with great accuracy. The average scatter of the reflected signal data for a change of 1 mmol/L is 0.1–0.15 dB, which is high for such a small change in glucose. The phantom created on the basis of polyurethane, graphite, and acetone shows a high similarity of the dielectric permeability with the mathematical calculations of biological media. The experimental data obtained using the manufactured sensor show a larger spread of the reflected signal with a small change in glucose level.

Much work is planned to verify the data obtained. In addition, it is necessary to improve the installation. It should be portable and mobile. That is, we will need to at least replace the bulky Agilent vector network analyzer with a compact version of the device. It will be necessary to develop the hardware and software of the device, which would allow for the collection and processing of data in real time. It will be necessary to conduct a series of preliminary measurements on volunteers to collect statistics indicating the presence of a correlation between glucose levels and the response to probing signals. We plan to conduct research on volunteers using the commercially available semi-non-invasive FreeStyle Libre blood glucose meter and at the same time using the developed sensor. Further work will be aimed at eliminating inaccuracies in the calculation and conducting experiments on humans with a comparison of the data obtained with the data of venous blood (laboratory analysis in a clinic). This approach will allow comparing the data on the level of sugar in venous blood and interstitial fluid and more accurately adjusting the developed sensor.

Author Contributions: Conceptualization, A.G. and K.Z.; methodology, A.G. and K.Z.; validation, A.G. and A.M.; formal analysis, A.M.; investigation, A.M.; resources, A.G.; data curation, A.G. and A.M.; writing, original draft preparation, A.G., K.Z. and A.M.; writing, review and editing, A.G. and K.Z.; visualization, A.M.; supervision, K.Z.; project administration, K.Z. All authors read and agreed to the published version of the manuscript.

Funding: This research was funded by the Russian Science Foundation, Grant No. 18-75-10101. The APC was funded by the Russian Science Foundation.

Institutional Review Board Statement: Not applicable.

Informed Consent Statement: Not applicable.

Data Availability Statement: Data sharing not applicable.

Conflicts of Interest: The authors declare no conflict of interest.

References

1. Jones, M.; Harrison, J.M. The future of diabetes technologies and therapeutics. *Diabetes Technol. Ther.* **2002**, *4*, 351–359. [CrossRef]
2. Forst, T.; Caduff, A.; Talary, M.; Weder, M.; Brändle, M.; Kann, P.; Flacke, F.; Friedrich, C.; Pfützner, A. Impact of environmental temperature on skin thickness and microvascular blood flow in subjects with and without diabetes. *Diabetes Technol. Ther.* **2006**, *8*, 94–101. [CrossRef]

3.   Hanlon, E.; Manoharan, R.; Koo, T.; Shafer, K.; Motz, J.; Fitzmaurice, M.; Kramer, J.; Itzkan, I.; Dasari, R.; Feld, M. Prospects for in vivo Raman spectroscopy. *Phys. Med. Biol.* **2000**, *45*, R1. [CrossRef]

4.   Caduff, A.; Hirt, E.; Feldman, Y.; Ali, Z.; Heinemann, L. First human experiments with a novel non-invasive, non-optical continuous glucose monitoring system. *Biosens. Bioelectron.* **2003**, *19*, 209–217. [CrossRef]

5.   Khalil, O.S. Spectroscopic and clinical aspects of noninvasive glucose measurements. *Clin. Chem.* **1999**, *45*, 165–177. [CrossRef]

6.   Heise, H. Non-invasive monitoring of metabolites using near-infrared spectroscopy: State of the art. *Horm. Metab. Res.* **1996**, *28*, 527–534. [CrossRef] [PubMed]

7.   Gebhart, S.; Faupel, M.; Fowler, R.; Kapsner, C.; Lincoln, D.; McGee, V.; Pasqua, J.; Steed, L.; Wangsness, M.; Xu, F.; et al. Glucose sensing in transdermal body fluid collected under continuous vacuum pressure via micropores in the stratum corneum. *Diabetes Technol. Ther.* **2003**, *5*, 159–166. [CrossRef]

8.   Lipson, J.; Bernhardt, J.; Block, U.; Freeman, W.R.; Hofmeister, R.; Hristakeva, M.; Lenosky, T.; McNamara, R.; Petrasek, D.; Veltkamp, D.; et al. Requirements for calibration in noninvasive glucose monitoring by Raman spectroscopy. *J. Diabetes Sci. Technol.* **2009**. [CrossRef] [PubMed]

9.   Roychoudhury, P.; Harvey, L.M.; McNeil, B. At-line monitoring of ammonium, glucose, methyl oleate and biomass in a complex antibiotic fermentation process using attenuated total reflectance-mid-infrared (ATR-MIR) spectroscopy. *Anal. Chim. Acta* **2006**, *561*, 218–224. [CrossRef]

10. Waynant, R.; Chenault, V. Overview of non-invasive fluid glucose measurement using optical techniques to maintain glucose control in diabetes mellitus. *IEEE LEOS Newsl.* **1998**, *12*, 3–6.

11. Khalil, O.S. Non-invasive glucose measurement technologies: An update from 1999 to the dawn of the new millennium. *Diabetes Technol. Ther.* **2004**, *6*, 660–697. [CrossRef]

12. Gabbay, R.A.; Sivarajah, S. Optical coherence tomography-based continuous noninvasive glucose monitoring in patients with diabetes. *Diabetes Technol. Ther.* **2008**, *10*, 188–193. [CrossRef]

13. Guo, X.; Mandelis, A.; Zinman, B. Noninvasive glucose detection in human skin using wavelength modulated differential laser photothermal radiometry. *Biomed. Opt. Express* **2012**, *3*, 3012–3021. [CrossRef] [PubMed]

14. Chretiennot, T.; Dubuc, D.; Grenier, K. Double stub resonant biosensor for glucose concentrations quantification of multiple aqueous solutions. In Proceedings of the 2014 IEEE MTT-S International Microwave Symposium (IMS2014), Tampa, FL, USA, 1–6 June 2014; pp. 1–4.

15. Hayashi, Y.; Brun, M.A.; Machida, K.; Lee, S.; Murata, A.; Omori, S.; Uchiyama, H.; Inoue, Y.; Kudo, T.; Toyofuku, T.; et al. Simultaneous assessment of blood coagulation and hematocrit levels in dielectric blood coagulometry. *Biorheology* **2017**, *54*, 25–35. [CrossRef] [PubMed]

16. Oliveira, B.L.; O'Loughlin, D.; O'Halloran, M.; Porter, E.; Glavin, M.; Jones, E. Microwave Breast Imaging: Experimental tumour phantoms for the evaluation of new breast cancer diagnosis systems. *Biomed. Phys. Eng. Express* **2018**, *4*, 025036. [CrossRef]

17. Gorst, A.; Zavyalova, K.; Shipilov, S.; Yakubov, V.; Mironchev, A. Microwave Method for Measuring Electrical Properties of the Materials. *Appl. Sci.* **2020**, *10*, 8936. [CrossRef]

18. Huang, S.Y.; Yoshida, Y.; Inda, A.J.G.; Xavier, C.X.; Mu, W.C.; Meng, Y.S.; Yu, W. Microstrip line-based glucose sensor for noninvasive continuous monitoring using the main field for sensing and multivariable crosschecking. *IEEE Sens. J.* **2018**, *19*, 535–547. [CrossRef]

19. Kim, S.; Kim, J.; Babajanyan, A.; Lee, K.; Friedman, B. Noncontact characterization of glucose by a waveguide microwave probe. *Curr. Appl. Phys.* **2009**, *9*, 856–860. [CrossRef]

20. Hu, S.; Nagae, S.; Hirose, A. Millimeter-wave adaptive glucose concentration estimation with complex-valued neural networks. *IEEE Trans. Biomed. Eng.* **2018**, *66*, 2065–2071. [CrossRef]

21. Omer, A.E.; Gigoyan, S.; Shaker, G.; Safavi-Naeini, S. WGM-based sensing of characterized glucose-aqueous solutions at mm-waves. *IEEE Access* **2020**, *8*, 38809–38825. [CrossRef]

22. Harnsoongnoen, S.; Wanthong, A. Coplanar waveguide transmission line loaded with electric-LC resonator for determination of glucose concentration sensing. *IEEE Sens. J.* **2017**, *17*, 1635–1640. [CrossRef]

 **biosensors**

MDPI

Article

Electronic Eye Based on RGB Analysis for the Identification of Tequilas

Anais Gómez, Diana Bueno and Juan Manuel Gutiérrez *

Bioelectronics Section, Department of Electrical Engineering, CINVESTAV-IPN, Mexico City 07360, Mexico; arochag@cinvestav.mx (A.G.); dbuenoh@ipn.mx (D.B.)
* Correspondence: mgutierrez@cinvestav.mx; Tel.: +52-55-5747-3800

Abstract: The present work reports the development of a biologically inspired analytical system known as Electronic Eye (EE), capable of qualitatively discriminating different tequila categories. The reported system is a low-cost and portable instrumentation based on a Raspberry Pi single-board computer and an 8 Megapixel CMOS image sensor, which allow the collection of images of Silver, Aged, and Extra-aged tequila samples. Image processing is performed mimicking the trichromatic theory of color vision using an analysis of Red, Green, and Blue components (RGB) for each image's pixel. Consequently, RGB absorbances of images were evaluated and preprocessed, employing Principal Component Analysis (PCA) to visualize data clustering. The resulting PCA scores were modeled with a Linear Discriminant Analysis (LDA) that accomplished the qualitative classification of tequilas. A Leave-One-Out Cross-Validation (LOOCV) procedure was performed to evaluate classifiers' performance. The proposed system allowed the identification of real tequila samples achieving an overall classification rate of 90.02%, average sensitivity, and specificity of 0.90 and 0.96, respectively, while Cohen's kappa coefficient was 0.87. In this case, the EE has demonstrated a favorable capability to correctly discriminated and classified the different tequila samples according to their categories.

Keywords: biologically inspired; electronic eye; optical methods; RGB analysis; tequila

Citation: Gómez, A.; Bueno, D.; Gutiérrez, J.M. Electronic Eye Based on RGB Analysis for the Identification of Tequilas. *Biosensors* **2021**, *11*, 68. https://doi.org/10.3390/bios11030068

Received: 29 January 2021
Accepted: 25 February 2021
Published: 2 March 2021

1. Introduction

Tequila is the traditional Mexican spirit made with agave *tequilana weber* (blue variety), which is grown in five states of Mexico, namely: Guanajuato, Michoacán, Nayarit, Tamaulipas, and Jalisco. Those geographical regions are established in the Protected Designation of Origin (PDO) [1], which guarantees both the manufacturing procedures and the quality necessary to comply with the strict export specifications from the United States [2] and the European Union [3].

Three main categories of tequila are recognized. The first category is Silver tequila. It is obtained directly from the distillation process without additives; it has a transparent appearance, not necessarily colorless proper to an unaged tequila. The second category is called Aged tequila, which means that the tequila has been aged at least two months using oak casks. This process produces a mellowed product with rich color and flavor. Finally, the third category is known as Extra-aged. This tequila is considered more sophisticated because it has been aged for at least one year in wood or oak recipients with V ≤ 600 L, which has enhanced its flavor with predominant woody notes in its color and aroma [1].

These spirits, whose world consumption ranks fourth after whiskey, vodka, and rum, have a significant presence in more than 120 countries, representing sales of more than 200 million liters per year [4]. Hence, quality control is increasingly important to know, characterize, and monitor its aging process, alcoholic content, and volatile composition that define each kind of tequila's flavor, color, and characteristic aroma.

Nowadays, several tests are carried out in the laboratory to analyze tequila, most of them performing conventional analytical methods such as UV–Vis spectrophotometry [5],

159

Raman spectroscopy [6], Gas Chromatography-Mass Spectrometry (GC-MS) [7], High-Performance Liquid Chromatography (HPLC) [8], Surface Plasmon Resonance (SPR) [9], and electrochemical analysis [10]. Nonetheless, despite their reliability and accuracy, they have several flaws related to long protocols demanding an analysis period from hours to days to carry out the tests, expensive equipment, and the need for technically qualified personnel, without forgetting that their use is confined to special installations with no online quality control possibilities.

Hence, there is an urgent need for fast, inexpensive, portable, and effective alternatives that achieve reliable, non-destructive analytical measurements. Today, biologically inspired analytical systems are beginning to play an important role in the food industry [11,12]. These technologies, sometimes defined as artificial senses, have the perspective of evaluating complex composition samples by emulating the human senses to determine their relevant characteristics. Essentially, the electronic eye has been designed to mimic human eyesight to analyze color and some other attributes related to the sample's appearance [13,14]. This task can be performed using computer vision, colorimetric, or spectrophotometric methods [15–17].

Usually, an electronic eye is built of technology capable of converting optical images into digital images, subsequently analyzed to identify particularities that allow the characterization of what is being observed, avoiding the subjective interpretation of a person [18]. In this sense, image sensors based on a Charge Coupled Device (CCD) or a Complementary Metal Oxide Semiconductor (CMOS) technology are commonly used. Both types of sensors are essentially made up of metal oxide semiconductors (MOS) distributed in a matrix form, and that each independently constitutes a pixel [19]. Once the images are collected, they are subjected to an image processing phase to improve their quality and extract characteristics requiring specific computational algorithms for their interpretation [20].

Electronic eyes have been proven to have an advantage in foodstuff analysis; their applications cover different food industry areas such as quality control, freshness assessment, shelf-life determination, process monitoring, and authenticity assessment [14]. Although the use of electronic eyes in the analysis of some liquids have widely described during the last two decades, these works were focused on samples such as orange juice [21], coffee [22], virgin olive oils [23], milk [24] and some semi-liquids like honey [25], and yogurt [26]. Only a few studies reported the analysis of alcoholic beverages mainly related to wine [27,28], beer, and vodka [29]. Comparatively, the tequila study still continues using conventional techniques instead of an electronic eye. Most tequila analyses have been focused on the determination of quality characteristics [30], sensory properties related to the distillation process [31], evaluation of volatile compounds [32], the relevance of the aging process [33], authenticity [34,35], and adulterations [36].

This work aims to establish the foundations for using an electronic eye in the qualitative identification of different kinds of tequila. The developed analytical system comprises an acquisition image platform that captures and digitalized images directly from tequila, coupled with a biomimetic processing stage based on the trichromatic theory of color vision [37]. In this context, RGB absorbances of images were evaluated and subsequently discriminated against using a Principal Component Analysis (PCA) and Linear Discriminant Analysis (LDA), with was possible to correctly classify the three main tequila categories coming from the Jalisco state region.

The paper is divided into five sections as follows: Section 1 introduction and state of art, Section 2 describes the materials and methods, including details of tequilas set under study, the electronic eye hardware design, experimental setup, image analysis, and data modeling of the proposed biomimetic vision system; Section 3 reports the experimental results obtained from the proposed RGB component analysis, followed by Section 4 where it discusses and compares the key results achieved with our work compared to those reported with conventional analytical techniques. Finally, Section 5 contains the conclusion together with some directions for future work.

2. Materials and Methods

2.1. Tequilas under Study

A total of 25 samples of different brands were acquired at the local supermarket, all of them with POD, made with 100% agave and certified by Consejo Regulador del Tequila (CRT, for its acronym in Spanish) to ensure their authenticity. These samples were chosen according to the main described categories and considering that they were made in the state of Jalisco. In this way, the formed set includes 8 Silver, 12 Aged, and 5 Extra-aged tequilas. Table 1 summarizes detailed information about the tequila samples used.

Table 1. Tequila samples under study grouped by category.

Type	Brand	Tag	Alcoholic Strength (vol %)
Silver	Hornitos	S1	38
	Orendain	S2	38
	Don Nacho	S3	38
	Corralejo	S4	38
	Dos Coronas	S5	38
	Sombrero Negro	S6	35
	Antigua Cruz	S7	40
	Herradura	S8	46
Aged	Hornitos	A1	38
	Jimador	A2	35
	Jarana	A3	35
	Don Nacho	A4	38
	Cabrito	A5	38
	Antigua Cruz	A6	40
	Don Julio	A7	40
	Dos coronas	A8	38
	Sombrero Negro	A9	38
	Reserva del Señor	A10	35
	Cazadores	A11	38
	Jose Cuervo	A12	38
Extra-aged	Corralejo	EA1	38
	Don Nacho	EA2	38
	Antigua Cruz	EA3	40
	Cazadores	EA4	38
	Hornitos	EA5	35

2.2. Electronic Eye System

A single-board computer Raspberry Pi (Model 3B+, Pencoed, Wales, UK) with a Raspbian operating system was chosen as a core for developing the Electronic Eye (EE) prototype. The light source was a white Light-Emitting Diode (LED) 2xLED (Flash Module Huawei LYA-L09, Shenzhen, Guangdong, China), and a camera module (Raspberry Pi Camera V2, Pencoed, Wales, UK) with an 8 Megapixels image sensor (Sony IMX219, Minato, Tokyo, Japan) to perform the image acquisition. It also has a 7-inch Liquid Crystal Display (LCD) (Hilitand hfpq73zx89, Shenzhen, Guangdong, China) that allows interaction with the equipment through a Graphical User Interface (GUI) created in MATLAB®2020a (MathWorks, Natick, MA, USA). The complete system is managed via Python IDLE 2.7 software, using specific routines programmed by the authors. Figure 1 shows a schematic diagram of the developed EE.

Different EE electronics parts are placed in an enclosure designed in SolidWorks 2019 and printed with a Da Vinci 3D 1.1 printer (Xyzprinting, New Taipei City, Taiwan) to operate as a PC peripheric. The design allows the light source location, camera module, and a disposable plastic UV-cuvette (BRAND, Wertheim, Germany) within a dark chamber. The cuvettes' filling volume has a range of 1.5 to 3.0 mL, with external dimensions of 4.5 mm × 23 mm that fits into an internal holder of the chamber, allowing it to be located at a fixed distance of 30 mm from the focal plane of the image sensor of the camera. The white LED was positioned in a centered zenith plane to improve accuracy and image acquisition (this position is widely used for samples with flat surfaces) [16]. In this way, white light can propagate from the source, passes from the chamber through the sample held in a cuvette, and reaches the image sensor avoiding possible external interference. At this point, it is possible to acquire the sample's corresponding digital image. The set of images captured by the EE system were saved automatically in a USB (Universal Serial Bus) device and processed offline employing the GUI designed for this purpose.

Figure 1. Schematic diagram of the developed Electronic Eye showing different electronic parts assembled and communication interface used.

2.3. Experimental Procedure

After opening each tequila bottle, the spirit was immediately taken. A sample volume of 1 mL of spirit was used directly without pretreatment to fill different UV-cuvettes free of dust and dirt to obtain trustworthy images. Additionally, a cuvette containing the same volume of deionized water was used as a blank solution. All experiments were carried out at room temperature (25 °C). The first measured sample with the EE corresponded to the blank solution to establish a system's reference signal. Subsequently, the UV-cuvettes containing different samples of tequila were measured one by one. The captured digitized images were recorded and stored using the programmed control software. During the entire experimental stage, it was ensured that the chamber remains closed during the image capture process to avoid the entry of external light and obtain good quality images.

Meanwhile, the white light source stayed on, waiting for the camera module to acquire the image and send it to the Raspberry Pi computer. Each sample was analyzed in triplicate, performing 10 repetitions each time to observe the repeatability and reproducibility of measures. The time to complete the measurement process by the EE system is 10 s.

2.4. Image Analysis

Digital images were obtained after placing a UV-cuvette with tequila sample in the lab-made EE system described above. In all cases, the camera settings used in our experiments were fixed (exposure time of 1/16 s, an aperture of f/2, and ISO 100). From the images captured by the EE of each tequila sample and the three categories involved, separate files were saved as a *jpeg* format on the Raspberry Pi memory; the average size per image is 2.7 MB (8 Megapixels resolution, 2592 × 1944 pixels). Although using compressed *jpeg* image format implies a loss of information regarding the *raw* format, some works have reported that the RGB obtained from them contained comparable information to those in large raw files [38,39]. Likewise, *jpeg* files retained the residing color information and allowed ease of handling due to the smaller file size, mainly when some multivariate calibration techniques were used to interpret them [39,40]. In our case, using the *jpeg* format also allowed efficient use of hardware resources (in terms of data storage and computational power requirements), as well as, this image format is closest to the images obtained by the human visual system since they are transformed using color-matching functions [41].

For the image analysis process, it is necessary to perform a preprocessing task that consists of selecting and clipping a region of interest (ROI). The ROI was chosen considering the viewing window of the UV-cuvette. This cropped area of the image and its relative position concerning the sample support is always constant. In this way, the complete set of images were cropped and saved as a separate file with a new dimension size of 1244 × 231 pixels.

Taking into account that digital images are a numeric representation of a two-dimensional collection of data, a digital image contains a fixed number of rows and columns of pixels where each pixel is specified for the red, blue, and green coordinates of a pixel array. This conceptualization of the image is related to the trichromatic theory of color vision based on the work of Maxwell, Young, and Helmholtz [37]. This theory states that there are three types of photoreceptors in the human eye, approximately sensitive to the red, green, and blue region of the spectrum, which are related to the three types of cone cells, generally referred to as L, M, and S (long, medium, and short wavelength sensitivity). These cells are responsible for the perception of colors; analogously, in the RGB color model, the image can be represented by the color's intensity, which indicates how much red, green, and blue is present in the image [42]. Hence, each component varies from zero to 255 [43]. If all the components are zero, the result is black color. In the opposite case, the result is a white color.

In the same way, considering that the obtained images are true-color images, it is possible to represent them as 3D matrices associated with RGB components. Making it possible to observe its tonal distribution through a histogram and evaluate its corresponding absorbance [44]. The critical steps followed for the EE acquisition and elaboration of RGB images' regions are illustrated in Figure 2.

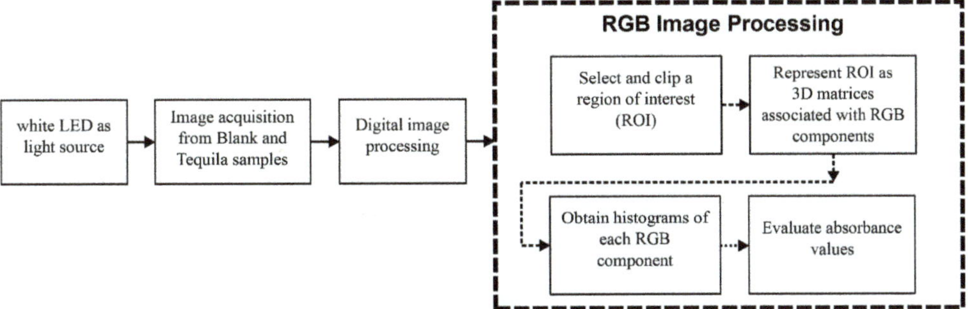

Figure 2. A generalized block diagram of acquisition and identification of image processing performed by the Electronic Eye.

The corresponding absorbances associated with the RGB components for the available image set were evaluated using the Lambert–Beer law. This law expresses the proportional relationship between the absorbance and the concentration of certain compounds present in the sample under analysis. The equation representing this law is a crucial element in evaluating the absorbance of a sample [45].

$$A_\lambda = -log\left(\frac{I_1}{I_0}\right) = \varepsilon b C \tag{1}$$

where A_λ is the absorbance defined via the incident intensity I_0 (incident light over the sample) and transmitted intensity I_1 (transmitted light that comes out of the sample), λ is the wavelength of the source light, C is the concentration of the absorbent sample expressed in *moles* $* L^{-1}$, b is the optical path (thickness of the cell), and ε is the molar absorptivity coefficient.

Similarly, it is possible to establish that (1) expresses the proportional relationship between the absorbance and the concentration of certain compounds present in the sample under analysis. Consequently, it was part of the implemented algorithms.

Experimentally, when light continues its path from the sample, passes through the camera lens, and reaches the image sensor, some light intensity is lost. This effect is because once a beam of light passes through the UV-cuvette made of transparent material containing the sample, its intensity varies due to the phenomena of absorbance, reflection, and transmission [46]. Therefore, it is possible to compare the light intensity transmitted by a standard (in our case, obtained by a blank solution) and the interest sample's light intensity. This procedure allows to obtain an experimental absorbance, as shown below in (2):

$$A_{\lambda experimental} = log\left(\frac{I_{solvent}}{I_{analyte\ solution}}\right) \tag{2}$$

where the experimental absorbance $A_{\lambda experimental}$ is evaluated by $I_{solvent}$ related to the blank solution (in this work it was used deionized water) considered a standard sample, and $I_{analyte\ solution}$ corresponding to each tequila sample to be analyzed.

2.5. Data Processing and Modelling

Data image processing and modeling were done using the specific routines written in MATLAB®2020a by the authors, based on already preprogrammed standard functions using Statistics and Machine Learning Toolbox (v11.7). Before carrying out any data processing and modeling task, it was decided to obtain information on the brightness and tonality characteristics of the acquired images to corroborate the equipment's optical adjustment. For this purpose, histograms of each RGB component were obtained for each available image. Subsequently, the experimental RGB absorbances were calculated (as described in Section 2.4). These calculated values were used as input for two different analysis methods: Principal Component Analysis (PCA) and Linear Discriminant Analysis (LDA). Considering that LDA is a supervised classification method, classification accuracy was evaluated using a Leave-One-Out Cross-Validation (LOOCV) procedure. This iterative method starts using as a training set all the available observations except one, which is excluded for use as validation.

As is known, PCA is an analysis method that depends on an orthogonal linear transformation, which allows summarizing almost all variance contained in a dataset on a fewer number of directions (PCs) with newer coordinates (scores) [47]. In most cases, PCA analysis allows showing clustering data according to their similarities, so it is possible to build a preliminary recognition model that shows the different classes involved according to the measurements made. Nevertheless, to perform a proper classification task, it is necessary to use a supervised learning approach. In this regard, LDA is one of the most used classification procedures with proved successful in many applications [48]. The idea behind LDA is to find a linear transformation that best discriminates among classes. This method operates maximizing between-class variability relative to within-class variability. In this manner, the classification is performed in the transformed space based on some metrics such as Euclidean distance. However, one of the most typical methods to implement is computing a scattering matrix, which must be non-singular. Nonetheless, this criterion cannot be applied when the matrix is singular. A situation that frequently occurs in applications using image databases for pattern recognition, where the number of measurements of each sample exceeds the number of samples in each class. To tackle this problem, it is possible to implement a two-stage approach based on PCA plus LDA. Considering that both methods project the data into a smaller subspace, PCA focused on finding the PCs that maximize the variance in the data set (without considering the class labels), while LDA finds the components that maximize between-class separation. Detailed information about this improved LDA method can be found in [49,50].

3. Results

3.1. RGB Image Processing

The experimental phase with the EE allowed capturing a total of 750 tequila images (10 photos for each sample of the 25 tequilas in triplicate). The selected ROI is automatically defined and fixed for all analyzed sample images from these data, as was described in Section 2.4. Figure 3 shows four representative samples and their corresponding captured images for one tequila sample per class plus the blank solution. The black dotted lines within the UV-cuvette image denote ROI selected image area. The histogram visualization shows the presence of reddish, greenish, and blueish pixels in association with the corresponding RGB components of the images. It is possible to show that both the distribution of these color components and their intensity is clearly different for each tequila type. Similarly, it can be assumed that the information captured in the images using fixed camera parameters (exposure time, aperture, and ISO) and under the same lighting conditions is representative to build a classifier model to identify different categories of tequila.

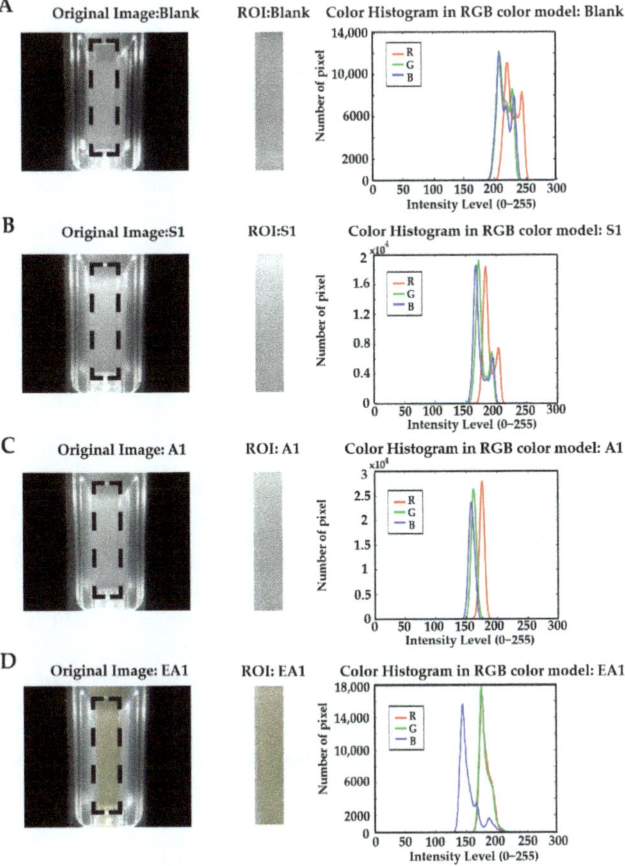

Figure 3. Representative images of (**A**) Blank solution, (**B**) Silver tequila, (**C**) Aged tequila, and (**D**) Extra-aged tequila samples. The image captured by the Electronic Eye (EE), region of interest (ROI) and Red, Green, and Blue (RGB) intensity histogram is noted from left to right.

As a reference, color has been one of the important factors in food quality measurement [39]. For this purpose, it is possible to use the RGB model because it is one of the best for detecting color variations of digital images. In this way, the acquired images were organized as a matrix of dimension 30×75, where the rows correspond to the total number of repetitions (3 tests with 10 repetitions for each test), and the columns represent the 25 tequila samples analyzed by triplicate. The intensities of RGB components are summarized in Table 2. It also integrated each RGB component's absorbance and samples' total absorbance, obtained through Equation (2).

Table 2. Electronic Eye RGB component's intensities and absorbances values for tequila samples. Furthermore, total absorbance values are reported.

	Average Components of RGB Pixel			Absorbance by Component			$Absorbance_\lambda$
Tag	**R**	**G**	**B**	**R**	**G**	**B**	
Blank	255	251 ± 4.0970	253 ± 3.4674	0	0.0026 ± 0.0040	0.0012 ± 0.0027	0.0016 ± 0.0038
S1	215 ± 0.6915	215 ± 1.1861	215 ± 1.0148	0.0729 ± 0.0014	0.0720 ± 0.0024	0.0729 ± 0.0021	0.0726 ± 0.0020
S2	216 ± 1.1059	216 ± 0.8137	216 ± 0.9248	0.0706 ± 0.0023	0.0696 ± 0.0017	0.0708 ± 0.0019	0.0703 ± 0.0019
S3	217 ± 1.7991	216 ± 1.7340	216 ± 1.6046	0.0685 ± 0.0036	0.0696 ± 0.0035	0.0711 ± 0.0033	0.0697 ± 0.0035
S4	225 ± 3.0820	227 ± 3.5519	227 ± 2.8720	0.0529 ± 0.0061	0.0487 ± 0.0070	0.0496 ± 0.0056	0.0504 ± 0.0062
S5	216 ± 1.1427	216 ± 1.1427	216 ± 1.1427	0.0705 ± 0.0023	0.0705 ± 0.0023	0.0705 ± 0.0023	0.0705 ± 0.0023
S6	215 ± 1.0417	215 ± 2.4542	215 ± 1.3113	0.0727 ± 0.0021	0.0717 ± 0.0050	0.0723 ± 0.0027	0.0722 ± 0.0033
S7	223 ± 1.4794	224 ± 2.2733	225 ± 2.8816	0.0556 ± 0.0029	0.0545 ± 0.0045	0.0530 ± 0.0056	0.0544 ± 0.0044
S8	223 ± 3.1220	224 ± 1.8144	224 ± 1.8144	0.0559 ± 0.0062	0.0543 ± 0.0035	0.0543 ± 0.0035	0.0549 ± 0.0044
A1	196 ± 1.3730	199 ± 1.2972	199 ± 1.3730	0.1118 ± 0.0031	0.1055 ± 0.0029	0.1053 ± 0.0030	0.1076 ± 0.0030
A2	211 ± 1.7682	213 ± 1.4794	208 ± 0.8193	0.0812 ± 0.0037	0.0755 ± 0.0031	0.0870 ± 0.0017	0.0813 ± 0.0028
A3	209 ± 0.6215	212 ± 0.7849	209 ± 3.6928	0.0855 ± 0.0013	0.0784 ± 0.0016	0.0850 ± 0.0076	0.0830 ± 0.0036
A4	219 ± 1.2243	222 ± 1.0417	211 ± 1.7207	0.0635 ± 0.0025	0.0576 ± 0.0021	0.0811 ± 0.0036	0.0674 ± 0.0027
A5	204 ± 0.7303	207 ± 0.4498	206 ± 0.6433	0.0955 ± 0.0016	0.0887 ± 0.0010	0.0910 ± 0.0014	0.0917 ± 0.0013
A6	207 ± 0.6915	210 ± 0.6288	206 ± 0.7849	0.0887 ± 0.0015	0.0817 ± 0.0013	0.0908 ± 0.0017	0.0871 ± 0.0015
A7	229 ± 0.6288	234 ± 0.5252	231 ± 0.6065	0.0459 ± 0.0012	0.0356 ± 0.0010	0.0406 ± 0.0012	0.0407 ± 0.0011
A8	212 ± 1.0807	217 ± 0.9965	204 ± 0.8023	0.0784 ± 0.0023	0.0680 ± 0.0020	0.0945 ± 0.0017	0.0803 ± 0.0020
A9	200 ± 1.8889	204 ± 1.2794	197 ± 2.4011	0.1048 ± 0.0042	0.0955 ± 0.0028	0.1113 ± 0.0054	0.1039 ± 0.0041
A10	223 ± 0.8996	224 ± 0.8996	211 ± 3.1397	0.0574 ± 0.0018	0.0537 ± 0.0018	0.0811 ± 0.0067	0.0641 ± 0.0033
A11	214 ± 1.3493	220 ± 0.8137	218 ± 1.3322	0.0748 ± 0.0028	0.0616 ± 0.0016	0.0654 ± 0.0027	0.0673 ± 0.0024
A12	217 ± 0.6288	222 ± 1.2576	213 ± 0.8841	0.0681 ± 0.0013	0.0590 ± 0.0025	0.0771 ± 0.0018	0.0681 ± 0.0019
EA1	207 ± 0.7611	207 ± 1.2015	180 ± 0.4983	0.0884 ± 0.0016	0.0887 ± 0.0026	0.1486 ± 0.0012	0.1086 ± 0.0018
EA2	206 ± 0.7240	210 ± 0.8193	191 ± 0.8683	0.0918 ± 0.0016	0.0836 ± 0.0017	0.1236 ± 0.0020	0.0997 ± 0.0018
EA3	221 ± 0.9248	224 ± 0.7303	208 ± 0.8469	0.0600 ± 0.0019	0.0537 ± 0.0014	0.0872 ± 0.0018	0.0670 ± 0.0086
EA4	215 ± 1.0283	220 ± 0.5509	209 ± 0.8137	0.0731 ± 0.0021	0.0628 ± 0.0011	0.0855 ± 0.0017	0.0738 ± 0.0016
EA5	211 ± 1.2576	206 ± 1.1059	169 ± 0.9685	0.0807 ± 0.0026	0.0912 ± 0.0024	0.1780 ± 0.0025	0.1166 ± 0.0025

It is possible to establish a relationship between the absorbance and the sample's content of each image provided by the Electronic Eye. According to the RGB model, an image's absorbance was calculated about each color component's average. As shown in Table 2, the average and standard deviation of each color intensity component were obtained together with their related absorbance from different tequila samples' images.

The Silver tequila sample's absorbance is 0.0644 ± 0.0034, while for Aged and Extra-aged tequila samples, the average absorbance is 0.0785 ± 0.0024 and 0.0931 ± 0.0019, respectively. The variability presented in the samples can be attributed to the characteristics of each brand's product, as well as to their particular aging process. Thus, the lowest absorbance values in Silver tequila are associated with its colorless and pure tone.

Depending on the tequila aging process, the tone can be yellowish for Aged tequilas or amber for Extra-aged tequilas. In this manner, while the intensity in the tequila tone increases, the absorbance values also increase.

Related to the RGB components' intensity, Silver tequila samples showed a prevalence of the three components. However, the Aged tequila samples predominate the red and blue components, whereas the blue component is more present and has the greatest intensity in the Extra-aged tequila samples. These differences have been associated with shades present in samples, despite being the same type of tequila, and these differences depend on the brand.

It is possible to observe that the similarity among obtained measurements for each tequila sample within the same class is minimal since the deviations are in the order of 0.0001–0.0005, demonstrating repeatability in the operativity of the designed EE.

To visualize the behavior of the RGB absorbances of the different tequila samples, radar plots were constructed. Figure 4 shows the RGB average absorbance of the complete set of tequilas grouped in each of the three categories under study. Here it is possible to observe some characteristic fingerprints for each type of tequila related to their optical properties. This evident pattern for each tequila class (i.e., Silver, Aged, and Extra-aged) will help interpret this information by the planned classifier models. The idea behind a pattern recognition process is to recognize the regularities present in data by a computational model that uses machine learning algorithms.

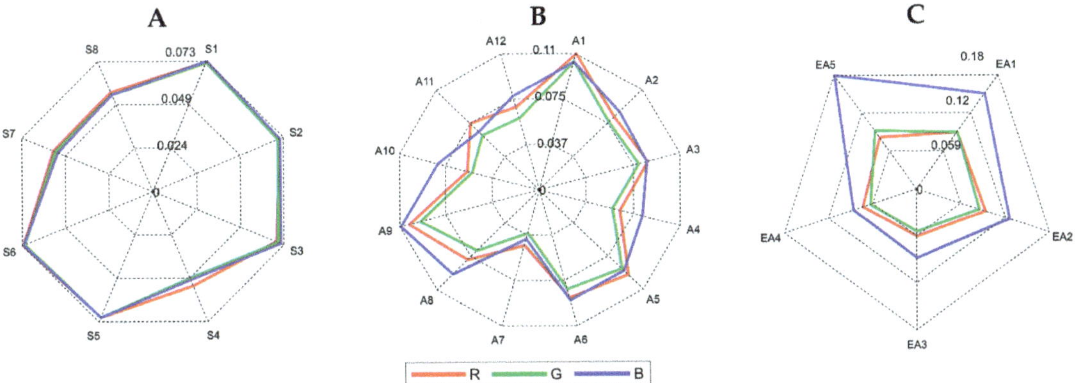

Figure 4. Radar plots for analyzed tequila samples with their respective absorbance values. (**A**) Silver, (**B**) Aged, and (**C**) Extra-aged.

3.2. EE Preliminary Recognition Model

Before modeling, RGB average absorbances were normalized to an interval of 0 to 1 to reduce illumination effects and for data treatment convenience. Afterward, a PCA analysis was done to build a preliminary recognition model, expecting to observe some sample clustering caused by the own absorbances and tequila class-related. The PCA plot with the three significant PCs is shown in Figure 5. Here the accumulated explained variance was ca. 99.96% with characteristic clusters that partially discriminate the different tequila kinds. That is, most of the Silver tequilas seem to be grouped in the upper right region of the plot, while the Aged tequilas are concentrated in the center, and the Extra-aged ones appear grouped in the left region. However, apart from the marked dispersion of these last two categories of tequilas, there is a clear overlap between some of their samples.

Although the aging mechanisms have been widely studied for different alcoholic beverages such as wine and spirits [51,52], there is still no scientific report that addresses it for tequila. Thus, considering that one of the physicochemical characteristics that are impacted during this process is the color, it is then possible to assume that the absorbances obtained with the electronic eye are also related to the aging of the analyzed tequila samples.

In this sense, the clustering regions observed in the PCA make sense when identifying that samples were grouped within the proper class. On the other hand, each cluster has a relationship with a different aging period. As a result, the dispersion present in the Aged and Extra-aged tequila cluster is clearly related to the aging times that each producer stipulates for their product. On the contrary, in the Silver tequilas cluster, the dispersion is minimal because these tequila samples do not have an aging process.

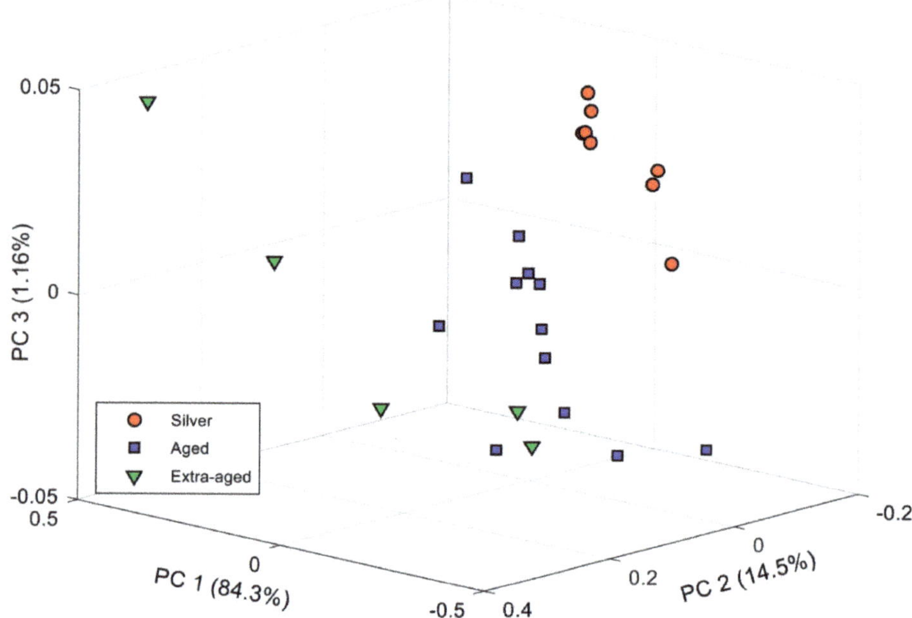

Figure 5. PCA score plot of the three first components obtained after analysis of tequila samples. As can be seen, some clustering is obtained according to different tequila classes.

Thus, it is highly probable that there are tequilas with different aging times within the set of tequila samples analyzed despite belonging to the same category. This may be because each tequila producer must comply with Mexican regulations to respect the minimum aging time. However, they can also establish longer aging periods without violating the standard's provisions to offer a product with better organoleptic characteristics than their competitions.

For this reason, to confirm these initial identifications seen by PCA, the next step was the use of LDA as a supervised pattern recognition method.

3.3. Tequila Categories Discrimination

Transformed data obtained by PCA were used as input information to perform LDA. Since this is a supervised method, classification success was evaluated using LOOCV. In this scheme, each sample is classified by means of the analysis function derived from the remaining samples (all cases except the case itself). This process was repeated as many times as the number of samples in the data set (i.e., 25 times), leaving out one different sample each time, considering it as a validation sample. With this approach, all samples are used once for validation. As can be observed in Figure 6, clear discrimination between the three categories of tequila was achieved. The clusters in the figure evidence that tequila samples are grouped according to their associated aged process. Although Silver tequilas are clearly grouped on the left region of the plot, the Aged and Extra-aged tequilas have class centroids located in the middle and right regions.

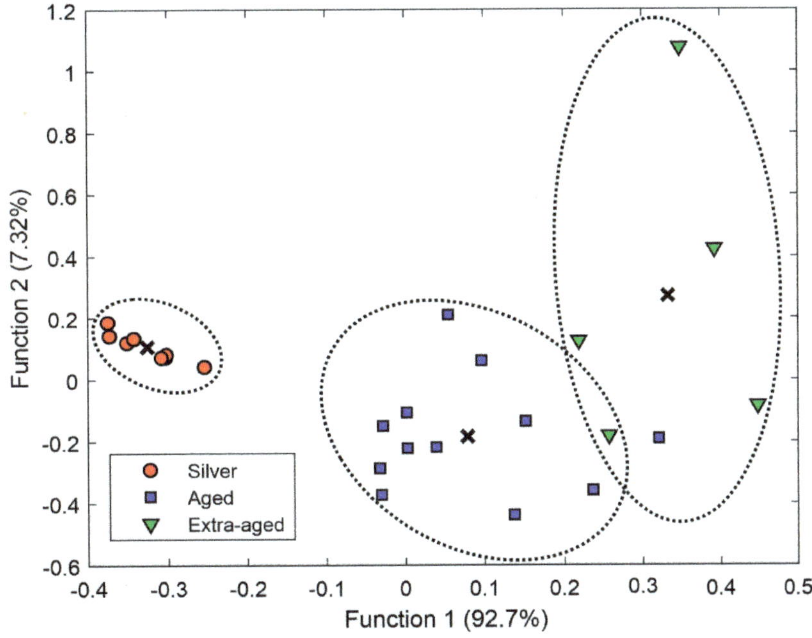

Figure 6. LDA score plot of the obtained functions after analysis of tequila samples, according to their category. Dotted lines represent classification clusters. In addition, the centroid of each class is plotted (x).

The average classification results obtained from the 25 LDA models built are reported in Table 3. Predictably from the LDA plot, the tequila samples managed to be correctly classified as Silver and Aged, reaching high classification rates (100% and 91.67%, respectively). In contrast, the Extra-aged class did not exceed 78.40% correct classification. The overall classification rate for the three classes was 90.02%. In order to evaluate the efficiency of the modeling, accuracy, precision, sensitivity, and specificity values were also calculated. It is possible to notice that sensitivity averaged for the three classes considered was 0.90, whereas specificity was 0.96.

Table 3. Average classification results of EE for the discrimination of different tequila samples according to expected categories employing PCA-LDA.

Tequila Category	Classification Rate (%)	Accuracy	Precision	Sensitivity	Specificity
Silver	100.00	1.00	1.00	1.00	1.00
Aged	91.67	0.92	0.91	0.92	0.92
Extra-aged	78.40	0.92	0.80	0.78	0.95
Average	90.02	0.94	0.90	0.90	0.96

Many studies have established that the overall classification rate is not the best criterion for measuring classifier performance where there is an imbalance in the number of samples per class [53]. In this direction, to corroborate that the results obtained from the LDA modeling are significant, it is necessary to use another criterion that reflects with more certainty the performance of the classifier in contexts of this imbalance. A well-known alternative measure to the accuracy is Cohen's kappa coefficient [54]. The fundamental idea for its calculation involves analyzing the differences between the reference data and the incoming data determined by the main diagonal of the confusion matrix, see definition (3).

$$\kappa = \frac{N \sum_{i=1}^{n} m_{i,i} - \sum_{i=1}^{n}(G_i C_i)}{N^2 - \sum_{i=1}^{n}(G_i C_i)} \tag{3}$$

where i is the class number, N is the total number of classified values compared to truth values, $m_{i,i}$ is the number of values belonging to the truth class i that have also been classified as class i (i.e., values found along the main diagonal of the confusion matrix), C_i is the total number of predicted values belonging to class i, and G_i is the total number of truth values belonging to class i.

Thus, kappa is an indicator that acquires values between 0 and 1, the first representing the absolute lack of agreement and the second, total agreement. According to their scheme, a value <0 indicates no agreement, 0–0.20 as slight, 0.21–0.40 as fair, 0.41–0.60 as moderate, 0.61–0.80 as substantial, and 0.81–1 as almost perfect agreement.

In this regard, kappa values were calculated for each of the 25 LDA models built considering the LOOCV process, obtaining an overall mean kappa coefficient of 0.87, which is defined as "perfect agreement". This finding indicates that this high agreement is related to reliable data. In other words, the RGB absorbances used to identify the tequila samples are representative enough to be modeled. Likewise, although the tequila classes are imbalanced, the LDA models do not privilege the Aged tequila class with the greatest number of samples over the Extra-aged tequila class with the least number of samples.

Additionally, from the obtained results, it is possible to confirm that even using a LOOCV does not produce an over-optimistic approach in the LDA classifiers performance.

4. Discussion

The results presented in Section 3 have provided some insight into the developed electronic eye's capabilities to authenticate the three categories of tequila: Silver (S), Aged (A), and Extra-aged (EA). First, from the preliminary recognition model using PCA, it is important to highlight the close relationship between tequilas' aging time and their clustering from the RGB absorbance analysis. This same aging effect in tequilas has been observed using more complex analytical methods such as HPLC [8]. This method is responsible for identifying and quantifying low molecular weight phenolic compounds acquired by tequila during the oak barrels' maturing process. Once characterized, they are related to the mentioned age classifications using analysis of variance (ANOVA) combined with discriminant analysis.

Other works instead deal the authentication of tequila recurring to methods of analysis as GC-MS [34], and UV-Vis [35] coupling some chemometric methods commonly based on LDA, Partial Least Squares Discriminant Analysis (PLS-DA), Multilayer Perceptron Artificial Neuronal Networks (MLP-ANN), and Support Vector Machines (SVM) to name a few. However, although these contributions differ from our study in factors such as the nature of analytical data obtained and the number of tequila samples analyzed, they represent the most recent state-of-the-art in identifying certified tequilas' three main categories of interest. Added to this, they report performance parameters like sensitivity and specificity of the classifier models they used, which allows direct comparisons with our results. In this way, Table 4 summarizes these parameters' comparison, including the analytical methods, classification models, and kinds of tequila reported by each research group.

Table 4. Comparison of the current study with representative publications dealing with tequila identification (S = Silver, A = Aged and EA = Extra-aged tequilas).

Reference	Analytical Method	Classification Model	Tequila Category	Sensitivity	Specificity
Ceballos-Magaña et al. [34]	GC-MS	LDA	S	0.66	0.75
			A	0.33	0.92
			EA	0.66	0.73
		MLP-ANN	S	1.00	1.00
			A	0.83	1.00
			EA	1.00	0.93
Pérez-Caballero et al. [35]	UV-Vis	PLS-DA	S	0.81	0.89
			A	0.71	0.88
			EA	1.00	0.93
		SVM	S	1.00	1.00
			A	1.00	0.99
			EA	0.96	1.00
This work	Electronic Eye	PCA-LDA	S	1.00	1.00
			A	0.92	0.92
			EA	0.78	0.95

In this way, it is clear that the model adopted in our study using PCA-LDA achieved superior performance in the individualized identification of classes (sensitivity for S = 1.00, A = 0.92, EA = 0.78 and specificity for S = 1.00, A = 0.92, EA = 0.95) than the LDA model (sensitivity for S = 0.66, A = 0.33, EA = 0.66 and specificity for S = 0.75, A = 0.92, EA = 0.73) reported by Ceballos-Magaña et al. [34], and the PLS-DA (sensitivity for S = 0.81, A = 0.71, EA = 1.00 and specificity for S = 0.89, A = 0.88, EA = 0.93) described by Pérez-Caballero et al. [35]. These results are remarkable because, in our study, a linear model was enough to identify tequilas from their RBG absorbances. In contrast, the authors mentioned above needed the use of models with non-linear strategies (e.g., MLP-ANN and SVM) that demand a high computational cost when performing their optimization process to tackle the classification problem properly.

On the other hand, if we compare the results obtained from the non-linear modeling of the PCA-LDA model described in the present work, the overall performance is competitive for the Silver and Aged tequila classes and limited for the Extra-aged class. Finally, the differentiation between non-aged tequilas and those with different maturity levels is closely related to the task of identifying mixed, fake, and adulterated tequilas. Taking into account that adulterations in tequila are also associated with practices such as dilution, the addition of alcohol or some prohibited substances, forbidden aging methods, or blending with lower quality tequila batches, these adulterations are closely related to changes in the UV-vis absorbance and, therefore, in samples' color [36,39]. Further work will attempt to include these kinds of samples applying the reported image processing procedure in order to find color variations (from RGB absorbances) to identify counterfeit tequilas.

5. Conclusions

An electronic eye based on lab-made instrumentation coupled with an image processing stage was developed to build a biologically inspired system capable of distinguishing between different tequila kinds, namely Silver, Aged, and Extra-aged. The system's repeatability was demonstrated by statistical analysis of the captured images using RGB information. Preliminary analysis employing PCA was relevant to observe data behavior and tequila class clustering mainly related to the aging process. LDA classifiers were built to recognize tequilas through the evaluated RGB absorbances using a LOOCV scheme to identify samples correctly.

Successful discrimination between tequilas was achieved by LDA, obtaining an overall classification rate of 90.02% for the three involved tequila classes mainly associated with their aging process. In the same way, the obtained sensitivity averaged was 0.90, whereas specificity was 0.96. Considering that the analyzed tequila samples are grouped in imbalanced classes, the kappa coefficient was calculated to corroborate that the performance measures were not over-optimistic. In this way, the kappa coefficient mean value was 0.87, which implies that models interpret reliable data without privileging any tequila class after adjustment.

These results show that the developed image analysis strategy based on obtained RGB information of compressed *jpeg* images, together with the PCA-LDA modeling stage, did not hamper the identification of tequilas by retaining enough color information of analyzed samples. Another notable point is that the method presented here agrees with the results reported by some previous studies that employ conventional analytical techniques such as UV-Vis and GC-MS combined with non-linear classification methods. In this sense, the developed electronic eye constitutes a reliable and easy-to-use tool that allows a quick and non-destructive analysis of tequilas to authenticate them according to the three main categories. Lastly, further research may be conducted to identifying fake or mixed tequilas applying the currently reported methodology based on color analysis.

Author Contributions: Conceptualization: J.M.G.; methodology and formal analysis: A.G.; validation: A.G. and D.B.; investigation: all authors; writing—original draft preparation, A.G.; writing—review and editing: D.B. and J.M.G. All authors have read and agreed to the published version of the manuscript.

Funding: This research received no external funding.

Institutional Review Board Statement: Not applicable.

Informed Consent Statement: Not applicable.

Data Availability Statement: Not applicable.

Acknowledgments: Authors would like to express their gratitude to the Mexican National Council of Science and Technology (CONACyT) for the financial support and fellowship for Anais Gómez.

Conflicts of Interest: The authors declare no conflict of interest.

References

1. Norma Oficial Mexicana NOM-006-SCFI-2012. Available online: http://www.dof.gob.mx/nota_detalle.php?codigo=5282165&fecha=13/12/2012 (accessed on 21 February 2021).
2. Electronic Code of Federal Regulations Title 27, 5.22(g). Available online: https://www.ecfr.gov/cgi-bin/text-idx?node=pt27.1.5&rgn=div5 (accessed on 21 February 2021).
3. Council of the European Union. Agreement between the European Community and the United Mexican States on the Mutual Recognition and Protection of Designations for Spirit Drinks. Available online: https://eur-lex.europa.eu/legal-content/EN/TXT/?uri=CELEX%3A21997A0611%2801%29 (accessed on 21 February 2021).
4. Consejo Regulador del Tequila. Available online: https://www.crt.org.mx/index.php/en/pages-2/proteccion-del-tequila-a-nivel-internacional (accessed on 21 February 2021).
5. Barbosa-García, O.; Ramos-Ortíz, G.; Maldonado, J.L.; Pichardo-Molina, J.L.; Meneses-Nava, M.A.; Landgrave, J.E.A.; Cervantes-Martínez, J. UV–vis absorption spectroscopy and multivariate analysis as a method to discriminate tequila. *Spectrochim. Acta Part A Mol. Biomol. Spectrosc.* **2007**, *66*, 129–134. [CrossRef]
6. Frausto-Reyes, C.; Medina-Gutiérrez, C.; Sato-Berrú, R.; Sahagún, L.R. Qualitative study of ethanol content in tequilas by Raman spectroscopy and principal component analysis. *Spectrochim. Acta Part A Mol. Biomol. Spectrosc.* **2005**, *61*, 2657–2662. [CrossRef]
7. De León-Rodríguez, A.; Escalante-Minakata, P.; Jiménez-García, M.I.; Ordoñez-Acevedo, L.G.; Flores, J.L.F.; Barba De La Rosa, A.P. Characterization of volatile compounds from ethnic Agave alcoholic beverages by gas chromatography-mass spectrometry. *Food Technol. Biotechnol.* **2008**, *46*, 448–455.
8. Muñoz-Muñoz, A.C.; Grenier, A.C.; Gutiérrez-Pulido, H.; Cervantes-Martínez, J. Development and validation of a High Performance Liquid Chromatography-Diode Array Detection method for the determination of aging markers in tequila. *J. Chromatogr. A* **2008**, *1213*, 218–223. [CrossRef] [PubMed]
9. Martínez-López, G.; Luna-Moreno, D.; Monzón-Hernández, D.; Valdivia-Hernández, R. Optical method to differentiate tequilas based on angular modulation surface plasmon resonance. *Opt. Lasers Eng.* **2011**, *49*, 675–679. [CrossRef]

10. Oliveira, P.R.; Lamy-Mendes, A.C.; Rezende, E.I.P.; Mangrich, A.S.; Marcolino Junior, L.H.; Bergamini, M.F. Electrochemical determination of copper ions in spirit drinks using carbon paste electrode modified with biochar. *Food Chem.* **2015**, *171*, 426–431. [CrossRef] [PubMed]

11. Kiani, S.; Minaei, S.; Ghasemi-Varnamkhasti, M. Fusion of artificial senses as a robust approach to food quality assessment. *J. Food Eng.* **2016**, *171*, 230–239. [CrossRef]

12. Tan, J.; Xu, J. Applications of electronic nose (e-nose) and electronic tongue (e-tongue) in food quality-related properties determination: A review. *Artif. Intell. Agric.* **2020**, *4*, 104–115. [CrossRef]

13. Orlandi, G.; Calvini, R.; Pigani, L.; Foca, G.; Vasile Simone, G.; Antonelli, A.; Ulrici, A. Electronic eye for the prediction of parameters related to grape ripening. *Talanta* **2018**, *186*, 381–388. [CrossRef]

14. Xu, C. Electronic eye for food sensory evaluation. In *Evaluation Technologies for Food Quality*; Zhong, J., Wang, X., Eds.; Woodhead Publishing: Cambridge, UK, 2019; pp. 37–59.

15. Wu, D.; Sun, D.W. Food colour measurement using computer vision. In *Instrumental Assessment of Food Sensory Quality*; Kilcast, D., Ed.; Woodhead Publishing: Cambridge, UK, 2013; pp. 165–195.

16. Wu, D.; Sun, D.-W. Colour measurements by computer vision for food quality control—A review. *Trends Food Sci. Technol.* **2013**, *29*, 5–20. [CrossRef]

17. Gomes, J.F.S.; Leta, F.R. Applications of computer vision techniques in the agriculture and food industry: A review. *Eur. Food Res. Technol.* **2012**, *235*, 989–1000. [CrossRef]

18. Ware, C. Color. In *Information Visualization*, 4th ed.; Ware, C., Ed.; Morgan Kaufmann: Burlington, MA, USA, 2021; pp. 95–141.

19. Jerram, P.; Stefanov, K. CMOS and CCD image sensors for space applications. In *High Performance Silicon Imaging*, 2nd ed.; Durini, D., Ed.; Woodhead Publishing: Cambridge, UK, 2020; pp. 255–287.

20. Patel, K.K.; Kar, A.; Jha, S.N.; Khan, M.A. Machine vision system: A tool for quality inspection of food and agricultural products. *J. Food Sci. Technol.* **2012**, *49*, 123–141. [CrossRef]

21. Fernández-Vázquez, R.; Stinco, C.M.; Meléndez-Martínez, A.J.; Heredia, F.J.; Vicario, I.M. Visual and instrumental evaluation of orange juice color: A consumers' preference study. *J. Sens. Stud.* **2011**, *26*, 436–444. [CrossRef]

22. Buratti, S.; Benedetti, S.; Giovanelli, G. Application of electronic senses to characterize espresso coffees brewed with different thermal profiles. *Eur. Food Res. Technol.* **2017**, *243*, 511–520. [CrossRef]

23. Apetrei, C.; Apetrei, I.M.; Villanueva, S.; de Saja, J.A.; Gutierrez-Rosales, F.; Rodriguez-Mendez, M.L. Combination of an e-nose, an e-tongue and an e-eye for the characterisation of olive oils with different degree of bitterness. *Anal. Chim. Acta* **2010**, *663*, 91–97. [CrossRef]

24. Figueroa, A.; Caballero-Villalobos, J.; Angón, E.; Arias, R.; Garzón, A.; Perea, J.M. Using multivariate analysis to explore the relationships between color, composition, hygienic quality, and coagulation of milk from Manchega sheep. *J. Dairy Sci.* **2020**, *103*, 4951–4957. [CrossRef] [PubMed]

25. Shafiee, S.; Minaei, S.; Moghaddam-Charkari, N.; Ghasemi-Varnamkhasti, M.; Barzegar, M. Potential application of machine vision to honey characterization. *Trends Food Sci. Technol.* **2013**, *30*, 174–177. [CrossRef]

26. Abildgaard, O.H.A.; Kamran, F.; Dahl, A.B.; Skytte, J.L.; Nielsen, F.D.; Thomsen, C.L.; Andersen, P.E.; Larsen, R.; Frisvad, J.R. Non-Invasive Assessment of Dairy Products Using Spatially Resolved Diffuse Reflectance Spectroscopy. *Appl. Spectrosc.* **2015**, *69*, 1096–1105. [CrossRef] [PubMed]

27. Martin, M.L.G.-M.; Ji, W.; Luo, R.; Hutchings, J.; Heredia, F.J. Measuring colour appearance of red wines. *Food Qual. Prefer.* **2007**, *18*, 862–871. [CrossRef]

28. Ouyang, Q.; Zhao, J.; Chen, Q. Instrumental intelligent test of food sensory quality as mimic of human panel test combining multiple cross-perception sensors and data fusion. *Anal. Chim. Acta* **2014**, *841*, 68–76. [CrossRef] [PubMed]

29. Benedetti, L.P.d.S.; dos Santos, V.B.; Silva, T.A.; Filho, E.B.; Martins, V.L.; Fatibello-Filho, O. A digital image-based method employing a spot-test for quantification of ethanol in drinks. *Anal. Methods* **2015**, *7*, 4138–4144. [CrossRef]

30. Villanueva-Rodríguez, S.J.; Rodríguez-Garay, B.; Prado-Ramírez, R.; Gschaedler, A. Tequila: Raw Material, Classification, Process, and Quality Parameters. In *Encyclopedia of Food and Health*; Caballero, B., Finglas, P.M., Toldrá, F., Eds.; Academic Press: Oxford, UK, 2016; pp. 283–289.

31. Prado-Ramírez, R.; Gonzáles-Alvarez, V.; Pelayo-Ortiz, C.; Casillas, N.; Estarrón, M.; Gómez-Hernández, H.E. The role of distillation on the quality of tequila. *Int. J. Food Sci. Technol.* **2005**, *40*, 701–708. [CrossRef]

32. Martín-del-Campo, S.T.; López-Ramírez, J.E.; Estarrón-Espinosa, M. Evolution of volatile compounds during the maturation process of silver tequila in new French oak barrels. *LWT* **2019**, *115*, 108386. [CrossRef]

33. López-Ramírez, J.E.; Martín-del-Campo, S.T.; Escalona-Buendía, H.; García-Fajardo, J.A.; Estarrón-Espinosa, M. Physicochemical quality of tequila during barrel maturation. A preliminary study. *Cyta-J. Food* **2013**, *11*, 223–233. [CrossRef]

34. Ceballos-Magaña, S.G.; de Pablos, F.; Jurado, J.M.; Martín, M.J.; Alcázar, Á.; Muñiz-Valencia, R.; Gonzalo-Lumbreras, R.; Izquierdo-Hornillos, R. Characterisation of tequila according to their major volatile composition using multilayer perceptron neural networks. *Food Chem.* **2013**, *136*, 1309–1315. [CrossRef]

35. Pérez-Caballero, G.; Andrade, J.M.; Olmos, P.; Molina, Y.; Jiménez, I.; Durán, J.J.; Fernandez-Lozano, C.; Miguel-Cruz, F. Authentication of tequilas using pattern recognition and supervised classification. *Trac Trends Anal. Chem.* **2017**, *94*, 117–129. [CrossRef]

36. Contreras, U.; Barbosa-García, O.; Pichardo-Molina, J.L.; Ramos-Ortíz, G.; Maldonado, J.L.; Meneses-Nava, M.A.; Ornelas-Soto, N.E.; López-de-Alba, P.L. Screening method for identification of adulterate and fake tequilas by using UV–VIS spectroscopy and chemometrics. *Food Res. Int.* **2010**, *43*, 2356–2362. [CrossRef]
37. Wyszecki, G.; Stiles, W. *Color Science: Concepts and Methods, Quantitative Data and Formulae*, 2nd ed.; Wiley-VCH: Weinheim, Germany, 2000; p. 968.
38. Foca, G.; Masino, F.; Antonelli, A.; Ulrici, A. Prediction of compositional and sensory characteristics using RGB digital images and multivariate calibration techniques. *Anal. Chim. Acta* **2011**, *706*, 238–245. [CrossRef]
39. Herrero-Latorre, C.; Barciela-García, J.; García-Martín, S.; Peña-Crecente, R.M. Detection and quantification of adulterations in aged wine using RGB digital images combined with multivariate chemometric techniques. *Food Chem. X* **2019**, *3*, 100046. [CrossRef]
40. Mutlu, A.Y.; Kılıç, V.; Özdemir, G.K.; Bayram, A.; Horzum, N.; Solmaz, M.E. Smartphone-based colorimetric detection via machine learning. *Analyst* **2017**, *142*, 2434–2441. [CrossRef] [PubMed]
41. Pennebaker, W.B.; Mitchell, J.L. *JPEG: Still Image Data Compression Standard*; Kluwer Academic Publishers: Drive Norwell, MA, USA, 1993; p. 638.
42. Tkalcic, M.; Tasic, J.F. Colour spaces: Perceptual, historical and applicational background. In Proceedings of the IEEE Region 8 EUROCON 2003. Computer as a Tool, Ljubljana, Slovenia, 22–24 September 2003; pp. 304–308.
43. Burger, W.; Burge, M.J. Color Images. In *Digital Image Processing: An Algorithmic Introduction Using Java*; Gries, D., Schneider, F.B., Eds.; Springer London: London, UK, 2016; pp. 291–328.
44. Burger, W.; Burge, M.J. Histograms and Image Statistics. In *Digital Image Processing: An Algorithmic Introduction Using Java*; Gries, D., Schneider, F.B., Eds.; Springer London: London, UK, 2016; pp. 37–56.
45. Oldham, K.B.; Parnis, J.M. Shining light on Beer's law. *ChemTexts* **2017**, *3*, 5. [CrossRef]
46. Corke, P. Light and Color. In *Robotics, Vision and Control: Fundamental Algorithms In MATLAB*, 2nd ed.; Springer International Publishing: Cham, Switzerland, 2017; pp. 287–318.
47. Jolliffe, I.T.; Cadima, J. Principal component analysis: A review and recent developments. *Philos. Trans. A Math. Phys. Eng. Sci.* **2016**, *374*, 20150202. [CrossRef]
48. Mitteroecker, P.; Bookstein, F. Linear Discrimination, Ordination, and the Visualization of Selection Gradients in Modern Morphometrics. *Evol. Biol.* **2011**, *38*, 100–114. [CrossRef]
49. Swets, D.L.; Weng, J.J. Using discriminant eigenfeatures for image retrieval. *IEEE Trans. Pattern Anal. Mach. Intell.* **1996**, *18*, 831–836. [CrossRef]
50. Belhumeur, P.N.; Hespanha, J.P.; Kriegman, D.J. Eigenfaces vs. Fisherfaces: Recognition using class specific linear projection. *IEEE Trans. Pattern Anal. Mach. Intell.* **1997**, *19*, 711–720. [CrossRef]
51. Carpena, M.; Pereira, A.G.; Prieto, M.A.; Simal-Gandara, J. Wine aging technology: Fundamental role of wood barrels. *Foods* **2020**, *9*, 1160. [CrossRef] [PubMed]
52. Delgado-González, M.J.; García-Moreno, M.V.; Sánchez-Guillén, M.M.; García-Barroso, C.; Guillén-Sánchez, D.A. Colour evolution kinetics study of spirits in their ageing process in wood casks. *Food Control* **2021**, *119*. [CrossRef]
53. Sharififar, A.; Sarmadian, F.; Malone, B.P.; Minasny, B. Addressing the issue of digital mapping of soil classes with imbalanced class observations. *Geoderma* **2019**, *350*, 84–92. [CrossRef]
54. Congalton, R.G. A review of assessing the accuracy of classifications of remotely sensed data. *Remote Sens. Environ.* **1991**, *37*, 35–46. [CrossRef]

Article

SPR-Optical Fiber-Molecularly Imprinted Polymer Sensor for the Detection of Furfural in Wine [†]

Maria Pesavento [1,*], Luigi Zeni [2], Letizia De Maria [3], Giancarla Alberti [1] and Nunzio Cennamo [2]

[1]   Department of Chemistry, University of Pavia, Via Taramelli n.12, 27100 Pavia, Italy;
       giancarla.alberti@unipv.it
[2]   Department of Engineering, University of Campania Luigi Vanvitelli, Via Roma n.29, 81031 Aversa, Italy;
       luigi.zeni@unicampania.it (L.Z.); nunzio.cennamo@unicampania.it (N.C.)
[3]   Ricerca sul Sistema Energetico-RSE S.p.A.-Via R. Rubattino n.54, 20134 Milano, Italy;
       letizia.demaria@rse-web.it
*   Correspondence: maria.pesavento@unipv.it; Tel.: +39-0382-987580
†   Presented at the 1st International Electronic Conference on Biosensors, 2–17 November 2020; Available online:
     https://iecb2020.sciforum.net/.

Abstract: A surface plasmon resonance (SPR) platform, based on a D-shaped plastic optical fiber (POF), combined with a biomimetic receptor, i.e., a molecularly imprinted polymer (MIP), is proposed to detect furfural (2-furaldheide, 2-FAL) in fermented beverages like wine. MIPs have been demonstrated to be a very convenient biomimetic receptor in the proposed sensing device, being easy and rapid to develop, suitable for on-site determinations at low concentrations, and cheap. Moreover, the MIP film thickness can be changed to modulate the sensing parameters. The possibility of performing single drop measurements is a further favorable aspect for practical applications. For example, the use of an SPR-MIP sensor for the analysis of 2-FAL in a real life matrix such as wine is proposed, obtaining a low detection limit of 0.004 mg L^{-1}. The determination of 2-FAL in fermented beverages is becoming a crucial task, mainly for the effects of the furanic compounds on the flavor of food and their toxic and carcinogenic effect on human beings.

Keywords: molecularly imprinted polymer (MIP); surface plasmon resonance (SPR); plastic optical fiber (POF); 2-furaldheide (2-FAL); beverages; optical chemical sensors

Citation: Pesavento, M.; Zeni, L.; De Maria, L.; Alberti, G.; Cennamo, N. SPR-Optical Fiber-Molecularly Imprinted Polymer Sensor for the Detection of Furfural in Wine . *Biosensors* **2021**, *11*, 72. https://doi.org/10.3390/bios11030072

Received: 3 February 2021
Accepted: 2 March 2021
Published: 5 March 2021

Publisher's Note: MDPI stays neutral with regard to jurisdictional claims in published maps and institutional affiliations.

1. Introduction

The need for low-cost and easy-to-use sensing systems for the rapid screening of food contaminants is constantly increasing. Traditional monitoring techniques are typically based on laboratory analyses of representative field-collected samples; this necessitates considerable time, effort, and expense and the sample composition may change before analysis. Alternatively, portable monitoring systems relying on sensing methods appear well suited to complement standard analytical methods and, also, can be permanently installed at the monitoring sites and can transmit the data remotely. Bio and chemo receptor-based sensors in optical fibers have been shown to be well suited in numerous applications [1–5]. In particular, the optical fiber sensing platforms based on surface plasmon resonance (SPR) [6–12] allow marker-free detection and have promising merits of low cost, high sensitivity, and small size. In general, the optical fiber for sensing application is either a glass or a plastic one. The prism-based Kretschmann and Otto configurations are the most commonly used in sensing to excite the SPR phenomenon. More recently, systems incorporating plastic optical fibers (POFs) have been introduced [12]. This upgrading makes it possible to reduce the SPR sensors' cost and dimensions by integrating the sensing platform with small optoelectronic devices (sources and detectors).

For low-cost sensing systems, POFs are especially advantageous due to their excellent flexibility, easy manipulation, great numerical aperture, large diameter, and the fact that plastic can withstand smaller bend radii than glass [12].

The SPR-POF platform developed by our research group is particularly convenient for sensing purposes when it is combined with receptors, as molecularly imprinted polymers (MIP) [13]. It is based on the SPR phenomenon taking place at a multilayer structure realized starting from a planar surface of exposed core POF, embedded in a resin block (D-shaped POF platform). The receptor layer is deposited on the gold-photoresist multilayer [13]. The flat shape of the sensing part is particularly suitable for measurements in a drop, for which no expensive and bulky flow-through cell is required.

The D-shaped POF sensing platform, developed by our research group, has been exploited in several application fields, employing different kinds of receptors, as aptamers, antibodies, metal ligands, and MIPs [13–20]. The MIPs are a class of artificial receptors whose concept dates back to the early 1930s [21–23]. Peculiar is the process of MIP preparation, which is based on a template-assisted synthesis [24]. The target molecule is dissolved in a liquid phase together with functional monomers able to coordinate around it both by covalent or non-covalent bonds. Next, the complex is polymerized in the presence of a cross-linking agent. Upon template removal, cavities are left on the polymeric material; they are complementary to the template in shape, size, and position of recognition sites. MIPs often possess recognition properties analogous to natural receptors but have the stability, ease of preparation, micromachining, integrability, and low cost of production, typical of synthetic materials [24–26].

MIPs are synthetic solids containing sites functionally and dimensionally complementary to the target molecular structure, similar to the receptor sites in bioreceptors. Moreover, it is important to emphasize that MIPs can be produced as layers in tight contact with the transducing surfaces [13,14]. It must be recognized that in this form, MIPs are different from the usual bioreceptors, for example, antibodies or aptamers, since their thickness can be higher than that of the bioreceptors' layers, usually constituted by only one or a few molecular layers. This aspect can be very important to reduce the so-called "bulk effect". In particular, the MIP's thickness can be modified in function of the real matrices in which the analyte is present; for instance, to exploit SPR-POF platforms to detect specific substances in power transformer oils, very thick MIP layers have been used [14].

MIPs have been found to have many interesting applications for selective separations and as biomimetic receptors in sensing devices, mainly for detection outside the laboratory, because of their robustness characteristics in different conditions (acidity, ionic strength, etc.), low cost, and fast development. In the D-shaped SPR-POF platform proposed by us, an MIP layer can be easily deposited by a drop coating and spinning procedure, as previously described in several cases [13,14,19,27].

In this work, the application of an SPR-POF platform with MIP as a specific receptor for the selective detection of 2-furaldheide (2-FAL) in aqueous solutions for food safety surveillance is presented. The 2-FAL detection in aqueous solutions or beverages, for instance, wine, is becoming a crucial task not only for its relevance in affecting the flavor and aroma [28], but also for its suspected toxic and carcinogenic effects on human beings [29–31]. Moreover, furanic compounds have been proposed to assess the aging of food and beverages due, for example, to inappropriate storing conditions [31–34].

An MIP for 2-FAL has been recently tested for sensing by electrochemical transduction by our group [35], taking advantage of the redox properties of 2-FAL. Here, the same MIP previously investigated is proposed with optical SPR transduction to improve the detection limits while maintaining similar characteristics of low cost and fast development of the biomimetic receptor. The portability of the apparatus is assured by the use of the POFs with a low dimension apparatus. In this study, the effect of the thickness of the MIP layer on the sensor response is investigated, particularly to modulate the SPR resonance wavelength according to the characteristics of the sample, mainly its refractive index (RI).

2. Materials and Methods

2.1. Chemicals and Instrumentation

Divynilbenzene (DVB, CAS N. 1321-74-0), methacrylic acid (MAA, CAS N. 79-41-4), 2-furaldehyde (2-FAL, CAS N. 98-01-1), and 2,2′-azobisisobutyronitrile (AIBN, CAS N. 78-67-1), were obtained from Sigma-Aldrich. MAA and DVB were purified with molecular sieves (Sigma-Aldrich cod. 208604, St. Louis, MO, USA) before use to remove stabilizers. All other chemicals were of analytical reagent grade. Pure water was obtained by a Milli-Q system (Merck Millipore, Billerica, MA, USA). Stock solutions of 2-FAL were prepared by weighing the liquid and dissolving it in pure water.

The wine-mimicking solution had the following composition: fructose 25 g/L, glucose 25 g/L, tartaric acid 5 g/L, glycerol 1.25 g/L, ethanol 180 mL/L (18% v/v). (pH = 3.3, $n = 1.3455$ RIU).

The white wine sample (WW) was a white wine in tetra pack purchased in a local supermarket.

The measurement apparatus consisted of a halogen lamp (HL–2000–LL, Ocean Optics) and a spectrometer connected to a PC (USB2000+UV–VIS spectrometer, Ocean Optics). The white light source presented an emission range from 360 nm to 1700 nm, whereas the spectrometer had a detection range from 350 nm to 1023 nm. The transmission spectra and data values were displayed online on the computer screen and saved by Spectra Suite software (Ocean Optics, Dunedin, FL, USA) [36]. An outline of the experimental setup based on spectral interrogation is reported in Figure 1a.

2.2. Preparation of the Specific MIP Layer

The prepolymeric mixture was composed of the reagents at molar ratio 1 (2-FAL):4 (MAA):40 (DVB) according to the method previously described [14]. The cross-linker divinylbenzene (DVB) was also used as the solvent in which the functional monomer (methacrylic acid, MAA) and the template, 2-FAL, were dissolved. The mixture was uniformly dispersed by sonication and de-aerated with nitrogen for 10 min. Then, the radical initiator AIBN (23 mg/mL of prepolymeric mixture) was added to the mixture.

The MIP layer was prepared directly over the flat part of the platform, dropping a small volume (about 50 μL) of the prepolymeric mixture on the platform maintained in a flat position with the help of the resin support. The prepolymeric mixture expanded spontaneously to cover the erased surface of the POF and the surface of the holder. The whole structure was spun at a given spin rate, typically at 1000 rpm for 2 min, and then placed in an oven for 16 h at 80 °C for the thermal polymerization in air [13,14,19]. Finally, the template and oligomeric polymer fragments were removed by repeated washings with 96% ethanol.

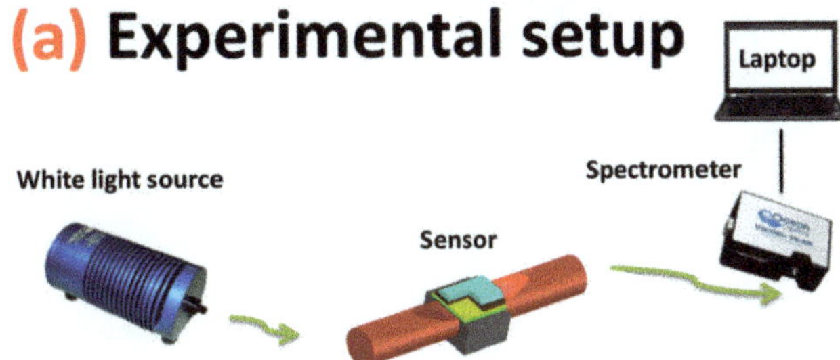

Figure 1. *Cont.*

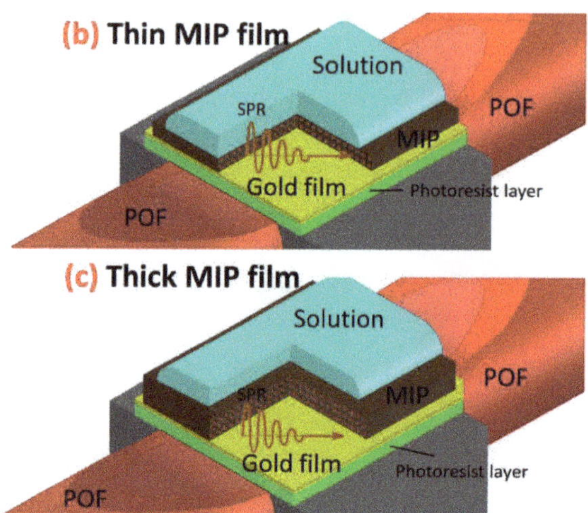

Figure 1. Optical-chemical sensor system: (**a**) outline of the experimental setup; (**b**) surface plasmon resonance- plastic optical fiber (SPR-POF) platform covered by a thin molecularly imprinted polymer (MIP) layer; (**c**) SPR-POF platform covered by a thick MIP layer. The grey block represents the resin support (1 cm × 1 cm × 1 cm) in which the POF is embedded.

2.3. Preparation of the Fiber Optic Platform

The optical platform was based on a multimode POF with a characteristic D-shaped sensing region, obtained by erasing the cladding and partially the core of the POF, held in a specially designed resin support, which produced a flat surface (see Figure 1) [13,14,36]. One half of the fiber was erased and the exposed POF core was 1 cm long. A multilayer structure was built up over the exposed core with a buffer layer (a photoresist of high refractive index with respect to the core, 1.5 µm thick), a thin metal film (gold, 60 nm thick) and, finally, an MIP layer as a specific chemical receptor for 2-FAL detection. Figure 1b,c show two typical sensing regions obtained by two MIP layers with different thickness.

In particular, the buffer layer (Microposit photoresist, MicroChemCorp., Westborough, MA, USA) was deposited on the exposed core, taking advantage of the flat shape by dropping and spinning at 6000 rpm. The so obtained layer was 1.5 µm thick and the gold layer was deposited over it by sputtering (SCD 500, Leica Microsystems, Wetzlar, Germany), forming a nanofilm 60 nm thick.

Figure 1b,c show a schematic view of SPR-POF probes, covered by MIP layers of different thicknesses, in which the penetration of the plasmonic wave in the liquid above, in the case of thin and thick MIP layers, is schematically displayed. The layers with different thickness could be obtained by spinning the prepolymeric mixture at different rates, or by depositing multiple layers of MIP.

2.4. Measurement in a Drop

The flat surface of the described optical platform makes it possible to perform the measurement in a drop simply deposited over the flat surface. The platform was fixed in a mini holder, which was purposely designed to keep the sensing surface, embedded in the resin block, in a flat position. A sample drop (50 µL) was deposited over the flat part of the sensor, allowed to expand over the sensing surface and the support, and equilibrated for 5 min. During this time, the drop over the surface maintained its shape due to the MIP surface's hydrophobicity, mainly constituted by DVB.

The concentration of 2-FAL in the sample only influenced the refractive index (RI) of the polymer in contact with the gold layer since the concentration of 2-FAL in the sample was too low to affect the solution's RI. Any RI variation related to the matrix, the so-called "bulk effect", had to be avoided or corrected.

As schematically shown in Figure 1b,c, this aspect is particularly relevant when very thin MIP layers are considered, i.e., thinner than the plasmonic wave penetration in the dielectric. In that case, the $\Delta\lambda$ must always be measured in solution with the same RI. If the sample and the standard solutions have the same RI, the spectra can be recorded directly in the drop of the liquid sample positioned over the sensing layer, but when RI is different, the solvent exchange method proposed for SPR measurements in serum ($n = 1.348$ RIU at 600 nm [37]) in flowing conditions [38] must be applied.

In the static condition required by the determination in a drop here considered, a small volume of sample (40 µL) was dropped over the flat part of the sensor and equilibrated for 10 min as in the direct measurement method, but without recording the spectrum and measuring $\Delta\lambda$. The sample solution was eliminated by suction, and the platform was washed with 40 µL of water. After the washing step, 40 µL of water was deposited over the MIP to record all the spectra with the same bulk liquid overlying the receptor layer. In this way, only the RI change induced by the binding of the analyte was detected, without any possible perturbation induced by the bulk refractive index of the examined sample. The transmission spectra in water were normalized to the spectrum obtained with the corresponding sensor (SPR-MIP) in air (reference spectrum).

As a matter of fact, in the reference spectrum acquired on SPR-MIP with a thin MIP layer, no plasmon resonance was excited in the operative refractive index range of the platform here described [36].

2.5. Standardization Curves

The standardization curves were obtained by plotting the variation of the resonance wavelength in the normalized transmission spectra ($\Delta\lambda$) vs. the concentration of 2-FAL. $\Delta\lambda$ was calculated respect to the resonance wavelength of a blank solution, i.e., a solution with the same composition of the sample but not containing the analyte of interest.

With the limited number of receptor sites in this kind of sensor, the response was linear only in very small concentration ranges. Therefore, the standardization curves were modeled by an equation deriving from the Langmuir adsorption isotherm [14,19,20], assuming that the signal is directly proportional to the amount of the template in the sensing layer ($\Delta\lambda = k\,g\,c_{\text{Aint}}$):

$$\Delta\lambda = \frac{k\,g\,c_{\text{int}}K_{\text{aff}}[A]}{1 + K_{\text{aff}}[A]} = \frac{\Delta\lambda_{\text{max}}K_{\text{aff}}[A]}{1 + K_{\text{aff}}[A]} \tag{1}$$

[A] is the concentration of the analyte in the sample solution. K_{aff} (in mg^{-1} L) is the affinity constant of the adsorption equilibrium, c_{int} is the concentration of the specific sites of the MIP (in mg g^{-1}), g is the polymer mass (in grams). $\Delta\lambda_{\text{max}} = k\cdot g\cdot c_{\text{int}}$ is the maximum $\Delta\lambda$ at high concentration of the analyte (i.e., when the analyte saturates all the specific sites).

Equation (1) can be applied as a standardization curve if the analyte concentration at the equilibrium in the sample, [A], is equal to the total one (c_A), i.e., when the concentration of the analyte adsorbed is negligible.

The parameters of Equation (1) were obtained by Solver, the Microsoft Excel add-in program.

Once the parameters are known, the concentration of the analyte can be evaluated in the whole detection range from the measured $\Delta\lambda$:

$$[A] = c_A = \frac{\Delta\lambda}{K_{\text{aff}}\cdot(k\cdot g\cdot c_{\text{int}} - \Delta\lambda)} \tag{2}$$

3. Results

3.1. Response of the SPR-Bare Platforms to the Refraction Index Variation

The response of the SPR platforms, not derivatized with MIP (SPR-bare), here considered, has been checked as previously proposed [36]. In particular, the resonance wavelengths of the platform in liquid media of different RI, measured by an Abbe refractometer, were registered. As an example, the spectra of water–glycerol solutions with different RI are reported in Figure 2a. The response curve $\Delta\lambda$ (referring to pure water) against RI is shown in Figure 2b.

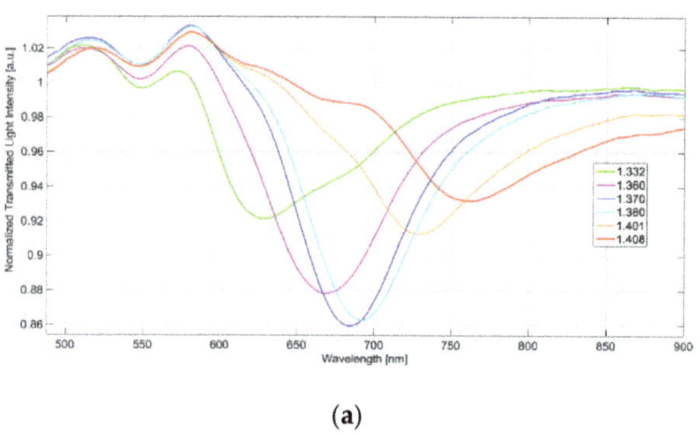

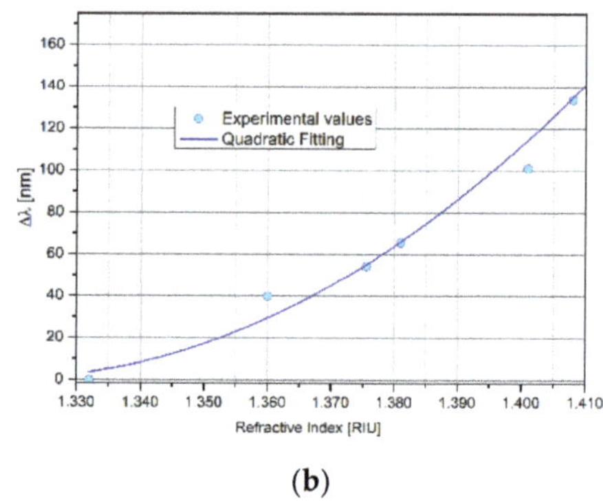

(a)   (b)

Figure 2. (a) Transmission spectra of the MIP-bare platform in water–glycerol solutions with different refractive index (RI). (b) Variation of the resonance wavelength with the RI of the overlying liquid.

The sensitivity ($\delta\Delta\lambda/\delta RI$) is significantly different from zero in the whole RI range, and it increases at increasing RI (Figure 2b). Nevertheless, for short RI ranges, the curve can be assumed to be linear. At increasing RI, the peak is shifted to higher wavelength values and becomes larger and flatter, making the acquisition of the minimum less accurate and precise. For this reason, the best RI of the dielectric layer in contact with the resonant surface must be a compromise between the highest sensitivity and precision.

3.2. MIP Layer Analysis

The SEM image reported in Figure 3 shows the gold surface modified with MIP polymerized in situ, where the presence of polymer aggregates of small particles can be seen. This SEM image is relative to a sensing platform in which the MIP layer has been obtained by spinning the prepolymeric mixture at high velocity (about 1000 rpm). It shows that at the micro level, the MIP film does not entirely cover the gold surface, so in some way it reproduces the surface derivatized with a bioreceptor, i.e. a large biomolecule.

Figure 4 reports the comparison between the SPR spectrum in water of a bare platform (SPR-bare) with that of the same platform modified with multiple layers of MIP (SPR-MIP). The MIP made the RI of the region overlying gold higher than pure water, as proven by the resonance wavelength shift towards higher values (red shift). The SPR spectra in Figure 4 are relative to a set of sensors prepared by depositing successively one, two, or three layers of MIP on equal platforms, as reported in Table 1. Water (n = 1.332 RIU) was the liquid over the platform. The example in Figure 4 shows that the resonance wavelength shift ($\Delta\lambda$) with respect to that of pure water depends on the amount of MIP deposited. The resonance wavelengths are reported in Table 1.

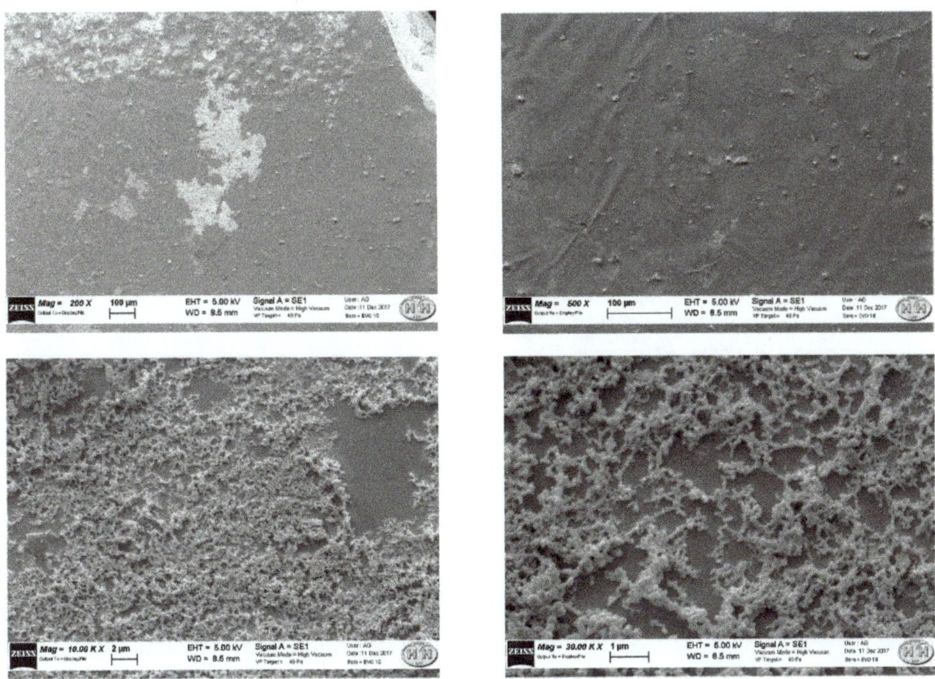

Figure 3. SEM images of different points at the surface of the sensing part.

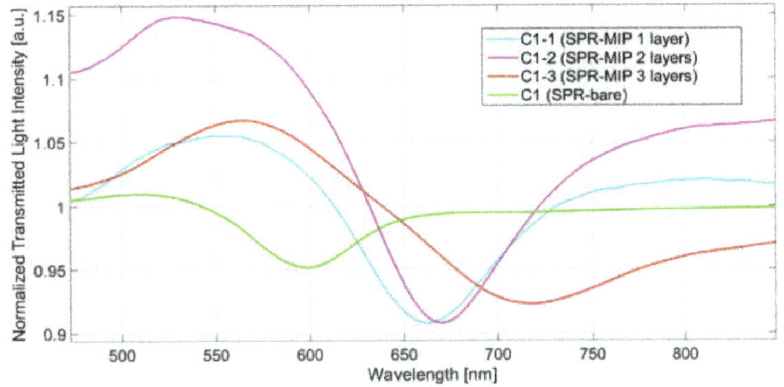

Figure 4. Transmission spectra of C1 (SPR-MIP) platform, with one, two, or three MIP layers, in water, normalized on the spectrum of the corresponding platforms in air.

Table 1. Resonance wavelength of sensors with multiple MIP layers. Formation of each MIP layer: 40 µL of prepolymeric mixture spun at 300 rpm. Normalization on the corresponding platform in air.

Type of Sensor	N. of MIP Layers	λ_{ris} in Water [nm]	$\Delta\lambda$ (SPR-MIP in Water Minus SPR-Bare in Water) [nm]
C1 (SPR-bare)	0	597.7	0
C1-1 (SPR-MIP)	1	663.2	65.5
C1-2 (SPR-MIP)	2	672.4	74.7
C1-3 (SPR-MIP)	3	718.3	120.6

By increasing the number of MIP layers, the resonance wavelength increases, and a shift to higher wavelengths (red shift) is observed (from 663.0 nm to 718.3 nm). Even with three layers deposited (C1-3), the resonance wavelength (718.3) is lower than the maximum wavelength useful for measurement.

In order to verify if the sensor with three layers of MIP is still sensitive to the RI of the overlying liquid, the SPR transmission spectra in water–glycerol solutions with different refractive indices, positioned like a drop over the MIP (bulk solution), are reported in Figure 5.

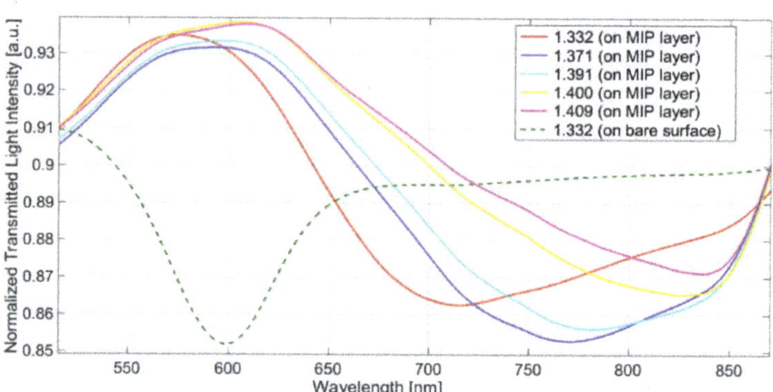

Figure 5. Comparison of the spectra of an SPR-MIP sensor (C1-3) in water–glycerol solutions with different refractive indices and the spectrum of an SPR-bare platform in water.

The resonance in water of the SPR-MIP platform is at 715 nm (see Figure 4), with a shift of 115 nm compared to the resonance of SPR-bare in water, indicating that the amount of MIP layer is relatively large. However, a shift to higher wavelengths is observed when liquids with RI higher than water substitute water. This behavior demonstrates that the MIP layer is not so thick as to completely include the plasmonic wave, as schematically illustrated in Figure 1b,c.

3.3. Sensor Response in Water and in Wine-Mimicking Solution

Water is a well-suited solvent for measurements with the platforms here described since its RI matches the operative range for the proposed SPR sensors [36]. The spectra can be recorded directly in the sample, since 2-FAL standards in water are easily prepared. When the 2-FAL concentration increases, the SPR wavelength is shifted to higher values, as seen in Figure 6a. As an example, a standardization curve of 2-FAL in water is shown in Figure 6b, and is obtained by reporting the wavelength variation $\Delta\lambda$ vs. 2-FAL concentration. The experimental data are well fitted by Equation (1), as expected when the sorption takes place according to the Langmuir model. A plateau corresponding to $\Delta\lambda_{max}$ is obtained for concentrations higher than 1 mg·L^{-1} due to the receptor's sites saturation. The parameters, evaluated by a non-linear regression method, are K_{aff}, 9.4 L·mg^{-1} (9.0 10^5 M^{-1}) and $\Delta\lambda_{max}$ = 3.3 (0.447) nm. The sensitivity at low concentration (i.e., in the linear part of the curve) is 31.6 nm·mg^{-1}·L. The saturation was reached at about 1 mg·L^{-1}.

The response of the sensor to 2-FAL concentration in wine-mimicking solution is reported in Figure 6c,d. In this case, the spectra were recorded in water, after equilibration with the sample and washing, according to the method of the solvent exchange.

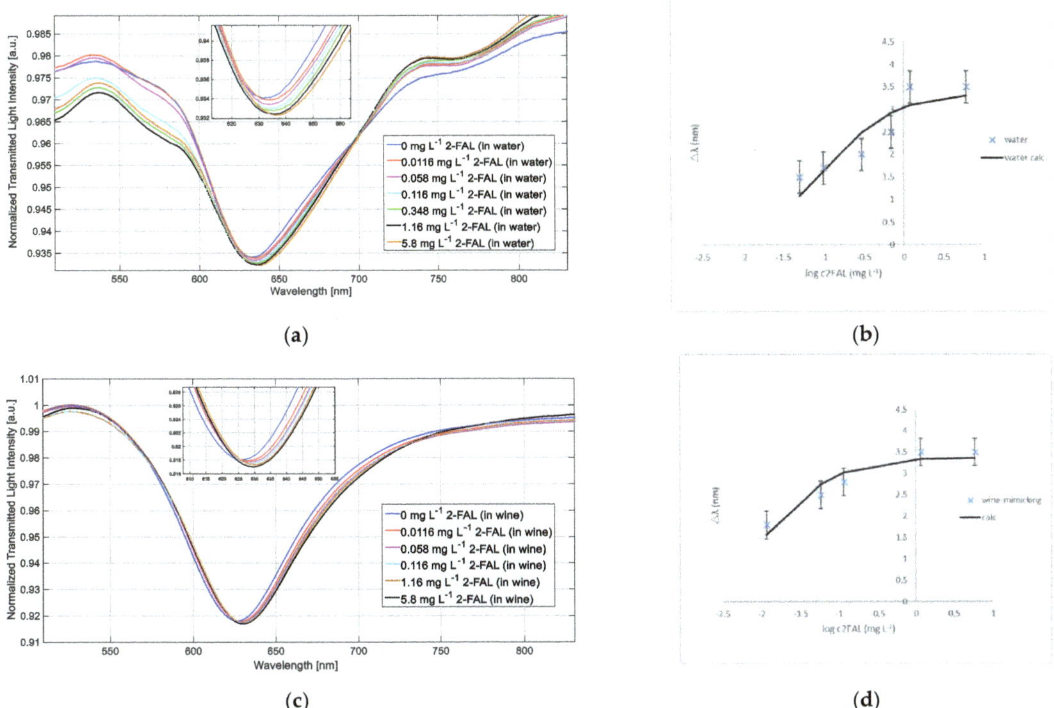

Figure 6. (**a**) SPR spectra of the SPR-MIP sensor at different 2-furaldheide (2-FAL) concentrations in water; (**b**) SPR wavelength shift vs. concentration of 2-FAL in a water solution and the calculated continuous curve; (**c**) SPR spectra of the SPR-MIP sensor at different 2-FAL concentrations in wine-mimicking solution; (**d**) SPR wavelength shift vs. concentration of 2-FAL in wine-mimicking solution and the calculated continuous curve. Vertical bars in (**b**,**d**) correspond to the standard deviations (error bars).

The dose–response curve ($\Delta\lambda$ vs. 2-FAL concentration) in wine-mimicking solution followed the Langmuir model (see Equation (1)) and the parameters obtained are: $K_{aff} = 75.6$ L·mg^{-1} and $\Delta\lambda_{max} = 3.3$ (0.247) nm. The sensitivity at low concentration is 254.9 nm·mg^{-1}·L, and the limit of detection (LOD) is 0.004 mg·L^{-1}. The saturation was reached at about 0.6 mg·L^{-1}.

The affinity constant is about 8 times higher in wine-mimicking solution than in pure water; consequently, the LOD is lower than in water.

3.4. Determination of 2-FAL in a White Wine Sample

The sample considered was a white wine in a tetra pack container (WW) purchased in a local supermarket, in which a concentration of 2-FAL of 0.1 ppm was found by an HPLC method. The SPR spectra were obtained as reported in the experimental part, using water as the liquid in which the spectra were finally recorded (bulk liquid). This method had to be applied since the exact RI of the white wine is unknown. Figure 7 shows the transmission spectra recorded in water of the considered WW sample, and with some standard additions of 2-FAL.

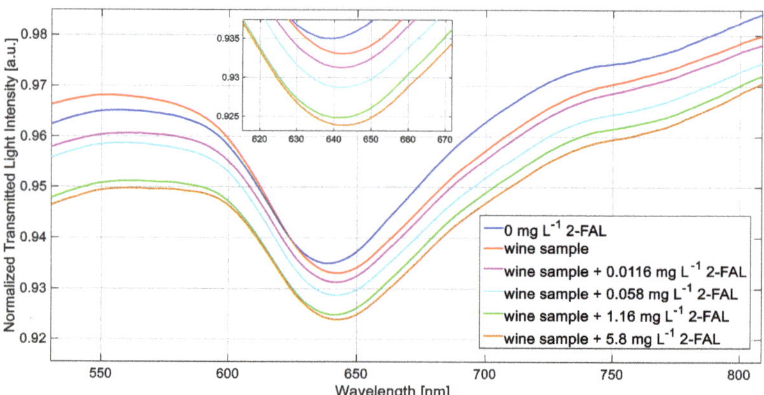

Figure 7. SPR spectra obtained by the SPR-MIP sensor in wine sample.

The variation of λ_{ris} from the blank (water) to the WW (2.8 nm) is due to the adsorption of 2-FAL on the MIP during the incubation with the sample. There is only a very slight variation of λ_{ris} in response to the standard additions of 2-FAL to the sample. This behavior indicates that the imprinted sites are almost saturated by the 2-FAL originally present in the WW considered. In fact, the concentration calculated from Equation (1), using the parameters evaluated in wine-mimicking solution, is 0.107 mg L^{-1}, near to the approximate saturation concentration, in acceptable agreement with the value found by the HPLC method (0.1 mg L^{-1}).

4. Discussion

In the SPR-POF sensing apparatus here considered, the operative RI ranges were from about 1.33 to 1.41 (see Figure 2). In aqueous matrices with low RI, an RI in this range is obtained when a thin MIP layer is deposited on the gold surface so that the plasmonic wave partially penetrates in the overlying aqueous medium, as shown in Figure 1b. The preparation of the MIP layer reported in the experimental part allowed this goal to be achieved, i.e., having an RI of the layer (MIP plus water solution) in the suitable range. The resonance wavelength shift of the sensor with MIP in water, with respect to that of the bare platform in water, was around 50–150 nm. Similar shifts were previously observed in sensors with MIP's receptor. For example, a value as high as 150 nm in sensors for TNT [13] and nicotine [19], and about 80 nm in a 2-FAL sensor [27] were obtained. Lower $\Delta\lambda$ was observed in the case of a monolayer of molecular receptor as in the case of an aptamer [17] or a metal ligand [16,19], confirming that the shift is higher when a higher amount of receptor is present at the resonant surface. As shown in the present investigation (Figure 2), the resonance wavelength should not exceed about 800 nm, so that an MIP layer that is not too thick must be used.

On the other hand, this makes the sensor sensitive to the RI of the overlying liquid, which thus must be carefully controlled.

For this reason, when measuring a real life sample, for example wine, using the sensing method proposed, any RI variation due to the matrix composition must be avoided. This aspect is relevant when thin MIP layers are considered, i.e., thinner than the plasmonic wave penetration in the dielectric. In that case, the $\Delta\lambda$ must always be measured in solution with the same RI. The matrix effects are avoided by equilibrating the sample with the sensor and then changing the sample with a solvent with a proper RI (for example, an aqueous buffer), for recording the spectrum and measuring $\Delta\lambda$. In the present work, to demonstrate the method's effectiveness, pure water was selected as the liquid for registering the spectrum for the analysis of 2-FAL in a white wine sample. An alternative approach could be applied, using a much thicker MIP layer; however, in this case, the RI of the dielectric over gold could not be suitable for the measurements since the resonance wavelength could be outside the operative range.

5. Conclusions

The proposed platform, based on D-shaped POF and an MIP as the receptor layer, has been demonstrated to be particularly suited for measurements in aqueous matrices since the RI of the layer in contact with the resonant surface (gold) matches the measurable RI range. On the other hand, this strongly depends on the MIP layer's thickness, at least for the MIP here considered.

It has been found that the "wine" matrix has a large effect on the SPR signal since the response of the sensor to 2-FAL is different in water and wine-mimicking solution. The affinity constant is one order of magnitude higher in wine than in water, and consequently, a lower LOD is obtained, reaching interesting values for the determination of 2-FAL in wine.

The method developed for the determination of 2-FAL in wine could have a high practical interest; on the one hand, for the possible toxic and carcinogenic effects of furanic aldehydes, particularly 2-FAL, on human beings and, on the other hand, for their impact on the aroma of food. Interestingly, the sensor here developed could be easily adapted to the analysis of other furanic compounds.

Author Contributions: Conceptualization, M.P. and N.C.; methodology, M.P., N.C., L.D.M., L.Z., G.A.; validation, M.P., N.C., L.D.M., L.Z., G.A.; formal analysis, M.P., N.C., L.D.M., L.Z., G.A.; investigation, M.P., N.C., L.D.M., L.Z., G.A.; data curation, M.P., N.C., L.D.M.; writing—original draft preparation, M.P., N.C., L.D.M., L.Z., G.A.; writing—review and editing, M.P., N.C., L.D.M., L.Z., G.A. All authors have read and agreed to the published version of the manuscript.

Funding: This research received no external funding.

Acknowledgments: This work was supported by the VALERE program of the University of Campania Luigi Vanvitelli (Italy).

Conflicts of Interest: The authors declare no conflict of interest.

References

1. Caucheteur, C.; Guo, T.; Albert, J. Review of plasmonic fiber optic biochemical sensors: Improving the limit of detection. *Anal. Bioanal. Chem.* **2015**, *407*, 3883–3897. [CrossRef]
2. Leung, A.; Shankar, P.M.; Mutharasan, R. A review of fiber-optic biosensors. *Sens Actuators B Chem.* **2007**, *125*, 688–703. [CrossRef]
3. Wang, X.D.; Wolfbeis, O.S. Fiber-Optic Chemical Sensors and Biosensors (2008–2012). *Anal. Chem.* **2013**, *85*, 487–508. [CrossRef]
4. Wang, X.D.; Wolfbeis, O.S. Fiber-Optic Chemical Sensors and Biosensors (2013–2015). *Anal. Chem.* **2016**, *88*, 203–227. [CrossRef]
5. Monk, D.J.; Walt, D.R. Optical fiber-based biosensors. *Anal. Bioanal. Chem.* **2004**, *379*, 931–945. [CrossRef] [PubMed]
6. Anuj, K.; Sharma, R.J.; Gupta, B.D. Fiber-optic sensors based on surface plasmon resonance: A comprehensive review. *IEEE Sens. J.* **2007**, *7*, 1118–1129.
7. Gupta, B.D.; Verma, R.K. Surface plasmon resonance-based fiber optic sensors: Principle, probe designs, and some applications. *J. Sens.* **2009**, *2009*, 979761. [CrossRef]
8. Trouillet, A.; Ronot-Trioli, C.; Veillas, C.; Gagnaire, H. Chemical sensing by surface plasmon resonance in a multimode optical fibre. *Pure Appl. Opt.* **1996**, *5*, 227–237. [CrossRef]
9. Piliarik, M.; Homola, J.; Manikova, Z.; Čtyroký, J. Surface Plasmon Resonance Sensor Based on a Single-Mode Polarization-Maintaining Optical Fiber. *Sens. Actuators B Chem.* **2003**, *90*, 236–242. [CrossRef]
10. Jorgenson, R.C.; Yee, S.S. A fiber-optic chemical sensor based on surface plasmon resonance. *Sens. Actuators B Chem.* **1993**, *12*, 213–220. [CrossRef]
11. Homola, J. Present and future of surface plasmon resonance biosensors. *Anal. Bioanal. Chem.* **2003**, *377*, 528–539. [CrossRef] [PubMed]
12. Cennamo, N.; Pesavento, M.; Zeni, L. A review on simple and highly sensitive plastic optical fiber probes for bio-chemical sensing. *Sens. Actuators B Chem.* **2021**, *331*, 129393. [CrossRef]
13. Cennamo, N.; D'Agostino, G.; Galatus, R.; Bibbò, L.; Pesavento, M.; Zeni, L. Sensors based on surface plasmon resonance in a plastic optical fiber for the detection of trinitrotoluene. *Sens. Actuators B* **2013**, *188*, 221–226. [CrossRef]
14. Cennamo, N.; De Maria, L.; Chemelli, C.; Profumo, A.; Zeni, L.; Pesavento, M. Markers detection in transformer oil by plasmonic chemical sensor system based on POF and MIPs. *IEEE Sens. J.* **2016**, *16*, 7663–7670. [CrossRef]

15. Cennamo, N.; Di Giovanni, S.; Varriale, A.; Staiano, M.; Di Pietrantonio, F.; Notargiacomo, A.; Zeni, L.; D'Auria, S. Easy to use plastic optical fiber-based biosensor for detection of butanal. *PLoS ONE* **2015**, *10*, e0116770. [CrossRef] [PubMed]
16. Cennamo, N.; Varriale, A.; Pennacchio, A.; Staiano, M.; Massarotti, D.; Zeni, L.; D'Auria, S. An innovative plastic optical fiber-based biosensor for new bio/applications. The Case of Celiac Disease. *Sens. Actuators B* **2013**, *176*, 1008–1014. [CrossRef]
17. Cennamo, N.; Alberti, G.; Pesavento, M.; D'Agostino, G.; Quattrini, F.; Biesuz, R.; Zeni, L. A Simple Small Size and Low Cost Sensor Based on Surface Plasmon Resonance for Selective Detection of Fe(III). *Sensors* **2014**, *14*, 4657–4661. [CrossRef]
18. Cennamo, N.; Pesavento, M.; Lunelli, L.; Vanzetti, L.; Pederzolli, C.; Zeni, L.; Pasquardini, L. An easy way to realize SPR aptasensor: A multimode plastic optical fiber platform for cancer biomarkers detection. *Talanta* **2015**, *140*, 88–95. [CrossRef] [PubMed]
19. Cennamo, N.; D'Agostino, G.; Pesavento, M.; Zeni, L. High selectivity and sensitivity sensor based on MIP and SPR in tapered plastic optical fibers for the detection of L-nicotine. *Sens. Actuators B* **2014**, *191*, 529–536. [CrossRef]
20. Pesavento, M.; Profumo, A.; Merli, D.; Cucca, L.; Zeni, L.; Cennamo, N. An Optical Fiber Chemical Sensor for the Detection of Copper(II) in Drinking Water. *Sensors* **2019**, *19*, 5246. [CrossRef] [PubMed]
21. Polyakov, M.V. Adsorption properties of silica gel and its structure. *Zhurnal Fiz. Khimii* **1931**, *6*, 799–805.
22. Wulff, G.; Sarhan, A. The Use of Polymers with Enzyme-Analogous Structures for the Resolution of Racemates. *Angew. Chem. Int. Ed.* **1972**, *11*, 341–344.
23. Arshady, R.; Mosbach, K. Synthesis of substrate-selective polymers by host-guest polymerization. *Makromol. Chem.* **1981**, *182*, 687–692. [CrossRef]
24. Uzun, L.; Turner, A.P.F. Molecularly-imprinted polymer sensors: Realising their potential. *Biosens. Bioelectron.* **2016**, *76*, 131–144. [CrossRef] [PubMed]
25. Li, S.; Ge, Y.; Piletsky, S.A.; Lunec, J. *Molecularly Imprinted Sensors, Overview and Applications*; Elsevier: Amsterdam, The Netherlands, 2012. [CrossRef]
26. Chen, L.; Wang, X.; Lu, W.; Wu, X.; Li, J. Molecular imprinting: Perspectives and applications. *Chem. Soc. Rev.* **2016**, *45*, 2137–2211. [CrossRef] [PubMed]
27. Zeni, L.; Pesavento, M.; Marchetti, S.; Cennamo, N. Slab plasmonic platforms combined with Plastic Optical Fibers and Molecularly Imprinted Polymers for chemical sensing. *Opt. Laser Technol.* **2018**, *107*, 484–490. [CrossRef]
28. Camara, J.S.; Alves, M.A.; Marques, J.C. Changes in volatile composition of Madeira wines during their oxidative ageing. *Anal. Chim. Acta* **2006**, *563*, 188–197. [CrossRef]
29. Abraham, K.; Gürtler, R.; Berg, K.; Heinemeyer, G.; Lampen, A.; Appel, K.E. Toxicology and risk assessment of 5-Hydroxymethylfurfural in food. *Mol. Nutr. Food Res.* **2011**, *55*, 667–678. [CrossRef] [PubMed]
30. Capuano, E.; Fogliano, V. Acrylamide and 5-hydroxymethylfurfural (HMF): A review on metabolism, toxicity, occurrence in food and mitigation strategies. *LWT Food Sci. Technol.* **2011**, *44*, 793–810. [CrossRef]
31. Islam, M.N.; Khalil, M.I.; Islam, M.A.; Gan, S.H. Toxic compounds in honey. *J. Appl. Toxicol.* **2014**, *34*, 733–742. [CrossRef] [PubMed]
32. Foo Wong, Y.; Makahleh, A.; Al Azzam, K.M.; Yahaya, N.; Saad, B.; Amrah Sulaiman, S. Micellar electrokinetic chromatography method for the simultaneous determination of furanic compounds in honey and vegetable oils. *Talanta* **2012**, *97*, 23–31. [CrossRef] [PubMed]
33. Cocchi, M.; Durante, C.; Lambertini, P.; Manzini, S.; Marchetti, A.; Sighinolfi, S.; Totaro, S. Evolution of 5-(hydroxymethyl)furfural and furfural in the production chain of the aged vinegar Aceto Balsamico Tradizionale di Modena. *Food Chem.* **2011**, *124*, 822–832. [CrossRef]
34. Verissimo, M.I.S.; Gamelas, J.A.F.; Evtuguin, D.V.; Gomes, M.T.S. Determination of 5-hydroxymethylfurfural in honey, using head space solid-phase microextraction coupled with a polyoxometalate-coated piezoelectric quartz crystal. *Food Chem.* **2017**, *220*, 420–426. [CrossRef] [PubMed]
35. Pesavento, M.; Marchetti, S.; De Maria, L.; Zeni, L.; Cennamo, N. Sensing by Molecularly Imprinted Polymer: Evaluation of the Binding Properties with Different Techniques. *Sensors* **2019**, *19*, 1344. [CrossRef]
36. Cennamo, N.; Massarotti, D.; Conte, L.; Zeni, L. Low cost sensors based on SPR in a plastic optical fiber for biosensor implementation. *Sensors* **2011**, *11*, 11752–11760. [CrossRef]
37. Liu, S.; Deng, Z.; Li, J.; Wang, J.; Huang, N.; Cui, R.; Zhang, Q.; Mei, J.; Zhou, W.; Zhang, C.; et al. Measurement of the refractive index of whole blood and its components for a continuous spectral region. *J. Biomed. Opt.* **2019**, *24*, 035003. [CrossRef] [PubMed]
38. Cennamo, N.; Chiavaioli, F.; Trono, C.; Tombelli, S.; Giannetti, A.; Baldini, F.; Zeni, L. A Complete Optical Sensor System Based on a POF-SPR Platform and a Thermo-Stabilized Flow Cell for Biochemical Applications. *Sensors* **2016**, *16*, 196. [CrossRef] [PubMed]

 *biosensors*

Article

Contactless Temperature Sensing at the Microscale Based on Titanium Dioxide Raman Thermometry [†]

Veronica Zani [1,2], **Danilo Pedron** [1,2], **Roberto Pilot** [1,2] and **Raffaella Signorini** [1,2,*]

[1] Department of Chemical Science, University of Padua, Via Marzolo 1, I-35131 Padova, Italy; veronica.zani@studenti.unipd.it (V.Z.); danilo.pedron@unipd.it (D.P.); roberto.pilot@unipd.it (R.P.)
[2] Consorzio INSTM, Via G. Giusti 9, I-50121 Firenze, Italy
* Correspondence: raffaella.signorini@unipd.it; Tel.: +39-049-8275118
[†] This paper is an extended version of our paper published in: Zani, V.; Pedron, D.; Pilot, R.; Signorini, R. Biocompatible Temperature Nanosensors Based on Titanium Dioxide. In Proceedings of the 1st International Electronic Conference on Biosensors, 2–17 November 2020.

Abstract: The determination of local temperature at the nanoscale is a key point to govern physical, chemical and biological processes, strongly influenced by temperature. Since a wide range of applications, from nanomedicine to nano- or micro-electronics, requires a precise determination of the local temperature, significant efforts have to be devoted to nanothermometry. The identification of efficient materials and the implementation of detection techniques are still a hot topic in nanothermometry. Many strategies have been already investigated and applied to real cases, but there is an urgent need to develop new protocols allowing for accurate and sensitive temperature determination. The focus of this work is the investigation of efficient optical thermometers, with potential applications in the biological field. Among the different optical techniques, Raman spectroscopy is currently emerging as a very interesting tool. Its main advantages rely on the possibility of carrying out non-destructive and non-contact measurements with high spatial resolution, reaching even the nanoscale. Temperature variations can be determined by following the changes in intensity, frequency position and width of one or more bands. Concerning the materials, Titanium dioxide has been chosen as Raman active material because of its intense cross-section and its biocompatibility, as already demonstrated in literature. Raman measurements have been performed on commercial anatase powder, with a crystallite dimension of hundreds of nm, using 488.0, 514.5, 568.2 and 647.1 nm excitation lines of the CW Ar^+/Kr^+ ion laser. The laser beam was focalized through a microscope on the sample, kept at defined temperature using a temperature controller, and the temperature was varied in the range of 283–323 K. The Stokes and anti-Stokes scattered light was analyzed through a triple monochromator and detected by a liquid nitrogen-cooled CCD camera. Raw data have been analyzed with Matlab, and Raman spectrum parameters—such as area, intensity, frequency position and width of the peak—have been calculated using a Lorentz fitting curve. Results obtained, calculating the anti-Stokes/Stokes area ratio, demonstrate that the Raman modes of anatase, in particular the E_g one at 143 cm^{-1}, are excellent candidates for the local temperature detection in the visible range.

Keywords: temperature sensor; Raman spectroscopy; anti-Stokes/Stokes spectra; titanium dioxide

Citation: Zani, V.; Pedron, D.; Pilot, R.; Signorini, R. Contactless Temperature Sensing at the Microscale Based on Titanium Dioxide Raman Thermometry . *Biosensors* **2021**, *11*, 102. https://doi.org/10.3390/bios11040102

Received: 10 February 2021
Accepted: 19 March 2021
Published: 31 March 2021

Publisher's Note: MDPI stays neutral with regard to jurisdictional claims in published maps and institutional affiliations.

1. Introduction

The determination of temperature with good accuracy and with nano/micro-spatial resolution (nanothermometry) has been matter of intense research efforts since it opens up new perspectives in different research fields like biomedicine, photonics and nanoelectronics [1–4]. For example, Okabe et al. [5] reported on the investigation of cell functions by mapping the intracellular temperature, Quintanilla et al. [6] engineered a probe for monitoring temperature during photothermal therapy, Santos et al. [7] worked on the early diagnosis of tumors, exploiting the different thermal dynamics of healthy and diseased tissues and Mi et al. mapped the temperature on a micro sized magneto-resistive device [8].

187

In optical nanothermometry, luminescence is currently the most widely employed detection technique. However, also Raman spectroscopy is emerging as a valuable tool for temperature measurements. Despite being an intrinsically weaker phenomenon than fluorescence (requiring longer integration times), advantages of Raman include the wide range of temperatures detectable, the ease of sample preparation and the ample availability of materials possessing a Raman spectrum [9–13]. Moreover, it is characterized by a great spatial resolution, in the order of the diffraction limit of the probe laser [14].

The Raman effect is the inelastic scattering of light and well-defined characteristics of the Raman spectrum, such as intensity, position in frequency and width of peak signals, are related to temperature. It follows that temperature can be measured from Raman spectra by determining the degree of the shift position of a defined peak at different temperatures, or by evaluating the broadening of its linewidth or by measuring the peak intensity ratio of the anti-Stokes signal to the Stokes signal [15].

A good Raman thermometer material should possess these properties: (a) a large Raman scattering cross-section (to reach high signal-to-noise ratios); (b) high-intensity Raman peaks at low Raman shifts (the upper limit, in frequency shift, depends on the working temperature and in general near room temperature it is about 600 cm^{-1} [13]), indeed the lower the Raman shift, the more sensitive is the peak intensity to temperature; (c) well-defined and distinguishable Raman peaks and (d) low absorbance at the excitation wavelength (to avoid heating mechanisms).

Temperature cannot be measured directly, signals, like frequency position or intensity of the anti-Stokes and Stokes Raman peaks, can be used as indication (Q), which is linked to temperature through a mathematical equation (the so called measurement model). Great care should also be devoted to the determination of the uncertainty of the obtained values, indicating the dispersion of values within which the true temperature value is expected to lie. The measured temperature will be accurate if it is very close to the true value, and if measured temperatures, acquired by replicate measurements on the same object, are close to each other. In addition, sensitivity and thermal resolution are two fundamental parameters to evaluate thermometry. Sensitivity (S) is defined as the derivative of the indication with respect to the temperature, $S = |\partial Q / \partial T|$ [9,16], while the thermal resolution is the smallest change in a temperature able to cause a perceptible change in the indication Q, calculated as the ratio of uncertainty (the standard deviation, σ) and sensitivity $\Delta T_{min} = \sigma / S$ [16].

When Raman thermometers are considered, it is also important to investigate the effect of the laser power on the local temperature, in order to avoid the heating of the sample due to the laser itself. For this purpose, it is interesting to examine the behavior of the system as a function of the laser intensity (when very high laser power intensities are achieved to obtain high intensity signals [17]). In general, the Raman signal depends on the third or the fourth power of the excitation frequency [18], so that it is expected to increase with decreasing excitation wavelength. That is the reason that many Raman measurements are conducted using shorter excitation wavelength; nonetheless, other factors have to be considered when choosing the proper laser frequency. Actually, it affects the depth of focus and the focal volume of the laser beam (the longer the wavelength the deeper light penetrates the sample), the spatial resolution and notably the photoluminescence background, which can be present also with transparent materials. Moreover, the presence of electronic transitions close to the excitation frequency has to be considered, as it causes an enormous enhancement of Raman signals [19], which is desirable to increase sensitivity, but may induce a local heating of the sample. When temperature is measured through the anti-Stokes/Stokes ratio, all these aspects have to be considered when choosing the material and/or the excitation wavelength, because a wavelength dependence of the Raman cross-section can determine an asymmetry between Stokes and anti-Stokes processes, resulting in anomalous anti-Stokes/Stokes ratios.

Studies on temperature with Raman measurements can be found in literature for silicon [14,20], gallium arsenide [21], gallium nitride [22] and graphene [23]. Titanium dioxide (TiO$_2$) has also been tested in few works as a Raman thermometer for titanium dioxide microparticles [24,25] and for thin films of titania used in solar cells [26].

Titanium dioxide's general features of chemical stability and nontoxicity make it a very interesting compound for various different applications, including photocatalysis [27], optical coatings, optoelectronic devices [28] and biomedicine [29]. It is a wide band gap insulator (3.0 eV [30–32]) and exists in nature in three different crystal structures: anatase, rutile and brookite. In particular, the anatase phase is exploited in photocatalysis, photochemical solar cells, optoelectronic devices and chemical sensors [33,34]. Titanium dioxide seems to fit perfectly all the requirements for a good thermometer material and has been chosen as Raman active thermometric material in our research.

The aim of this study is to obtain a protocol for temperature determination, with a high spatial resolution, of the order of the micro-nanometer dimension, exploiting Raman spectroscopy on anatase powder. As multiple signals are present in the Raman spectrum of Titanium dioxide, the choice of the actual Raman mode to be used has been performed on the base of its sensitivity to temperature. The ratio between Stokes and anti-Stokes signals of the same Raman mode has been investigated as a function of temperature (T), excitation wavelength (λ_{exc}) and input power. The control of the temperature is obtained by using a temperature controller, which is assumed also as reference for the determination of the absolute temperature. The performances of the temperature sensor are examined in the wavelength range 488.0–647.1 nm, to individuate the best excitation wavelength in terms of reaching the highest sensitivity, and in the temperature range 283–323 K, which is important for biological applications. The work will demonstrate that a different calibration constant is necessary for different wavelengths and Raman modes. The calibration constants, determined with this work, have been tested on a titanium dioxide based *Test Sample*, obtaining results with high sensitivity and low uncertainty and open the way to the use, in the future, of titanium dioxide-based new biosensors.

2. Materials and Methods

Raman measurements have been performed using a micro-Raman setup in a back-scattering geometry; the principal elements of the setup are showed in Figure 1.

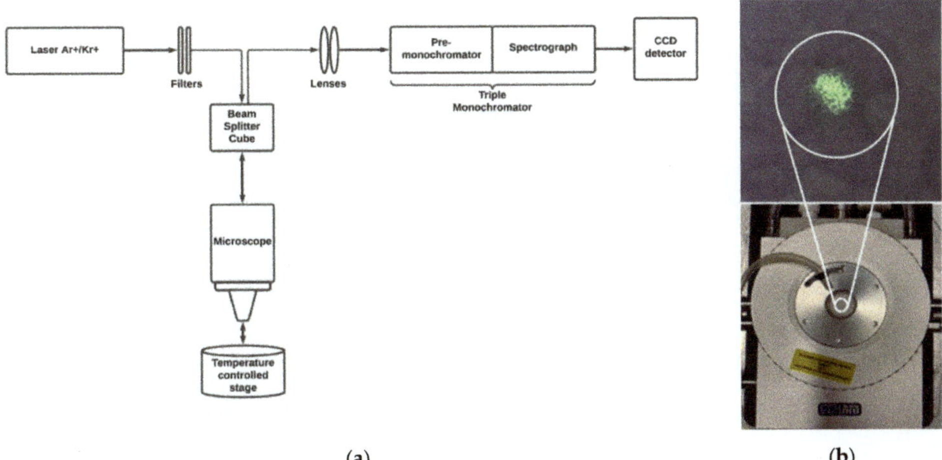

Figure 1. Experimental micro-Raman setup (**a**) and Linkam THMS600/720 temperature-controlled stage with zoom on the sample inserted, showing the laser spot (**b**).

The system is equipped with a CW Ti:Sapphire Laser, tunable in the range 675–1000 nm (MKS Instruments, Spectra Physics, 3900S, Santa Clara, CA, USA) and pumped by a CW

Optically Pumped Semiconductor Laser (Coherent, Verdi G7, Santa Clara, CA, USA), and an Ar^+/Kr^+ gas laser (Coherent, Innova 70, Santa Clara, CA, USA) providing the lines at 488.0, 514.5, 530.8, 568.2 and 647.1 nm. The laser beam is coupled to a microscope (Olympus BX 40, Tokyo, Japan) and focused on the sample by $100\times$, $50\times$ or $20\times$ objectives (Olympus SLMPL, Tokyo, Japan). The Raman scattering is collected into the slit of a three-stages subtractive spectrograph (Jobin Yvon S3000, Horiba, Kyoto, Japan) by means of a set of achromatic lenses. The spectrograph is made up of a double monochromator (Jobin Yvon, DHR 320, Horiba, Kyoto, Japan), working as a tunable filter rejecting elastic scattering, and a spectrograph (Jobin Yvon, HR 640, Horiba, Kyoto, Japan). The Raman signal is detected by a liquid nitrogen-cooled CCD (Jobin Yvon, Symphony 1024×256 pixels front illuminated). When an entrance slit of 50 μm is used, a precision of 0.6 cm^{-1} in the determination of the peak position is obtained, with this experimental set-up.

A temperature-controlled stage (Linkam, THMS600/720, Tadworth, UK) is used to change and control the temperature of the sample, by means of a liquid nitrogen reservoir and heating resistances, giving a control of 0.1 K on the temperature inside the sample chamber. The sample is inserted into the temperature controller stage, and uniformly heated or cooled to reach the desired temperature, with a rate of 5 K/min and a thermalization time of at least 30 min. Before starting the experiment, a procedure of purging air from the stage chamber with nitrogen is performed; by this way the air in the chamber is eliminated and an inert static nitrogen atmosphere is realized, allowing fast temperature variations. Once the thermalization process is done, consecutive Stokes and anti-Stokes measurements are conducted to measure the local temperature of the sample. The wavelength incident on the sample is properly chosen in order to avoid sample heating (by using photons less energetic than the band gap of sample). Raman spectra have been collected in the visible range, by exciting at 488.0, 514.5, 568.2 and 647.1 nm, at different temperatures, ranging from 283 to 323 K. Measurements are repeated, at each wavelength and temperature, to obtain a consistent set of data (from 5 to 10 measurements at each temperature), by collecting the Raman signal in different positions of the sample.

Measurements of the anti-Stokes/Stokes ratio have been collected also at different laser powers, in the range 0.1–20 mW, to individuate the region where the signal is independent from the power, and the local temperature is not influenced by the presence of a laser beam. An input power of few mW has been used for temperature measurements.

All Raman spectra have been collected using a $20\times$ objective (Olympus) with numerical aperture (N.A.) of 0.4 and a working distance of 12 mm, under these conditions the spot of the laser on the sample is expected to be approximately 1.5 μm (nearly equal for all the lines of the Ar^+/Kr^+ laser).

The sample is titanium dioxide, a commercial anatase powder (Sigma Aldrich, Merck KGaA, St. Louis, MO, USA), with a crystallite dimensions of ~200 nm; it possesses a band gap of 3.4 eV [32]. Titanium dioxide has been used as pristine powder inserted in the temperature stage and as powder pressed on KBr pellet sample (herein called *Test Sample*), with a final thickness of few hundreds μm.

3. Results

3.1. TiO$_2$ Raman Spectra at 488.0 nm

The Stokes Raman spectrum of anatase powder, recorded at room temperature, using a laser at 488.0 nm with an input power of 1.72 mW, is reported in Figure 2. The spectrum clearly shows an intense peak centered at 143 cm^{-1} and four peaks, at 197, 397, 515 and 640 cm^{-1}, with lower intensity.

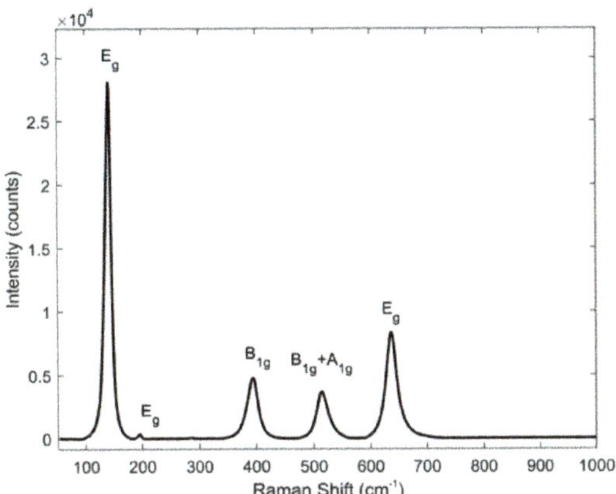

Figure 2. Stokes Raman spectrum of titanium dioxide anatase excited at 488.0 nm, with input power of 1.72 mW, with the indication of the Raman mode symmetry.

Anatase crystals are characterized by 15 optical modes at the Γ point of the Brillouin zone, described with the following irreducible representation of the normal vibrational modes [28,35]:

$$1A_{1g} + 1A_{2u} + 2B_{1g} + 1B_{2u} + 3E_g + 2E_u$$

Among these, only six modes, A_{1g}, $2B_{1g}$ and $3E_g$ are the Raman-active ones (reported in Figure 2 and in Table 1); the Raman spectrum shows only five peaks since the B_{1g} and A_{1g} modes, at 512 and 518 cm^{-1} respectively, are not distinguishable at the experimental conditions, due to their intrinsic amplitude [28,36].

Table 1. Experimental and literature [28] TiO$_2$ anatase Raman-active modes, excited at 514.5 nm.

Symmetry	Experimental [cm^{-1}]	Literature [cm^{-1}]
E_g	143.0	143
E_g	197.8	198
B_{1g}	394.8	395
B_{1g}	516.1	512
A_{1g}	516.1	518
E_g	638.4	639

The experimental frequencies, reported in the central column of Table 1, corrected by using the cyclohexane frequencies as calibration frequency, are comparable to data reported in literature [28].

The Stokes (positive Raman shift) and anti-Stokes (negative Raman shift) Raman spectra are reported in Figure 3, where it is possible to observe also the zooms of the low intensity peaks. By observing Stokes and anti-Stokes data, it turned out that the best peak for temperature monitoring is the E_g mode at 143 cm^{-1}. It is well defined, very intense even at low laser powers and highly sensitive to temperature, as will be demonstrated later, thanks to its low Raman shift.

3.2. Raman Spectra of TiO$_2$ as A Function of Temperature, Excitation Wavelength and Input Power

Stokes and anti-Stokes Raman spectra of anatase have been collected in the temperature range of 283–323 K, by exciting at 488.0, 514.5, 568.2 and 647.1 nm, using an input power of 1.47, 1.20, 2.20 and 5.86 mW, respectively. The Raman spectra collected at room temperature, at different excitation wavelengths, are reported in Figure 4a, while the 143 cm^{-1} E$_g$ mode, excited at 514.5 nm, at different temperatures, is illustrated in Figure 4b.

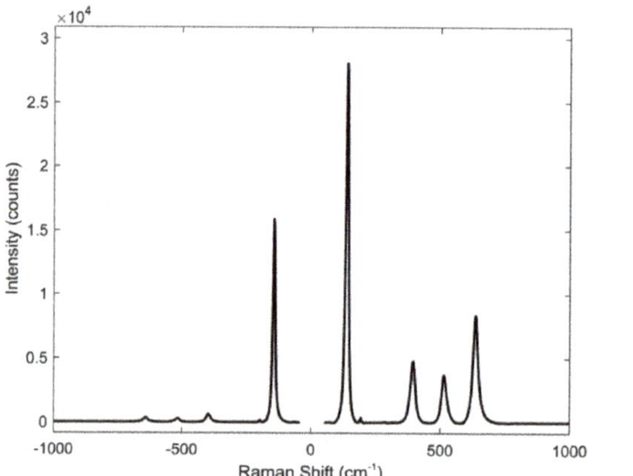
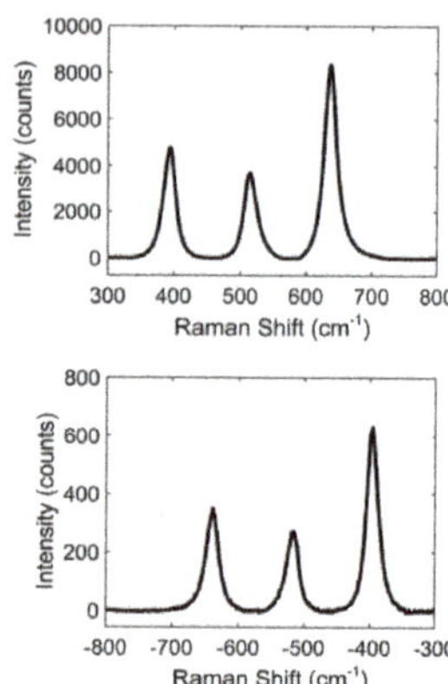

Figure 3. Stokes (positive Raman shift) and anti-Stokes (negative Raman shift) Raman spectrum of TiO$_2$ anatase recorded at 488.0 nm, with input power of 1.6 mW. In the right panel the zooms of the three less intense peaks are displayed, Stokes in the upper part, anti-Stokes in the lower one.

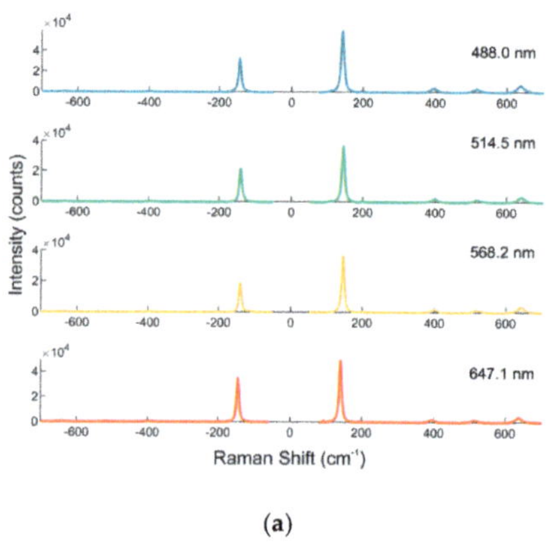

(a)

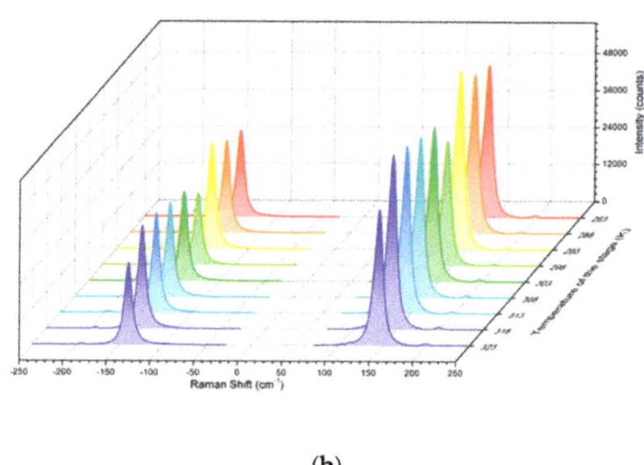

(b)

Figure 4. (a) Experimental anti-Stokes and Stokes spectra of TiO2 collected at room temperature and different excitation wavelengths (488.0 nm blue line, 514.5 nm green line, 568.2 nm yellow line and 647.1 nm red line); (b) Stokes and anti-Stokes spectra of the 143 cm^{-1} E$_g$ mode, collected at 488.0 nm as function of temperature (from 283 to 323 K, with 5 K increment step).

All Raman spectra have been analyzed with Matlab, using a Lorentz fitting, to obtain the Raman spectrum parameters, such as frequency position, width, intensity and area of the peak (see Appendix A).

An example of the data, obtained from the analysis of the Raman spectra centered on the 143 cm^{-1} peak, by exciting at 514.5 nm at different temperatures, is reported in Table 2, where data from the Stokes spectra are reported together with the area of the anti-Stokes peaks, which allows to calculate the anti-Stokes/Stokes ratio, a parameter important for the determination of the local temperature. The errors of the peak positions and widths are experimental errors, directly derived from the characteristics of the experimental set-up; the errors on the area have been derived from the fitting (see the example in Appendix A); for the anti-Stokes/Stokes ratio data, the error propagation has been considered [12].

Table 2. Frequency position and width of the Stokes peak at 143 cm^{-1}, Stokes and anti-Stokes areas and anti-Stokes/Stokes ratios for the temperature range 283–323 K, explored during the calibration procedure at 514.5 nm.

T [K]	Peak Center [cm^{-1}]	Peak Width [cm^{-1}]	Stokes Area 10^5	Anti-Stokes Area 10^5	Anti-Stokes/Stokes Ratio
283	142.4 ± 0.6	10.7 ± 0.6	8.45 ± 0.32	4.11 ± 0.08	0.487 ± 0.021
288	142.3 ± 0.6	11.1 ± 0.6	9.29 ± 0.37	4.55 ± 0.09	0.490 ± 0.022
293	142.4 ± 0.6	11.4 ± 0.6	10.68 ± 0.44	5.30 ± 0.11	0.497 ± 0.023
298	143.0 ± 0.6	10.7 ± 0.6	7.00 ± 0.23	3.54 ± 0.06	0.505 ± 0.019
303	143.4 ± 0.6	10.8 ± 0.6	8.68 ± 0.29	4.41 ± 0.08	0.508 ± 0.019
308	143.5 ± 0.6	11.0 ± 0.6	9.16 ± 0.30	4.71 ± 0.08	0.515 ± 0.019
313	143.8 ± 0.6	11.0 ± 0.6	9.63 ± 0.32	5.04 ± 0.09	0.523 ± 0.020
318	144.3 ± 0.6	10.9 ± 0.6	8.99 ± 0.28	4.75 ± 0.08	0.530 ± 0.019
323	144.9 ± 0.6	10.8 ± 0.6	7.63 ± 0.23	4.15 ± 0.07	0.543 ± 0.019

Corresponding to the increasing in temperature, both the peak frequency position and the anti-Stokes/Stokes ratio clearly show an increment, from 142.5 to 145 cm^{-1} and 0.48 and 0.54, respectively, while the peak width varies in between 10.7 and 11.4 cm^{-1}.

In order to test whether the Raman spectra, and the corresponding parameters, are perturbed by the laser irradiation, it is also necessary to investigate the effect of increasing incident laser power by keeping constant all other variables, like excitation wavelength. The laser power was changed in a range between 0.1 and 18 mW (depending on the laser wavelength); the experimental results (peak position, peak width (FWHM, i.e., full width half maximum), peak area and anti-Stokes/ Stokes ratio), for the 143 cm^{-1} E$_g$ mode, at 514.5 nm, are shown in Figure 5.

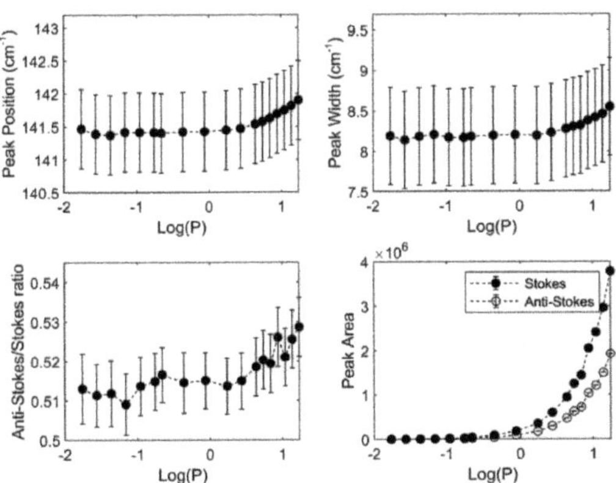

Figure 5. Peak position, peak width, anti-Stokes/ Stokes ratio and peak area for the 143 cm^{-1} E$_g$ mode of anatase as function of the laser power incident on the sample; λ_{exc} = 514.5 nm. The dashed lines are only for eyes guidance. All data are reported against the laser Power (mW) in a logarithmic scale.

It is evident that parameters are not increasing with the laser power only in the small range, 0–2 mW, while at higher input power they increase with the increase of the laser power. In order to assume that the temperature of the sample is not influenced by the presence of the laser irradiating the sample, it is necessary to keep the laser power at values lower than 2 mW.

3.3. Test Sample Measurement

A validation of the method can be done by evaluating anti-Stokes and Stokes intensities in various positions of the *Test Sample*, thus verifying if the parameters measured all over the sample are uniform or not, and comparable with the results previously obtained. The calculated anti-Stokes/Stokes ratios are reported in Table 3.

Table 3. Repeated measurements of the anti-Stokes/Stokes ratio at room temperature collected at 488.0, 514.5, 568.2 and 647.1 nm using a laser power of 1–2 mW depending on the excitation wavelength used. In the final row, mean values of the anti-Stokes/Stokes ratios and the standard deviations are reported.

	Anti-Stokes/Stokes Experimental Ratio			
Measurements	**@ 488.0 nm**	**@ 514.5 nm**	**@ 568.2 nm**	**@ 647.1 nm**
1	0.490 ± 0.010	0.5026 ± 0.007	0.477 ± 0.006	0.609 ± 0.008
2	0.486 ± 0.009	0.5012 ± 0.007	0.471 ± 0.006	0.602 ± 0.008
3	0.487 ± 0.010	0.5028 ± 0.007	0.477 ± 0.006	0.610 ± 0.008
4	0.483 ± 0.009	0.5030 ± 0.008	0.474 ± 0.006	0.609 ± 0.007
5	0.489 ± 0.009	0.5058 ± 0.007	0.478 ± 0.006	0.607 ± 0.007
6	0.485 ± 0.009	0.5034 ± 0.007	0.471 ± 0.006	0.597 ± 0.008
7	0.488 ± 0.009	0.5017 ± 0.007	0.476 ± 0.006	0.609 ± 0.008
8	0.486 ± 0.009	0.5034 ± 0.007	0.474 ± 0.006	0.605 ± 0.007
9	0.488 ± 0.009	0.5006 ± 0.007	0.477 ± 0.006	0.608 ± 0.007
10	0.486 ± 0.009	0.5040 ± 0.007	0.474 ± 0.006	0.606 ± 0.007
Mean $\pm\ \sigma$	0.487 ± 0.002	0.503 ± 0.001	0.475 ± 0.002	0.606 ± 0.004

4. Discussion

Raman spectroscopy allows for temperature measurements with two methods: (a) Since the anti-Stokes and Stokes Raman intensities are proportional to the populations of their respective initial vibrational states, described by the Boltzmann distribution, the T of the sample can be estimated from the ratio of the Stokes and anti-Stokes intensities [37–39]; (b) the frequency position, the intensity or the shape of a (Stokes) Raman band may change as a function of temperature: The increase in temperature is expected to loosen chemical bonds (and hence to decrease the frequency of the mode), and/or induce more significant structural changes in the material under investigation [40–42]. Within this work, the local temperature has been investigated considering Stokes and anti-Stokes Raman spectra.

The anti-Stokes/Stokes ratio allows to determine the local temperature T of the sample, through the relation:

$$\rho = \frac{I_{aS}}{I_S} = C \cdot \frac{(\nu_0 + \nu_m)^3}{(\nu_0 - \nu_m)^3} \exp\left(-\frac{h\nu_m}{k_B T}\right) \tag{1}$$

where ν_m is the frequency of the vibrational mode m considered, ν_0 the laser frequency, h the Planck's constant, k_B the Boltzmann constant and C is the calibration constant. A frequency dependence to the third power of the anti-Stokes/Stokes ratio is needed as the detection system is based on photon counting (CCD), whereas if the detection is energy-based, a fourth power dependence is more appropriate [40]. The calibration constant is related to the experimental setup, in particular to parameters such as the polarization of the incident laser and the CCD and grating efficiency. The determination of the calibration constant, at each working excitation wavelength, is a key point to the determination of the local temperature.

4.1. Determination of the Calibration Constant

Raw Raman spectra were analyzed to obtain the parameters, such as area, intensity, frequency position and width of the peak. In particular, the anti-Stokes area of the E_g mode was divided by that of the Stokes signal to obtain the experimental anti-Stokes/Stokes ratio, ρ (reported as example in Table 2—Section 3.2 for the 514.5 nm excitation wavelength).

With the Raman scattering cross-section being nearly constant over the wavelength range explored, normalized intensities still differ depending on the CCD and grating efficiency (instrumental response function), which is particular of the instrumentation set-up used.

Equation (1) was fitted to the experimental values, collected at different temperatures and defined excitation wavelength, leaving C as a free parameter.

Figure 6 reports the anti-Stokes/Stokes ratio obtained from the calculated area of the anatase Raman modes, at 143, 397, 515 and 640 cm^{-1}, in the temperature range 283–323 K, excited at 514.5 nm, and the curve resulting from the fitting with Equation (1) (the dashed lines). The corresponding calibration constants are calculated to be 0.9605, 0.9411, 0.9461 and 0.9491, respectively.

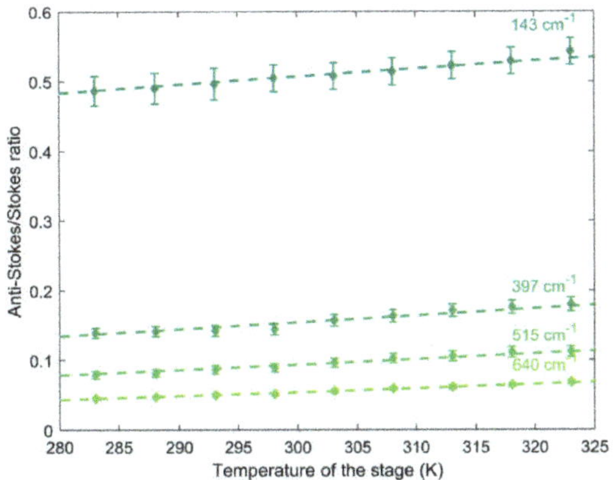

Figure 6. Experimental anti-Stokes/Stokes ratios in the range 283–323 K for the Raman modes of anatase, at 143 cm^{-1} (circles), 397 cm^{-1} (down-pointing triangles), 515 cm^{-1} (squares) and 640 cm^{-1} (up-pointing triangles) collected at λ_{exc} = 514.5 nm.

The anti-Stokes/Stokes experimental ratios of the 143 cm^{-1} Raman mode, measured at different excitation wavelengths as a function of the temperature imposed by the thermostat are reported in Table 4 and depicted in Figure 7, together with the curve obtained from the fitting and the corresponding calculated calibration constants, reported in Table 5. The curve is well fitted to experimental data, as the R2 parameters range from a minimum of 0.84, at 647 nm, to a maximum of 0.95, at 514.5 nm.

Table 4. Anti-Stokes/Stokes experimental ratios at different excitation wavelengths, as function of the temperature imposed by the thermostat.

	Anti-Stokes/Stokes Experimental Ratio vs. Temperature			
T [K]	**@ 488.0 nm**	**@ 514.5 nm**	**@ 568.2 nm**	**@ 647.1 nm**
283	0.468 ± 0.023	0.487 ± 0.021	0.461 ± 0.007	0.582 ± 0.012
288	0.469 ± 0.023	0.490 ± 0.022	0.471 ± 0.008	0.591 ± 0.014
293	0.477 ± 0.023	0.497 ± 0.023	0.471 ± 0.008	0.603 ± 0.014
298	0.481 ± 0.022	0.505 ± 0.019	0.468 ± 0.009	0.605 ± 0.013
303	0.486 ± 0.021	0.508 ± 0.019	0.473 ± 0.009	0.602 ± 0.015
308	0.498 ± 0.020	0.515 ± 0.019	0.486 ± 0.009	0.616 ± 0.015
313	0.506 ± 0.018	0.523 ± 0.020	0.494 ± 0.008	0.614 ± 0.015
318	0.516 ± 0.016	0.530 ± 0.019	0.501 ± 0.008	0.633 ± 0.015
323	0.522 ± 0.016	0.543 ± 0.019	0.504 ± 0.008	0.624 ± 0.015

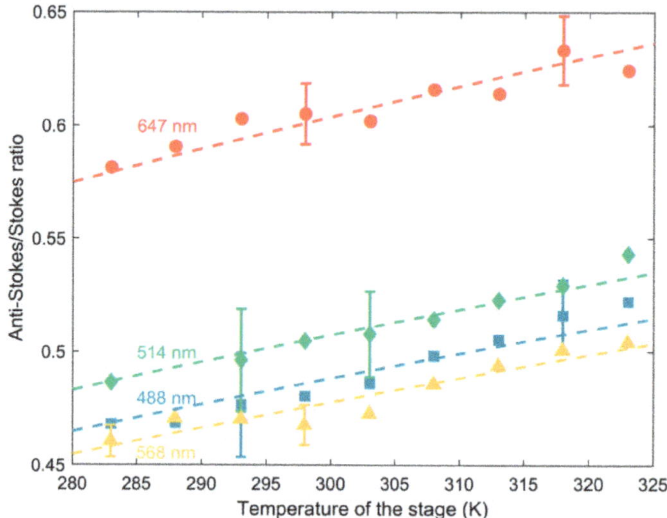

Figure 7. Experimental anti-Stokes/Stokes ratios of the 143 cm^{-1} E$_g$ mode of anatase as a function of the temperature, and corresponding calibration curves; data are collected by exciting at different wavelengths: 488.0 (blue squares and dashed line), 514.5 (green diamonds and dashed line), 568.2 (yellow triangles and dashed line) and 647.1 nm (red circles and dashed line).

Table 5. Calibration constants for the 143 cm^{-1} Raman mode at four different wavelengths.

Excitation Wavelength [nm]	Calibration Constant	R^2
488.0	0.931 ± 0.009	0.91
514.5	0.961 ± 0.006	0.95
568.2	0.904 ± 0.007	0.90
647.1	1.135 ± 0.009	0.84

As we can see qualitatively in Figures 6 and 7 and quantitatively in Tables 4 and 5, the calibration constant C is different depending on the excitation wavelength and the frequency of the Raman mode, due to the difference in response of the optical components of the experimental set-up to the wavelength. For this reason, it is necessary to individuate a calibration constant for a defined Raman mode, at a specific excitation wavelength.

4.2. Comparison between the Four Raman Modes of Titanium Dioxide

In order to individuate the best Raman mode, to obtain the more efficient signal in the determination of the local temperature, it is necessary to compare the behavior of the four TiO$_2$ Raman modes as a function of the temperature variation. Figure 8a shows, for example, the theoretical anti-Stokes/Stokes ratios of the Titania modes, calculated with excitation wavelength at 514.5 nm, in the temperature range of interest, 283–323 K. The 143 cm^{-1} E$_g$ anti-Stokes/Stokes ratio shows the highest value, in comparison with the other Raman modes, it presents values in the range 0.50–0.55, while the 397 cm^{-1} mode, the 515 cm^{-1} and the 640 cm^{-1} modes present values in the ranges 0.15–0.19, 0.08–0.12 and 0.05–0.07, respectively. The total variations of the ratio, in the whole T range ($\Delta\rho$), decreases from 0.0476, with the 143 cm^{-1} mode, to 0.0233, with the 640 cm^{-1} mode. At the excitation wavelengths of 488.0, 568.2 and 647.1 nm the behavior and the values obtained are comparable to that obtained at 514.5 nm.

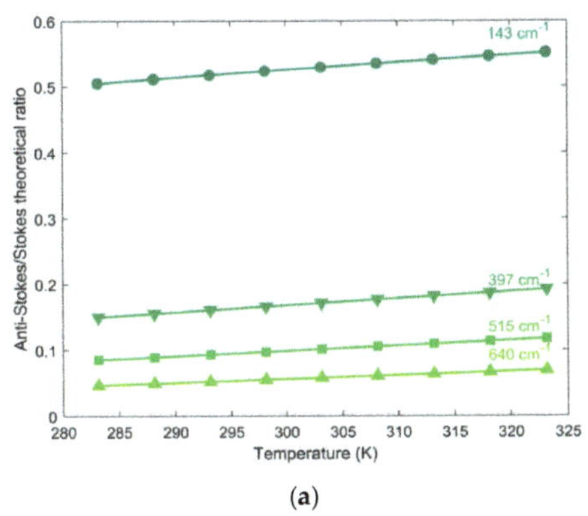 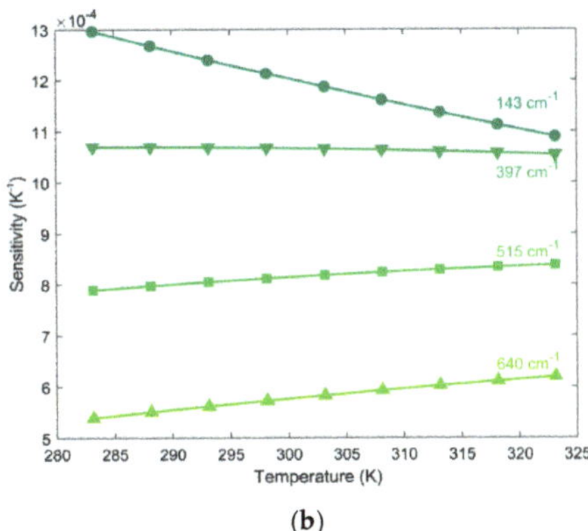

(a)   **(b)**

Figure 8. (a) Theoretical behavior and (b) sensitivity of the anti-Stokes/Stokes ratio of the four anatase modes, at 143, 397, 515, and 640 cm^{-1}, calculated with λ_{exc} = 514.5 nm, in the range 283–323 K.

To evaluate the efficiency in temperature detection, one of the most used figures of merit of thermometry is the sensitivity, calculated, for Raman measurements, through the following Equation [9,16]:

$$S = \left| \frac{\partial \rho}{\partial T} \right| = \left| -\frac{(v_0 + v_m)^3}{(v_0 - v_m)^3} \frac{h v_m}{kT^2} \exp\left(-\frac{h v_m}{kT}\right) \right| \tag{2}$$

where ρ is the anti-Stokes over Stokes ratio.

The derivative with respect to the temperature, has been evaluated at 514.5 nm, for all the Raman modes of anatase. The results, plotted in Figure 8b, show an increasing in sensitivity corresponding to the decreasing in frequency of the Raman modes and a decrease of the sensitivity of all Raman mode as temperature increases, also already observed in literature [43]. This can be attributed to the fact that at high temperature the differences between the population of the ground state and the first vibrational excited state are smaller than at room temperature.

In particular, it is possible to evaluate the sensitivity, at 300 K close to the room temperature, and to compare the obtained values for different wavelengths and different Raman modes of titanium dioxide. The outcome, shown in Table 6, is that the sensitivity at constant excitation wavelength decreases with increasing frequency of the Raman mode and for a given Raman mode it increases with increasing excitation wavelength. Corresponding to the 143 cm^{-1} E$_g$ mode a thermal resolution in the range 1.2 (@ 514.5 nm) to 3.4 K (@ 647.1 nm) have been calculated. All these experimental data are comparable with those already reported in literature [9].

Table 6. Sensitivity at 300 K of the anti-Stokes/Stokes ratio to temperature, calculated for the 143, 397, 515 and 640 cm^{-1} modes of anatase.

	Sensitivity at 300 K [10^{-3} K^{-1}]			
Excitation Wavelength [nm]	**Raman Mode**			
	143 cm^{-1}	**397 cm^{-1}**	**515 cm^{-1}**	**640 cm^{-1}**
488.0	1.20	1.06	0.81	0.57
514.5	1.20	1.07	0.81	0.57
568.2	1.22	1.08	0.83	0.59
647.1	1.23	1.10	0.85	0.60

The behavior of the theoretical ratio can be compared to the experimental anti-Stokes/Stokes ratios for the four Raman modes, reported in Figure 6. It is evident that there is a good agreement, confirming the 143 cm^{-1} Raman mode, the lowest in Raman frequency, the more sensitive to Temperature variation. Moreover, the sensitivity calculated starting from the experimental data overlaps well the theoretical ones.

4.3. Validation of the Method and Temperature Determination: Test Sample

By using the calibration constants, it is possible to determine the local temperature. In Figure 9a, the temperature derived from repeated measurements, performed on a different region of the *test sample*, of the anti-Stokes/Stokes ratio for the 143 cm^{-1} mode of anatase at 514.5 nm, at 297 K, is reported; the figure also shows the mean temperature calculated for these measurements and the standard deviation. In this context, the standard deviation of temperatures, rather than the errors derived from each measurement, is reported, as it is preferable when repeated measures are conducted; moreover, the two values have the same order of magnitude (few kelvins). Detailed values are reported in Table 7, and it turns out that the data show an excellent overlap between the expected and measured data, with a deviation as low as 3 K, maximum. Moreover, the results obtained by changing the input power, reported in Figure 9b, confirm that the local sample temperature is not affected by the laser power, when an incident power of few mW is used, while at higher input powers the laser is heating the sample, at all the used wavelengths.

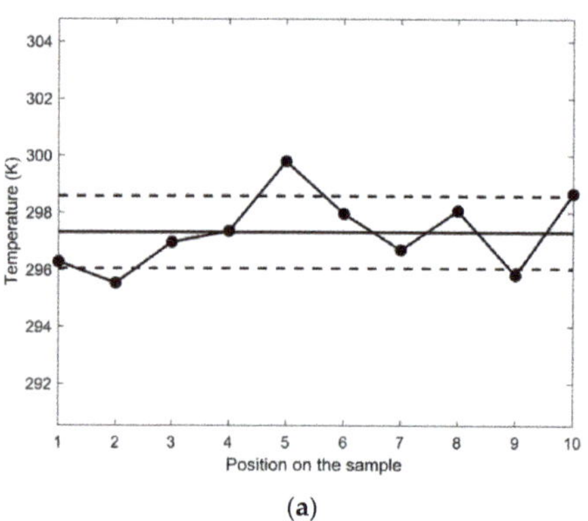
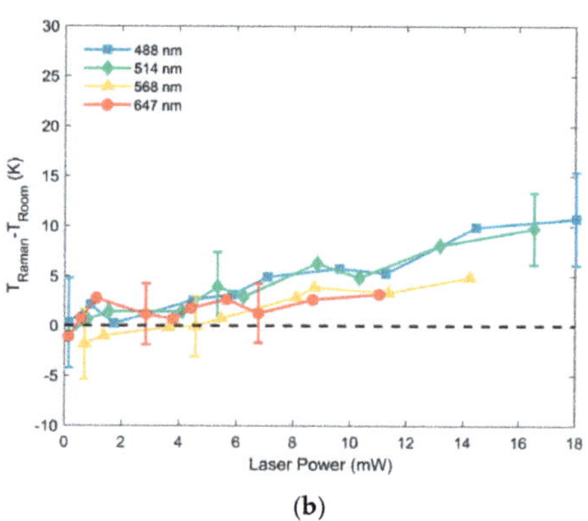

(a)   (b)

Figure 9. (a) Temperature determined from the anti-Stokes/Stokes values of the 143 cm^{-1} E$_g$ mode at λ_{exc} = 514.5 nm of the *Test Sample*, with the mean value (solid line) and the standard deviation (dashed lines); (b) difference between the temperature derived from the anti-Stokes/Stokes ratio (corrected with the proper calibration constant) for the 143 cm^{-1} E$_g$ mode of anatase as function of the laser power incident on the sample at 488.0 (blue squares), 514.5 (green diamonds), 568.2 (yellow triangles) and 647.1 nm (red circles). The dashed line represents the room temperature.

Table 7. Results of repeated measurements at four different excitation wavelengths conducted at room temperature.

Excitation Wavelength [nm]	Laser Power [mW]	Room Temperature [K]	Averaged Raman Temperature [K]	Root Mean Square Deviation [K]
488.0	1.72	297.0	298	2
514.5	1.52	297.0	297	1
568.2	1.35	296.0	296	2
647.1	1.09	298.0	299	3

These results confirm the validation of the method thus verifying that the temperature measured all over the sample is uniform.

It is possible to conclude that the Raman modes of anatase, in particular the E$_g$ one at 143 cm^{-1}, are excellent candidate for the local temperature detection in the visible range.

However, the need remains to investigate the behavior at longer wavelengths, towards the near IR, where the biological window is located.

The performances obtained with this TiO_2 based Raman thermometer are compatible with data reported in literature [9,44]; the lower thermal resolution, with respect to fluorescent thermometer, is compensated with the wider wavelength working range.

5. Conclusions

Temperature is an important parameter influencing physical, chemical and biological processes: For this reason, the investigation of new materials, with enhanced performances, together with the definition of the more performing experimental protocol is a hot topic in the nanothermometry field. The experimental work reported in this article will contribute to the development of a new Raman based biocompatible nanothermometer, by investigating the optical performances of titanium dioxide, as anatase, with Raman technique.

The spectroscopic characterization of titanium dioxide has been carried out in the visible range, at 488.0, 514.5, 568.2 and 647.1 nm, and the Raman-active modes have been investigated to find the more performing one, as temperature sensor. Both Stokes and anti-Stokes spectra were collected at different temperature, input power and wavelengths, to investigate the temperature range, the temperature resolution, the eventual self-heating (due to the input laser power) of the sample and to identify the working range of the nanothermometer. A key point for the identification of the local temperature is the calibration of the experimental set-up, which allows defining the best experimental protocol. The calibration procedure has been conducted by controlling the sample temperature with a temperature-controlled stage and exploring the Raman signals in the temperature range of 283–323 K (with 5 K increment), as it is of interest for biological applications. The obtained values of the anti-Stokes/Stokes ratio allow the determination of the calibration constant, specified for all anatase Raman modes at each excitation wavelength. The calibration constant permits to determine the sample local temperature and to identify the power range where the local temperature is not affected by the laser power. Working with an incident laser power higher than 2 mW the sample experiences self-heating, while at lower power samples do not experience any self-heating. The validation of the proposed protocol has been finally achieved with the analysis of the Raman spectra of the *Test Sample*. Repeated anti-Stokes and Stokes measurements, have been performed on various positions of the sample at room temperature (~297 K), with an incident laser power of 1.5 mW. An excellent agreement between the temperature derived from the anti-Stokes/Stokes ratios and the expected temperature was found, with a standard deviation of repeated temperature measurements calculated to be in between 1 and 3 K, for the most intense peak, located at 143 cm^{-1}, which has been demonstrated to be the most sensitive to temperature. This titanium dioxide mode seems to be an excellent candidate for the local temperature detection in the visible range from 488.0 to 647.1 nm, reaching the highest sensitivity in the red region.

Author Contributions: Conceptualization, D.P. and R.S.; Methodology, R.P., V.Z., D.P., R.S.; Software, V.Z., R.P.; Validation, V.Z., D.P., R.S. and R.P.; Formal analysis, V.Z.; Investigation, V.Z.; Resources, R.S. and R.P.; Data curation, V.Z.; Writing—original draft preparation, V.Z. and R.S.; Writing—review and editing, V.Z., D.P., R.P. and Signorini R; Supervision, R.S. and D.P.; Project administration, R.S.; Funding acquisition, R.S. and R.P. All authors have read and agreed to the published version of the manuscript.

Funding: This research was funded by Chemical Science Department of University od Padova, project P-DiSC#10BIRD2019-UNIPD.

Conflicts of Interest: The authors declare no conflict of interest.

Appendix A.

The fitting procedures performed are reported in the following, by outlining the expression used and showing an example of result and the corresponding error.

Appendix A.1. Fitting of Raman Signals

The Raman spectra have been analyzed with Matlab, using a Lorentz fitting, to obtain the Raman spectrum parameters, such as frequency peak position, width, intensity and area of the peak. The fitting of a spectrum, collected under excitation at 514.5 nm, is reported in Figure A1, and the corresponding parameters obtained from the fitting are reported in Table A1. The goodness of the Lorentz fitting is evaluated with the R^2 parameter, known as the coefficient of determination in statistics, it is calculated from the sum of squared errors. It indicates the proportionate amount of variation in the response variable explained by the independent variables in the model. The larger it is, the more variability is explained by the model.

Table A1. Fitting model, parameters and goodness of the fit of TiO_2 Raman spectrum at 514.5 nm.

Fitting Model		$y=y_0+\frac{2A}{\pi}\cdot\frac{w}{4(x-x_0)^2+w^2}$	
Parameters		**Parameter Value**	**Confidence Interval (95%)**
Offset	y_0	2.946×10^{-6}	$(-55.56, 55.56)$
Area	A	3.506×10^5	$(3.464 \times 10^5, 3.549 \times 10^5)$
Width	w	8.186	$(8.062, 8.31)$
Peak position	x_0	143.2	$(143.2, 143.3)$
Goodness of Fit			
R^2		0.99	

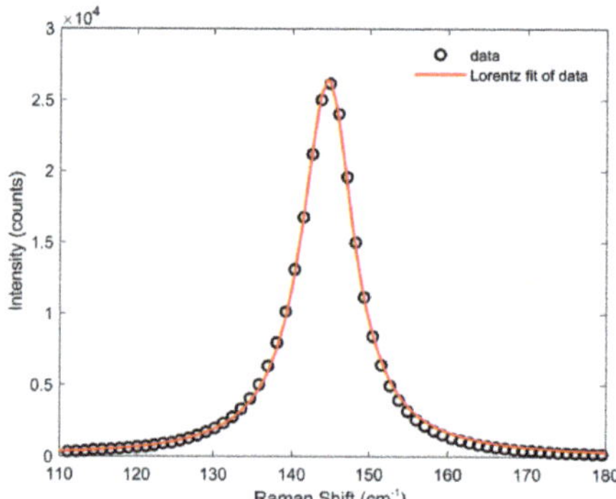

Figure A1. Experimental data collected at λ_{exc} = 514.5 nm (open circles) fitted with a Lorentzian curve (red line).

Appendix A.2. Fitting of Experimental Anti-Stokes/Stokes Ratios as Function of Temperature

The calibration coefficients have been obtained by fitting anti-Stokes over Stokes ratio data, at each excitation wavelength, using the equation:

$$\rho = C\cdot\frac{(v_0+v)^3}{(v_0-v)^3}\exp\left(-\frac{hv}{kT}\right)$$

We substitute $v_0 = 5.83\cdot10^{14}\ s^{-1}$ as λ_{exc} = 514.5 nm , $v = 4.29\cdot10^{12}\ s^{-1}$ (frequency of the 143.0 cm^{-1} Raman mode), and the values of the Planck and Boltzmann constants. Temperatures are those imposed by the temperature-controlled stage. C is left as a free parameter. The value, obtained with data reported in Figure A2, $C = 0.961 \pm 0.006$ is the calibration constant for the 143 cm^{-1} Raman mode and λ_{exc} = 514.5 nm). The value of the R^2 for this fitting is 0.95.

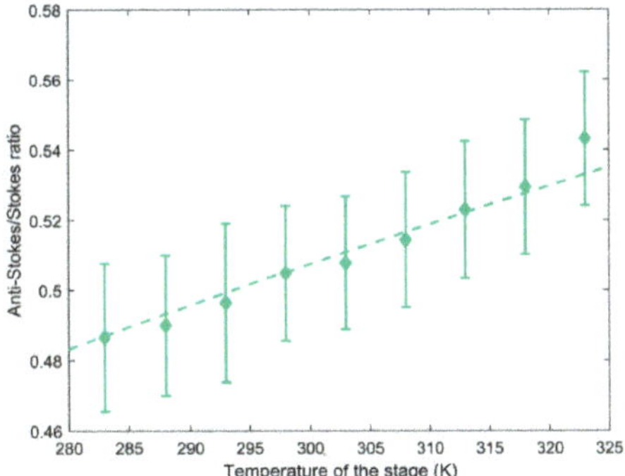

Figure A2. Experimental anti-Stokes/Stokes ratios (green diamonds) of the 143 cm^{-1} E$_g$ mode of anatase as a function of the temperature, and corresponding calibration curves (dashed line); data are collected by exciting at 514.5 nm.

Appendix A.3. Theoretical Anti-Stokes/Stokes Ratio Behavior as Function of Temperature

The determination of the local temperature using the strength of a Raman band at the Stokes and anti-Stokes position is based on the Boltzmann distribution of the ground and first excited state populations. At room temperature a significant difference between the Stokes and anti-Stokes signal strengths are expected, since the population of the ground state will be higher than that of the excited one, for an oscillator with energy proportional to 140 cm^{-1}, as can be observed from Figure 2. It is also evident that the higher the energy of the vibrational mode is, the lower the population of the excited state and therefore the weaker the signal strength of the anti-Stokes Raman band will be. At increasing temperature, the population of the excited state increases, thus increasing the intensity of the anti-Stokes band and decreasing the intensity of the corresponding Stokes band. The dependence of this variations is strictly related to the population distribution, related to the Boltzmann distribution, which consider an exponential dependence with temperature.

Figure A3 reports the behavior of the Stokes over anti-Stokes area ratio of the Anatase Raman modes as a function of the temperature, only on a large temperature range the exponential behavior is evident, while by considering the small range the behavior seems to be linear, but it is necessary to consider an exponential dependence due to the underlined process.

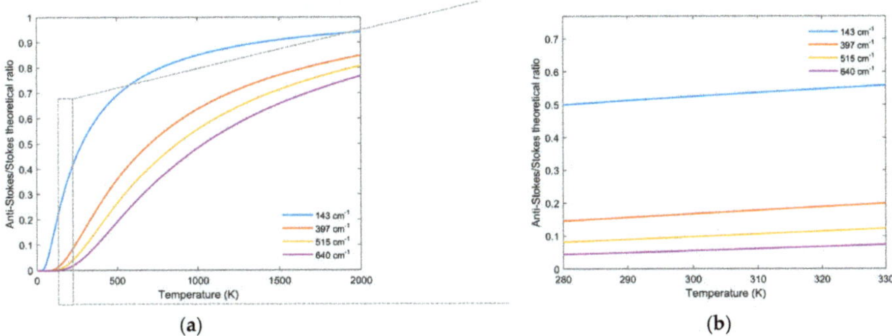

(a) (b)

Figure A3. Theoretical anti-Stokes/Stokes ratio evaluated at λ_{exc} = 514.5 nm, for all the four Raman modes of titanium dioxide, in a temperature range 0–2200 K (**a**) and in the range of interest (**b**).

Theoretical models predict a direct dependence of the Stokes and anti-Stokes Raman intensity on the temperature, but experimental data show a direct correlation with temperature of the anti-Stokes/Stokes ratio. This fact, already observed in literature [10,45–48], can be ascribed also to small variations of experimental parameters, like the intensity of the laser, the effective focal volume, linked to the position of the focus on the sample, to the inhomogeneity of the sample, which can lead to a number of different active molecules in the focal volume. All these experimental parameters will influence the absolute intensity of the Raman bands. By performing the ratio between anti-Stokes and Stokes bands a sort of internal normalization is obtained.

References

1. Zhou, H.; Sharma, M.; Berezin, O.; Zuckerman, D.; Berezin, M.Y. Nanothermometry: From Microscopy to Thermal Treatments. *Eur. J. Chem. Phys. Phys. Chem.* **2016**, *17*, 27–36. [CrossRef] [PubMed]
2. Zhou, J.; Del Rosal, B.; Jaque, D.; Uchiyama, S.; Jin, D. Advances and Challenges for Fluorescence Nanothermometry. *Nat. Methods* **2020**, *17*, 967–980. [CrossRef]
3. Bednarkiewicz, A.; Marciniak, L.; Carlos, L.D.; Jaque, D. Standardizing Luminescence Nanothermometry for Biomedical Applications. *Nanoscale* **2020**, *12*, 14405–14421. [CrossRef] [PubMed]
4. Jauffred, L.; Samadi, A.; Klingberg, H.; Bendix, P.M.; Oddershede, L.B. Plasmonic Heating of Nanostructures. *Chem. Rev.* **2019**, *119*, 8087–8130. [CrossRef]
5. Okabe, K.; Inada, N.; Gota, C.; Harada, Y.; Funatsu, T.; Uchiyama, S. Intracellular Temperature Mapping with a Fluorescent Polymeric Thermometer and Fluorescence Lifetime Imaging Microscopy. *Nat. Commun.* **2012**, *3*, 1–9. [CrossRef] [PubMed]
6. Quintanilla, M.; García, I.; de Lázaro, I.; García-Alvarez, R.; Henriksen-Lacey, M.; Vranic, S.; Kostarelos, K.; Liz-Marzán, L.M. Thermal Monitoring during Photothermia: Hybrid Probes for Simultaneous Plasmonic Heating and near-Infrared Optical Nanothermometry. *Theranostics* **2019**, *9*, 7298–7312. [CrossRef]
7. Santos, H.D.; Ximendes, E.C.; del Iglesias-de la Cruz, M.C.; Chaves-Coira, I.; del Rosal, B.; Jacinto, C.; Monge, L.; Rubia-Rodríguez, I.; Ortega, D.; Mateos, S.; et al. In Vivo Early Tumor Detection and Diagnosis by Infrared Luminescence Transient Nanothermometry. *Adv. Funct. Mater.* **2018**, *28*, 1–10. [CrossRef]
8. Mi, C.; Zhou, J.; Wang, F.; Lin, G.; Jin, D. Ultrasensitive Ratiometric Nanothermometer with Large Dynamic Range and Photostability. *Chem. Mater.* **2019**, *31*, 9480–9487. [CrossRef]
9. Quintanilla, M.; Liz-Marzan, L.M. Guiding Rules for Selecting a Nanothermometer. *Nano Today* **2018**, *19*, 126–145. [CrossRef]
10. McCreery, R.L. *Raman Spectroscopy for Chemical Analysis*; John Wiley & Sons: Hoboken, NJ, USA, 2000.
11. Griffiths, J.E.; Malyj, M. Stokes/Anti-Stokes Raman Vibrational Temperatures: Reference Materials, Standard Lamps, and Spectrophotometric Calibrations. *Appl. Spectrosc.* **1983**, *37*, 315–333. [CrossRef]
12. Beechem, T.E.; Serrano, J.R. Raman Thermometry of Microdevices: Choosing a Method to Minimize Error. *Spectroscopy* **2011**, *26*, 36–44.
13. Tuschel, D. Raman Thermometry: Understanding the Mathematics to Better Design Raman Measurements. *Spectroscopy* **2019**, *34*, 8–13.
14. Serrano, J.R.; Phinney, L.M.; Kearney, S.P. Micro-Raman Evaluation of Polycrystalline Silicon MEMS Devices. *Surf. Eng. Manuf. Appl.* **2006**. [CrossRef]
15. Beechem, T.; Christensen, A.; Graham, S.; Green, D. Micro-Raman Thermometry in the Presence of Complex Stresses in GaN Devices. *J. Appl. Phys.* **2008**, *103*, 1–8. [CrossRef]
16. Dramićanin, M.D. Sensing Temperature via Downshifting Emissions of Lanthanide-Doped Metal Oxides and Salts. A Review. *Methods Appl. Fluoresc.* **2016**, *4*, 1–23. [CrossRef] [PubMed]
17. Kip, B.J.; Meier, R.J. Determination of the Local Temperature at a Sample during Raman Experiments Using Stokes and Anti-Stokes Raman Bands. *Appl. Spectrosc.* **1990**, *44*, 707–711. [CrossRef]
18. Tuschel, D. Selecting an Excitation Wavelength for Raman Spectroscopy. *Spectroscopy* **2016**, *31*, 14–23.
19. Adar, F. Resonance Enhancement of Raman Spectroscopy: Friend or Foe? *Spectroscopy* **2013**, *28*, 6.
20. Abel, M.R.; Graham, S.; Serrano, J.R.; Kearney, S.P.; Phinney, L.M. Raman Thermometry of Polysilicon Microelectromechanical Systems in the Presence of an Evolving Stress. *J. Heat Transf. Trans. Asme* **2007**, *129*, 329–334. [CrossRef]
21. Sarua, A.; Bullen, A.; Haynes, M.; Kuball, M. High-Resolution Raman Temperature Measurements in GaAs p-HEMT Multifinger Devices. *IEEE Trans. Electron Devices* **2007**, *54*, 1838–1842. [CrossRef]
22. Ahmad, I.; Kasisomayajula, V.; Holtz, M.; Berg, J.M.; Kurtz, S.R.; Tigges, C.P.; Allerman, A.A.; Baca, A.G. Self-Heating Study of an AlGaN/GaN-Based Heterostructure Field-Effect Transistor Using Ultraviolet Micro-Raman Scattering. *Appl. Phys. Lett.* **2005**, *86*, 1–3. [CrossRef]
23. Calizo, I.; Balandin, A.A.; Bao, W.; Miao, F.; Lau, C.N. Temperature Dependence of the Raman Spectra of Graphene and Graphene Multilayers. *Nano Lett.* **2007**, *7*, 2645–2649. [CrossRef] [PubMed]
24. Rassat, S.D.; Davis, E.J. Temperature-Measurement of Single Levitated Microparticles Using Stokes Anti-Stokes Raman Intensity Ratios. *Appl. Spectrosc.* **1994**, *48*, 1498–1505. [CrossRef]

25. Lundt, N.; Kelly, S.T.; Rodel, T.; Remez, B.; Schwartzberg, A.M.; Ceballos, A.; Baldasseroni, C.; Anastasi, P.A.F.; Cox, M.; Hellman, F.; et al. High Spatial Resolution Raman Thermometry Analysis of TiO2 Microparticles. *Rev. Sci. Instrum.* **2013**, *84*, 1–7. [CrossRef] [PubMed]

26. Gallardo, J.J.; Navas, J.; Zorrilla, D.; Alcantara, R.; Valor, D.; Fernandez-Lorenzo, C.; Martin-Calleja, J. Micro-Raman Spectroscopy for the Determination of Local Temperature Increases in TiO_2 Thin Films Due to the Effect of Radiation. *Appl. Spectrosc.* **2016**, *70*, 1128–1136. [CrossRef]

27. Peral, J.; Domenech, X.; Ollis, D.F. Heterogeneous Photocatalysis for Purification, Decontamination and Deodorization of Air. *J. Chem. Technol. Biotechnol.* **1997**, *70*, 117–140. [CrossRef]

28. Giarola, M.; Sanson, A.; Monti, F.; Mariotto, G.; Bettinelli, M.; Speghini, A.; Salviulo, G. Vibrational Dynamics of Anatase TiO2: Polarized Raman Spectroscopy and Ab Initio Calculations. *Phys. Rev. B* **2010**, *81*, 1–7. [CrossRef]

29. Yin, Z.F.; Wu, L.; Yang, H.G.; Su, Y.H. Recent Progress in Biomedical Applications of Titanium Dioxide. *Phys. Chem. Chem. Phys.* **2013**, *15*, 4844–4858. [CrossRef]

30. Pascual, J.; Camassel, J.; Mathieu, H. Resolved Quadrupolar Transition in TiO2. *Phys. Rev. Lett.* **1977**, *39*, 1490–1493. [CrossRef]

31. Mo, S.-D.; Ching, W.Y. Electronic and Optical Properties of Three Phases of Titanium Dioxide: Rutile, Anatase, and Brookite. *Phys. Rev. B* **1995**, *51*, 13023–13032. [CrossRef]

32. Mohamad, M.; Ul Haq, B.; Ahmed, R.; Shaari, A.; Ali, N.; Hussain, R. A Density Functional Study of Structural, Electronic and Optical Properties of Titanium Dioxide: Characterization of Rutile, Anatase and Brookite Polymorphs. *Mater. Sci. Semicond. Process.* **2015**, *31*, 405–414. [CrossRef]

33. Fujishima, A.; Honda, K. Electrochemical Photolysis of Water at a Semiconductor Electrode. *Nature* **1972**, *238*, 37–38. [CrossRef]

34. Michael, G. The Artificial Leaf, Molecular Photovoltaics Achieve Efficient Generation of Electricity from Sunlight. *Comments Inorg. Chem.* **2006**, *12*, 93–111. [CrossRef]

35. Ohsaka, T.; Fujio, I.; Yoshinori, F. Raman Spectrum of Anatase, TiO_2. *J. Raman Spectrosc.* **1978**, *7*, 321–324. [CrossRef]

36. Du, Y.L.; Deng, Y.; Zhang, M. S Variable-temperature Raman scattering study on anatase titanium dioxide nanocrystals. *J. Phys. Chem. Solids* **2006**, *67*, 2405–2408. [CrossRef]

37. Brites, C.D.; Balabhadra, S.; Carlos, L.D. Lanthanide-Based Thermometers: At the Cutting-Edge of Luminescence Thermometry. *Adv. Opt. Mater.* **2019**, *7*, 1801239. [CrossRef]

38. Maher, R.; Galloway, C.; Le Ru, E.; Cohen, L.; Etchegoin, P. Vibrational Pumping in Surface Enhanced Raman Scattering (SERS). *Chem. Soc. Rev.* **2008**, *37*, 965–979. [CrossRef]

39. Kim, M.M.; Giry, A.; Mastiani, M.; Rodrigues, G.O.; Reis, A.; Mandin, P. Microscale Thermometry: A Review. *Microelectron. Eng.* **2015**, *148*, 129–142. [CrossRef]

40. Tuschel, D.; Adar, F. Molecular Spectroscopy Workbench Raman Thermometry. *Spectroscopy* **2016**, *31*, 8–13.

41. Freitag, M.; Steiner, M.; Martin, Y.; Perebeinos, V.; Chen, Z.; Tsang, J.C.; Avouris, P. Energy Dissipation in Graphene Field-Effect Transistors. *Nano Lett.* **2009**, *9*, 1883–1888. [CrossRef]

42. Smith, J.D.; Cappa, C.D.; Drisdell, W.S.; Cohen, R.C.; Saykally, R.J. Raman Thermometry Measurements of Free Evaporation from Liquid Water Droplets. *J. Am. Chem. Soc.* **2006**, *128*, 12892–12898. [CrossRef]

43. LaPlant, F.; Laurence, G.; BenAmotz, D. Theoretical and Experimental Uncertainty in Temperature Measurement of Materials by Raman Spectroscopy. *Appl. Spectrosc.* **1996**, *50*, 1034–1038. [CrossRef]

44. Brites, C.D.S.; Lima, P.P.; Silva, N.J.O.; Angel Millan, A.; Amaral, V.S.; Palacio, F.; Carlos, L.D. Thermometry at the nanoscale. *Nanoscale* **2012**, *4*, 4799–4829. [CrossRef] [PubMed]

45. Venkateswarlu, K. Temperature Dependence of the Intensities of Raman Lines. *Nature* **1947**, *159*, 96–97. [CrossRef]

46. Long, D.A. *Raman Spectroscopy*; McGraw-Hill: New York, NY, USA, 1976.

47. Albrecht, A.C. On the Theory of Raman Intensities. *J. Chem. Phys.* **1961**, *34*, 1476–1484. [CrossRef]

48. Weber, W.H.; Merlin, R. *Raman Scattering in Materials Science*; Springer: Berlin/Heidelberg, Germany, 2000.

biosensors

Article

Detection of Sub-Nanomolar Concentration of Trypsin by Thickness-Shear Mode Acoustic Biosensor and Spectrophotometry

Ivan Piovarci [1], Sopio Melikishvili [1], Marek Tatarko [1], Tibor Hianik [1,*] and Michael Thompson [2,*]

[1] Department of Nuclear Physics and Biophysics, Faculty of Mathematics, Physics and Informatics, Comenius University, Mlynska dolina F1, 84248 Bratislava, Slovakia; piovarci6@uniba.sk (I.P.); s.melikishvili@gmail.com (S.M.); marek.tatarko@fmph.uniba.sk (M.T.)

[2] Lash Miller Laboratories, Department of Chemistry, University of Toronto, Toronto, ON M5S 3H6, Canada

[*] Correspondence: tibor.hianik@fmph.uniba.sk (T.H.); m.thompson@utoronto.ca (M.T.)

Abstract: The determination of protease activity is very important for disease diagnosis, drug development, and quality and safety assurance for dairy products. Therefore, the development of low-cost and sensitive methods for assessing protease activity is crucial. We report two approaches for monitoring protease activity: in a volume and at surface, via colorimetric and acoustic wave-based biosensors operated in the thickness-shear mode (TSM), respectively. The TSM sensor was based on a β-casein substrate immobilized on a piezoelectric quartz crystal transducer. After an enzymatic reaction with trypsin, it cleaved the surface-bound β-casein, which increased the resonant frequency of the crystal. The limit of detection (LOD) was 0.48 ± 0.08 nM. A label-free colorimetric assay for trypsin detection has also been performed using β-casein and 6-mercaptohexanol (MCH) functionalized gold nanoparticles (AuNPs/MCH-β-casein). Due to the trypsin cleavage of β-casein, the gold nanoparticles lost shelter, and MCH increased the attractive force between the modified AuNPs. Consequently, AuNPs aggregated, and the red shift of the absorption spectra was observed. Spectrophotometric assay enabled an LOD of 0.42 ± 0.03 nM. The Michaelis–Menten constant, K_M, for reverse enzyme reaction has also been estimated by both methods. This value for the colorimetric assay (0.56 ± 0.10 nM) is lower in comparison with those for the TSM sensor (0.92 ± 0.44 nM). This is likely due to the better access of the trypsin to the β-casein substrate at the surface of AuNPs in comparison with those at the TSM transducer.

Keywords: trypsin; β-casein; AuNPs; acoustic wave biosensor; colorimetric assay

Citation: Piovarci, I.; Melikishvili, S.; Tatarko, M.; Hianik, T.; Thompson, M. Detection of Sub-Nanomolar Concentration of Trypsin by Thickness-Shear Mode Acoustic Biosensor and Spectrophotometry. *Biosensors* **2021**, *11*, 117. https://doi.org/10.3390/bios11040117

Received: 11 February 2021
Accepted: 6 April 2021
Published: 11 April 2021

Publisher's Note: MDPI stays neutral with regard to jurisdictional claims in published maps and institutional affiliations.

1. Introduction

Peptidases, more frequently referred to as proteases, are a group of enzymes that irreversibly hydrolyze a peptide bond in an amino acid sequence through the nucleophilic attack and subsequent hydrolysis of a tetrahedral intermediate. They play critical roles in biological and physiological processes such as blood clotting, digestion, and a variety of cellular activities [1,2]. Proteases are highly involved in the dairy industry as well, where their activity is directly linked to the shelf life of dairy products [3]. Owing to their specificity, protease activity-based biosensors are used in various diseases diagnostics [4–6]. For example, pancreatic diseases such as cystic fibrosis, acute pancreatitis, or the acute phase of chronic pancreatitis are associated with the increased trypsin level of 2.1–71.42 nM in the serum of patients [7,8]. In the healthy physiological condition, the concentration of trypsin varies in magnitude. Additionally, levels of trypsin differ between serum and intestinal levels. For serum levels for fasting individuals, the concentration of trypsin was measured from 4 to 20 nM [9,10]. The intestinal level of trypsin depends on the location in the intestine and ranges from 4 to 30 μM, which is much higher than in serum [11].

Moreover, the inhibitors of these proteases are successfully employed as therapeutic agents [2,12,13].

Trypsin is an extremely important serine protease of the chymotrypsin family. It is produced in the pancreas and it plays crucial roles in the small intestine. Trypsin catalyzes the hydrolysis of consumed proteins and activates protease proenzymes as part of the digestive system. It is highly specific toward the cleavage of peptide bonds at the carboxyl side of lysine or arginine. Trypsin is often used as a model protease because it is inexpensive and readily available [14–16]. Standard assays for the detection of proteases such as trypsin usually utilize fluorogenic and chromogenic substrates. Those assays are useful, practical, and highly sensitive. However, spectroscopic assays are incapable of measuring protease activity in highly colored and turbid samples such as cells, tissue lysates, or milk. Therefore, the development of a new label-free method for detecting protease activity without interruption from impurity inclusions is needed [1,15,17].

The thickness-shear mode (TSM) acoustic wave biosensor may present an attractive platform for the development of cost-effective and highly sensitive techniques for trypsin detection. The use of TSM devices is a well-known method for the detection of mass changes due to depositions or chemical/biochemical reactions on its surface. It is also an established method for detecting changes in the viscoelastic properties of the contacting material. Therefore, the TSM biosensor is a sensitive tool for the study of molecular interactions on surfaces [18]. Moreover, the coupling of a flow injection analysis (FIA) system to a TSM sensor device permits the monitoring of kinetic processes that take place at the surface of the sensor [19]. The TSM device applies a high-frequency AC voltage across an AT-cut quartz crystal on which, due to the piezoelectric effect, an acoustic shear wave is generated and propagated through the sensing layer perpendicular to the surface of the crystal [20]. It has a low noise level and higher Q-factor in clinical liquids such as tissue fluids and serum. Compared to other common biosensing technologies, TSM electroacoustic resonators have the combined advantages of high sensitivity and low cost, label-free detection of analyte, and simple operation without the requirement of bulky detection systems [21]. Moreover, in contrast with traditional quartz crystal microbalance (QCM) techniques, the analysis of complex impedance spectra allows for the receipt of information about changes in the properties of layers even with the adsorption of relatively small molecules that do not contribute to the mass but only to the viscoelastic properties of the layer [22]. The multi-harmonic QCM method has previously been applied for the detection of plasmin and trypsin at the surface of β-casein layers [23]. This method allows the detection of these proteases at the sub-nM level. However, the possible contribution of viscoelastic effects has not been analyzed.

In addition to the acoustic methods also the colorimetric assay based on gold nanoparticles (AuNPs) for protease detection is of increased interest. AuNPs have attracted tremendous interest because of their optical and electronic properties, which are tunable by changing the size, shape, surface chemistry, or aggregation state. Colloidal AuNPs have a distinctive red color, which arises from the tiny dimensions of the AuNPs. The changes in the UV–vis spectra of the resultant colloids are measured to investigate the size effect of AuNPs on the surface plasmon resonance (SPR). Interestingly, the red color of citrate-stabilized AuNPs turns to blue when they are aggregated [24]. This approach has been widely applied to various methods for colorimetric detection of analytes via the aggregation of AuNPs [25], including those of trypsin detection [26].

In this work, we designed an analytical method based on the TSM biosensor for the real-time and label-free detection of trypsin. Using TSM frequency responses, we studied the assembly and stability of self-assembled β-casein layers on a quartz crystal electrode and measured the dynamics of TSM response and changes in motional resistance during casein cleaving by the protease.

Additionally, we compared the sensitivity of the TSM method with another label-free assay of the protease activity by employing AuNPs coated by β-casein and 6-mercaptohexanol (MCH). Unlike the surface-sensitive TSM biosensor, the suggested approach was volume-sensitive, thus allowing us to monitor tryptic activity in the reaction mixture. We used an approach developed by Chuang et al. [26]. However, instead of gelatin, β-casein was used as a substrate for trypsin. β-casein adsorbed on AuNPs kept the modified nanoparticles stably suspended in solution.

Considering the results obtained, we believe that the proposed approaches constitute rapid, cost-efficient, sensitive and useful tools for protease analysis. This paper is an extension of a conference paper published in the 1st International Electronic Conference on Biosensors [27].

2. Materials and Methods

2.1. Reagents

Ultrapure water obtained by reverse osmosis (Thermo Scientific, Waltham, MA, USA, ρ = 18.2 MΩ cm) was used for the preparation of all aqueous solutions. As a medium, 10 mM, pH 7.4 phosphate-buffered saline (PBS) was used (10 mM Na_2HPO_4, 2 mM KH_2PO_4, 2.7 mM KCl and 137 mM NaCl), prepared from tablets (Sigma-Aldrich, Darmstadt, Germany, Cat. No. P4417). In the experiments, trypsin from bovine pancreas ($\geq$90%, $\geq$7500 BAEE units/mg solid, Sigma-Aldrich, Darmstadt, Germany, Cat. No. T9201) served as a model protease. The concentration of stock bovine β-casein ($\geq$98%, Sigma-Aldrich, Darmstadt, Germany, Cat. No. C6905) solutions, prepared in PBS, was 0.5 mg/mL. 11-Mercaptoundecanoic acid (MUA, Sigma-Aldrich, Cat. No. 450561), *N*-(3-dimethylaminopropyl)-*N'*-ethylcarbodiimide (EDC, $\geq$98%, Sigma-Aldrich, Cat. No. E6383), and N–Hydroxysuccinimide (NHS, Sigma-Aldrich, Darmstadt, Germany, Cat. No. 130672) were employed for β-casein immobilization. The chemicals needed to prepare the gold nanoparticles, such as auric acid ($HAuCl_4$), sodium citrate, and 6-mercapto-1-hexanol (MCH), were purchased from Sigma-Aldrich (Darmstadt, Germany). All experiments were carried out at 20 °C.

2.2. Cleaning and Modification of Gold Electrode-Coated Quartz Crystals

Symmetric gold electrode-coated quartz discs (Total Frequency Control, Storrington, UK, working area, 0.2 cm^2) with a fundamental frequency of 8 MHz were cleaned in a basic Piranha solution (29% NH_3, 30% H_2O and H_2O_2 with volumetric 1:5:1 ratio, respectively) for 25 min. After this treatment, the crystals were washed three times with deionized water and stored in ethanol. After drying in a flow of nitrogen, the TSM crystals were immersed in 2 mM MUA and were incubated for 16 h to form a self-assembled monolayer. After this step, the crystals were rinsed several times with deionized water and dried under nitrogen, followed by incubation for 20 min in a 20 mM EDC and 50 mM NHS mixture in order to activate the carboxylic groups of MUA for the further immobilization of bovine β-casein on the gold electrode of the quartz sensor. The scheme of modification of the TSM transducer as well as the cleavage of β-casein by trypsin is shown in Figure 1.

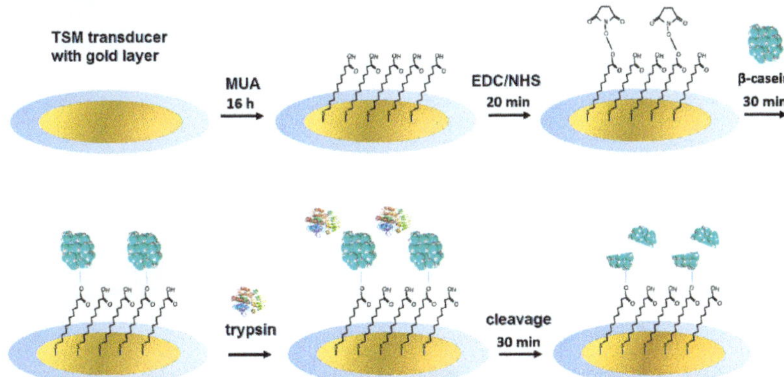

Figure 1. The scheme for modification of the gold layer on a TSM transducer and the cleavage of β-casein by trypsin.

2.3. TSM Measurements

AT-cut 8.0 MHz gold electrode-coated quartz crystals, modified on one side by MUA with activated carboxylic groups by NHS/EDC as described above, were incorporated into a home-built flow-through thickness shearing mode (TSM) acoustic wave device sensor system. The setup and general configuration of the flow-through system is described in reference [19]. One side of the crystal was exposed to liquid, the other one was exposed to air. The liquid was introduced using a syringe pump (Genie Plus, Torrington, CT, USA). Runs were performed with the crystals in the vertical position and at ambient temperature (approximately 20 °C). The modified crystal was secured in the holder using two O-rings. The gold electrodes were kept in contact with the gold leads in the holder. Resonance frequency, f, and motional resistance, R_m, were determined based on the Butterworth–van Dyke (BVD) model of a quartz crystal resonator [19]. The resonant frequency represents the energy storage and reflects the mass changes of the oscillating layer, while R_m is related to the dissipation of energy and provides evidence of changes in the shearing viscosity of the layer [22]. The measuring procedure was as follows. Each slide was flushed through with PBS at a flow rate of 50 µL/min until a stable baseline was achieved (45 min), using the flow-through injection system. This step was necessary to remove any weakly adsorbed molecules at the surface of the TSM transducer. Next, the pump was momentarily stopped. The β-casein solution (0.5 mg/mL in PBS) was slowly introduced to the sample, while the PBS was exchanged out in order to minimize pressure effects to the system. β-casein was introduced at a rate of 50 µL/min for approximately 45 min. Once again, the pump was momentarily stopped, and the sample input tube was slowly placed back into the PBS solution. The PBS was re-introduced at a rate of 50 µL/min to remove any loosely bound casein until a stable baseline was achieved. Changes of the resonant frequency and motional resistance were recorded. For proteolysis measurements, solutions with various concentrations of trypsin in PBS (0.1, 0.5, 1, 5, 10, and 20 nM) were flowed over TSM crystals with an immobilized β-casein layer at a flow rate of 50 µL/min. Trypsin and β-casein solutions were freshly prepared before each experiment.

2.4. Synthesis and Modification of AuNPs

AuNPs were prepared using a modified citrate method described in reference [28]. Briefly, 100 mL of $HAuCl_4$ (0.01%) was heated to boiling under vigorous stirring, which was followed by the addition of 5 mL of sodium tris-citrate solution (1%). The solution was left boiling while stirring until it turned a deep red. Then, we let the AuNPs solution cool down and stored it in the dark. In order to modify the gold nanoparticles with casein, we added 2 mL of 0.1 mg/mL β-casein to 18 mL of the AuNPs solution. After 2 h of incubation at room temperature without stirring, the gold nanoparticles were further incubated with 200 µL of 1 mM MCH overnight for approximately 18 h. The scheme of modification of AuNPs is showed in Figure 2.

2.5. Sprectrophotometric Assay

For the colorimetric assay, we prepared 0.95 mL of AuNPs. Trypsin was dissolved in deionized water, and 0.05 mL of trypsin from the stock solution was added to each cuvette. The concentration of trypsin in cuvettes was 0.1, 0.5, 1, 5, and 10 nM at 1 mL total volume of solution. We also used a reference cuvette where only 0.05 mL of protease-free water was added to the AuNPs solution. We measured the spectra of the AuNPs before trypsin addition (0 min), just after trypsin addition (approximately 1 s), and then every 15 min up to 60 min. The measurement was repeated 3 times. We multiplied the value of absorbance at time t = 0 by the dilution factor to correct the changes in absorbance intensity caused by the initial protease addition. Absorbance was measured by UV-1700 spectrophotometer at a temperature of around 20 °C and in the wavelength range of 220–800 nm (Shimadzu, Kyoto, Japan).

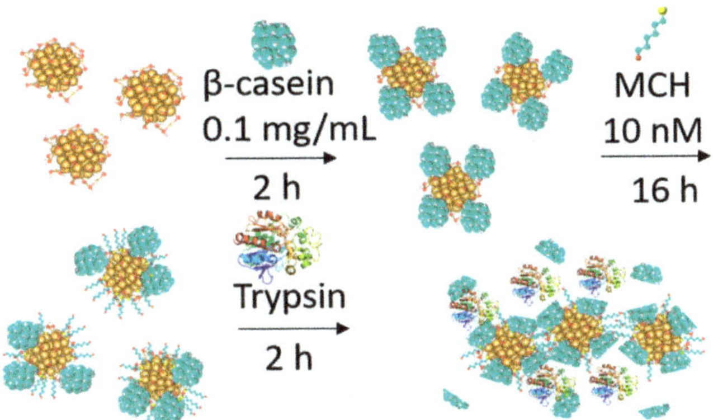

Figure 2. The scheme for modification of gold nanoparticles (AuNPs) by β-casein and by 6-mercapto-1-hexanol (MCH) as well as the cleavage of β-casein by trypsin. Before enzymatic digestion, functionalized AuNPs were stable due to steric stabilization. After the AuNPs were subjected to protease cleavage, the casein was removed from the surface of AuNPs/MCH/β-casein. This caused the destabilization of the NPs, followed by their aggregation.

2.6. Analysis of Casein Adsorption and Hydrolysis Processes

The surface concentration (Γ_{QCM}, ng/cm^2) of the adsorbed β-casein layer on the TSM transducer was determined by a modified Sauerbrey Equation (1) as follows:

$$\Gamma_{QCM} = \frac{-A\sqrt{\mu\rho}\Delta f}{2f_0^2},\qquad(1)$$

where A is the area of the electrode, ρ = 2.648 g/cm^3 is the density of quartz, μ = 2.947 × 10^{11} g/cms^2 is the shear modulus of AT-cut crystal, and f_0 is fundamental resonant frequency [29]. The Sauerbrey equation is strongly valid for thin rigid films at the surface of quartz crystal in vacuum. However, in a liquid, the viscoelastic contribution can affect the frequency changes. Through analysis of the motional resistance, R_m, it is possible to estimate whether the mass or viscosity is dominant in frequency changes. It has been shown that the slope of $|\Delta f/\Delta R_m|$ can be used for quantitative estimation whether the changes in frequency can be attributed to mass or to viscosity effects. For ideal rigid films, the ΔR_m values are practically zero. This means that $|\Delta f/\Delta R_m|$ parameters higher than a certain critical value can be assigned to the mass effect [30]. According to the calculations made in ref. [30] for the AT cut quartz crystal with fundamental frequency f_0 = 8 MHz, $|\Delta f/\Delta R_m|$ = 10.37 Hz/Ω.

The frequency changes following the addition of the trypsin were normalized to the changes of the resonant frequency caused by adsorption of the β-casein at the surface of TSM crystal. This allowed consideration of a possible variation in the properties of the β-casein layers that were subsequently cleaved by trypsin. The normalized frequency changes were expressed as $\Delta f_N = (\Delta f_{TRY}/\Delta f_{casein}) \times 100(\%)$, where Δf_{TRY} are changes in frequency following the addition of trypsin at certain concentration of the protease and Δf_{casein} are changes in frequency caused by the formation of a β-casein layer.

An inverse Michaelis–Menten (MM) model [31] was used to describe the dependence of the normalized frequency changes vs. concentration of trypsin at fixed concentration of the β-casein at the surface of TSM transducer:

$$\Delta f_N = (\Delta f_N)_{max}\frac{C_{TRY}}{K_M + C_{TRY}}\qquad(2)$$

where $(\Delta f_N)_{max}$ is the maximal change of the frequency that corresponds to the maximum rate of enzyme reaction achieved by the system happening at saturating enzyme concentration, C_{TRY} is the concentration of trypsin, K_M is the reverse Michalis–Menten constant that is equal to the trypsin concentration that achieves half of maximum rate. The hydrolysis of β-casein in a volume was modeled with an inverse MM Equation (3) as well

$$\frac{A_0 - A_{15}}{A_0}100 = v_{max}\frac{C_{TRY}}{K_M + C_{TRY}},\tag{3}$$

where A_0 is the absorbance of AuNPs before exposure to trypsin, A_{15} is the absorbance after 15 min of exposure to trypsin, and $v_{max} = [100 \times (A_0 - A_{15})/A_0]_{max}$ represents the maximum rate achieved by the system.

2.7. Data Analysis

Origin version 7.5 software (Microcal Software Inc., Northampton, MA, USA) was used for curve-fitting and data analysis. Data were obtained from a minimum of 3 independent experiments.

3. Results and Discussion

3.1. Development of Acoustic Biosensor for the Detection of Trypsin Activity at Surfaces

For the detection of trypsin activity at surfaces, it is crucial to optimize the methods of preparation of the protein layers that serve as a substrate for the protease of interest. The preparation of the protein layers on the surface of the transducers is a common application of acoustic biosensors. For instance, the preparation of casein layers is attractive for future applications in the pharmaceutical and food industries [32].

In this study, we have monitored the activity of trypsin at various concentrations (from 0.1 to 20 nM) in the hydrolysis of a β-casein layer immobilized onto a gold surface by a carboxylate terminated self-assembled monolayer (SAM) of MUA using a TSM technique. MUA strongly binds to gold through thiol groups in a high level of molecular dimension order, forming a stable SAM [33]. The formation of the SAM itself enables the coupling of activated carboxylic groups with free amino groups in the β-casein, which is an effective method for immobilizing proteins on a gold surface [34–36].

Figure 3 illustrates typical kinetic changes of the frequency, Δf, and motional resistance, ΔR$_m$, obtained during the TSM experiment. The TSM crystal covered by the MUA layer activated by EDC/NHS established in a flow cell has been first washed by PBS. As soon as the stable baseline was established, the β-casein dissolved in PBS in a concentration of 0.5 mg/mL has been added. The sharp drop of the resonant frequency and an increase of the motional resistance were observed, indicating the adsorption of the β-casein to the quartz crystal/liquid interface. The washing of the surface by PBS resulted in only a slight increase of the frequency, which is evidence of removal of weakly adsorbed β-casein molecules from the surface. Thus, the frequency did not recover to the original value obtained when the crystals were exposed to the buffer. This suggests that there were two modes of casein binding to the MUA surface, a tightly bound layer and a weakly bound layer, and that only loosely bound casein layers were removed during the PBS washing [37]. Since the increase in resonant frequency after PBS washing was so small, we can speculate that β-casein adsorbed on the MUA formed a stable immobilized layer, which makes this result attractive for its potential applications in biosensors for the detection of protease activity.

The resulting frequency shift after the adsorption of the β-casein to the surfaces of the crystal was around −199.43 Hz. Furthermore, the buffer was changed to a 20 nM trypsin solution. The frequency increased asymptotically to reach a stable value, indicating that the proteolysis process occurred. Washing of the surface by PBS did not result in significant changes of frequency and motional resistance, which is evidence that the cleaved peptide residues were removed from the surface in a flow mode during the application of trypsin. The kinetics of the changes of the resonant frequency and motional resistance were recorded for different trypsin concentrations, each one with a new quartz crystal and a newly adsorbed β-casein layer.

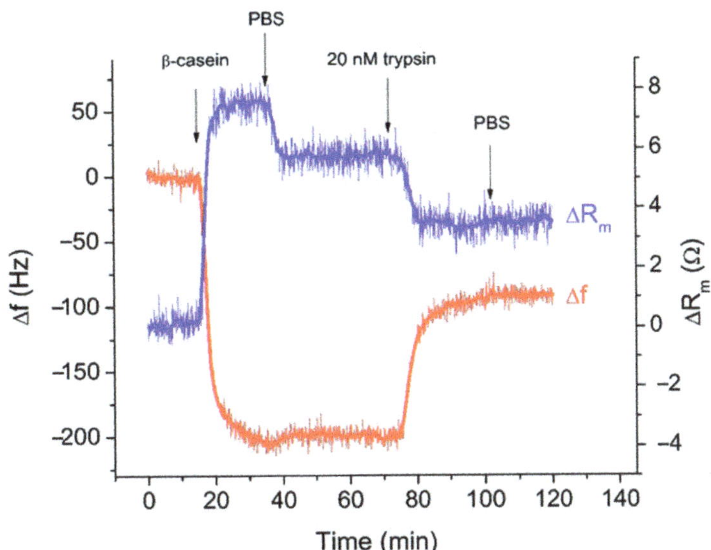

Figure 3. Typical kinetics of the changes of resonant frequency, Δf, and motional resistance, ΔR_m, of the thickness-shear mode (TSM) transducer for various modifications. The additions of β-casein, trypsin, and washing of the surface by phosphate-buffered saline (PBS) are shown by arrows.

Earlier works indicated that the Sauerbrey Equation (1) can be applied to obtain a rough estimate for the surface concentration of the adsorbed β-casein layer [23,38,39], which is valid only for the specific case of a crystal being loaded with rigid, well-adhered layers in air with a minor contribution to the surface viscosity [19,40,41]. As we mentioned in Section 2.6, the contribution of viscosity into the frequency changes can be estimated from the ratio $|\Delta f/\Delta R_m|$. At the highest concentration of trypsin (20 nM) studied and at the steady-state conditions (Figure 1), $|\Delta f/\Delta R_m| = 199.43\ \text{Hz}/7.4\ \Omega = 26.95\ \text{Hz}/\Omega$. This value is much higher than the threshold value (10.37 Hz/Ω). This means that the changes of frequency are related mainly to the changes of the mass.

Therefore, with an awareness of the limitations stated above, Equation (1) can be used to estimate the amount of proteins on the surface (Γ_{QCM}, ng/cm^2) [38]. The average value of the frequency shift after the adsorption of the β-casein to hydrophilic surfaces was -165.26 ± 47.7 Hz. Using this value, as well as A = 0.2 cm^2 for the area of the electrode of an AT-cut quartz crystal ($f_0 = 8$ MHz fundamental resonant frequency), a surface concentration of 228.1 ± 65.8 ng/cm^2 was obtained for β-casein.

This is in good agreement with earlier experimental works based on ellipsometry that reported 200–300 ng/cm^2 for a full-coverage monolayer of β-casein [42–44]. Furthermore, QCM studies by Tatarko et al. estimated a mass density of 350 ng/cm^2 for the immobilized β-casein monolayer [23]. These results support the interpretation that the surface concentration of β-casein obtained by TSM measurements corresponds to monolayer formation.

Based on the kinetic curves obtained for the concentration range of trypsin 0.1–20 nM, we prepared a plot of the frequency and motional resistance changes as a function of trypsin concentration (Figure 4). It can be seen that the frequency changes increase with increasing the trypsin concentration and started to saturate at $C_{TRY} > 10$ nM. In contrast with frequency, R_m decreased with increasing the concentration of the protease, which is evidence of dominant mass changes.

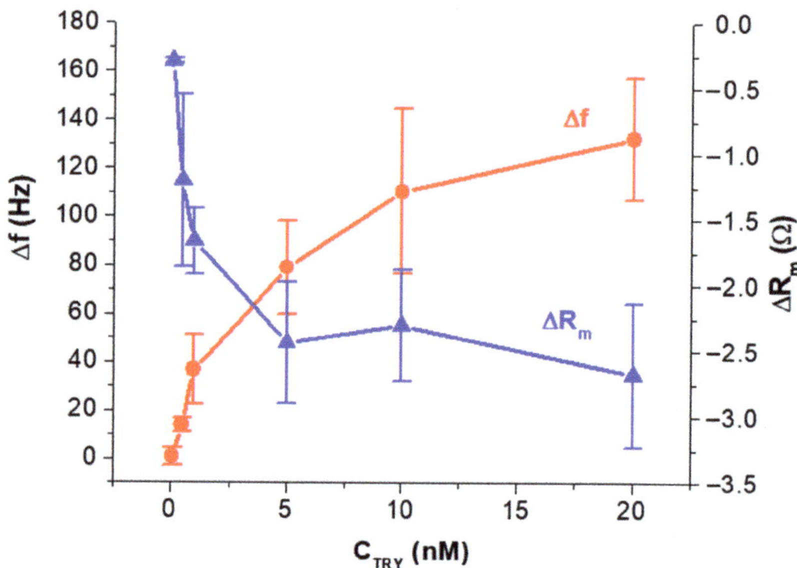

Figure 4. Plots of changes of frequency, Δf, and motional resistance, ΔR_m, vs. trypsin concentration (C_{TRY}). Statistically, a value for the standard deviation was obtained from three independent experiments at each trypsin concentration.

For practical purposes, for the detection of trypsin in food or in other biological samples such as blood or blood plasma, it is convenient to analyze the effect of trypsin on the cleavage of β-casein by changes of resonant frequency of the quartz crystal under steady-state conditions. In order to minimize the effect of variation of the properties of β-casein layers at the monolayer of 11-mercaptoundecanoic acid (MUA) on the resonant frequency, we plotted the normalized frequency changes: $\Delta f_N = (\Delta f_{TRY} / \Delta f_{casein}) \times 100\%$ vs. concentration of trypsin, C_{TRY} (Δf_{TRY} is the frequency change corresponded to the cleavage of β-casein layer after incubation with a certain concentration of trypsin and Δf_{casein} is the frequency changes corresponded to the adsorption of β-casein at the MUA layer before trypsin addition). This dependence shown on Figure 5 can be fitted by the Langmuir isotherm (see Section 2.6 and Equation (2)).

The fitting of calibration plots yielded $(\Delta f_N)_{max} = 70.36 \pm 4.60$ and $K_M = 0.92 \pm 0.44$ nM. The limit of detection (LOD) has been determined from the linear part of the dependence presented in Figure 5 using the 3.3(SD)/S rule (SD is standard deviation at the lowest concentration of trypsin, S is the slop of the linear dependence) as LOD = 0.48 ± 0.08 nM. Thus, in the presence of 20 nM trypsin, almost 70% of the casein layer is removed due to protease cleavage. This value is close to the maximum cleavage obtained by fitting the Langmuir isotherm. It can be assumed that due to the restricted access of the trypsin to the casein layer at the surface of the TSM transducer, around 30% of the casein remained at the surface after protease cleavage.

β-casein interacts with the immobilized MUA layer preferably with N-terminus. This part of the protein contains most of the charge [45]. It also contains numerous free amino groups that amino-reactive MUA can bind. β-casein is composed of 209 amino acids starting with arginine at the N-end (Arg1) [46]. The immediate binding of Arg1 to MUA is possible. Following the addition of trypsin, the cleavage of available peptide bonds toward the C-terminus of lysine and partially arginine occurs. These cleavage sites for trypsin are mostly identical to that of the plasmin [47]. The only unique cleavage site for trypsin is located between Arg202-Gly203, near the C-terminus [48]. The most common hydrolysis takes place at Lys28-Lys29, Lys105-His106, and Lys107-Glu108 with the subsequent cleavage at Lys97-Ala98, Lys99-Glu100, and Lys113-Tyr114 [49]. The cleavage of these peptide bonds should cause release of the β-casein fragments (or so called γ-casein fragments) that corresponds to up to 88% of the β-casein molecular weight. This ratio can be affected by the β-casein assembly on the MUA layer and thus the availability of such bonds to the trypsin. Considering that approximately 70% of casein fragments are released from the sensing, we can speculate that the closest site for its cleavage by trypsin at the MUA layer is probably after Lys48. According to the ExPASy Peptide Cutter tool [50], the cleavage of β-casein by trypsin at Lys48 is highly probable.

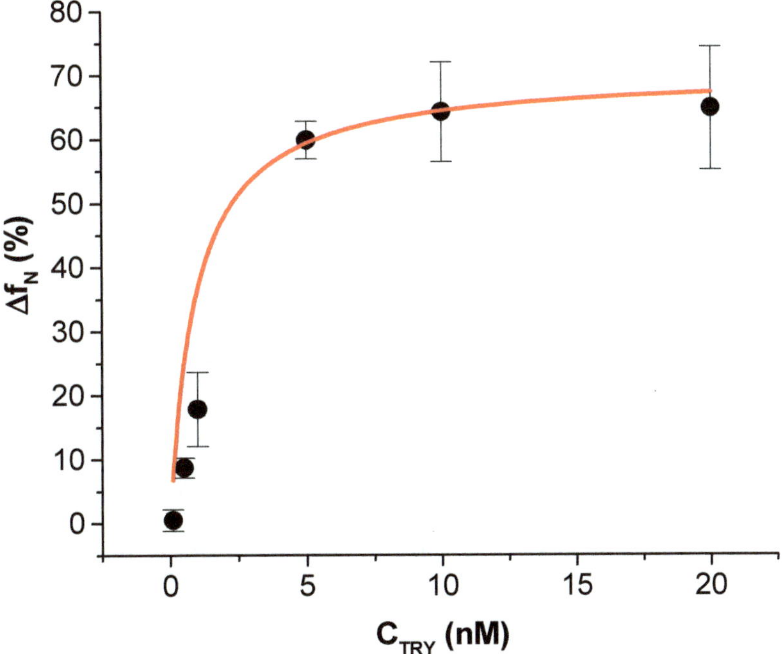

Figure 5. Plot of the normalized changes of the resonant frequency Δf_N vs. trypsin concentrations, C_{TRY}. Standard deviation values are obtained from three independent experiments. The red line is the fit according to the Langmuir isotherm (Equation (2)) with accuracy $R^2 = 0.99$.

In the paper by Chen et al. [51], the detection of trypsin activity based on the electrochemical method has been reported. They applied a gold working electrode modified with a short peptide substrate conjugated with graphene oxide (GO) and the thionine redox label. The incubation of the sensor with trypsin for 2 h resulted in cleavage of the peptide substrate, removal of the redox probe, and a decrease of the current amplitude. Although the authors reported a lower detection limit, down to 0.05 nM, and a high selectivity to trypsin, this method has some drawbacks. First, the biosensor was based on the peptide substrate labeled by the graphene oxide (GO)–thionine conjugate, which is not available commercially. The peptide–GO–thionine conjugates are more expensive in comparison with the β-casein used in our work. Therefore, this limits the practical application of such an electrochemical sensor. Moreover, unlike the label-free approach presented in our work, the method by Chen et al. cannot monitor the trypsin activity in real time, because this activity was detected only after 2 h of incubation of the trypsin with the peptide-modified electrode. It should be also mentioned that commercially available enzyme-linked immunosorbent assay (ELISA) kits for trypsin possess also high selectivity and sensitivity similar to the work of Chen et al. (down to 0.012 nM) [52]. However, those kits require expensive antibodies, and detection is carried out in several steps. Furthermore, ELISA does not allow monitoring of the kinetics of the trypsin activity. The acoustic sensor developed by us is sufficiently sensitive (LOD of 0.48 ± 0.08 nM) to detect such dangerous diseases as cystic fibrosis, acute pancreatitis, or the acute phase of chronic pancreatitis that are characterized by raised concentration of trypsin in blood in the range of 2.1–71.4 nM [7,8]. In contrast with ELISA, the TSM biosensor is label-free, straightforward, and facile regarding the evaluation of the response. In addition, the TSM method can be used in samples that are not transparent.

3.2. Sprectophtometric Assay of Protease Activity

In colorimetric sensor applications, AuNPs are most widely used due to their high stability, facile synthesis, excellent biocompatibility, and strong surface plasmon resonance effect. This effect can be utilized to produce visual color changes in a process termed the colorimetric method [53,54]. Here, we report the results of a simple colorimetric assay based on the optical properties of functionalized AuNPs (Figure 2). The purpose of this study was a comparison of the sensitivity of surface-based (TSM) and volume-sensitive methods of trypsin activity detection. We used a slightly modified version of the method reported by Chuang et al. [26]. However, instead of gelatin, β-casein has been used as a substrate for trypsin digestion. Briefly, for the protease assay, AuNPs were first modified by β-casein and subsequently with MCH. The molecules of MCH are chemisorbed to the AuNPs through a thiol group (-SH) substitution and the hydroxyl group (-OH) exposed on the AuNPs surface enhances the attraction force between AuNPs. Additionally, MCH molecules on the AuNPs act as blockers, while covering the surface area of the AuNPs that are not conjugated with casein and blocking adsorption of the protease on the surface of the AuNPs [26]. The addition of MCH to the AuNPs–β-casein solution led to a color change from wine-red to violet. When trypsin digests the casein at AuNPs/MCH–casein, NPs aggregated due to the removal of the protective layer of casein and the color change from violet to blue occurred within minutes; then, the solution became colorless.

The absorption spectra of AuNPs in the absence of β-casein (black curve), presence of β-casein (red curve), presence of β-casein and MCH (blue curve), and AuNPs with chemisorbed MCH (magenta curve) are shown in Figure 6. The absorption peak of pure AuNPs is centered at 520 nm as expected. This indicates that the gold colloids are not aggregated but well dispersed as individual particles [55]. After the modification of AuNPs with β-casein, the position of the maximum absorption of AuNPs shifted from 520 to 525 nm, which indicates the formation of bioconjugates [55]. The shift is identical with those reported in [26] for AuNPs modified by gelatin. The red shift in the position of the plasmon absorption band is produced by a perturbation in the dielectric constant around the nanoparticles due to the chemisorption of β-casein molecules [56]. No significant broadening of the spectrum was observed after the β-casein adsorption process, which indicates that the separation distance between AuNPs is higher than their radii, and that AuNPs do not experience aggregation into larger nanoparticles upon the adsorption of β-casein [55]. Further modification with MCH resulted in a significant red shift around 60 nm accompanied by the broadening of the spectrum. This broadening is indicative of an aggregation of nanoparticles This is due to the replacement of the β-casein protective layer with MCH, which in turn makes the nanoparticles closer to each other [57,58]. The modification of AuNPs with MCH resulted in a significant red shift, indicating strong aggregation of the nanoparticles.

Additionally, a less expressed maximum at 280 nm is observed after the modification of AuNPs by β-casein. This is due to the absorption of β-casein's amino acids at this wavelength. The amplitude of this peak decreases after the chemisorption of MCH, which is probably due to the removal of weakly adsorbed casein molecules from the surface of AuNPs. Furthermore, we carried out a quantitative analysis of trypsin activity via the UV-vis spectroscopy method. For this purpose, trypsin was added to the AuNPs solution. We recorded the changes of absorbance spectra of the AuNPs suspension during the trypsin cleavage at 0 min, 0.01 min, 15 min, 30 min, 45 min, and 60 min. Figure 7 illustrates the changes in spectra over time in a 10 nM concentration of trypsin. A substantial red shift (up to 640 nm) of the spectra and a decrease in absorbance with time was observed at this concentration of trypsin, due to trypsin-induced aggregation caused by the cleavage of the AuNPs' protective shell as well as the MCH induced increase of attractive force between the AuNPs. Moreover, the absorbance spectra showed a decrease in the absorption spectra at 280 nm when the AuNPs/MCH–casein was digested by trypsin. It can also be seen that the absorbance decreased with time. The absorbance also started to decrease after maximum shifting. Our results are in good agreement with those previously reported by Chuang et al., whose work served as our inspiration to design a colorimetric assay based on an AuNPs/MCH-protein platform. Chuang et al. demonstrated that protein modified AuNPs aggregation after treatment with protease can be successfully monitored via the red shift of absorption spectra [26].

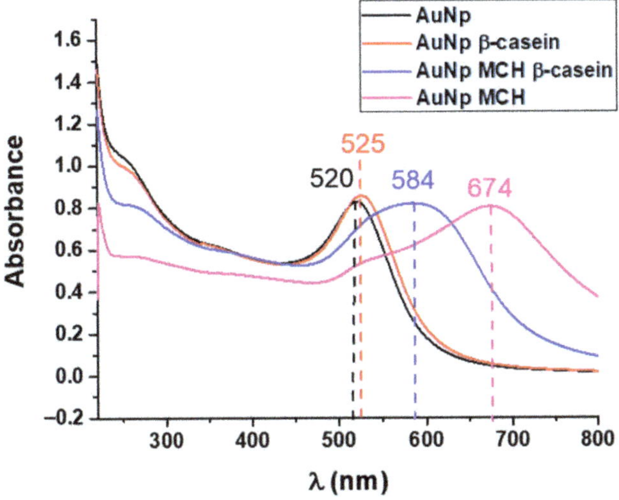

Figure 6. UV-vis absorption spectra of gold nanoparticles (AuNPs): bare (black), modified by β-casein (red), and subsequently modified by 6-mercapto-1-hexanol (MCH) (blue) as well as AuNPs modified by MCH (magenta).

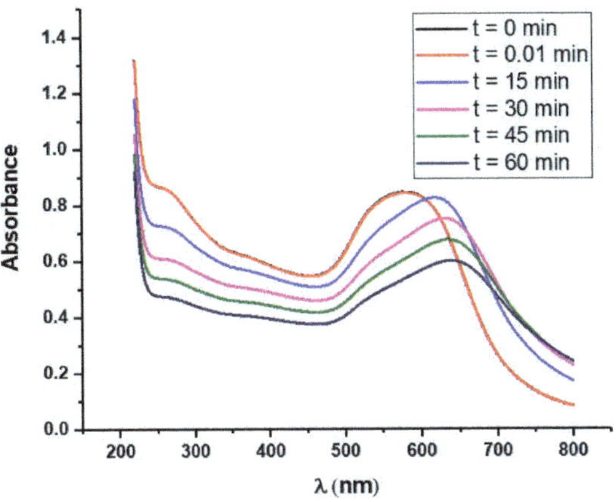

Figure 7. UV-vis absorption spectra of β-casein and MCH-conjugated AuNPs treated with 10 nM trypsin at different time points. Note that at time 0 and 0.01 min, the spectra are almost identical.

Trypsin at concentrations ranging from 0.1 to 10 nM was used in the study to estimate the detection limit of the optical AuNPs assay. In order to construct the calibration curve, we have plotted the changes in relative values of absorbance measured at around 640 nm after 15 min of trypsin exposure against the trypsin concentration (Figure 8a). As in the case of the analysis of trypsin activity via the TSM method, we were able to use an inverse Michaelis–Menten (MM) model expressed by Equation (3) to analyze the obtained calibration curve for trypsin activity in volume.

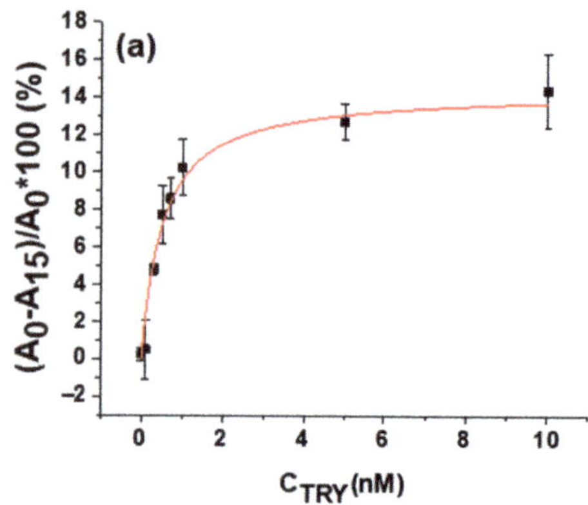
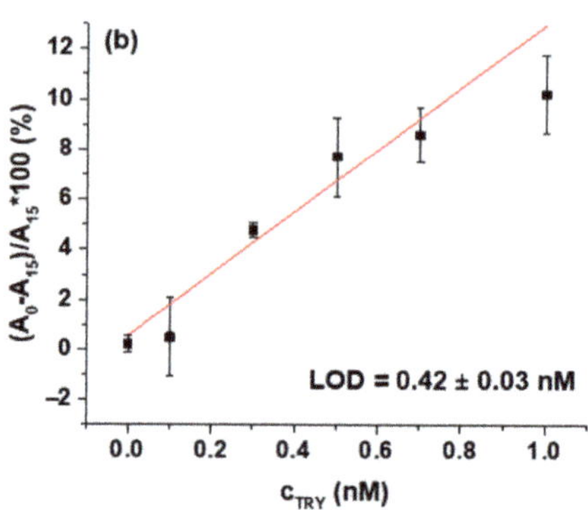

Figure 8. Calibration plots of colorimetric assay. (**a**) Changes in relative values of absorbance after β-casein and MCH functionalized gold nanoparticles (AuNPs/MCH-β-casein) exposure to trypsin (A_0—exposure time 0 min, A_{15}—exposure time 15 min.) vs. concentration of trypsin (C_{TRY}). Symbols are experimental data, and the red line is the best fit of Equation (3). (**b**) Linear part of the calibration curve for calculation of the limit of detection (LOD). Values are means ± SD (n = 3). Red line is the linear regression fit.

The fitting of calibration plots with the MM model yielded v_{max} = 14.98 ± 0.81% and K_M = 0.56 ± 0.10 nM. As was already mentioned, in the inverse MM model, the roles of the enzyme and substrate are swapped, and the concentration of the enzyme is changed while the substrate is presented in excess.

To calculate the LOD, we used only part of the calibration curve from 0 to 5 nM, where the dependence was almost linear. The obtained LOD was 0.42 ± 0.03 nM, according to the rule 3.3 (SD)/S, where SD is the standard deviation of the sample with the lowest concentration and S is slope calculated from the fit of the linear part of the calibration curve [59]. The results are shown in Figure 8b.

It is interesting to compare the properties of the AuNPs assay and the TSM method used to detect trypsin activity (Table 1). On one hand, both methods successfully detected protease activity at the sub-nM level, within a similar time range in a real-time mode. However, it should be noted that a major drawback of the AuNPs assay is that the method is of limited application in a turbid medium. On the other hand, unlike the TSM method, detection using the AuNPs assay can be carried out in only one step, as the signal detection simply involves the direct measurement of the absorbance values at A_{640}. It is also interesting to compare the reverse Michaelis–Menten constants for both methods. As can be seen in Table 1, a lower K_M value has been obtained for the AuNPs-based colorimetric assay. This can be attributed to trypsin's better access to the β-casein substrate. Certainly, the β-casein layer is formed at MUA monolayers by covalent binding of the casein hydrophilic amino groups. Thus, the cleavage sites are closer to the quartz crystal surface with limited access to the trypsin. In addition, due to covalent binding of casein molecules at the self-assembled MUA, the casein layer is compactly packed, which creates additional restriction of access of trypsin to the cleavage sites. A similar conclusion was also obtained for chymotrypsin detection [57]. In contrast, at AuNPs, the casein is physically adsorbed at the gold surface. This means that casein molecules are randomly oriented, which make the access of trypsin to the casein cleavage sites more advantageous.

Table 1. Comparison of TSM biosensor and AuNPs platform (colorimetric biosensor) used for detection of trypsin.

Parameters	TSM Biosensor	AuNPs Assay
Detection time	30 min	30 min
K_M	0.92 ± 0.44 nM	0.56 ± 0.10 nM
Detection limit	0.48 ± 0.08 nM	0.42 ± 0.03 nM
Signal detection	Acoustic wave at surface	UV-vis absorbance in a volume

We should also mention that in contrast with the colorimetric method, the acoustic TSM technique is sensitive to air bubbles presented in the sample and to the pressure changes caused by handling of the flow cell. Air bubbles are more prone to growth at the hydrophilic interface, which likely altered to a hydrophobic case upon the adsorption of the casein layer. Therefore, special care can be taken in avoiding this effect, for example by degassing the sample before starting the experiments.

Finally, we briefly discuss the most often used techniques employed for trypsin detection. The advantages and disadvantages of these techniques as well as their LOD are summarized in Table 2. Nowadays, researchers' efforts are focused on the development of simple and rapid biosensors for the sensitive determination of trypsin because traditional methods such as enzyme-linked immunosorbent assay (ELISA), gelatin-based film technique, and high-performance liquid-chromatography (HPLC) are time-consuming and require specialized instruments and trained personnel [60]. Moreover, those methods do not allow for the monitoring of the kinetics of protease activity. Recently, many efforts have been reported regarding trypsin determination. Several biosensors based on fluorescent, electrochemical, and colorimetric methods have been developed to detect trypsin [15,26,60,61]. Fluorescence-based homogeneous assays are the most popular ones for trypsin activity monitoring due to their simple processes, high sensitivity, and convenient operation. These methods usually need peptide-based molecular probes containing fluorochrome and quencher pairs to monitor specific proteases by fluorescence resonance energy transfer. Nevertheless, these labeled fluorogenic substrates are expensive and are difficult to synthesize [62]. Colorimetry is another method reported for the detection and screening of trypsin [63]. It is the simplest, less expensive, and most widely used method. It can be directly observed with the naked eye or accurately quantified via UV-vis spectrophotometer. Unfortunately, the colorimetric method is limited to only optically transparent liquids. Electrochemical methods are rather sensitive. However, they require the conjugation of a specific peptide substrate by redox probes, longer incubation time with protease, and cannot be used for measuring the kinetics of enzyme reaction [51]. Acoustic methods are among the most effective and promising approaches for the detection of trypsin activity. Their advantage lies in high sensitivity, which reaches levels comparable to state-of-the-art techniques such as ELISA. Since most of the biochemical samples are acoustically transparent, measurements can be performed in a wider range of solutions without the need for a chemical probe, as well as in opaque and high-concentration samples that are difficult to measure with optical methods. Most recently, we successfully demonstrated the feasibility of a volume-sensitive acoustic method for the detection of proteolytic activity of trypsin [64]. However, it is worth noting a possible limitation of the proposed acoustic methods: namely, the influence of air bubbles and temperature stability. This limitation must be addressed in future research.

Table 2. Comparison of the most used analytical methods for trypsin determination.

Method	Advantages	Disadvantages	LOD, nM	References
ELISA	High selectivity and sensitivity	Requires expensive antibodies, the kinetics of trypsin activity cannot be measured	0.012	[42]
Fluorescent assay	High sensitivity, operates in real-time mode	Fluorogenic substrates are expensive and difficult to be synthesized.	3.8–29	[15,61]
Colorimetric assay	Simple, inexpensive, and sensitive, enables real-time detection of trypsin activity	Limited to only optically transparent liquids	0.19 0.42 ± 0.03	[63] This work
Electrochemical sensor	High sensitivityy	Necessity to use peptide substrate conjugated with graphene oxide and thionine	0.05	[51]
Acoustic TSM sensor	High sensitivity, capable of real-time monitoring of kinetics of the trypsin mediated cleavage	Measurements are sensitive to air bubbles presented in the sample	0.2 0.48 ± 0.08	[23] This work
High-resolution ultrasonic spectroscopy	High sensitivity, capable of real-time monitoring of kinetics of the trypsin mediated cleavage	Measurements are sensitive to air bubbles presented in the sample	~1.0	[64]

4. Conclusions

We have shown that β-casein forms a stable monolayer via an 11-mercaptoundecanoic acid (MUA) cross-linker at the gold surface of a piezoelectric transducer. The TSM sensor based on a β-casein layer enabled a detection limit of 0.48 ± 0.08 nM for trypsin. The cleavage of β-casein resulted in an increase of resonant frequency and a decrease of motional resistance. Furthermore, we compared the results obtained by the TSM method with a colorimetric assay for quantifying trypsin activity in a volume. This assay was based on AuNPs modified by β-casein and MCH and on the phenomena of surface plasmon resonance (SPR) and yielded a detection limit of 0.42 ± 0.03 nM, which is comparable with the LOD obtained from TSM experiments. We also analyzed the Michaelis–Menten constants, K_M, for reverse enzymatic reaction and showed that the K_M value for the colorimetric assay (0.56 ± 0.10 nM) is lower in comparison with that obtained in the case of the TSM method (0.92 ± 0.44 nM). This has been explained by better access of trypsin to the β-casein in a volume. The TSM method is useful for the study of the kinetics of the protease's activity, which is not possible via conventional ELISA or HPLC methods. The obtained results can be considered as a first step toward the application of a TSM sensor and colorimetric assays based on β-casein for the label-free detection of trypsin activity. For practical application in medical diagnostics, both acoustic and optical methods need additional validation in complex biological fluids such as blood or blood plasma. In addition, the sensitivity of the TSM method can be improved by the application of hydrophobic substrates for casein immobilization. We anticipate in this case that the detection limit can be improved at least five times. The improved sensitivity of detection is important for working with diluted bilogical samples in order to minimize the matrix effect.

Author Contributions: Investigation, validation, formal analysis, writing—original draft preparation, I.P.; investigation, validation, formal analysis, writing—original draft preparation, S.M.; investigation and formal analysis, writing—review and editing, M.T. (Marek Tatarko); conceptualization, writing—original draft preparation, writing—review and editing, supervision, project administration, funding acquisition, T.H.; methodology, project administration, writing—review and editing, M.T. (Michael Thompson). All authors have read and agreed to the published version of the manuscript.

Funding: This work was supported by European Union's Horizon 2020 research and innovation programme under the Marie Sklodowska-Curie grant agreement No. 690,898 and by Science Grant Agency VEGA, project No. 1/0419/20.

Institutional Review Board Statement: Not applicable.

Informed Consent Statement: Not applicable.

Data Availability Statement: Not applicable.

Conflicts of Interest: The authors declare no conflict of interest.

References

1. Gemene, K.L.; Meyerhoff, M.E. Detection of protease activities by flash chronopotentiometry using a reversible polycation-sensitive polymeric membrane electrode. *Anal. Biochem.* **2011**, *416*, 67–73. [CrossRef]
2. Siklos, M.; Aissa, B.; Thatcher, G.R.J. Cysteine proteases as therapeutic targets: Does selectivity matter? A systematic review of calpain and cathepsin inhibitors. *Acta Pharm. Sin. B* **2015**, *5*, 506–519. [CrossRef]
3. Glantz, M.; Rosenlow, M.; Lindmark-Månsson, H.; Johansen, L.B.; Hartmann, J.; Hojer, A.; Waak, E.; Lofgren, R.; Saeden, K.H.; Svensson, S.; et al. Impact of protease and lipase activities on quality of Swedish raw milk. *Int. Dairy J.* **2020**, *107*, 104724. [CrossRef]
4. Verdoes, M.; Verhels, S.H.L. Detection of protease activity in cells and animal. *Biochim. Biophys. Acta (BBA) Proteins Proteom.* **2016**, *1864*, 130–142. [CrossRef] [PubMed]
5. Ku, M.; Hong, Y.; Heo, D.; Lee, E.; Hwang, S.; Suh, J.-S.; Yang, J. In vivo sensing of proteolytic activity with an NSET-based NIR fluorogenic nanosensor. *Biosens. Bioelectr.* **2016**, *77*, 471–477. [CrossRef]
6. Buss, C.G.; Dudani, J.S.; Akana, R.T.K.; Fleming, H.E.; Bhatia, S.N. Protease activity sensors noninvasively classify bacterial infections and antibiotic responses. *Ebiomedicine* **2018**, *38*, 248–256. [CrossRef]
7. Sharma, H.; Vyas, R.K.; Vyas, S. Role of serum trypsin level in diagnosis and prognosis of pancreatitis and compared with healthy subjects of rajasthan. *Am. J. Biochem.* **2018**, *8*, 93–99.
8. Heinrich, H.C.; Gabbe, E.E.; Ičagić, F. Immunoreactive serum trypsin in diseases of the pancreas. *Klin. Wochenschr.* **1979**, *57*, 1237–1238. [CrossRef]
9. Lake-Bakaar, G.; McKavanagh, S.; Redshaw, M.; Wood, T.; Summerfield, J.A.; Elias, E. Serum immunoreactive trypsin concentration after a lunch meal. Its value in the diagnosis of pancreatic disease. *J. Clin. Pathol.* **1979**, *32*, 1003–1008. [CrossRef]
10. Artigas, J.M.; Garcia, M.E.; Faure, M.R.; Gimeno, A.M. Serum trypsin levels in acute pancreatic and non-pancreatic abdominal conditions. *Postgrad. Med. J.* **1981**, *57*, 219–222. [CrossRef] [PubMed]
11. Borgstrom, B.; Dahlqvist, A.; Lundh, G.; Sjovall, J. Studies of intestinal digestion and absorption in the human. *J. Clin. Investig.* **1957**, *36*, 1521–1536. [CrossRef] [PubMed]
12. Mumtaz, T.; Qindeel, M.; Rehman, A.; Tarhini, M.; Ahmed, N.; Elaissari, A. Exploiting proteases for cancer theranostic through molecular imaging and drug delivery. *Int. J. Pharm.* **2020**, *587*, 119712. [CrossRef]
13. Dunn, D.T.; Stöhr, W.; Arenas-Pinto, A.; Tostevin, A.; Mbisa, J.L.; Paton, N.I. Next generation sequencing of HIV-1 protease in the PIVOT trial of protease inhibitor monotherapy. *J. Clin. Virol.* **2018**, *101*, 63–65. [CrossRef] [PubMed]
14. Kahler, U.; Kamenik, A.S.; Waibl, F.; Kraml, J.; Liedl, K.R. Protein-protein binding as a two-step mechanism: Preselection of encounter poses during the binding of BPTI and trypsin. *Biophys. J.* **2020**, *119*, 652–666. [CrossRef]
15. Hou, S.; Feng, T.; Zhao, N.; Zhang, J.; Wang, H.; Liang, N.; Zhao, L. A carbon nanoparticle-peptide fluorescent sensor custom-made for simple and sensitive detection of trypsin. *J. Pharm. Anal.* **2020**, *10*, 482–489. [CrossRef]
16. Sato, D.; Kato, T. Novel fluorescent substrates for detection of trypsin activity and 541 inhibitor screening by self-quenching. *Bioorg. Med. Chem. Lett.* **2016**, *26*, 5736–5740. [CrossRef]
17. Sao, K.; Murata, M.; Fujisaki, Y.; Umezaki, K.; Mori, T.; Niidome, T.; Katayama, Y.; Hashizume, M. A novel protease activity assay using a protease-responsive chaperone protein. *Biochem. Biophys. Res. Commun.* **2009**, *383*, 293–297. [CrossRef] [PubMed]
18. Sakti, S.P.; Lucklum, R.; Hauptmann, P.; Bühling, F.; Ansorge, S. Disposable TSM-biosensor based on viscosity changes of the contacting medium. *Biosens. Bioelectr.* **2001**, *16*, 1101–1108. [CrossRef]
19. Cavic, B.A.; Thompson, M. Interfacial nucleic acid chemistry studied by acoustic shear wave propagation. *Anal. Chim. Acta* **2002**, *469*, 101–113. [CrossRef]
20. Poturnayova, A.; Karpisova, I.; Castillo, G.; Mezo, G.; Kocsis, L.; Csámpai, A.; Keresztes, Z.; Hianik, T. Detection of plasmin based on specific peptide substrate using acoustic transducer. *Sens. Actuators B Chem.* **2016**, *223*, 591–598. [CrossRef]
21. Liu, J.; Chen, D.; Wang, P.; Song, G.; Zhang, X.; Li, Z.; Wang, Y.; Wang, J.; Yang, J. A microfabricated thickness shear mode electroacoustic resonator for the label-free detection of cardiac troponin in serum. *Talanta* **2020**, *2015*, 120890. [CrossRef] [PubMed]
22. Šnejdárková, M.; Poturnayová, A.; Rybár, P.; Lhoták, P.; Himl, M.; Flídrová, K.; Hianik, T. High sensitive calixarene-based sensor for detection of dopamine by electrochemical and acoustic methods. *Bioelectrochemistry* **2010**, *80*, 55–61. [CrossRef]
23. Tatarko, M.; Muckley, E.S.; Subjakova, V.; Goswami, M.; Sumpter, B.G.; Hianik, T.; Ivanov, I.N. Machine learning enabled acoustic detection of sub-nanomolar concentration of trypsin and plasmin in solution. *Sens. Actuators B Chem.* **2018**, *272*, 282–288. [CrossRef]

24. Lerdsri, J.; Chananchana, W.; Upan, J.; Sridara, T.; Jakmunee, J. Label-free colorimetric aptasensor for rapid detection of aflatoxin B1 by utilizing cationic perylene probe and localized surface plasmon resonance of gold nanoparticles. *Sens. Actuators B Chem.* **2020**, *320*, 128356. [CrossRef]

25. Borghei, Y.-S.; Hosseinkhani, S. Colorimetric assay of apoptosis through in-situ biosynthesized gold nanoparticles inside living breast cancer cells. *Talanta* **2020**, *208*, 120463. [CrossRef] [PubMed]

26. Chuang, Y.-C.; Li, J.-C.; Chen, S.-H.; Liu, T.-Y.; Kuo, C.-H.; Huang, W.-T.; Lin, C.-S. An optical biosensing platform for proteinase activity using gold nanoparticles. *Biomaterials* **2010**, *31*, 6087–6095. [CrossRef] [PubMed]

27. Melikishvili, S.; Hianik, T.; Thompson, M. Detection of sub-nanomolar concentration of trypsin by thicken-shear mode (TSM) acoustic wave biosensor. *Proceedings* **2020**, *60*, 6.

28. Kimling, J.; Maier, M.; Okenve, B.; Kotaidis, V.; Ballot, H.; Plech, A. Turkevich method for gold nanoparticle synthesis revisited. *J. Phys. Chem. B* **2006**, *110*, 15700–15707. [CrossRef]

29. Sauerbrey, G. Verwendung von schwingquarzen zur wagung dunnerschichten und zur mikrowagung. *Z. Phys.* **1959**, *155*, 206–222. [CrossRef]

30. Rehman, A.; Zeng, X. Monitoring the cellular binding events with quartz crystal microbalance (QCM) biosensors. In *Biosensors and Biodetection. Methods in Molecular Biology*; Prickril, B., Rasooly, A., Eds.; Humana Press: New York, NY, USA, 2017; Volume 1572, pp. 313–326.

31. Kari, J.; Andersen, M.; Borch, K.; Westh, P. An inverse michaelis-menten approach for interfacial enzyme kinetics. *Catalysis* **2017**, *7*, 4904–4914. [CrossRef]

32. Dizon, M.; Tatarko, M.; Hianik, T. Advances in analysis of milk proteases activity at surfaces and in a volume by acoustic methods. *Sensors* **2020**, *20*, 5594. [CrossRef]

33. Ahmadab, A.; Moore, E. Electrochemical immunosensor modified with self-assembled monolayer of 11-mercaptoundecanoic acid on gold electrodes for detection of benzo[a]pyrene in water. *Analyst* **2012**, *137*, 5839–5844. [CrossRef] [PubMed]

34. Huenerbein, A.; Schmelzer, C.E.H.; Neubert, R.H.H. Real-time monitoring of peptic and tryptic digestions of bovine -casein using quartz crystal microbalance. *Anal. Chim. Acta* **2007**, *584*, 72–77. [CrossRef] [PubMed]

35. Yao, J.; Lin, C.; Tao, T.; Lin, F. The effect of various concentrations of papain on the properties and hydrolytic rates of β-casein layers. *Colloids Surf. B Biointerfaces* **2013**, *101*, 272–279. [CrossRef]

36. Murray, B.S.; Cros, L. Adsorption of β-lactoglobulin and β-casein to metal surfaces and their removal by a non-ionic surfactant, as monitored via a quartz crystal microbalance. *Colloids Surf. B Biointerfaces* **1998**, *10*, 227–241. [CrossRef]

37. Ozeki, T.; Verma, V.; Uppalapati, M.; Suzuki, Y.; Nakamura, M.; Catchmark, J.M.; Hancock, W.O. Surface-bound casein modulates the adsorption and activity of kinesin on SiO₂ surfaces. *Biophys. J.* **2009**, *96*, 3305–3318. [CrossRef] [PubMed]

38. Goda, T.; Miyahara, Y. Interpretation of protein adsorption through its intrinsic electric charges: A comparative study using a field-effect transistor, surface plasmon resonance, and quartz crystal microbalance. *Langmuir* **2012**, *28*, 14730–14738. [CrossRef] [PubMed]

39. Pérez-Fuentes, L.; Drummond, C.; Faraudo, J.; Bastos-González, D. Adsorption of milk proteins (β-casein and β-lactoglobulin) and BSA onto hydrophobic surfaces. *Materials* **2017**, *10*, 893. [CrossRef] [PubMed]

40. Románszki, L.; Tatarko, M.; Jiao, M.; Keresztes, Z.; Hianik, T.; Thompson, M. Casein probe–based fast plasmin determination in the picomolar range by an ultra-high frequency acoustic wave biosensor. *Sens. Actuators B Chem.* **2018**, *275*, 206–214. [CrossRef]

41. Miodek, A.; Poturnayová, A.; Šnejdárková, M.; Hianik, T.; Korri-Youssoufi, H. Binding kinetics of human cellular prion detection by DNA aptamers immobilized on a conducting polypyrrole. *Anal. Bioanal. Chem.* **2013**, *405*, 2505–2514. [CrossRef]

42. Nylander, T.; Wahlgren, N.M. Competitive and sequential adsorption of β-casein and β-lactoglobulin on hydrophobic surfaces and the interfacial structure of β-casein. *J. Colloid Interface Sci.* **1994**, *162*, 151–162. [CrossRef]

43. Nylander, T.; Tiberg, F.; Wahlgren, N.M. Evaluation of the structure of adsorbed layers of β-casein from ellipsometry and surface force measurements. *Int. Dairy J.* **1999**, *9*, 313–317. [CrossRef]

44. Krisdhasima, V.; Vinaraphong, P.; McGuire, J. Adsorption kinetics and elutability of α-lactalbumin, β-casein, β-lactoglobulin, and bovine serum albumin at hydrophobic and hydrophilic interfaces. *J. Colloid Interface Sci.* **1993**, *161*, 325–334. [CrossRef]

45. Evers, C.H.J.; Andredsson, T.; Lund, M.; Skepo, M. Adsorption of unstructured protein β-casein to hydrophobic and charged surfaces. *Langmuir* **2012**, *28*, 11843–11849. [CrossRef]

46. Eskin, N.A.M.; Goff, H.D. Milk. In *Biochemistry of Foods*, 3rd ed.; Eskin, N.A.M., Shaidi, F., Eds.; Academic Press: Cambridge, MA, USA, 2013; pp. 187–214.

47. Kelly, A.L.; McSweeney, P.L.H. Indigenous proteinases in milk. In *Advanced Dairy Chemistry—1 Proteins*, 3rd ed.; Fox, P.F., McSweeney, P.L.H., Eds.; Springer: New York, NY, USA, 2003; pp. 495–521.

48. Bumberger, E.; Belitz, H.D. Bitter taste of enzymic hydrolysates of casein. I. Isolation, structural and sensorial analysis of peptides from tryptic hydrolysates of beta-casein. *Z. Lebensmittel-Unters. Forsch.* **1993**, *197*, 14–19. [CrossRef] [PubMed]

49. Rauh, V.M.; Johansen, L.B.; Ipsen, R.; Paulsson, M.; Larsen, L.B.; Hammershøj, M. Plasmin activity in UHT milk: Relationship between proteolysis, age gelation, and bitterness. *J. Agricult. Food Chem.* **2014**, *62*, 6852–6860. [CrossRef]

50. Gasteiger, E.; Hoogland, C.; Gattiker, A.; Duvaud, S.; Wilkins, M.R.; Appel, R.D.; Bairoch, A. Protein identification and analysis tools on the ExPASy server. In *The Proteomics Protocols Handbook*; Walker, J.M., Ed.; Humana Press: Totowa, NJ, USA, 2005; pp. 571–607.

51. Chen, G.; Shi, H.; Ban, F.; Zhang, Y.; Sun, L. Determination of trypsin activity using a gold electrode modified with a nanocover composed of graphene oxide and thionine. *Microchim. Acta* **2015**, *182*, 2469–2476. [CrossRef]
52. Trypsin ELISA Kit. Available online: https://assets.thermofisher.com/TFS-Assets/LSG/manuals/EH468RB.pdf (accessed on 27 February 2021).
53. Akshaya, K.; Arthi, C.; Pavithra, A.J.; Poovizhi, P.; Shilpa Antinate, S.; Hikku, G.S.; Jeyasubramanian, K.; Murugesan, R. Bioconjugated gold nanoparticles as an efficient colorimetric sensor for cancer diagnostics. *Photodiagnosis Photodyn. Ther.* **2020**, *30*, 101699. [CrossRef] [PubMed]
54. Lapenna, A.; Dell'Aglio, M.; Palazzo, G.; Mallardi, A. "Naked" gold nanoparticles as colorimetric reporters for biogenic amine detection. *Colloids Surf. A Physicochem. Eng. Asp.* **2020**, *600*, 124903. [CrossRef]
55. Liu, Y.; Guo, R. The interaction between casein micelles and gold nanoparticles. *J. Colloid Interface Sci.* **2009**, *332*, 265–269. [CrossRef]
56. Lee, S.; Pérez-Luna, V.H. Dextran-gold nanoparticle hybrid material for biomolecule immobilization and detection. *Anal. Chem.* **2005**, *77*, 7204–7211. [CrossRef]
57. Piovarci, I.; Hianik, T.; Ivanov, I.N. Detection of chymotrypsin by optical and acoustic methods. *Biosensors* **2021**, *11*, 63. [CrossRef] [PubMed]
58. Lang, N.J.; Liu, B.; Zhang, X.; Liu, J. Dissecting colloidal stabilization factors in crowded polymer solutions by forming self-assembled monolayers on gold nanoparticles. *Langmuir* **2013**, *29*, 6018–6024. [CrossRef] [PubMed]
59. Mei, H.; Chu, H.; Chen, W.; Xue, F.; Liu, J.; Xu, H.; Zhang, R.; Zheng, L. Ultrasensitive one-step rapid visual detection of bisphenol A in water samples by label-free aptasensor. *Biosens. Bioelectron.* **2013**, *39*, 26–30. [CrossRef] [PubMed]
60. Lin, Y.; Shen, R.; Liu, N.; Yi, H.; Dai, H.; Lin, J. A highly sensitive peptide-based biosensor using NiCo2O4 nanosheets and g-C3N4 nanocomposite to construct amplified strategy for trypsin detection. *Anal. Chim. Acta* **2018**, *1035*, 175–183. [CrossRef] [PubMed]
61. Duan, X.; Li, N.; Wang, G.; Su, X. High sensitive ratiometric fluorescence analysis of trypsin and dithiothreitol based on WS2 QDs. *Talanta* **2020**, *219*, 121171. [CrossRef]
62. Hu, L.; Han, S.; Parveen, S.; Yuan, Y.; Zhang, L.; Xu, G. Highly sensitive fluorescent detection of trypsin based on BSA-stabilized gold nanoclusters. *Biosens. Bioelectron.* **2012**, *32*, 297–299. [CrossRef]
63. Zhang, L.; Du, J. A sensitive and label-free trypsin colorimetric sensor with cytochrome c as a substrate. *Biosens. Bioelectron.* **2016**, *79*, 347–352. [CrossRef]
64. Melikishvili, S.; Dizon, M.; Hianik, T. Application of high-resolution ultrasonic spectroscopy for real-time monitoring of trypsin activity in β-casein solution. *Food Chem.* **2021**, *337*, 127759. [CrossRef]

 *biosensors* 

Article

Fast and Sensitive Determination of the Fungicide Carbendazim in Fruit Juices with an Immunosensor Based on White Light Reflectance Spectroscopy [†]

Georgios Koukouvinos [1,‡], Chrysoula-Evangelia Karachaliou [2,*,‡], Ioannis Raptis [3,4], Panagiota Petrou [1], Evangelia Livaniou [2,*] and Sotirios Kakabakos [1]

[1] Immunoassay/Immunosensors Lab, Institute of Nuclear & Radiological Sciences & Technology, Energy & Safety, National Centre for Scientific Research "Demokritos", P.O. Box 60037, 15310 Agia Paraskevi, Greece; geokoukoubinos@yahoo.gr (G.K.); ypetrou@rrp.demokritos.gr (P.P.); skakab@rrp.demokritos.gr (S.K.)

[2] Immunopeptide Chemistry Lab, Institute of Nuclear & Radiological Sciences & Technology, Energy & Safety, National Centre for Scientific Research "Demokritos", P.O. Box 60037, 15310 Agia Paraskevi, Greece

[3] Institute of Nanoscience and Nanotechnology, National Centre for Scientific Research "Demokritos", P.O. Box 60037, 15310 Agia Paraskevi, Greece; i.raptis@inn.demokritos.gr

[4] ThetaMetrisis S.A., Polydefkous 14, 12243 Egaleo, Greece

* Correspondence: xrisak15@rrp.demokritos.gr or xrisak15@hotmail.com (C.-E.K.); livanlts@rrp.demokritos.gr (E.L.)

[†] This paper is the extended version of a paper (Koukouvinos, G.; Karachaliou, C.-E.; Kakabakos, S.; Livaniou, E. "A White Light Reflectance Spectroscopy Label-Free Biosensor for the Determination of Fungicide Carbendazim") published in: Proceedings of the 1st International Electronic Conference on Biosensors, 2–17 November 2020.

[‡] These authors contributed equally to this work.

Citation: Koukouvinos, G.; Karachaliou, C.-E.; Raptis, I.; Petrou, P.; Livaniou, E.; Kakabakos, S. Fast and Sensitive Determination of the Fungicide Carbendazim in Fruit Juices with an Immunosensor Based on White Light Reflectance Spectroscopy. *Biosensors* **2021**, *11*, 153. https://doi.org/10.3390/bios 11050153

Received: 22 March 2021
Accepted: 10 May 2021
Published: 13 May 2021

Publisher's Note: MDPI stays neutral with regard to jurisdictional claims in published maps and institutional affiliations.

Abstract: Carbendazim is a systemic benzimidazole-type fungicide with broad-spectrum activity against fungi that undermine food products safety and quality. Despite its effectiveness, carbendazim constitutes a major environmental pollutant, being hazardous to both humans and animals. Therefore, fast and reliable determination of carbendazim levels in water, soil, and food samples is of high importance for both food industry and public health. Herein, an optical biosensor based on white light reflectance spectroscopy (WLRS) for fast and sensitive determination of carbendazim in fruit juices is presented. The transducer is a Si/SiO_2 chip functionalized with a benzimidazole conjugate, and determination is based on a competitive immunoassay format. Thus, for the assay, a mixture of an in-house developed rabbit polyclonal anti-carbendazim antibody with the standards or samples is pumped over the chip, followed by biotinylated secondary antibody and streptavidin. The WLRS platform allows for real-time monitoring of biomolecular interactions carried out onto the Si/SiO_2 chip by transforming the shift in the reflected interference spectrum caused by the immunoreaction to effective biomolecular adlayer thickness. The sensor is able to detect 20 ng/mL of carbendazim in fruit juices with high accuracy and precision (intra- and inter-assay CVs $\leq$ 6.9% and $\leq$9.4%, respectively) in less than 30 min, applying a simple sample treatment that alleviates any "matrix-effect" on the assay results and a 60 min preincubation step for improving assay sensitivity. Excellent analytical characteristics and short analysis time along with its small size render the proposed WLRS immunosensor ideal for future on-the-spot determination of carbendazim in food and environmental samples.

Keywords: white light reflectance spectroscopy; real-time immunosensor; ELISA; pesticides; carbendazim; fruit juices

1. Introduction

Fungal contamination causes significant damages to the crops for human consumption every year, resulting in poor yield, deficient food quality, and huge economic loss. To circumvent these problems, the use of fungicides has been intensified over the last decades [1].

Carbendazim (methyl 1H-benzimidazol-2-ylcarbamate) is a synthetic, systemic, broad-spectrum, benzimidazole-type fungicide used worldwide as pre- and post-harvest treatment to control fungi that compromise the quality of various food commodities such as vegetables, fruits, cereals, and seeds [2,3]. Despite the unquestionable benefits regarding crop yield, carbendazim is a major pollutant, which induces acute and chronic effects on humans and livestock. In this context, carbendazim has been documented to induce infertility, embryotoxicity, teratogenicity, hepatocellular dysfunction, endocrine-disrupting effects, disruption of hematological functions, and mutagenicity [2]. Additionally, the World Health Organization has classified carbendazim as a possible human carcinogen [4]. Due to its aforementioned severe toxicities and its persistence in food and the environment, carbendazim has been officially banned in most of the European Union countries, USA, and Australia. However, its production and use in various formulations are still permitted in some countries, such as UK, Portugal, India, China, and Brazil [5], raising a growing concern for its adverse effects on the health of both humans and animals. To protect consumers from food products contaminated with carbendazim, regulatory authorities have established maximum residue limits (MRLs) for this pesticide in several matrices (fresh fruits and vegetables, oil seeds, cereals, spices, etc.). For example, the carbendazim MRL in fruit juices is 200 ppb and the MRL sum of benzimidazole pesticides has been set at 500 ppb by European Union [6,7]. The presence of carbendazim in fruit juice products has raised many concerns due to their worldwide popularity and the fact that children are their primary consumers [8].

The protection of the public health from pesticide residues in food requires, in addition to relevant legislation, accurate analytical methodologies. Determination of carbendazim is routinely performed by instrumental analytical techniques, mainly high-performance liquid chromatography coupled to mass spectroscopy or ultraviolet spectroscopy [9,10]. Alternatively, immunochemical techniques have been developed and used, offering low-cost and short-time analyses, simple assay protocols, and minimum sample pretreatment, and high-throughput sample analysis capacity. In this context, classic enzyme-linked immunosorbent assays (ELISA) [11] and immunochromatographic strips [12] have been reported in the literature, while immunosensors have lately attracted much attention due to their simplicity, rapidity, portability, and potential for the point-of-need application [13].

In this work, a real-time immunosensor based on white light reflectance spectroscopy (WLRS) is employed for the accurate, fast, and sensitive determination of carbendazim in fruit juice samples with the potential for use at the point-of-need. The transducer is a Si chip with a 1-μm thick SiO_2 layer on top and it is transformed to a versatile biosensing element through immobilization of a suitable recognition molecule. The optical set-up includes a white light source, a reflection probe consisting of a bundle of seven optical fibers; six at the periphery of the probe and one at the center, and a spectrometer. The six fibers at the periphery of the reflection probe guide the light from the source to the chip surface, while the seventh central fiber collects the light reflected by the chip and guides it to the spectrometer. As the light strikes the chip surface vertically, it is reflected by the silicon surface and by the transparent materials adlayers (silicon dioxide and biomolecular layer) of different refractive index. This way interference takes place at each wavelength resulting in an interference spectrum that is collected by the central fiber of the reflection probe. The increase of the biomolecular adlayer thickness due to binding reactions taking place onto the chip surface causes a shift of the interference spectrum. The software receives the interference spectrum from the embedded spectrometer and the effective biomolecular adlayer thickness (that is the signal of the WLRS sensor) is determined implementing the Levenberg–Marquart algorithm [14]. As this conversion is done by the software in real-time, the evolution of the effective biomolecular adlayer thickness in the course of the binding reactions occurring on a biochip surface could be monitored in real-time (Scheme 1). Thus, the WLRS biosensing platform allows for the label-free, real-time monitoring of biomolecular interactions carried out onto the Si/SiO_2 chip with a detectable effective adlayer thickness <0.1 nm. A presentation of WLRS set-up and operation principle is presented in Scheme 1. The WLRS sensing principle has been successfully applied to the quantitative determination of both high and low molecular weight analytes into a plethora of matrices, after proper biofunctionalization of the sensing surface [14–16].

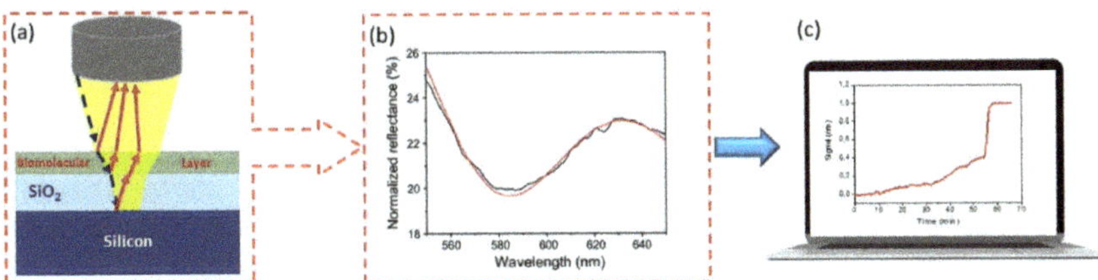

Scheme 1. WLRS set-up and operation principle: (**a**) a schematic of the incident light beam (blue dotted line) reflection at the layers of different refractive index of a WLRS chip (red lines); (**b**) typical reflectance spectrum (black line) and its fitting by the sensor software (red line); (**c**) a depiction of the real-time signal monitoring resulting by the recorded spectrum processing by dedicated software.

For carbendazim determination, a competitive immunoassay involving the delivery of a mixture of standards or fruit juice samples with a carbendazim-specific antibody to a benzimidazole conjugate-modified chip was conducted, as it is schematically depicted in Figure 1a. The primary immunoreaction, i.e., the competitive reaction between the immobilized onto the chip benzimidazole moieties and carbendazim in the standards or samples for the limited binding sites of the carbendazim-specific antibody [17], was followed by two signal enhancement steps. The first step included reaction with a biotiny-lated secondary antibody and the second one with streptavidin so as to further increase the thickness of the adlayer formed. The implementation of these two reactions aimed at the increase of the effective biomolecule adlayer thickness, i.e., the sensor signal, thus increasing the detection sensitivity. The benzimidazole conjugate [17] used for chip coating is also schematically presented in Figure 1b. Initially, a lysine-core peptidyl moiety was prepared using a fluorenylmethoxycarbonyl (Fmoc) solid phase peptide synthesis strategy previously described by us [18] with slight modifications. This moiety was then functionalized with 3-maleimidopropionic acid [19]. A benzimidazole derivative, 2-mercaptobenzimidazole, was then coupled to the 3-maleimidopropionic acid-functionalized peptidyl moiety through its thiol functional group. Notably, 2-mercaptobenzimidazole was used here because it is structurally similar to, but less toxic than, carbendazim. All assay parameters were optimized to achieve the highest possible detection sensitivity and the shortest assay duration. A simple sample preparation procedure was also developed to demonstrate the analysis of several commercially available fruit juice samples without any detectable "matrix-effect". The accuracy of measurements with the proposed methodology was evaluated through recovery experiments using carbendazim-spiked samples prepared in commercially available fruit juices. Finally, the potential of regeneration and re-use of the biofunctionalized sensor chips was investigated, as a means to reduce the total cost of analysis. The novelty of the present work is mainly based on the combined use of an in-house prepared antibody for carbendazim recognition [17] and a benzimidazole derivative as a coating conjugate [17–19] in the WLRS immunoassay for carbendazim in different fruit juices.

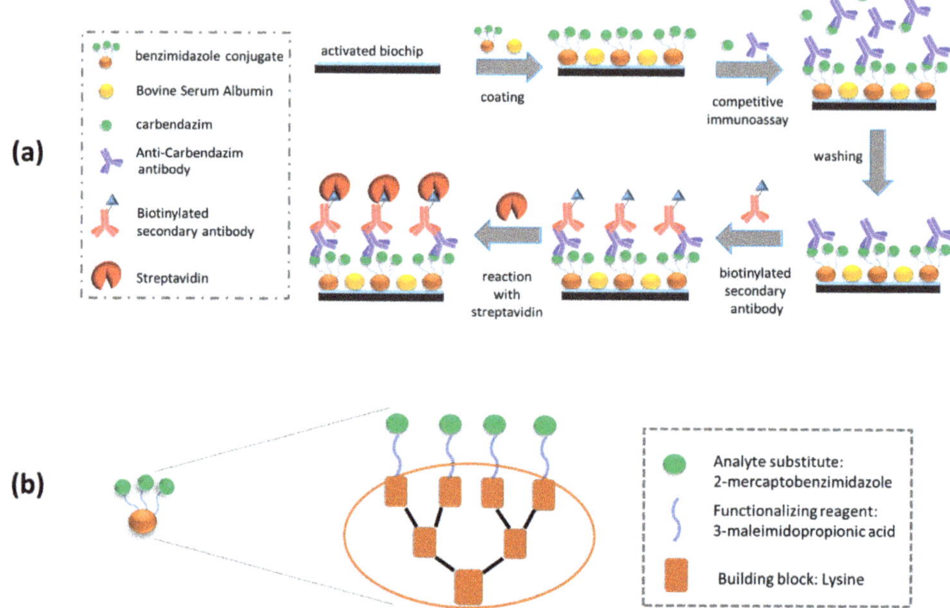

Figure 1. Schematic representation of (**a**) assay procedure for the detection of carbendazim with the WLRS sensor and (**b**) benzimidazole conjugate structure.

2. Materials and Methods

2.1. Reagents and Materials

The rabbit polyclonal antibody for carbendazim (primary antibody) and the benzimidazole conjugate used for surface functionalization were in-house developed, as previously described [17]. Biotinylated goat anti-rabbit IgG (secondary antibody), streptavidin, streptavidin-horseradish peroxidase conjugate (streptavidin-HRP), 3,3′,5,5′-tetramethylbenzidine (TMB), carbendazim pestanal® , highly pure ethanol, acetone, isopropanol, and (3-aminopropyl) triethoxysilane (APTES) were from Sigma-Aldrich (St. Louis, MO, USA). Bovine serum albumin (BSA) was purchased from Acros Organics (Geel, Belgium). Polytetrafluoroethylene (PTFE) syringe filters of 0.45 µm pore size were the product of Membrane Solutions (Auburn, WA, USA). All other chemicals were purchased from Merck (Darmstadt, Germany). High-binding 96-well polystyrene microtiter plates (Code No. 3590) were purchased from Corning-Costar (Corning, NY, USA). Four-inch Si wafers (< 100 >) were purchased from Si-Mat Germany (Kaufering, Germany). These wafers were sequentially sonicated in acetone and isopropanol before a 1000-nm thick SiO_2 layer was thermally oxidized on them at 1100 °C using the clean room facility at the Institute of Nanoscience and Nanotechnology of NCSR "Demokritos". Then, the wafers were diced to chips with dimensions of 5 mm × 15 mm.

2.2. Preparation of Carbendazim Standard Solutions and Fruit Juice Samples

A 5 mg/mL carbendazim stock solution in absolute ethanol was prepared and stored at −20 °C. Standard solutions ranging from 20 ng/mL to 20 µg/mL prepared in 10 mM phosphate buffer, pH 7.4, containing 0.9% (*w/v*) NaCl, and 0.4% (*w/v*) BSA (assay buffer), were kept at −20 °C for up to 2 months. Fruit juices used in this study were purchased from local markets. In particular the following juices have been purchased: Amita orange and orange–lemon–carrot juice, and nine fruits Amita Motion juice from Coca-Cola 3E SA (Maroussi, Greece); orange, orange–apple–carrot, and nine fruits juice from Olympos Greek

Dairies SA (Larissa, Greece); Eviva orange (short and long self-life) and lemon juice from Lidl Hellas (Sindos, Greece); Marata orange juice from Sklavenitis SA (Peristeri, Greece). All juices were separately filtered through a 0.45 μm-PTFE filter and the pH of the filtrate was adjusted to 7.4 ± 0.2 with 1 M NaOH solution.

2.3. Carbendazim-ELISA Assay

An indirect competitive ELISA was set up to conduct carbendazim detection. In brief, ELISA wells were incubated overnight in 100 μL of a 1 μg/mL benzimidazole conjugate in 0.05 M carbonate buffer, pH 9.2 (coating buffer). The plates were then washed three times with 10 mM phosphate buffered saline (PBS), pH 7.4, containing 0.05% (*v/v*) Tween-20 (PBS-T) using an ELISA plate washer (DIA Source), and incubated with 10 mM PBS, pH 7.4, containing 2% (*w/v*) BSA (200 μL/well), at room temperature for 1 h to block any remaining binding sites. After three washes with PBS-T, to each well were added 100 μL of 1:1 (*v/v*) mixtures of carbendazim standards (0.02–20 μg/mL in assay buffer) or samples and the anti-carbendazim antibody solution (2 μg/mL in assay buffer), which have been preincubated at room temperature for 1 h. After incubation at 37 °C for 90 min, the wells were washed three times with PBS-T, and incubated at 37 °C for 1 h with the biotinylated secondary antibody diluted 1:2000 in assay buffer (100 μL/well). The wells were then washed again with PBS-T and incubated at 37 °C for 45 min with a 750 ng/mL streptavidin-HRP solution in assay buffer (100 μL/well). The wells were washed again with PBS-T and incubated at room temperature for 20 min in 100 μL of chromogenic HRP substrate (TMB/H_2O_2). Finally, 50 μL of a 2 M aqueous sulfuric acid solution were added per well to terminate color development, and the absorbance of the wells at 450 nm (A_{450}) was measured using a microtiter plate reader (Sirio S, Seak). An absorbance percentage was then calculated by dividing the absorbance of each standard (A_x) to that of zero standard (A_0). An absorbance percentage versus carbendazim concentration calibration plot was then constructed.

2.4. Evaluation of Primary Antibody—Specificity with the Carbendazim-ELISA Assay

The specificity of the primary antibody was evaluated through cross-reactivity studies with the pesticides carbaryl, imazalil, atrazine, and paraquat as follows. ELISA microwells were coated, blocked and washed as described in Section 2.3. Then, the wells were incubated at 37 °C for 90 min in 100 μL of a 1:1 (*v/v*) mixture (preincubated at room temperature for 1 h) of carbendazim standard solutions or standard solutions containing 0.02–5.0 μg/mL of each cross-reactant in assay buffer, and a 2 μg/mL anticarbendazim antibody solution in assay buffer. Microwells were washed, incubated with biotinylated secondary antibody, streptavidin-HRP, and TMB/H_2O_2 solution as described in Section 2.3. Finally, the absorbance was read at 450 nm, the calibration plots were constructed and the percent cross-reactivity (%CR) with each pesticide tested was determined according to the equation:

$$\%CR = (IC_{50} \text{ carbedazim}/IC_{50} \text{ cross-reactant pesticide}) \times 100$$

where IC_{50} carbedazim and IC_{50} cross-reactant represent the concentrations of carbendazim and cross-reactant in question, respectively, which provided 50% signal drop with respect to zero standard.

2.5. WLRS Instrumentation

WLRS instrumentation involves an FR-Pro tool operating in the 450–720 nm spectral range (ThetaMetrisis SA; Egaleo, Greece). The tool is equipped with a stabilized visible/near infrared light source and a high-performance miniaturized spectrometer tuned to provide very high optical resolution in the dedicated spectral range. The reflection probe delivers the incident light emitted by the source to the biofunctionalized chip surface through six fibers (diameter 400 μm) positioned to its periphery and collects the reflected light directing it to the spectrometer through a seventh fiber (diameter 400 μm) positioned at the center of the probe. The sensing surface consists of a transparent SiO_2 film over a Si reflecting substrate covered by a custom-designed microfluidic cell (Jobst Technologies GmbH; Freiburg, Germany) providing the fluidic connections to the solutions. The FR-Pro tool was accompanied by a dedicated software that evaluates the initial thickness of the SiO_2/protein adlayer and transforms in real-time the spectral shift due to the binding reactions in effective thickness of biomolecular adlayer expressed in nm. In more detail, a reference [$R(\lambda)$) and a dark spectrum $(D(\lambda)]$ were acquired prior to the continuous recording of the reflectance spectrum [$S(\lambda)$], and the absolute reflectance spectrum is calculated by Equation (1):

$$R(\lambda) = \frac{S(\lambda) - D(\lambda)}{R(\lambda) - D(\lambda)} \tag{1}$$

The normalized spectrum is then processed applying Levenberg–Marquart algorithm to calculate the thickness of the biomolecular adlayer, d_1, from the shift in the interference spectrum wavelength, $\delta\lambda$, according to Equation (2):

$$\delta\lambda = r_1 \frac{1 - r_2^2}{(r_1 + r_2)(1 + r_1 r_2)} \frac{n_1 d_1}{n_2 d_2} \lambda_{0m} \tag{2}$$

where, r_1 and r_2, and n_1 and n_2 are the Fresnel coefficients and refractive indices of the biomolecular and the silicon dioxide layer, respectively, d_1 and d_2 are the thickness of the two layers, and λ the wavelength.

2.6. Biochip Preparation and Biosensor Assay Performance

Chips were first cleaned and hydrophilized by O_2 plasma treatment (10 mTorr) for 30 s in a reactive ion etcher. Then, they were immersed in a 2% (*v/v*) aqueous APTES solution for 20 min, gently washed with distilled water, dried under a nitrogen (N_2) stream and cured by heating at 120 °C for 20 min. Chips were kept in a desiccator at room temperature for at least 48 h prior to use. For chip biofunctionalization, the benzimidazole conjugate (500 µg/mL, in coating buffer) was deposited on the chips and incubated at room temperature overnight. The following day, chips were rinsed with 10 mM PBS, pH 7.4 (washing buffer), blocked through immersion in 10 mM PBS, pH 7.4, containing 2% (*w/v*) BSA, for 3 h, rinsed with washing buffer and distilled water, dried with N_2, and used for the assay (Figure 1). Prior to assay, each biofunctionalized chip was assembled with the fluidic module, placed on the docking station and equilibrated with assay buffer to acquire a stable baseline. For the assay, 1:1 (*v/v*) mixtures of standards (0.02–20 µg/mL in assay buffer) or samples with the rabbit anticarbendazim antibody (2 µg/mL in assay buffer), preincubated for 60 min, were passed over the chip for 18 min, followed by a 1:200 dilution of biotinylated anti-rabbit IgG antibody solution for 7 min, and a 10 µg/mL streptavidin solution in assay buffer for 3 min. The flow rate throughout the assay was 50 µL/min. Finally, the biochip was regenerated by running a 0.1 M glycine-HCl buffer, pH 2.5, for 3 min, followed by re-equilibration with assay buffer. The calibration plot was constructed by plotting the effective thickness of the built-up biomolecular adlayer (signal) corresponding to different standards, S_x, expressed as percentage of the zero standard signal, S_0 (maximum signal), against the carbendazim concentration in the standard solutions.

3. Results and Discussion

3.1. ELISA Assay for Carbendazim

Prior to the development of the carbendazim WLRS immunosensor assay, an ELISA in microtiter plates was set up to evaluate the basic immunoreagents used on the sensor platform, e.g., specificity of the anti-carbendazim antibody, as well as to determine optimal conditions for each assay step, e.g., optimum assay buffer. Moreover, the results obtained from the analysis of fruit juice samples with the ELISA assay were compared with those of the WLRS immunosensor.

3.1.1. Assay Optimization

For the development of the competitive indirect ELISA for determination of carbendazim, a benzimidazole conjugate was used as a solid-phase reagent in combination with a rabbit polyclonal anticarbendazim antibody, both developed in-house as previously described [17]. The optimum concentration of the conjugate for coating was determined by preliminary titration experiments using mixtures of the anti-carbendazim polyclonal antibody with carbendazim standards prepared in assay buffer. As shown in Figure S1a, zero standard signal values in the range 1.0–1.5 (which are considered optimum for an ELISA) were received for the following combinations of conjugate/antibody concentrations 0.5/4.0, 1.0/2.0, and 2.5/1.0 in μg/mL. These combinations were tested further regarding not only the zero standard signal but also the signal received in presence of 200 ng/mL carbendazim. From Figure S1b, where the percent absorbance values are presented, it can be concluded that the combination that provided the highest percent signal drop in presence of carbendazim was 1.0 μg/mL benzimidazole conjugate and 2.0 μg/mL anticarbendazim antibody. Thus, this combination was adopted in the final ELISA protocol. Different assay buffers were then tested including a 10 mM PBS buffer, pH 6.5, 10 mM PBS, pH 7.4, 50 mM Tris-HCl buffer, pH 7.8, and 50 mM Tris-HCl buffer, pH 8.25, and the results obtained are shown in Scheme S1 (see Supplementary Material). All buffers contained 0.4% BSA. It was found that 10 mM PBS, pH 7.4, containing 0.4% (*w/v*) BSA, was the optimum buffer for the immunoreaction between the primary antibody and the antigen since it provided the highest signals for zero standard and detection sensitivity compared to the other buffers tested. The implementation of a preincubation step of the antibody with the standards prior to addition onto the biofunctionalized wells was investigated as a means to improve assay detection sensitivity. As shown in Figure S3, the assay sensitivity, expressed by the slope of the linear segment covering the two standards of the lowest concentration in the calibration plot, ranged from −0.19 [dS/(ng/mL)] for 0 min, −0.26 [dS/(ng/mL)] for 30 min, −0.42 [dS/(ng/mL)] for 60 min, and −0.38 [dS/(ng/mL)] for 120 min. These results indicate considerably improved assay sensitivity when a preincubation up to 60 min was used. However, longer preincubation did not result in any additional improvement. Thus, 60-min pre-incubation was adopted in the final protocol. Under optimal conditions, the detection limit of the assay (LoD, calculated as the carbendazim concentration corresponding to percent signal value equal to 100-3SD of 16 measurements of zero standard) was 20 ng/mL. The quantitation limit (LoQ) was calculated as the carbendazim concentration for which the mean absorbance value + 3SD is equal or lower than the mean zero standard value—3SD (LoD), and was 50 ng/mL. The dynamic range of the assay was from 50 ng/mL to 2 μg/mL.

The ELISA assay presented here differs from that previously described [17] in the configuration followed. More specifically, a biotinylated secondary antibody along with streptavidin-HRP have been used, instead of an HRP-labelled secondary antibody, while TMB (instead of 2,2′-azino-bis(3-ethylbenzothiazoline-6-sulfonic acid) diammonium salt—ABTS) has been employed as a chromogen (Scheme S1). Moreover, a preincubation step was adopted so as to further increase the detection sensitivity.

3.1.2. Evaluation of the Anti-Carbendazim Antibody Specificity

The cross-reactivity of the anti-carbendazim antibody with four commonly reported pesticides in fruit juices [20–22], including carbaryl, imazalil, atrazine, and paraquat was tested. In Figure 2, the calibration plots obtained for each one of the four pesticides are provided. As a rule, compounds that do not provide at least 50% decrease in signal under the above described conditions, are not considered as cross-reacting materials. Thus, as shown, no cross-reactivity with any pesticide tested could be detected.

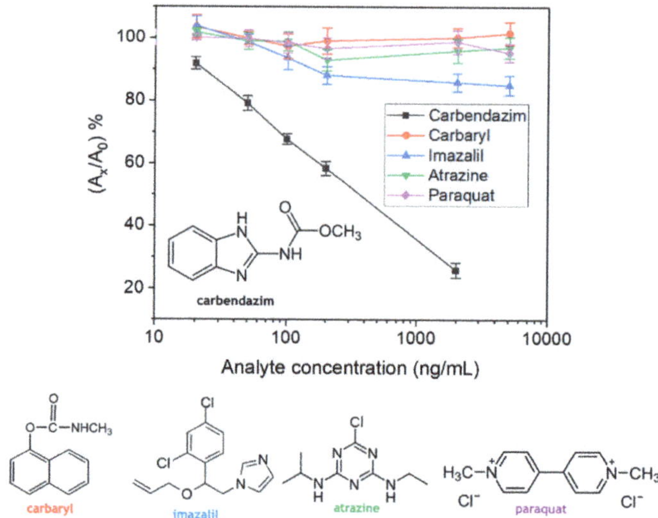

Figure 2. Calibration plots obtained with carbendazim (black squares), carbaryl (red circles), imazalil (blue up triangle), atrazine (green down triangle), and paraquat (purple rhombus) standards with concentrations ranging from 0 to 5000 ng/mL. All standards were preincubated for 60 min with the anti-carbendazim antibody prior to addition in the microwells. Each point is the mean value of 3 replicates $\pm$ SD.

3.2. WLRS Assay

3.2.1. Assay Optimization

The transfer of the carbendazim assay to the WLRS platform was based on results obtained in ELISA-optimization experiments, e.g., in terms of the composition of the assay buffer and the duration of the preincubation step. However, there were some parameters that also needed optimization. Due to the competitive nature of the assay, the two most crucial parameters to optimize were the concentration of benzimidazole conjugate used for coating and the concentration of anti-carbendazim antibody, as their combination determines the highest signal (zero standard signal) and the assay sensitivity. Hence, different concentrations of the benzimidazole conjugate (100–1000 µg/mL) in combination with different concentrations of the anti-carbendazim antibody (0.5–5.0 µg/mL) were tested. As shown in Figure S4, for all antibody concentrations tested, zero standard signal values increased as the concentration of benzimidazole-conjugate used for coating inreased and maximum signal plateau values were obtained for concentrations $\geq$500 µg/mL. The increase in the signal observed as the concentration of both benzimidazole-conjugate and anti-carbendazim antibody increases, depicts the increase in the biomolecular adlayer thickness due to the antigen–antibody reaction taking place onto the chip surface. Regarding the antibody concentration, it was found that the signal increased almost linearly up to an antibody concentration of 2 µg/mL, whereas higher antibody concentrations provided only marginal signal (10%) increase. In addition, as it can be concluded from Figure S5, the assay sensitivity deteriorated by approximately 50% when the anti-carbendazim antibody concentration increased from 2 to 4 µg/mL. Thus, a 2 µg/mL anti-carbendazim antibody concentration was adopted in the final assay protocol.

Another parameter to optimize was the duration of the whole assay, in order to achieve a fast and reliable assay. As shown in Figure 3, the primary immunoreaction reached a signal plateau after 30 min, the secondary immunoreaction in 20 min, and the reaction with streptavidin in 3 min. The real-time sensor response showed that the reaction with biotinylated secondary antibody and the subsequent reaction with streptavidin resulted in considerable signal enhancement. In particular, the reaction with the secondary antibody increased by 2.5-fold the signal obtained from the primary immunoreaction, whereas the reaction with streptavidin further increased the signal by approximately 3-times, leading to an overall 7.5-fold increase in the signal of the primary immunoreaction. The signal increase observed in the two reactions that follow the primary immunoreaction is ascribed to the fact that more than one biotinylated secondary antibody molecules bind per immunosorbed primary antibody, since the secondary antibody is a polyclonal one and has been raised against the whole anti-rabbit IgG molecule and it is expected, therefore, to bind to several epitopes of the primary antibody molecule. Similarly, the secondary antibody has several biotin-moieties and thus, more than one streptavidin molecule could bind per biotinylated secondary antibody molecule, providing considerable signal enhancement. Due to this accumulation of multiple molecules per immunosorbed primary antibody molecule, the effective biomolecular adlayer thickness, and consequently the WLRS sensor signal, is considerably increased.

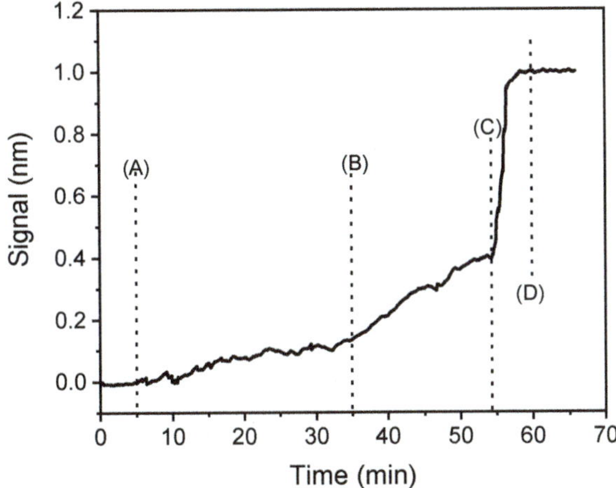

Figure 3. Real time response obtained from a biochip functionalized with 500 µg/mL of benzimidazole conjugate upon running over the chip: assay buffer (from start to point A); a 1:1 (*v/v*) mixture of zero standard with a 2 µg/mL rabbit anti-carbendazim antibody solution (A–B); a 1:200 diluted solution of a biotinylated secondary antibody (B–C); a 10 µg/mL streptavidin solution (C–D).

The signal increase achieved by the implementation of the two signal enhancement steps allowed the reduction of the whole assay time. More specifically, the duration of the primary immunoreaction was set at 18 min, the reaction with the secondary antibody at 7 min, and that with streptavidin at 3 min, resulting in a total assay time of 28 min. This reduction in assay time reduced also the maximum signal obtained by 50%, however, the signal received with the shorter assay protocol was adequate for the performance of the assay. This was confirmed by the good discrimination of the real-time sensor responses obtained for carbendazim standards with concentrations ranging from 0 to 20 µg/mL as shown in Figure 4a. In addition, Figure 4b depicts the relevant calibration plot.

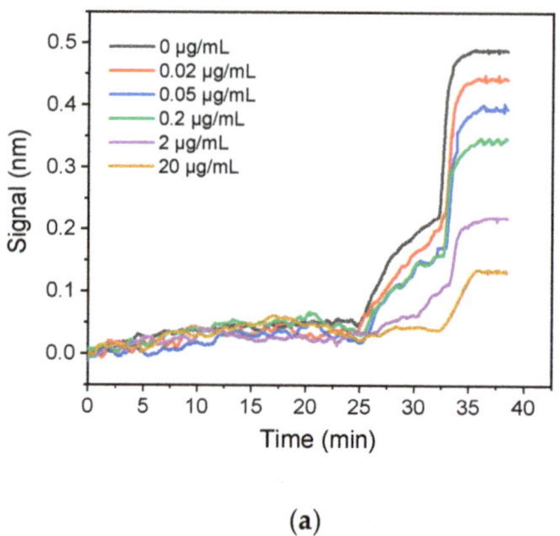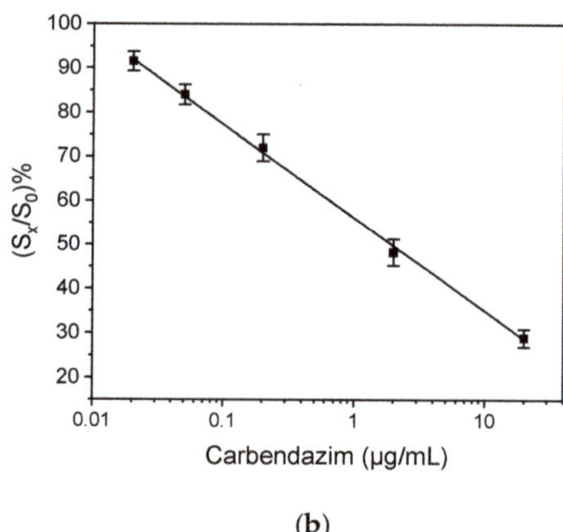

(**a**) (**b**)

Figure 4. (**a**) Real-time responses obtained from biochips functionalized with 500 µg/mL benzimidazole conjugate upon running sequentially: assay buffer; a 1:1 *v/v* mixture of carbendazim standards (0–20,000 ng/mL) prepared in assay buffer with a 2 µg/mL anti-carbendazim antibody solution in the same buffer; a 1:200 diluted solution of biotinylated secondary antibody in assay buffer; a 10 µg/mL streptavidin solution. (**b**) Typical calibration plot for carbendazim. Each point represents the mean value of 4 runs ± SD.

The LoD of the proposed immunosensor was evaluated as described in Section 3.1.1 for the respective ELISA assay and was found to be 20 ng/mL. The assay dynamic range was 50 ng/mL–20 µg/mL.

3.2.2. Optimization of Sample Preparation Procedure

The developed immunosensor was applied to carbendazim detection in fruit juices. The acidic pH and the pulp in fruit juices are reportedly impacting the performance of immunoassays by affecting the antibody–antigen binding, resulting in the so-called "matrix-effect" [23,24]. Thus, it is necessary to perform a sample preparation procedure that minimises interferences before analysing fruit juice samples. In this respect, filtration through a 0.45 µm PTFE membrane syringe filter was performed to remove the pulp, and then the pH of the filtrate was adjusted to 7.4 ± 0.2 with addition of 1 M NaOH solution (without significantly changing the sample volume). In addition, to investigate the possible matrix effect of fruit juice samples to assay performance, the signal obtained by undiluted and 2- to 20-fold diluted samples was compared to that of zero standard signal in buffer. As shown in Figure 5a for an orange juice sample, the undiluted sample, which has been used after filtration and pH neutralization, provided the same signal with the assay buffer indicating that there was no need for additional sample dilution with assay buffer in order to alleviate the matrix effect. This is supported by the fact that the undiluted sample provided statistically the same zero standard values with orange juice diluted 2–20 times with assay buffer. Similar results were obtained using juices from other fruits, e.g., lemon, and from juices made from a combination of different fruits. Thus, all juices involved in the study were analyzed without dilution after filtration and pH adjustment. Further verification that there was not any matrix-effect in the assay performance was provided by the fact that the carbendazim calibration plots obtained in assay buffer and a suitably treated, commercial orange juice were superimposed, as presented in Figure 5b.

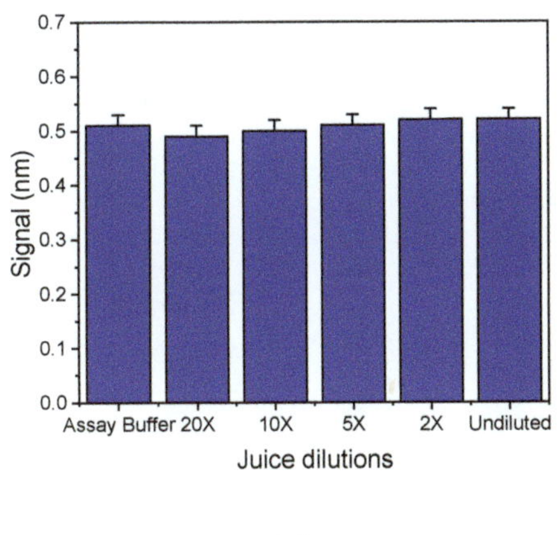
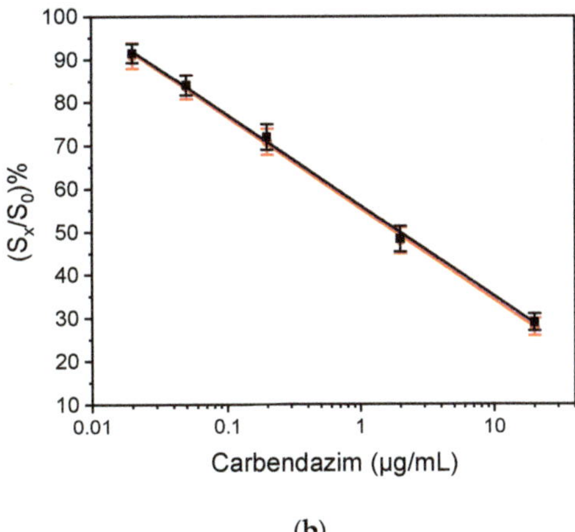

Figure 5. (**a**) Zero standard signal corresponding to assay buffer and filtered and pH-adjusted juice, undiluted and diluted $2\times$ to $20\times$ with assay buffer. Each bar represents the mean value of 3 independent measurements $\pm$ SD. (**b**) Typical calibration plots obtained with carbendazim standards prepared in assay buffer (black squares) and treated orange juice (red circles). Each point represents the mean value of 4 runs $\pm$ SD.

3.2.3. Accuracy and Precision of the Developed ELISA and Sensor Assays

The accuracy of the measurements performed with the developed immunosensor was evaluated through recovery experiments using three commercially available juices prepared from different fruits. To this end, three fruit juices previously analyzed with the carbendazim ELISA and found not to contain any detectable carbendazim were fortified with concentrations of the pesticide that corresponded to three concentration levels, i.e., 100, 500, and 1000 ng/mL. The fortified samples were analyzed in triplicates both with the developed biosensor and the ELISA assay, prior to and after the addition of carbendazim. The results obtained from the analysis of the spiked samples are shown in Table 1, while no carbendazim could be detected in any of the samples prior to the addition of the pesticide. The percent recovery was calculated as the percent ratio of the carbendazim concentration determined in the spiked samples to that expected based on the amount spiked. The recovery values determined with the two methods are presented in Table 1. As shown, the recovery values obtained with the ELISA and the biosensor assays ranged from 90 to 110% and 89 to 110%, respectively, demonstrating the high accuracy of the assays developed. Furthermore, there was a very good agreement of the values determined with the immunosensor to those determined for the same samples with the ELISA. A paired *t*-test confirmed that there was not statistically significant difference between the results of the two methods ($p < 0.05$).

Table 1. Recovery values of carbendazim in commercial fruit juices spiked with the indicated concentrations of the pesticide.

Juice Sample	Spiked Level (ng/mL)	Determined Concentration (ng/mL)		% Recovery	
		WLRS	ELISA	WLRS	ELISA
Orange	1000	1052 ± 53	1097 ± 41	105	110
	500	522 ± 32	540 ± 22	104	108
	100	98 ± 4	106 ± 3	98.0	106
Orange–apple–carrot	1000	973 ± 57	1010 ± 37	97.3	101
	500	547 ± 25	451 ± 26	109	90.2
	100	89 ± 25	91 ± 4	89.0	91.0
Lemon	1000	951 ± 61	896 ± 32	95.1	89.6
	500	487 ± 28	496 ± 18	97.4	99.2
	100	110 ± 7	113 ± 6	110	113

3.2.4. Analysis of Commercially Available Fruit Juices

A survey on carbendazim residues in commercial packages of fruit juices was performed. Ten different juices of four different brands made from orange, lemon, or combination of different fruits were purchased from local supermarkets and analyzed by the WLRS sensor. The results are presented in Table 2. None of the products tested contained detectable amounts of carbendazim. This finding was confirmed by analysing the same samples with the carbendazim ELISA.

Table 2. Carbendazim residues in commercially available fruit juices. In the parentheses the trade name and storage conditions recommended by the manufacturers are provided.

Sample Number/Name (Storage)	Amount Detected (ng/mL)
1/ Orange juice (Amita, RT)	<LoD
2/ Orange juice (Eviva, RT)	<LoD
3/ Orange juice (Marata, RT)	<LoD
4/ Orange-lemon-carrot (Amita, RT)	<LoD
5/ 9 fruits Motion (Amita, RT)	<LoD
6/ Orange juice (Olympos, 4 °C)	<LoD
7/ Orange juice (Eviva, 4 °C)	<LoD
8/ Orange-apple-carrot (Olympos, 4 °C)	<LoD
9/ 9 Fruits (Olympos, 4 °C)	<LoD
10/ Lemon juice (Eviva, 4 °C)	<LoD

3.2.5. Regeneration of Biochips

The stability of the developed immunosensor response to sequential assay/regeneration cycles was also determined as a means to exploit the use of a single biochip for analysis of several samples thus reducing the analysis cost. A low-pH buffer is often used in affinity-based biosensor assays for surface regeneration purposes, to quantitatively remove the antibody from the coating analyte, without affecting the coating analyte structure. The regeneration was achieved by running a 0.1 M glycine-HCl buffer, pH 2.5, for 3 min after completion of the assay. Figure 6 shows the zero standard responses obtained from a single biochip in 15 assay/regeneration cycles performed over a period of three days. As shown, for up to 12 assay/regeneration cycles, all values consistently fell within the mean value ± 2SD range, which demonstrates the potential reuse of biosensor.

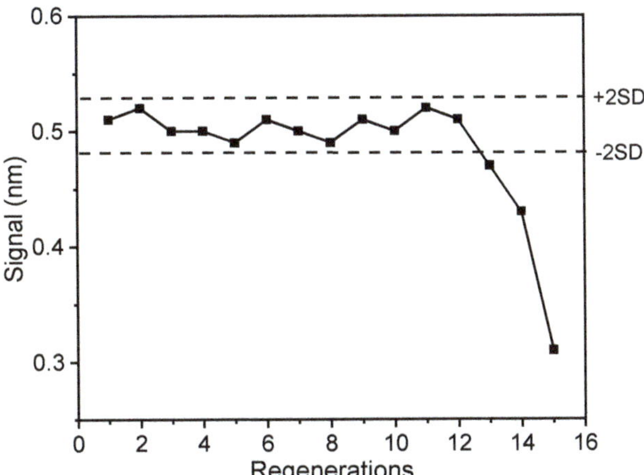

Figure 6. Signal responses obtained from a single biochip for 15 regeneration/assay cycles. Dashed lines represent the mean value ± 2SD limits.

3.3. Comparison with Other Biosensors

In the last decade carbendazim has received particular attention from the research community and many efforts from research groups all over the world have been carried out towards the development of biosensors for carbendazim determination in various food commodities. The majority of the reported carbendazim biosensors are electrochemical [25–31]. Only few optical sensors for carbendazim detection have been reported in the literature. These include a sensor based on surface-enhanced Raman scattering technology, which involves cyclodextrin inclusion complexes on gold nanorods as recognition element [32], a sensor based on luminescence resonance energy transfer from aptamer-labeled upconversion nanoparticles to manganese dioxide nanosheets that act as an acceptor [33], and a surface plasmon resonance-based immunosensor [34]. The immunosensor proposed herein is the first optical sensor based on white light reflectance spectroscopy, dedicated to the determination of carbendazim in foodstuff.

With respect to actual analysis time of 28 min, our sensor is considered among the fastest sensors reported for carbendazim detection in foodstuff generally; even if the separate, 60-min preincubation step is taken into account, the proposed sensor can be still considered as an analytical tool capable of determining carbendazim very quickly. In terms of analytical sensitivity, the LoD of the developed method (20 ng/mL) is well below the current European Union regulatory limit of 200 ng/mL for carbendazim in fruit juices. It is noteworthy that in our case, as in the aforementioned SPR sensor [34], a signal enhancement step was introduced after the primary immunoreaction in order to generate a measurable response while keeping the analysis time as short as possible. Overall, the developed sensor could be regarded as a very fast and sensitive real-time biosensing platform for the detection of carbendazim, which could be considered as label-free, since it is not based on any typical label, such as a fluorophore or an enzyme, for the development of an optical signal. The sensor is accompanied by a rather simple sample preparation protocol that can be easily reproduced at the point-of-need as it does not require any special equipment. In addition, the sample preparation procedure could be applied to juices prepared from different fruits without any noticeable effect in the immunosensor performance by the difference in the sample matrix, which is very important for future on-site analysis of different juices.

4. Conclusions

In conclusion, a WLRS-based biosensing platform for the real-time immunochemical determination of carbendazim in fruit juices was successfully developed. The proposed sensor allowed for the sensitive and fast quantification of carbendazim levels down to 20 ng/mL within less than 30 min. Moreover, the fact that a single chip functionalized with the benzimidazole conjugate could be regenerated and reused for at least 12 times without any effect onto the signal received provides a further advantage. In summary, excellent analytical characteristics and short analysis time combined with the small size of the analytical device render the proposed WLRS biosensor ideal for future on-the-spot determination of carbendazim in food and environmental samples.

Supplementary Materials: The following are available online at https://www.mdpi.com/article/10.3390/bios11050153/s1, Scheme S1. TMB peroxidase substrate reaction. Figure S1. (a) Absorbance values at 450 nm received for zero carbendazim standard from wells coated with benzimidazole-conjugate concentration 0.5 (black squares), 1 (red circles), 2.5 (blue up triangles), or 5 μg/mL (green down triangles) when assayed with anti-carbendazim antibody concentrations ranging from 0.5 to 8 μg/mL. Each point is the mean value of four wells ± SD. (b) Percent absorbance values obtained for the zero carbendazim standard (orange bars) and a standard containing 200 ng/mL carbendazim (green bars) using different combinations of benzimidazole conjugate for well coating and anti-carbendazim antibody. Each point is the mean value of four wells ± SD. Figure S2. Absorbance values at 450 nm received for zero carbendazim standard (orange bars) and a standard containing 200 ng/mL carbendazim (green bars) using the following assay buffers: 10 mM PBS buffer, pH 6.5; 10 mM PBS, pH 7.4; 50 mM Tris-HCl buffer, pH 7.8; 50 mM Tris-HCl buffer, pH 8.25. All buffers contained 0.4% BSA. Each point is the mean value of four wells ± SD. Figure S3. Carbendazim calibration plots obtained without preincubation of carbendazim standards with the anti-carbendazim antibody solution (green down triangles) or with preincubation for 30 min (black squares), 60 min (red circles), and 120 min (blue up triangles). Each point is the mean value of four wells ± SD. Figure S4. Signals received for zero carbendazim standard from WLRS chips coated with benzimidazole conjugate concentrations ranging from 100 to 1000 μg/mL when assayed with anti-carbendazim antibody solutions of 0.5 (black squares), 1 (red circles), 2 (blue up triangles), or 4 μg/mL (green down triangles). Each point is the mean value of measurements obtained by three chips ± SD. Figure S5. Calibration plots obtained from chips coated with 500 μg/mL of benzimidazole conjugate and assayed with anti-carbendazim antibody solutions with concentration 2 (black squares) or 4 μg/mL (red circles). Each point is the mean value of three measurements ± SD.

Author Contributions: Conceptualization, G.K., C.-E.K., E.L. and S.K.; methodology, G.K. and C.-E.K.; investigation, G.K. and C.-E.K.; data curation, G.K. and C.-E.K.; resources, I.R., P.P.; funding acquisition, E.L. and S.K.; supervision, E.L. and S.K.; writing—original draft preparation, C.-E.K.; writing—review and editing, G.K., C.-E.K., P.P., E.L., and S.K. All authors have read and agreed to the published version of the manuscript.

Funding: This research is cofinanced by Greece and the European Union (European Social Fund—ESF) through the Operational Program "Human Resources Development, Education and Lifelong Learning 2014–2020" in the context of the project "Development of immunochemical analytical methodology, especially an ELISA and a biosensor assay, for determining benzimidazole-type pesticides" (MIS 5047811).

Institutional Review Board Statement: Animal immunization was performed in accordance with the Presidential Decree 56/2013 for the Protection of Animals used for Scientific Purposes (Directive 2010/63/EU) and approved by the local committee of the Animal House (Institute of Biosciences & Applications, NCSR "Demokritos") which is a certified installation (EL 25 BIOexp 039, Prefecture of Attica) and the Prefecture of Attica, Division of Agriculture and Veterinary Medicine (license No 247871/08-04-2020).

Informed Consent Statement: Not applicable.

Data Availability Statement: The data presented in this study are available on request from the corresponding author. The data are not publicly available due to privacy issues.

Conflicts of Interest: The authors declare no conflict of interest. The funders had no role in the design of the study; in the collection, analyses, or interpretation of data; in the writing of the manuscript, or in the decision to publish the results.

References

1. Karlsson, I.; Friberg, H.; Steinberg, C.; Persson, P. Fungicide Effects on Fungal Community Composition in the Wheat Phyllosphere. *PLoS ONE* **2014**, *9*, e111786. [CrossRef] [PubMed]
2. Singh, S.; Singh, N.; Kumar, V.; Datta, S.; Wani, A.B.; Singh, D.; Singh, K.; Singh, J. Toxicity, monitoring and biodegradation of the fungicide carbendazim. *Environ. Chem. Lett.* **2016**, *14*, 317–329. [CrossRef]
3. Tortella, G.; Mella-Herrera, R.; Sousa, D.; Rubilar, O.; Briceño, G.; Parra, L.; Diez, M. Carbendazim dissipation in the biomixture of on-farm biopurification systems and its effect on microbial communities. *Chemosphere* **2013**, *93*, 1084–1093. [CrossRef]
4. Goodson, W.H.; Lowe, L.; Carpenter, D.O.; Gilbertson, M.; Ali, A.M.; Salsamendi, A.L.D.C.; Lasfar, A.; Carnero, A.; Azqueta, A.; Amedei, A.; et al. Assessing the carcinogenic potential of low-dose exposures to chemical mixtures in the environment: The challenge ahead. *Carcinogenesis* **2015**, *36*, S254–S296. [CrossRef]
5. Stahel, W.R. Reuse Is the Key to the Circular Economy. Available online: http://ec.europa.eu/environment/ecoap/about-eco-innovation/experts-interviews/reuse-is-the-key-to-the-circular-economy_en.htm (accessed on 12 March 2020).
6. Grujic, S.; Radišić, M.; Vasiljevic, T.; Lausevic, M. Determination of carbendazim residues in fruit juices by liquid chromatography-tandem mass spectrometry. *Food Addit. Contam.* **2005**, *22*, 1132–1137. [CrossRef] [PubMed]
7. EU Database. Available online: https://eur-lex.europa.eu/eli/reg/2011/559/oj (accessed on 2 November 2020).
8. Heyman, M.B.; Abrams, S.A.; Gastroenterology, H.S.O.; Committee on Nutrition. Fruit Juice in Infants, Children, and Adolescents: Current Recommendations. *Pediatrics* **2017**, *139*, e20170967. [CrossRef] [PubMed]
9. Hiemstra, M.; De Kok, A. Comprehensive multi-residue method for the target analysis of pesticides in crops using liquid chromatography–tandem mass spectrometry. *J. Chromatogr. A* **2007**, *1154*, 3–25. [CrossRef]
10. Chayata, H.; Lassalle, Y.; Nicol, É.; Manolikakes, S.; Souissi, Y.; Bourcier, S.; Gosmini, C.; Bouchonnet, S. Characterization of the ultraviolet–visible photoproducts of thiophanate-methyl using high performance liquid chromatography coupled with high resolution tandem mass spectrometry—Detection in grapes and tomatoes. *J. Chromatogr. A* **2016**, *1441*, 75–82. [CrossRef]
11. Gough, K.C.; Jarvis, S.; Maddison, B.C. Development of competitive immunoassays to hydroxyl containing fungicide metabolites. *J. Environ. Sci. Health Part B* **2011**, *46*, 581–589. [CrossRef]
12. Guo, L.; Wu, X.; Liu, L.; Kuang, H.; Xu, C. Gold Nanoparticle-Based Paper Sensor for Simultaneous Detection of 11 Benzimidazoles by One Monoclonal Antibody. *Small* **2017**, *14*, 1701782. [CrossRef]
13. Reynoso, E.C.; Torres, E.; Bettazzi, F.; Palchetti, I. Trends and Perspectives in Immunosensors for Determination of Currently-Used Pesticides: The Case of Glyphosate, Organophosphates, and Neonicotinoids. *Biosensor* **2019**, *9*, 20. [CrossRef] [PubMed]
14. Koukouvinos, G.; Petrou, P.; Goustouridis, D.; Misiakos, K.; Kakabakos, S.; Raptis, I. Development and Bioanalytical Applications of a White Light Reflectance Spectroscopy Label-Free Sensing Platform. *Biosensors* **2017**, *7*, 46. [CrossRef] [PubMed]
15. Anastasiadis, V.; Koukouvinos, G.; Petrou, P.S.; Economou, A.; Dekker, J.; Harjanne, M.; Heimala, P.; Goustouridis, D.; Raptis, I.; Kakabakos, S.E. Multiplexed mycotoxins determination employing white light reflectance spectroscopy and silicon chips with silicon oxide areas of different thickness. *Biosens. Bioelectron.* **2020**, *153*, 112035. [CrossRef] [PubMed]
16. Stavra, E.; Petrou, P.S.; Koukouvinos, G.; Economou, A.; Goustouridis, D.; Misiakos, K.; Raptis, I.; Kakabakos, S.E. Fast, sensitive and selective determination of herbicide glyphosate in water samples with a White Light Reflectance Spectroscopy immunosensor. *Talanta* **2020**, *214*, 120854. [CrossRef] [PubMed]
17. Zikos, C.; Evangelou, A.; Karachaliou, C.-E.; Gourma, G.; Blouchos, P.; Moschopoulou, G.; Yialouris, C.; Griffiths, J.; Johnson, G.; Petrou, P.; et al. Commercially available chemicals as immunizing haptens for the development of a polyclonal antibody recognizing carbendazim and other benzimidazole-type fungicides. *Chemosphere* **2015**, *119*, S16–S20. [CrossRef]
18. Papasarantos, I.; Klimentzou, P.; Koutrafouri, V.; Anagnostouli, M.; Zikos, C.; Paravatou-Petsotas, M.; Livaniou, E. Solid-Phase Synthesis of a Biotin Derivative and its Application to the Development of Anti-Biotin Antibodies. *Appl. Biochem. Biotechnol.* **2009**, *162*, 221–232. [CrossRef] [PubMed]
19. Bayer, E.A.; Zalis, M.G.; Wilchek, M. 3-(N-maleimido-propionyl) biocytin: A versatile thiol-specific biotinylating reagent. *Anal. Biochem.* **1985**, *149*, 529–536. [CrossRef]
20. Ortelli, D.; Edder, P.; Corvi, C. Pesticide residues survey in citrus fruits. *Food Addit. Contam.* **2005**, *22*, 423–428. [CrossRef]
21. De Souza, D.; Machado, S. Electrochemical detection of the herbicide paraquat in natural water and citric fruit juices using microelectrodes. *Anal. Chim. Acta* **2005**, *546*, 85–91. [CrossRef]
22. Zhao, B.; Feng, S.; Hu, Y.; Wang, S.; Lu, X. Rapid determination of atrazine in apple juice using molecularly imprinted polymers coupled with gold nanoparticles-colorimetric/SERS dual chemosensor. *Food Chem.* **2019**, *276*, 366–375. [CrossRef]
23. Skerritt, J.H.; Rani, B.E.A. Detection and removal of sample matrix effects in agrochemical immunoassays. In *Immunoassays for Residue Analysis*; ACS Symposium Series; American Chemical Society: Washington, DC, USA, 1996; Volume 621, pp. 29–43.
24. Moreno, M.-J.; Plana, E.; Montoya, A.; Caputo, P.; Manclús, J.J. Application of a monoclonal-based immunoassay for the determination of imazalil in fruit juices. *Food Addit. Contam.* **2007**, *24*, 704–712. [CrossRef]

25. Feng, S.; Li, Y.; Zhang, R.; Li, Y. A novel electrochemical sensor based on molecularly imprinted polymer modified hollow N, S-Mo2C/C spheres for highly sensitive and selective carbendazim determination. *Biosens. Bioelectron.* **2019**, *142*, 111491. [CrossRef] [PubMed]
26. Barboza, A.D.M.; Da Silva, A.B.; Da Silva, E.M.; De Souza, W.P.; Soares, M.A.; De Vasconcelos, L.G.; Terezo, A.J.; Castilho, M. A biosensor based on microbial lipase immobilized on lamellar zinc hydroxide-decorated gold nanoparticles for carbendazim determination. *Anal. Methods* **2019**, *11*, 5388–5397. [CrossRef]
27. Liao, X.; Huang, Z.; Huang, K.; Qiu, M.; Chen, F.; Zhang, Y.; Wen, Y.; Chen, J. Highly Sensitive Detection of Carbendazim and Its Electrochemical Oxidation Mechanism at a Nanohybrid Sensor. *J. Electrochem. Soc.* **2019**, *166*, B322–B327. [CrossRef]
28. Eissa, S.; Zourob, M. Selection and Characterization of DNA Aptamers for Electrochemical Biosensing of Carbendazim. *Anal. Chem.* **2017**, *89*, 3138–3145. [CrossRef] [PubMed]
29. Dong, Y.; Yang, L.; Zhang, L. Simultaneous Electrochemical Detection of Benzimidazole Fungicides Carbendazim and Thiabendazole Using a Novel Nanohybrid Material-Modified Electrode. *J. Agric. Food Chem.* **2017**, *65*, 727–736. [CrossRef] [PubMed]
30. Arruda, G.J.; De Lima, F.; Cardoso, C.A.L. Ultrasensitive determination of carbendazim in water and orange juice using a carbon paste electrode. *J. Environ. Sci. Health Part B* **2016**, *51*, 534–539. [CrossRef] [PubMed]
31. Razzino, C.A.; Sgobbi, L.F.; Canevari, T.C.; Cancino, J.; Machado, S.A. Sensitive determination of carbendazim in orange juice by electrode modified with hybrid material. *Food Chem.* **2015**, *170*, 360–365. [CrossRef]
32. Strickland, A.D.; Batt, C.A. Detection of Carbendazim by Surface-Enhanced Raman Scattering Using Cyclodextrin Inclusion Complexes on Gold Nanorods. *Anal. Chem.* **2009**, *81*, 2895–2903. [CrossRef]
33. Ouyang, Q.; Wang, L.; Ahmad, W.; Rong, Y.; Li, H.; Hu, Y.; Chen, Q. A highly sensitive detection of carbendazim pesticide in food based on the upconversion-MnO2 luminescent resonance energy transfer biosensor. *Food Chem.* **2021**, *349*, 129157. [CrossRef]
34. Li, Q.; Dou, X.; Zhao, X.; Zhang, L.; Luo, J.; Xing, X.; Yang, M. A gold/Fe$_3$O$_4$ nanocomposite for use in a surface plasmon resonance immunosensor for carbendazim. *Microchim. Acta* **2019**, *186*, 313. [CrossRef] [PubMed]

 *biosensors*

Article

Comparison of Leading Biosensor Technologies to Detect Changes in Human Endothelial Barrier Properties in Response to Pro-Inflammatory TNFα and IL1β in Real-Time [†]

James J. W. Hucklesby [1,2,*], Akshata Anchan [2,3], Simon J. O'Carroll [3,4], Charles P. Unsworth [5], E. Scott Graham [2,3,‡] and Catherine E. Angel [1,*,‡]

1   School of Biological Sciences, Faculty of Science, University of Auckland, Auckland 1010, New Zealand
2   Department of Molecular Medicine and Pathology, Faculty of Medical and Health Sciences, University of Auckland, Auckland 1010, New Zealand; a.anchan@auckland.ac.nz (A.A.); s.graham@auckland.ac.nz (E.S.G.)
3   Centre for Brain Research, University of Auckland, Auckland 1010, New Zealand; s.ocarroll@auckland.ac.nz
4   Department of Anatomy and Medical Imaging, Faculty of Medical and Health Sciences, University of Auckland, Auckland 1010, New Zealand
5   Department of Engineering Science, Faculty of Engineering, University of Auckland, Auckland 1010, New Zealand; c.unsworth@auckland.ac.nz
*   Correspondence: james.hucklesby@auckland.ac.nz (J.J.W.H.); c.angel@auckland.ac.nz (C.E.A.)
†   Presented at the 1st International Electronic Conference on Biosensors, 2–17 November 2020; Available online: https://iecb2020.sciforum.net/. Presentation published in Proceedings.
‡   These authors contributed equally to this work.

Citation: Hucklesby, J.J.W.; Anchan, A.; O'Carroll, S.J.; Unsworth, C.P.; Graham, E.S.; Angel, C.E. Comparison of Leading Biosensor Technologies to Detect Changes in Human Endothelial Barrier Properties in Response to Pro-Inflammatory TNFα and IL1β in Real-Time . *Biosensors* **2021**, *11*, 159. https://doi.org/10.3390/bios11050159

Received: 22 April 2021
Accepted: 13 May 2021
Published: 18 May 2021

Publisher's Note: MDPI stays neutral with regard to jurisdictional claims in published maps and institutional affiliations.

Abstract: Electric Cell-Substrate Impedance Sensing (ECIS), xCELLigence and cellZscope are commercially available instruments that measure the impedance of cellular monolayers. Despite widespread use of these systems individually, direct comparisons between these platforms have not been published. To compare these instruments, the responses of human brain endothelial monolayers to TNFα and IL1β were measured on all three platforms simultaneously. All instruments detected transient changes in impedance in response to the cytokines, although the response magnitude varied, with ECIS being the most sensitive. ECIS and cellZscope were also able to attribute responses to particular endothelial barrier components by modelling the multifrequency impedance data acquired by these instruments; in contrast the limited frequency xCELLigence data cannot be modelled. Consistent with its superior impedance sensing, ECIS exhibited a greater capacity than cellZscope to distinguish between subtle changes in modelled endothelial monolayer properties. The reduced resolving ability of the cellZscope platform may be due to its electrode configuration, which is necessary to allow access to the basolateral compartment, an important advantage of this instrument. Collectively, this work demonstrates that instruments must be carefully selected to ensure they are appropriate for the experimental questions being asked when assessing endothelial barrier properties.

Keywords: ECIS; xCELLigence; cellZscope; hCMVEC; endothelial cell; impedance sensing

1. Introduction

Impedance sensing is a label-free, real-time technique used to monitor cellular function. First pioneered by Giaever and Keese, impedance sensing exposes live cells to very small electrical currents across a range of frequencies [1,2]. By measuring the impedance that the cells provide to this current, we can accurately measure the responses of the cells in real-time. As no labelling is required, measurements are non-invasive and can be carried out over extended periods to give high-resolution information in real-time [3,4]. Furthermore, this information is inherently quantitative and thus can be readily analysed statistically [5,6]. Mathematical models can also be applied to this data to allow the exploration of various cellular parameters that cannot be measured directly [7]. These advantages have triggered the broad adoption of impedance sensing in a wide variety of applications, with a range of custom instruments having been developed [8–10]. However, the adoption of these systems has been limited, as the construction of customised specialist instrumentation is technically challenging. In contrast, commercially available instruments provide a turnkey solution to accessing impedance sensing. There are, however, only a few commercially available instruments including the Electrical Cell-Substrate Impedance Sensing (ECIS), xCELLigence and cellZscope platforms [11–13]. Despite the widespread use of these platforms individually to assess endothelial barriers [3,14–18] a systematic comparison of each platform's capacity to resolve changes in endothelial barrier properties has not been conducted. Therefore, in this paper, the ability of these instruments to detect changes in endothelial barrier properties in response to TNFα and IL1β were compared.

Giaever and Keese's original Electric Cell-substrate Impedance Sensing (ECIS) invention has since been commercialized by Applied BioPhysics [11]. One such instrument is the ECIS ZΘ, which can be configured to measure cellular impedance in 96-well plates with gold electrodes fabricated directly onto the base of each well that has a growth area of 0.32 cm^2 (Figure 1 and Supplementary Table S1). Impedance and phase measurements at frequencies ranging from 10 Hz to 10^5 Hz are collected by the instrument (Supplementary Table S1). Subsequently, these can be modelled computationally to indicate biologically relevant cellular parameters. Three key values are generated: Rb, Cm and Alpha (Supplementary Table S1). Rb represents the cell–cell contacts, such as those formed by junctional molecules; Cm represents the membrane resistance of the cells; whilst Alpha represents the basolateral adhesion, which is influenced by both the distance between the cells and the underlying substrate and the presence of any junctional molecules bridging this interface [7]. Together, these values allow for the in-depth analysis of biological responses [19].

More recently, ACEA Biosciences (now part of Agilent) released the xCELLigence instrument [12]. Much like ECIS, this instrument uses gold electrodes fabricated directly onto the base of wells in a 96-well plate; each well has a growth area of 0.196 cm^2 (Figure 1 and Supplementary Table S1). However, this instrument only collects impedance measurements at three frequencies, 10, 25 and 50 kHz (Supplementary Table S1). Although modelling cellular parameters is theoretically possible using three frequency measurements, the limited range of readings makes any results unreliable.

Finally, cellZscope is the most recent addition to the market, and is able to measure impedance across a Transwell filter with a cell growth area of 0.33 cm^2 (Supplementary Table S1) [13]. The Transwell is seated in a stainless steel pot that acts as an electrical conductor. A second electrode suspended over the cells makes contact with the media in the apical chamber, completing the circuit and allowing impedance to be measured (Figure 1 and Supplementary Table S1). Like ECIS, phase and impedance data are collected at a range of frequencies from 1 Hz to 100 kHz and hence, can also be modelled (Supplementary Table S1). This results in the calculation of transepithelial-endothelial electrical resistance (TER) as a measurement of the cell–cell junctional interactions, and C_{CL} as a measure of cell layer capacitance (Supplementary Table S1) [13]. An equivalent of the Alpha value generated by the ECIS instrument is not included in this model, as the porous nature of the Transwells means that this parameter is not physically present and therefore not appropriate to infer.

Despite numerous studies using these instruments, direct comparisons between them have not been conducted. This is a critical lack of knowledge, as the inferences from the data collected from all three instruments are regularly used together to interrogate cellular responses [17,20–23]. Therefore, in this paper, we analyse the similarities and differences between these three commercially available instruments. The hCMVEC cell line was chosen due to its low overall resistance, which, although characteristic of brain microvascular endothelial cell lines [23], dictates the use of more sophisticated and more sensitive instrumentation [19].

The inflammatory cytokines TNFα and IL1β were selected for these experiments due to their well-defined biphasic response in this cell line. The response of hCMVECs to IL1β

and TNFα has been explored at a molecular level and has been well-characterized using impedance instruments [17]. These responses are ideal for this study, as the cytokines first cause a decrease in resistance, followed by a substantial increase for an extended period. Therefore, both decreases and increases in resistance can be examined with the same stimulus. The transient initial decrease in resistance also showcases the high time resolution of impedance sensing, by highlighting a response that could easily go undetected between the time points of a traditional end-point assay [24]. For this study, TNFα and IL1 β concentrations were selected to provide a robust biphasic response with which to test the impedance instruments [17].

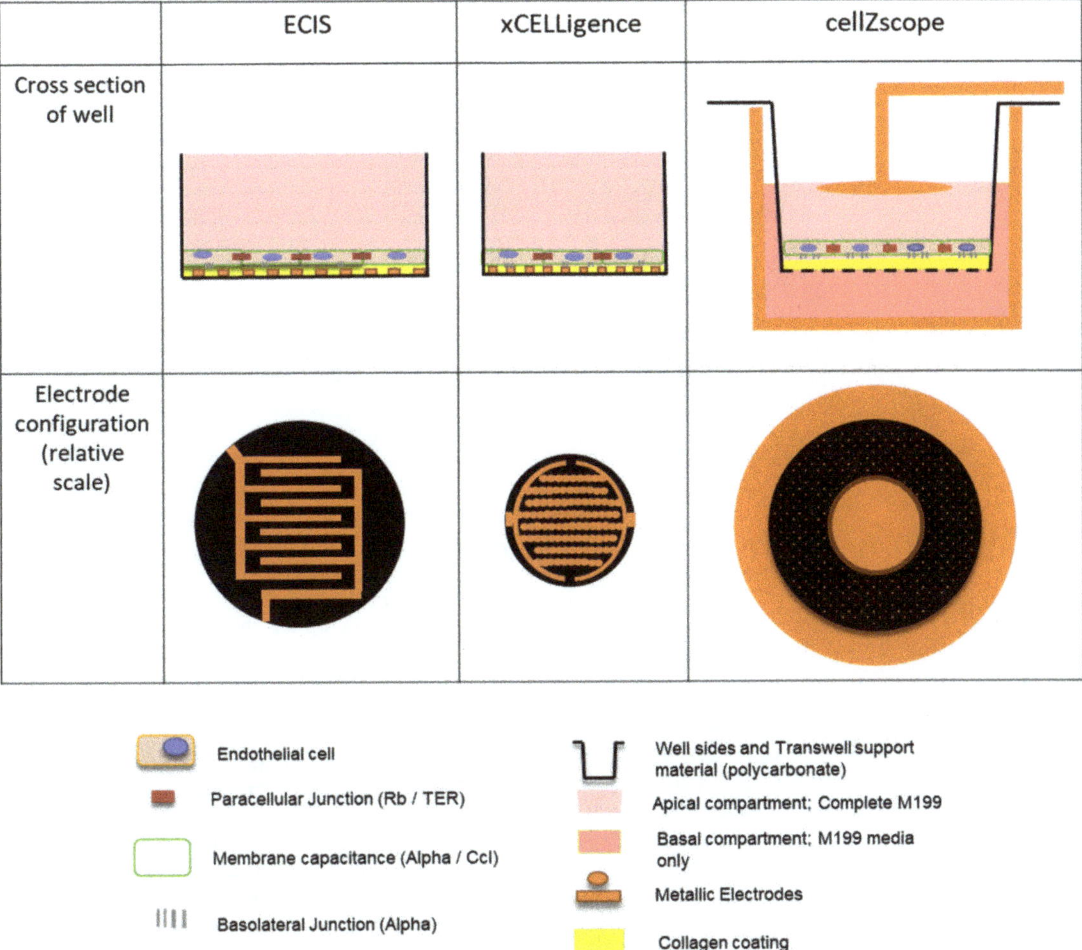

Figure 1. The electrode arrays used most widely in the ECIS (96W20idf plate), xCELLigence (E-plate) and cellZscope instruments differ in their electrode configuration. Both the ECIS and xCELLigence electrodes have a similar interdigitating electrode configuration, which covers a high proportion of the bottom of the well. Hence, their electrodes are directly coated with collagen and in intimate contact with the endothelial cells. In contrast, one of the cellZscope electrodes lines the lower compartment of the Transwell, whilst the second is suspended above the cell monolayer. Therefore, these electrodes are not in direct contact with the endothelial cells.

We evaluated two key parameters of the data produced: the difference in magnitude at key points in time, and the profile of the temporal measurements resulting in different curve shapes. A difference in magnitude is informative, straightforward to interpret and correlates with traditional single-time point assays [19,20]. The second characteristic, the profile of the temporal measurements or shape of the curve, is also useful. Even if two responses have the same magnitude at a key time point, they may reach that point in a very different way. This characteristic was analysed using cross-correlation with no lag, which generates a single value between 1 and −1 for each pair of curves. A value of 1 represents identical curves, 0 shows no correlation between the curves and −1 represents curves with a mirror image opposing profile or inverse correlation [25]. By assessing the magnitude and temporal profile of the response in concert, we are able to rigorously compare the measurements from all three instruments.

In this study, we ran the same experiment simultaneously on the three impedance-sensing instruments. We show that, although the instruments' temporal impedance measurements have similar profiles, they differ in magnitude, demonstrating significant differences in sensitivity. Furthermore, the modelled data reinforces the differences in sensitivity between the instruments and reveals changes in endothelial barrier properties that were not evident from the overall impedance measurements. Together, this demonstrates the importance of selecting the most appropriate instrument for a particular study.

2. Materials and Methods

2.1. Culture of Human Brain Endothelial Cells

Human cerebral microvascular endothelial cells (hCMVECs) were purchased from Applied Biological Materials Inc (cat# T0259). hCMVECs were cultured in 75 cm^2 (T75) Nunc flasks (cat# 156499) with M199 medium containing 10% FBS, 1 µg/mL hydrocortisone, 3 ng/mL hFGF, 1 ng/mL hEGF, 10 µg/mL heparin, 2 mM GlutaMAX and 80 µM dibutyryl-cAMP (later referred to as complete M199 medium) at 37 °C, with 5% CO_2 and 100% humidity. For both hCMVEC maintenance and experiments, culture vessels were coated with 1 µg/cm^2 collagen I dissolved in 0.02 M acetic acid for 1 h at room temperature, before being washed 3 times with sterile MilliQ water and seeding the hCMVECs. To passage the hCMVECs, T75 flasks were washed twice with 4 mL pre-warmed PBS before being incubated with 4 mL pre-warmed TrypLE for 5 min at 37 °C. The TrypLE activity was then neutralized with 4 mL complete M199 and the cells were centrifuged at $100\times g$ for 5 min, counted, and seeded for experiments. All experiments used hCMVECs between passages 11 to 16. All impedance instruments and experimental hCMVEC cultures were kept in dedicated incubators at 37 °C, with 5% CO_2 and 100% humidity.

2.2. Impedance Sensing Experiments

ECIS: 96W20idf plates were treated with 10 mM cysteine for 15 min to clean the electrode and standardize the electrode impedance (as per manufacturers' instructions). The wells were then coated with collagen as described above. The hCMVECs were seeded in 200 µL complete M199 medium. The ECIS machine was run continuously in multi-frequency mode using the default frequency spectra (Supplementary Table S1).

xCELLigence: E-plates (96 wells) were coated with collagen as described above. Complete M199 was added to each well and calibration was conducted. Cells were seeded in 122 µL Complete M199. Impedance was measured at 10, 25 and 50 kHz (Supplementary Table S1).

CellZscope: before the experiment, cellZscope components were cleaned with MilliQ water, 70% ethanol, and then MilliQ water again. The pots and dipping electrodes were autoclaved, whilst the remainder of the Cell Module was sterilised with 70% ethanol. Before coating, the Cell Module was assembled under sterile conditions, and each of the stainless steel pots was flooded with 900 µL basal M199 media. The assembled module was then placed in the cell culture incubator to equilibrate for at least one hour. Transwells (Corning; 6.5 mm insert, 0.4 µm pore size) were coated from the apical side, as previously described in Section 2.1. The hCMVECs were then seeded into the apical chamber in 200 µL complete M199 medium. Transwells were then transferred into the Cell Module, taking care not to trap any bubbles underneath the membrane. The Cell Module was then placed in the instrument, and the spectra were acquired at the highest resolution between 1 and 100 kHz (Supplementary Table S1). Measurements were made every 15 min, the fastest rate possible at these frequency settings.

2.3. Treatment with Inflammatory Cytokines

After seeding, the cells were cultured for 48 h to allow the barrier to fully develop and impedance to stabilise. On the day of treatment a $5\times$ stock of TNFα and IL1β in complete M199 was prepared; once added to the corresponding culture wells this provided a final concentration of 500 pg/mL of TNFα or 500 pg/mL IL1β. For the control treatment, the $5\times$ stock consisted of complete M199 with an equivalent amount of MilliQ water (henceforth labelled as the control). Each instrument was then paused, and the $5\times$ stock was gently introduced to the middle of the well or apical chamber. The cultures were then returned to the respective instrument and the measurements resumed. Cell monitoring continued on all instruments for a further 27 h.

2.4. Data Analysis

Modelling was conducted against a cell-free well in the same experiment using software provided by the vendor for each instrument; ECIS Software (V 1.1.252, Applied Biophysics), RTCA Software (V 2.0.0.2301, AECA Biosciences Inc.) and cellZscope (V 4.3.4, nanoAnalytics) for ECIS, xCELLigence and the cellZscope respectively.

Graphs were generated using ggplot2 version 3.3.2 [26]. All experiments were conducted in triplicate and the mean $\pm$ standard error of the mean (SEM) from three independent experiments were plotted.

RStudio (version 1.1.414, RStudio, Inc., Boston, MA, USA) and vascr (developed by J. Hucklesby) [6] were used to generate the cross-correlation values. vascr uses the ccf function in the stats package to run the underlying cross-correlation analysis. No lag value was applied. Temporal response profiles for each experiment were generated by averaging measurements from three technical replicates. Cross-correlation results show the mean $\pm$ SEM of the values derived from the two temporal response profiles being compared, each of which includes data from three independent experiments.

3. Results and Discussion

To assess the comparability of the three instruments, we first collected the impedance spectra of a confluent hCMVEC monolayer at 5 and 47 h, after the seeding of either 250,000 cells/cm^2, 62,500 cells/cm^2 or media only (Figure 2). As all three platforms were seeded simultaneously using the same preparation of cells, we can directly compare the measurements collected.

The dataset in Figure 2 clearly demonstrates the quantity of data collected by each instrument and the relative concordance of the data obtained from these three instruments. The cellZscope captured 34 data points, compared to 9 from the ECIS instrument, and only 3 from the xCELLigence platform. For modelling to be accurately conducted, we require data showing the impedance response of cells over a large frequency range [2]; hence the small number of data points spanning a narrow frequency range that were acquired using xCELLigence cannot be accurately used to model different endothelial barrier properties. Therefore, only the overall impedance change obtained using the xCELLigence can be assessed, a value that incorporates several undistinguishable cellular parameters, limiting the interpretation of these data. The capacity of the ECIS and cellZscope instruments to model the data they acquire over a larger frequency range will be explored later in this study.

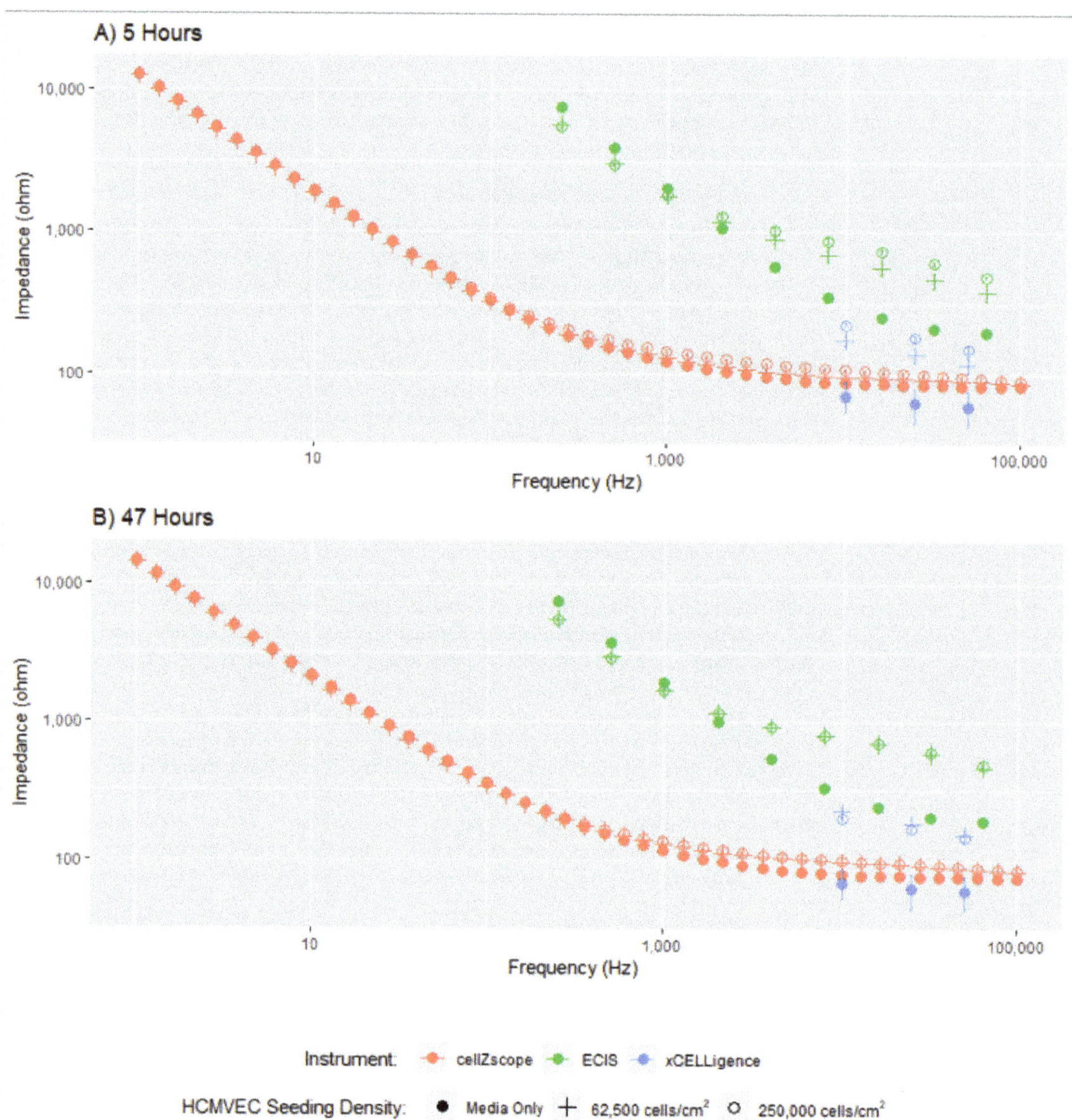

Figure 2. Impedance spectra of hCMVECs obtained using the ECIS, cellZscope and xCELLigence instruments. The impedance spectra of two different initial cell seeding densities were assessed at 5 and 47 h. Each point represents the mean ± SEM at each frequency where the impedance was measured. These data are derived from three independent experiments, each of which was conducted in triplicate.

Despite variation in magnitudes between all three instruments, the cell impedance spectra generated by each instrument follow a similar trend, indicating that similar cellular characteristics are being measured. As expected, the impedance data for cell-seeded wells is higher than the media-only controls, indicating the cell monolayer has been detected. The difference between cell-seeded wells and the media-only control is subtle for the cellZscope data. This may be partly because the cells are not in direct contact with the electrodes, as they are in the ECIS and xCELLigence instruments (Figure 1 and Supplementary Table S1), meaning the current may be less concentrated when it passes through the cells resulting in a more subtle response. Collectively, the differences in magnitude observed between these data are likely due to variabilities in electrode area and configuration between the instruments (Figure 1 and Supplementary Table S1), which affects the absolute value of the impedance measured.

The dataset in Figure 2 also provides some insight into the temporal changes in impedance for the cells assessed using each of these three instruments. At the 5-h time point, there is a clear difference in impedance between the two cell seeding densities, however, at 47 h this difference is no longer apparent. This is because endothelial cells rapidly proliferate until they come into contact with neighbouring cells and form a monolayer, at which point proliferation slows and a mature monolayer forms [27]. By 47 h, the cells seeded at the lower density had sufficient time to proliferate and form a monolayer similar to the monolayer formed earlier by the cells seeded at a higher density.

To further explore the effect of cell seeding density on hCMVEC proliferation and monolayer formation, the measured impedances and modelled barrier properties were assessed over 48 h (Figure 3). Data from all three instruments showed that cells seeded at the higher density exhibited a high level of impedance at the start, which declined slowly and to varying extents during the 48-h period. In contrast, cells seeded at the lower density started with a lower impedance value that slowly increased and plateaued by approximately 40 h. Interestingly, the data acquired using the ECIS and cellZscope showed that the impedance values of the two seeding densities overlapped by 48 h, consistent with the endothelial cell growth properties discussed earlier. However, the data generated using xCELLigence showed that the impedance of the cells seeded at the higher density dropped below those seeded at the lower density by 48 h.

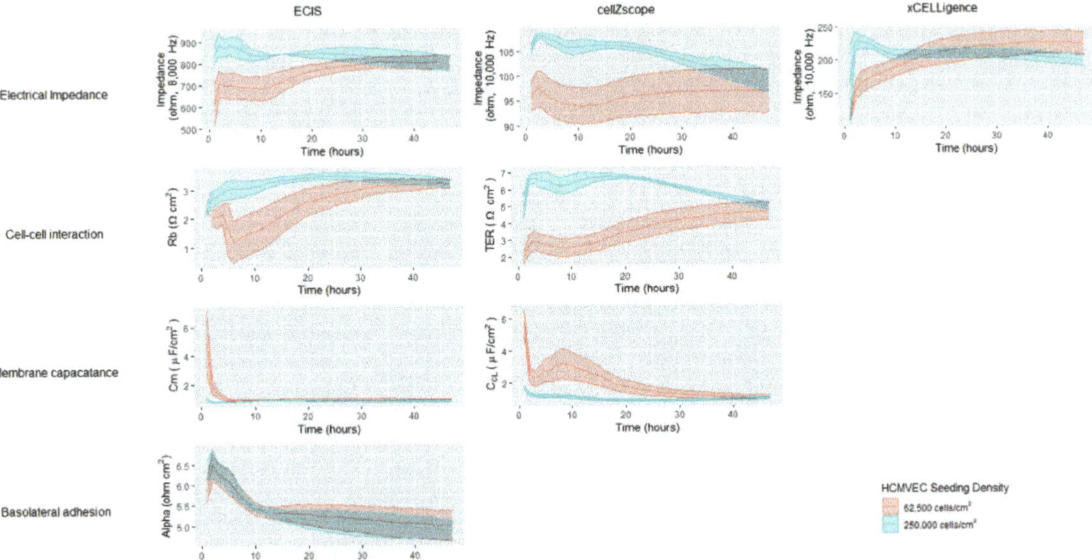

Figure 3. Temporal profile of impedance and modelled endothelial barrier properties of hCMVECs monitored over 48 h by ECIS, cellZscope and xCELLigence instruments. hCMVECs were initially seeded at either 62,500 cells/cm^2 or 250,000 cells/cm^2 and incubated for 48 h until confluent. Ribbon plots show the mean ± SEM of three independent experiments, each of which was conducted in triplicate.

Impedance measurements provide useful insight into overall cellular and monolayer properties, however modelling impedance data acquired over a large frequency range can provide valuable information regarding more distinct cellular and barrier properties. As mentioned earlier it is not appropriate to model xCELLigence data that is generated using a limited frequency range. It is possible however to model the multifrequency data generated using the ECIS and cellZscope instruments (Supplementary Table S1). ECIS and cellZscope can provide Rb and TER values respectively, that use the entire impedance spectra to infer the extent of the cell-cell interactions that have formed (Figure 3). Modelled data for the lower seeding density from both instruments showed an increase in cell-cell interactions during the 48-h period; interestingly the cell–cell interactions modelled from the cellZscope data continued to increase after the impedance had begun to plateau (Figure 3). This shows that a plateau in impedance does not necessarily mean that the cell-cell interactions are fully formed. Membrane capacitance values were also modelled using the ECIS and cellZscope data; overall the hCMVECs exhibited low-level membrane capacitance (Figure 3). The cell membrane capacitance appeared to stabilise within the first 10 h on the ECIS instrument, however, it took a full 40 h for the cell membrane capacitance to stabilise on the Transwells in the cellZscope. ECIS was the only instrument capable of generating data that could infer the basolateral adhesive properties of the cells. This is because the cells are grown on a solid substrate, directly beneath which lie the electrodes. Meaning alpha, which represents basolateral adhesion, can be calculated as illustrated in the equivalent circuit diagram in Supplementary Table S1. In contrast, in the cellZscope the electrodes are not in direct contact with the cells or the solid substrate they are grown on and hence the basolateral adhesion can't be modelled (Figure 1 and Supplementary Table S1). The ECIS profiles of the basolateral adhesive properties of both cell-seeding densities were similar, both declining and stabilising at approximately 12 h (Figure 3). Collectively the data generated by all three instruments using the lower cell seeding density indicates that a stable confluent monolayer had formed by 48 h; this consistency between instruments reflects their similar trends in impedance spectra (Figure 2). These data demonstrate the utility of these three instruments to assess cell growth and monolayer barrier properties, however, it does not interrogate their ability to assess temporal changes in response to biological stimuli.

Next, each system's capacity to detect temporal cellular changes in response to a biological stimulus was tested by treating, the confluent cell monolayers formed at 48 h with the proinflammatory cytokines TNFα and IL1β (Figure 4). The impedance data presented in Figure 4 has been normalized at one hour before treatment and is presented as a change in impedance, to allow direct comparisons between the instruments to be made. Finally, cross-correlation analysis was conducted between each treatment for all instruments; this analysis will test each instrument's ability to discern between different temporal response profiles by comparing the curve shapes.

Each of the instruments were able to detect impedance differences between the control, TNFα and IL1β treatments, however, the ECIS instrument appeared to be the most sensitive showing the largest difference between treatments (Figure 4). The ECIS data showed that both IL1β and TNFα induced an initial rapid reduction in impedance, followed by a sustained increase that slowly declined after 70 h. These trends were apparent for both cell-seeding densities. Similar trends were also observed for the xCELLigence data, however, there was a reduced magnitude in both the responses detected and the differences between treatments, when compared with the ECIS data. The similar trends observed using these two instruments could be attributed to the similar interdigitating electrode configuration and the high proportion of electrode coverage on the bottom of the wells (Figure 1, Supplementary Table S1), meaning that both instruments can detect changes in endothelial monolayer impedance throughout a large proportion of the well. The ECIS instrument's superior ability to resolve the temporal profiles of each of the proinflammatory treatments from the control was reflected by corresponding low cross-correlation values.

In contrast, the higher cross-correlation values obtained for the analogous xCELLigence data reinforce this instrument's reduced resolving capacity.

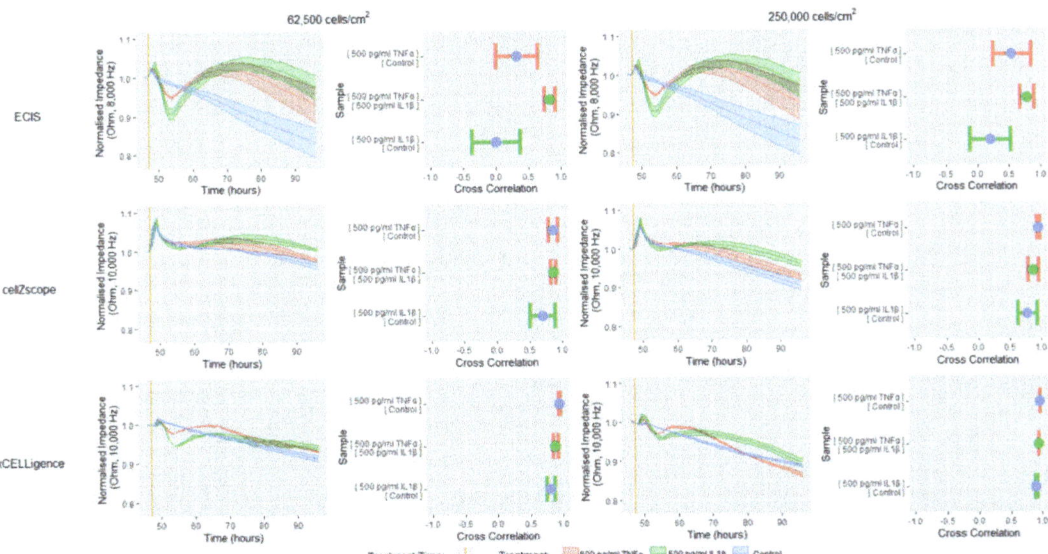

Figure 4. ECIS has a greater capacity to distinguish between different temporal impedance response profiles than cellZscope or xCELLigence. hCMVECs were seeded at either 62,500 cells/cm^2 or 250,000 cells/cm^2 and incubated for 48 h until confluent. The cells were then treated with TNFα or IL1β, and monitored using ECIS, cellZscope or xCELLigence for a further 48 h. The impedance data has been normalized at one hour before treatment and is presented as a change in impedance to allow direct comparisons between instruments to be made. Ribbon plots show the mean ± SEM of three independent experiments. Cross correlation results show the mean ± SEM of the values derived from the two temporal response profiles being compared, each of which includes data from three independent experiments. Cross-correlation is expressed as a value between 1 and −1, where 1 represents identical curves, 0 shows no correlation between the curves and −1 represents curves with a mirror image opposing profile or inverse correlation. N.B. Each treatment is indicated by the same colour in both the ribbon plots and cross correlation plots.

In contrast to ECIS and xCELLigence, the impedance data obtained using the cellZscope did not show a substantial difference between treatments during the initial 5 h, just the peak associated with adding a treatment (Figure 4). Thereafter, however, there was a slight increase in impedance following treatment with both TNFα and IL1β for both cell seeding densities, which slowly declined after approximately 65 h. Despite the subtle differences in cellZscope's temporal profiles, the cross-correlation data indicated that IL1β appeared to influence the impedance of the endothelial monolayer to a greater extent than TNFα, a trend that was consistent for all three instruments. The reduced magnitude of the differences in the impedance temporal profiles observed with cellZscope, when compared with ECIS and xCELLigence may be a result of its distinct electrode configuration and the fact that the electrodes are not in direct contact with the cells or the substrate they are grown on (Figure 1, Supplementary Table S1). Collectively, these data demonstrate that the ECIS platform has a superior capacity to distinguish between the temporal impedance profile of a control endothelial monolayer and endothelial monolayers responding to TNFα or IL1β, when compared with the cellZscope and xCELLigence platforms.

To reveal the endothelial cellular and monolayer properties that are causing the temporal changes in impedance in response to TNFα or IL1β, we next modelled the data generated using the ECIS and cellZscope (Figure 5, equivalent circuits in Supplementary Table S1). These experiments were conducted using the lower cell seeding density as the data in Figure 4, and demonstrated that the magnitudes of the impedance responses were similar for each cell seeding density tested. The data in Figure 5 has been normalized and is presented as a change in either impedance, cell-cell interactions or membrane capacitance, to allow direct comparisons between instruments to be made.

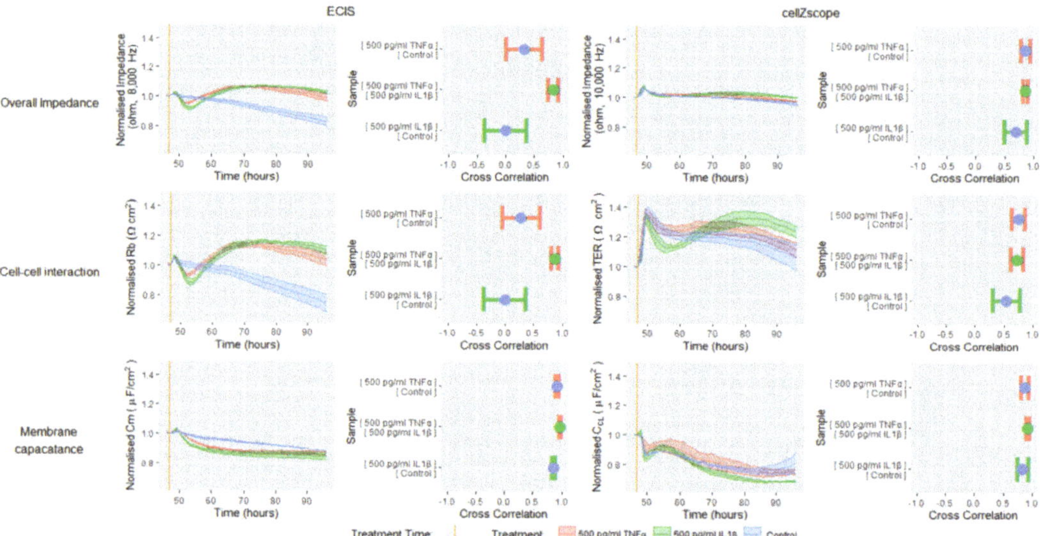

Figure 5. Modelling impedance data generated by ECIS or cellZscope reveals changes in endothelial barrier properties in response to TNFα or IL1β. hCMVECs were seeded at 62,500 cells/cm² and incubated for 48 h until confluent. The cells were then treated with TNFα, IL1β or a vehicle, and the temporal profile of impedance was monitored using ECIS or cellZscope for a further 48 h. The impedance data were modelled to provide a temporal profile of cell-cell interactions (Rb) and membrane capacitance (Cm). All data has been normalized at one hour before treatment and is presented as a change in impedance, cell-cell interactions or membrane capacitance, to allow direct comparisons between instruments to be made. Ribbon plots show the mean ± SEM of three independent experiments. Cross-correlation results show the mean ± SEM of the values derived from the two temporal response profiles being compared, each of which includes data from three independent experiments. Cross-correlation is expressed as a value between 1 and −1, where 1 represents identical curves, 0 shows no correlation between the curves and −1 represents curves with a mirror image opposing profile or inverse correlation. N.B. Each treatment is indicated by the same colour in both the ribbon plots and cross-correlation plots.

The modelled Rb and TER values in Figure 5 represent the level of interaction that exists between neighbouring endothelial cells in a monolayer, for example, the junctional molecules and cell-cell contacts [7]. The TNFα and IL1β induced impedance profiles measured by ECIS and cellZscope were mirrored by the modelled Rb and TER profiles respectively, indicating that the cell-cell interactions between the hCMVECs contribute substantially to the overall impedance measurement. Both the ECIS and cellZscope data in Figure 5 indicate that IL1β stimulates an initial weakening followed by a sustained strengthening of cell-cell interactions, relative to the control. In contrast, the effect of TNFα on cell–cell interactions differ between instruments; ECIS shows that TNFα stimulates an initial weakening followed by a sustained strengthening of cell–cell interactions, relative to the control; whilst the cellZscope profile infers that TNFα does not weaken cell-cell interactions below that of the control, and the subsequent strengthening is slight, relative to the control.

The impedance data can also be modelled to indicate the capacitance of the endothelial membrane layer. The magnitude of the changes in capacitance in response to IL1β and TNFα for both instruments was small relative to the changes in cell-cell interaction, indicating that these cytokines influenced capacitance to a lesser extent than they did the cell-cell interactions. Interestingly, the capacitance profiles in response to TNFα and IL1β differs for each instrument; ECIS shows that both proinflammatory cytokines induce a reduction in capacitance relative to the control that stabilises at approximately 65 h, whereas the cellZscope indicates that neither TNFα or IL1β stimulate a change in capacitance relative to the control until approximately 65 h when IL1β induces a decline in capacitance.

Collectively, the magnitude of the differences in measured impedance and modelled data, between treated and untreated endothelial monolayers was greater for ECIS than it was for the cellZscope data. This increased sensitivity meant that the ECIS system was able to definitively distinguish between both proinflammatory treatments and the control temporal profile, which was reinforced by the low cross-correlation values generated from these comparisons. These observations confirmed earlier findings showing that both IL1β and TNFα can influence cell-cell interactions that contribute to endothelial monolayer impedance [17]. The reduced sensitivity of the cellZscope meant it was only able to distinguish between the IL1β and control profiles; therefore it appears that this platform may not be able to definitively resolve subtle changes in endothelial monolayer properties, such as the lesser TNFα response in this study. As mentioned previously, the reduced sensitivity of the cellZscope may result from the electrodes being distant from the monolayer culture and measuring the impedance of the culture as a whole (Supplementary Table S1). In contrast, the ECIS electrodes span a large proportion of the culture surface (i.e., 3.985mm^2, Supplementary Table S1) and are therefore in direct contact with a high proportion of the cell monolayer and its underlying substrate. Collectively this widespread electrode coverage and direct culture contact likely contributes to the enhanced sensitivity of the ECIS platform.

Despite cellZscope's reduced sensitivity, it is important to note that this platform provides access to the basolateral compartment, thereby enabling studies that either need to apply treatments from beneath the monolayer, want to generate and study stratified cultures or wish to assess the transport of cells or molecules across the monolayer. The cellZscope's 24 well capacity also provides considerable scope for assessing multiple treatments in these types of studies. Although, if access to the basolateral compartment is not required, either the xCELLigence or ECIS platforms 96 well arrays may be more attractive for large-scale experiments.

4. Conclusions

The data presented in this study highlights that the instrument used to assess changes in endothelial cell monolayer properties should be carefully selected, to ensure it is appropriate for the experimental questions being addressed. Although both the ECIS and xCELLigence platforms can facilitate large-scale screening on 96 well plates with similar electrode configurations, the ECIS platform is more sensitive than xCELLigence when detecting impedance changes in response to a stimulus. Furthermore, ECIS can acquire data at multiple frequencies, which can be modelled to identify which of the endothelial barrier components contributing to impedance are being affected, something xCELLigence is unable to do because of its limited frequency acquisition range. The cellZscope instrument also acquires impedance data at multiple frequencies which can be modelled to identify changes in particular endothelial barrier components, however, the reduced sensitivity of this platform relative to ECIS means that subtle changes in endothelial monolayer properties may not be resolved to the same extent as they can be by ECIS technology. The reduced sensitivity of the cellZscope platform could be due to its distinct electrode configuration that allows access to the basolateral compartment, which is essential for certain types of experimental approaches. Ultimately, the choice of platform hinges on: (1) whether access to the basolateral compartment is required, (2) if the researcher wishes to identify which endothelial monolayer properties are being influenced and (3) whether a high degree of sensitivity is required to detect subtle changes.

Supplementary Materials: The following is available online at https://www.mdpi.com/article/10.339 0/bios11050159/s1, Supplementary Table S1: ECIS, xCELLigence and cellZscope instrument parameters.

Author Contributions: Conceptualization: J.J.W.H., C.E.A. and E.S.G.; methodology: J.J.W.H., A.A., S.J.O., C.E.A., and E.S.G.; software: J.J.W.H.; formal analysis: J.J.W.H., C.P.U., C.E.A., and E.S.G.; resources: S.J.O., C.E.A., and E.S.G.; data curation: J.J.W.H., C.P.U., and E.S.G.; writing—original draft preparation: J.J.W.H., A.A., C.E.A. and E.S.G.; writing—review and editing: J.J.W.H., A.A., S.J.O., C.P.U., C.E.A., and E.S.G.; supervision: C.E.A. and E.S.G.; project administration: J.J.W.H., C.E.A. and E.S.G.; funding acquisition: J.J.W.H., A.A., C.E.A. and E.S.G. All authors have read and agreed to the published version of the manuscript.

Funding: J.J.W.H. was supported by an Auckland Medical Research Foundation Doctoral Scholarship, and A.A. was supported by a Neurological Foundation Doctoral Scholarship. The instrument purchase was supported by the New Zealand Lottery Health Fund (E.S.G.; xCELLigence and S.O./E.S.G.; ECIS ZΘ). The cellZscope was purchased with funding from the University of Auckland (E.S.G. and S.O.). The consumables support was provided by the University of Auckland Faculty Research Development Fund (C.E.A.).

Data Availability Statement: Not applicable.

Acknowledgments: We thank the University of Auckland Statistical Consulting Centre for their input and verification of the statistical analysis conducted in this paper. We also thank Jo Dodd for her generous guidance and technical support.

Conflicts of Interest: The authors declare no conflicts of interest. There was no involvement of the funders in any role pertaining to the choice of the research project; the design of the study; in the collection, analyses or interpretation of data; in the writing of the manuscript; or in the decision to publish the results.

References

1. Giaever, I.; Keese, C.R. Micromotion of mammalian cells measured electrically. *Proc. Natl. Acad. Sci. USA* **1991**, *88*, 7896–7900. [CrossRef] [PubMed]
2. Wegener, J.; Keese, C.R.; Giaever, I. Electric cell-substrate impedance sensing (ECIS) as a noninvasive means to monitor the kinetics of cell spreading to artificial surfaces. *Exp. Cell Res.* **2000**, *259*, 158–166. [CrossRef]
3. Johnson, R.H.; Kho, D.T.; O'Carroll, S.J.; Angel, C.E.; Graham, E.S. The functional and inflammatory response of brain endothelial cells to Toll-Like Receptor agonists. *Sci. Rep.* **2018**, *8*. [CrossRef] [PubMed]
4. Keese, C.R.; Bhawe, K.; Wegener, J.; Giaever, I. Real-time impedance assay to follow the invasive activities of metastatic cells in culture. *Biotechniques* **2002**, *33*, 842–850. [CrossRef] [PubMed]
5. Anchan, A.; Kalogirou-Baldwin, P.; Johnson, R.; Kho, D.T.; Joseph, W.; Hucklesby, J.; Finlay, G.J.; O'Carroll, S.J.; Angel, C.E.; Graham, E.S. Real-Time Measurement of Melanoma Cell-Mediated Human Brain Endothelial Barrier Disruption Using Electric Cell-Substrate Impedance Sensing Technology. *Biosensors* **2019**, *9*, 56. [CrossRef]
6. Anchan, A.; Martin, O.; Hucklesby, J.J.W.; Finlay, G.; Johnson, R.H.; Robilliard, L.D.; O'Carroll, S.J.; Angel, C.E.; Graham, E.S. Analysis of melanoma secretome for factors that directly disrupt the barrier integrity of brain endothelial cells. *Int. J. Mol. Sci.* **2020**, *21*, 8193. [CrossRef]
7. Lo, C.M.; Keese, C.R.; Giaever, I. Impedance analysis of MDCK cells measured by electric cell-substrate impedance sensing. *Biophys. J.* **1995**, *69*, 2800–2807. [CrossRef]
8. Soscia, D.A.; Lam, D.; Tooker, A.C.; Enright, H.A.; Triplett, M.; Karande, P.; Peters, S.K.G.; Sales, A.P.; Wheeler, E.K.; Fischer, N.O. A flexible 3-dimensional microelectrode array for: In vitro brain models. *Lab. Chip.* **2020**, *20*, 901–911. [CrossRef]
9. Yúfera, A.; Rueda, A.; Muñoz, J.M.; Doldán, R.; Leger, G.; Rodríguez-Villegas, E.O. A tissue impedance measurement chip for myocardial ischemia detection. *IEEE Trans. Circuits Syst. I Regul. Pap.* **2005**, *52*, 2620–2628. [CrossRef]
10. Pérez, P.; Huertas, G.; Maldonado-Jacobi, A.; Martín, M.; Serrano, J.A.; Olmo, A.; Daza, P.; Yúfera, A. Sensing Cell-Culture Assays with Low-Cost Circuitry. *Sci. Rep.* **2018**, *8*, 8841. [CrossRef]
11. ECIS Z-Theta—Applied BioPhysics. Available online: https://www.biophysics.com/ztheta.php (accessed on 13 October 2020).
12. Agilent. Continuously Monitor Live Cell Changes with xCELLigence RTCA SP. Available online: https://www.agilent.com/en/ product/cell-analysis/real-time-cell-analysis/rtca-analyzers/xcelligence-rtca-sp-single-plate-741232#literature (accessed on 13 October 2020).
13. CellZscope 2—nanoAnalytics EN. Available online: https://www.nanoanalytics.com/en/products/cellzscope/cellzscope2 .html#cZs_applications (accessed on 13 October 2020).
14. Maeda, H.; Hashimoto, K.; Go, H.; Miyazaki, K.; Sato, M.; Kawasaki, Y.; Momoi, N.; Hosoya, M. Towards the development of a human in vitro model of the blood–brain barrier for virus-associated acute encephalopathy: Assessment of the time- and concentration-dependent effects of TNF-α on paracellular tightness. *Exp. Brain Res.* **2021**, *239*, 451–461. [CrossRef]

15. Maherally, Z.; Fillmore, H.L.; Ling Tan, S.; Fei Tan, S.; Jassam, S.A.; Quack, F.I.; Hatherell, K.E.; Pilkington, G.J. Real-time acquisition of transendothelial electrical resistance in an all-human, in vitro, 3-dimensional, blood–brain barrier model exemplifies tight-junction integrity. *FASEB J.* **2018**. [CrossRef]
16. Leo, L.M.; Familusi, B.; Hoang, M.; Smith, R.; Lindenau, K.; Sporici, K.T.; Brailoiu, E.; Abood, M.E.; Brailoiu, G.C. GPR55-mediated effects on brain microvascular endothelial cells and the blood–brain barrier. *Neuroscience* **2019**, *414*, 88–98. [CrossRef]
17. O'Carroll, S.J.; Kho, D.T.; Wiltshire, R.; Nelson, V.; Rotimi, O.; Johnson, R.; Angel, C.E.; Graham, E.S. Pro-inflammatory TNFα and IL-1β differentially regulate the inflammatory phenotype of brain microvascular endothelial cells. *J. Neuroinflamm.* **2015**, *12*, 131. [CrossRef]
18. Silwedel, C.; Haarmann, A.; Fehrholz, M.; Claus, H.; Speer, C.P.; Glaser, K. More than just inflammation: Ureaplasma species induce apoptosis in human brain microvascular endothelial cells. *J. Neuroinflamm.* **2019**, *16*, 1–13. [CrossRef]
19. Robilliard, L.D.; Kho, D.T.; Johnson, R.H.; Anchan, A.; O'Carroll, S.J.; Graham, E.S. The importance of multifrequency impedance sensing of endothelial barrier formation using ECIS technology for the generation of a strong and durable paracellular barrier. *Biosensors* **2018**, *8*, 64. [CrossRef]
20. Bischoff, I.; Hornburger, M.C.; Mayer, B.A.; Beyerle, A.; Wegener, J.; Fürst, R. Pitfalls in assessing microvascular endothelial barrier function: Impedance-based devices versus the classic macromolecular tracer assay. *Sci. Rep.* **2016**, *6*, 23671. [CrossRef] [PubMed]
21. Morgan, K.; Gamal, W.; Samuel, K.; Morley, S.D.; Hayes, P.C.; Bagnaninchi, P.; Plevris, J.N. Application of Impedance-Based Techniques in Hepatology Research. *J. Clin. Med.* **2019**, *9*, 50. [CrossRef] [PubMed]
22. Hillger, J.M.; Lieuw, W.L.; Heitman, L.H.; IJzerman, A.P. Label-free technology and patient cells: From early drug development to precision medicine. *Drug Discov. Today* **2017**, *22*, 1808–1815. [CrossRef] [PubMed]
23. Eigenmann, D.E.; Xue, G.; Kim, K.S.; Moses, A.V.; Hamburger, M.; Oufir, M. Comparative study of four immortalized human brain capillary endothelial cell lines, hCMEC/D3, hBMEC, TY10, and BB19, and optimization of culture conditions, for an in vitro blood-brain barrier model for drug permeability studies. *Fluids Barriers CNS* **2013**, *10*, 33. [CrossRef] [PubMed]
24. Kho, D.; MacDonald, C.; Johnson, R.; Unsworth, C.; O'Carroll, S.; Mez, E.; Angel, C.; Graham, E. Application of xCELLigence RTCA Biosensor Technology for Revealing the Profile and Window of Drug Responsiveness in Real Time. *Biosensors* **2015**, *5*, 199–222. [CrossRef] [PubMed]
25. Venables, W.N.; Ripley, B.D. Time Series Analysis. In *Modern Applied Statistics with S*; Springer: Cham, Switzerland, 2002; pp. 387–418.
26. Wickham, H. *ggplot2*; Use R! Springer International Publishing: Cham, Switzerland, 2016; ISBN 978-3-319-24275-0.
27. Schwartz, S.M.; Gajdusek, C.M. Contact inhibition in the endothelium. In *Biology of Endothelial Cells*; Springer: Boston, MA, USA, 1984; pp. 66–73.

 **biosensors**

MDPI

Article

Surface Plasmon Resonance Assay for Label-Free and Selective Detection of HIV-1 p24 Protein

Lucia Sarcina [1], Giuseppe Felice Mangiatordi [2], Fabrizio Torricelli [3], Paolo Bollella [1], Zahra Gounani [4], Ronald Österbacka [4], Eleonora Macchia [4,*] and Luisa Torsi [1,4,5]

[1] Dipartimento di Chimica, Universita' degli Studi di Bari A. Moro, 70125 Bari, Italy; lucia.sarcina@uniba.it (L.S.); paolo.bollella@uniba.it (P.B.); luisa.torsi@uniba.it (L.T.)
[2] CNR–Institute of Crystallography, 70125 Bari, Italy; giuseppe.mangiatordi@ic.cnr.it
[3] Dipartimento di Ingegneria dell'Informazione, Università degli Studi di Brescia, 25123 Brescia, Italy; fabrizio.torricelli@unibs.it
[4] Physics, Faculty of Science and Engineering, Åbo Akademi University, 20500 Turku, Finland; zahra.gounani@abo.fi (Z.G.); ronald.osterbacka@abo.fi (R.Ö.)
[5] CSGI (Centre for Colloid and Surface Science), 70125 Bari, Italy
* Correspondence: eleonora.macchia@abo.fi

Abstract: The early detection of the human immunodeficiency virus (HIV) is of paramount importance to achieve efficient therapeutic treatment and limit the disease spreading. In this perspective, the assessment of biosensing assay for the HIV-1 p24 capsid protein plays a pivotal role in the timely and selective detection of HIV infections. In this study, multi-parameter-SPR has been used to develop a reliable and label-free detection method for HIV-1 p24 protein. Remarkably, both physical and chemical immobilization of mouse monoclonal antibodies against HIV-1 p24 on the SPR gold detecting surface have been characterized for the first time. The two immobilization techniques returned a capturing antibody surface coverage as high as $(7.5 \pm 0.3) \times 10^{11}$ molecule/cm^2 and $(2.4 \pm 0.6) \times 10^{11}$ molecule/cm^2, respectively. However, the covalent binding of the capturing antibodies through a mixed self-assembled monolayer (SAM) of alkanethiols led to a doubling of the p24 binding signal. Moreover, from the modeling of the dose-response curve, an equilibrium dissociation constant K_D of 5.30×10^{-9} M was computed for the assay performed on the SAM modified surface compared to a much larger K_D of 7.46×10^{-5} M extracted for the physisorbed antibodies. The chemically modified system was also characterized in terms of sensitivity and selectivity, reaching a limit of detection of (4.1 ± 0.5) nM and an unprecedented selectivity ratio of 0.02.

Keywords: HIV-1 p24 protein; surface plasmon resonance; surface modifications; label-free detection

Citation: Sarcina, L.; Mangiatordi, G.F.; Torricelli, F.; Bollella, P.; Gounani, Z.; Österbacka, R.; Macchia, E.; Torsi, L. Surface Plasmon Resonance Assay for Label-Free and Selective Detection of HIV-1 p24 Protein. *Biosensors* **2021**, *11*, 180. https://doi.org/10.3390/bios11060180

Received: 5 May 2021
Accepted: 1 June 2021
Published: 3 June 2021

Publisher's Note: MDPI stays neutral with regard to jurisdictional claims in published maps and institutional affiliations.

1. Introduction

One of the main features of a biosensing platform is combining a high sensitivity with selectivity in the binding interactions between immobilized biorecognition species and the target analyte [1,2]. Relevantly, the design of a high throughput and reliable transducing interface in biosensors plays a pivotal role in the positive outcome of the assay. Indeed, the immobilization of bioreceptors to a surface always results in the reduction or loss of mobility. Consequently, to prevent any partial or complete loss of bioactivity, arisen from random orientation or structural deformations, bioreceptors should be attached onto surfaces without affecting conformation and functions. Indeed, the biosensor analytical figures of merit might be strongly influenced by the parameter related to the immobilization process itself [3,4].

Many efforts have been made to study suitable immobilization techniques of biorecognition elements on metal surfaces [5–8]. Some advantages may arise from the stable anchoring of biomolecules by covalent immobilization by forming chemical bonds between complementary functional groups present on the biomolecules and on the solid surface, compared to their direct adsorption on sensor surfaces [3]. For instance, by using antibody fragments or protein G mediated immobilization, a more efficient capture of the bio-recognition element has been observed, thus improving the sensitivity of immunosensing platforms [9,10]. On the other hand, physical immobilization is particularly suited to deposit biorecognition elements on various surfaces. Indeed, it does not require any additional coupling reagents or chemical modification of the biomolecules, therefore being cost-effective and faster than other immobilization techniques. Nevertheless, the resulting biofilms usually lack homogeneity, and the long-term stability of the device needs to be assessed [11]. In the present work, HIV p24 antibodies (anti-p24) by physisorption and chemical deposition through self-assembled monolayers on a 0.42 cm^2 wide gold detecting interface were characterized with surface plasmon resonance (SPR) for the first time. In particular, the detection efficacies toward human immunodeficiency virus (HIV-1) p24 capsid proteins were compared utilizing the SPR real-time monitoring of the bio-affinity reactions.

The HIV-1 p24 protein is one of the most important biomarkers for the timely and accurate diagnosis of HIV infection due to its presence in the serum or plasma as early as 4–11 days after infection, while only by weeks 3–12 of infection do the HIV host antibodies generally become detectable [12,13]. Therefore, tests that detect the p24 antigen generally allow for the timely detection of HIV infection than the ones based on host antibodies to HIV [14]. Remarkably, blood serum from individuals recently infected with HIV contains from 10 to 30,000 virions per mL, resulting in an estimated concentration of the p24 capsid antigen in the femtoMolar range (fM, 10^{-15} M) [15]. The study of new platforms for the early detection of HIV infection, through an anti-p24 biofunctionalized detecting interface, is of great interest [16], especially from the perspective of developing disposable tests, suitable as fast screening platforms, in the early stage of infection [17–19].

To this aim, multi-parameter SPR is herein proposed for the real-time study of biological interaction occurring at the biofunctionalized detecting surface [20,21] as well as a reliable and label-free detection method, achieving limits of detection comparable to the label-needing enzyme-linked immunosorbent assay (ELISA) gold standard [22]. In particular, the binding affinity constants were evaluated for both the immobilization strategies, achieving an equilibrium dissociation constant K_D of 5.30×10^{-9} M for the assay performed on the SAM modified surface, compared to a K_D of 7.46×10^{-5} M for that with physisorbed antibodies. This evidence suggests a reduced ligand affinity for the physiosorbed anti-p24 binding sites. Remarkably, the selectivity of the SPR assay in the presence of interferent species has been evaluated. Notably, the human C-reactive protein (CRP) was cross-tested for the first time, demonstrating the selectivity of the immunosensor for p24 detection, achieving an unprecedented selectivity ratio—computed as the ratio between the SPR angle-shifts—as low as 0.02. Moreover, a limit of detection (LOD) of (4.1 ± 0.5) nM was also demonstrated, falling in the same range of the LOD gathered with the label-needing ELISA gold standard and being one order of magnitude lower than the state-of-the-art limit of detection reported for HIV-1 p24 direct SPR assays. Remarkably, this study provides important pieces of information for a reliable and optimized biofunctionalization strategy suitable for further developing a wide-field bioelectronic sensor [19] to accomplish an efficient pre-symptomatic diagnosis of diseases caused by HIV infections.

2. Materials and Methods

Mouse monoclonal antibodies to HIV-1 p24 (anti-p24) and the recombinant HIV-1 p24 capsid protein (p24, molecular weight 26 kDa), expressed in *Escherichia coli*, were purchased from Abcam (Cambridge, UK). Human C-reactive protein (CRP, molecular weight 118 kDa) was purchased from Sigma-Aldrich-Darmstadt, Germany. 3-Mercaptopropionic acid (3MPA) (98%), 11-mercaptoundecanoic acid (11MUA), ethanolamine hydrochloride (EA), 1-ethyl-3-(3-dimethylamino-propyl)carbodiimide (EDC), N-hydroxysulfosuccinimide sodium salt (NHSS), and bovine serum albumin (BSA, molecular weight 66 kDa) were purchased from Sigma-Aldrich and used without further purification. A phosphate buffered saline (PBS, phosphate buffer 0.01 M, KCl 0.0027 M, NaCl 0.137 M, Sigma-Aldrich) tablet was dissolved in 200 mL HPLC water and used upon filtration on a Corning 0.22 μm polyethersulfone membrane. 2-(N-morpholino)ethane-sulfonic acid (MES) was purchased from Sigma-Aldrich; a 0.1 M buffer solution was prepared and adjusted at pH 4.8–4.9 with sodium hydroxide solution (NaOH 1 M).

The glass sensor slides (SPR Navi-200) were provided with a 50 nm gold layer on a 2 nm layer of chromium adhesion promoter. They were used after a dip cleaning in a NH_3 aq./H_2O_2 aqueous solution (1:1:5 v/v) at 80–90 °C for 10 min, then rinsed with water, dried with nitrogen, and treated for 10 min in a UV–ozone cleaner.

A BioNavis-200 multi-parameter surface plasmon resonance (MP-SPR) NaviTM instrument, in the Kretschmann configuration, was used. The SPR modulus was equipped with two laser sources (at 670 and 785 nm wavelengths) and they were both set at 670 nm, scanning an angular range of 50.29–77.93 degrees (SPR liquid range). A single channel cell was used, provided with an internal volume of 100 μL. The injections in the SPR cell were made manually with a 1 mL sterile syringe, with no automatic flow-rate setting.

All data were analyzed with Origin2018 graphing software by OriginLab Corporation.

3. Results and Discussion

3.1. Gold Layer Bio-Modification

The characterization of both physiosorbed and covalently immobilized anti-p24 capturing antibodies on a gold surface was assessed via surface plasmon resonance (SPR) characterization. To this aim, a multi-parameter SPR (MP-SPR) Navi 200-L apparatus in the Kretschmann configuration was used [23]. In Figure 1, a schematic of the apparatus is shown. Two laser beams, inspecting two different sample areas, pass through the high refractive index material (the prism), and are totally reflected at the low refractive index material (i.e., at the prism–metal layer interface) [23]. The presence of the noble metal thin film (50 nm gold layer) causes partial loss of the reflected light by exciting the metal surface electrons. This produces an evanescent wave that propagates along the interface between the dielectric (sample medium) and the metal layer [24,25]. The so-called surface plasma wave is mostly confined at the metal–dielectric boundary and decreases exponentially into both media, with higher field concentration in the dielectric [25]. Thus, the technique is very sensitive to any variation in the local refractive index on this surface, where biomolecule interactions can be inspected.

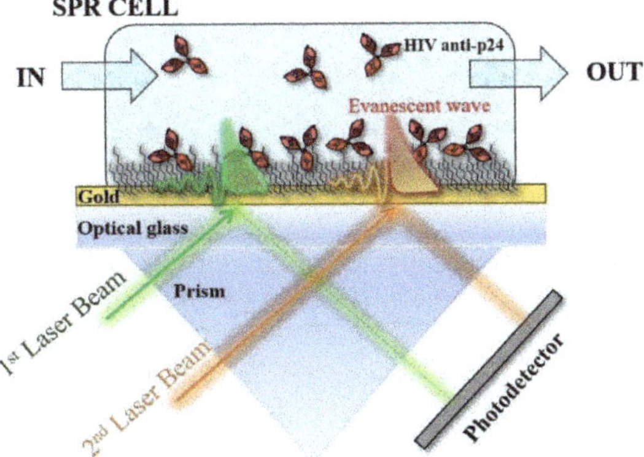

Figure 1. SPR apparatus in the Kretschmann configuration. During the biofunctionalization of gold, the green and orange laser beams (λ = 670 nm) sampled the surface in two points and serves to monitor the layer homogeneity.

The scanning of the SPR resonance angle allows the real-time monitoring of each specimen approaching the metal surface from the dielectric medium. Hence, this setup was used first to characterize the efficacy of the immobilization strategies proposed to deposit the capturing antibodies on the detecting surface, and then to study the interaction of the bio-recognition element with its cognate ligands (vide infra). The biofunctionalized SPR slide holding an area of 0.42 cm^2 was inspected in two different spots by two laser sources, both set at 670 nm (green and orange arrows in Figure 1) to estimate the layer homogeneity.

Two immobilization strategies were compared for the biofunctionalization of gold electrodes with the anti-p24 antibodies against the HIV-1 p24 capsid protein. The former consists of a physisorption of the bio-recognition elements directly on bare gold; the latter involved the chemical modification of the surface utilizing self-assembled monolayers (SAMs) of alkylthiols.

The biolayer formation on surfaces can be monitored with SPR in real-time. Thus, as shown in Figure 2, the physisorption of anti-p24 was performed in situ by scanning the plasmon peak angle vs. time. Once the baseline was established in PBS, the solution of anti-p24 antibodies at a concentration of 50 µg/mL was injected into the cell. The contact with the surface was kept for two hours, after which the equilibrium was reached between molecules deposited on gold and those in the bulk solution [26]. Then, by rinsing the cell with PBS, the unbounded residues of anti-p24 are removed. The angular shift ($\Delta\theta_{SPR}$) recorded upon anti-p24 physisorption is reported on the sensogram in Figure 2. Moreover, the subsequent deposition of BSA (100 µg/mL in PBS) on the same surface has been performed to prevent non-specific binding.

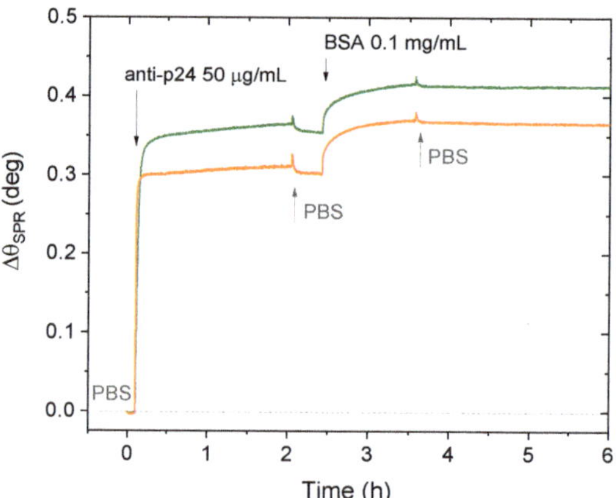

Figure 2. SPR sensogram of the anti-p24 physisorption on the gold surface and the subsequent BSA blocking. Green and orange curves refer to the surface inspection performed by the two laser beams in two different points of the sample points.

The surface coverage of physisorbed anti-p24 can be determined by using Feijter's Equation (1) [27,28]:

$$\Gamma = d \cdot (n - n_0) \cdot (dn/dC)^{-1} \tag{1}$$

where Γ, expressed in ng cm^{-2}, is the surface coverage; d is the thickness of the biolayer deposited on the gold surface; $(n - n_0)$ is the difference between the refractive index of the layer and the one of the bulk medium; and dn/dC is the so-called refractive index increment of the adsorbed biolayer [28]. Deriving the equation further to consider the instrument response, the difference in refractive index returns Equation (2):

$$(n - n_0) = \Delta\theta_{SPR} \cdot k \tag{2}$$

where $\Delta\theta_{SPR}$ is the experimental angular shift, and k is the wavelength dependent sensitivity coefficient. For laser beams with $\lambda = 670$ nm and thin layers (d < 100 nm), the following approximations hold true: (i) $dn/dC \approx 0.182$ cm^3 g^{-1}, (ii) k·d $\approx 1.0 \cdot 10^{-7}$ cm·deg [29].

Therefore, under these assumptions, and by including Equation (2) in Equation (1), the surface coverage Γ can be expressed as a function of the experimental angular shift ($\Delta\theta_{SPR}$, deg) [7,30]:

$$\Gamma = \Delta\theta_{SPR} \cdot 550 \; (ng/cm^2), \tag{3}$$

which can also be expressed in the number of molecules per cm^2, by considering the molecular weight of the species.

The average value of $\Delta\theta_{SPR}$, measured from the two curves shown in Figure 2, was used to calculate the surface coverage of the physisorbed anti-p24 antibodies. An experimental value of $\Delta\theta_{SPR} = (0.33 \pm 0.04)$ deg was registered for the physisorbed anti-p24 antibodies. Hence, by using Equation (3), the surface coverage was calculated, obtaining an average value of (181 ± 20) ng/cm^2, corresponding to $(7.5 \pm 0.3) \times 10^{11}$ molecules/cm^2. Additionally, the BSA blocking step on the sensor surface determined a further variation in the SPR angle, with a shift of $\Delta\theta_{SPR} = 0.059 \pm 0.002$ deg, corresponding to $(1.2 \pm 0.3) \times 10^{11}$ molecules/cm^2 BSA molecules. The resulting values for the surface coverage of anti-p24 and BSA layers are reported as the average among the surface coverages evaluated on two different replicates and four different sampled areas. The error bars were estimated as the relative standard deviation.

On the other hand, the chemical bonding of anti-p24 antibodies on the detecting interface foresees the chemical modification of the gold surface with a self-assembled monolayer (SAM) of mixed alkanethiols, prior to the bio-layer formation [30,31]. The mixed SAM with different chain lengths is preferable for anchoring large biomolecules like antibodies, since it provides improved accessibility for protein binding due to reduced steric hindrance [32]. To this aim, gold-coated glass slides were immersed, immediately after cleaning, in the thiol solution. A mixture of 11MUA: 3MPA (1:10 molar ratio) in ethanol was used at a final concentration of 10 mM. The sample, immersed in the thiol solution, was left overnight in a nitrogen atmosphere at room temperature. Afterward, the slide was rinsed in ethanol and mounted in the SPR sample holder.

To achieve the bio-conjugation of antibodies on the SAM, the established EDC/NHSS coupling method was used [33,34]. A scheme of the procedure is depicted in Figure 3 and extensively discussed elsewhere [30]. Briefly, the carboxylic terminal groups of the chemical SAM are converted into intermediate reactive species (NHSS, N-hydroxysulfosuccinimide esters) that react with the amine groups of the antibody, anti-p24, at a concentration of 50 µg/mL, achieving its covalent coupling. Then, the ethanolamine saturated solution (EA, at concentration 1 M) is injected to deactivate the unreacted esters in an inactive hydroxyethyl amide. Finally, to cover possible voids on the SAM and to prevent non-specific binding, a BSA solution 100 µg/mL in PBS was used [35].

The real-time monitoring of the anti-p24 chemical bonding on the mixed-SAM is reported in the sensogram of Figure 4a, showing the variation in the SPR signal for each biofunctionalization step. It is worth mentioning that any possible change in the refractive index due to the buffer composition could lower the SPR signal. To this end, to make sure to control any possible effect due to the different refractive index of the solvent involved in the biofunctionalization, a stable baseline has been recorded before each step. Specifically, for the covalent binding of the bio-recognition elements, the baseline level was established in PBS, as indicated from the bottom-up arrows in the sensogram. The succeeding injections of ethanolamine and BSA were also performed in the same buffer. In Table 1, all the details on the reagents injected and the time of exposure are reported as well as the angular shift, $\Delta\theta_{SPR}$, measured after the binding of anti-p24, the deactivation with EA, and the adsorption of BSA.

As reported in Table 2, the contact of the activated SAM with the antibodies was kept for two hours, the time required to observe the saturation of the bio-recognition elements, which reached equilibrium on the SAM modified surface. The biolayer homogeneity was also assessed for this immobilization strategy. Thus, the SPR angular shift is reported as the average signal of four replicate experiments while the error bars were evaluated as the relative standard deviation. The sensogram portion corresponding to the anti-p24 binding is highlighted in Figure 4b. Here, the variation in the SPR signal with a value of 0.13 ± 0.02 deg can be appreciated.

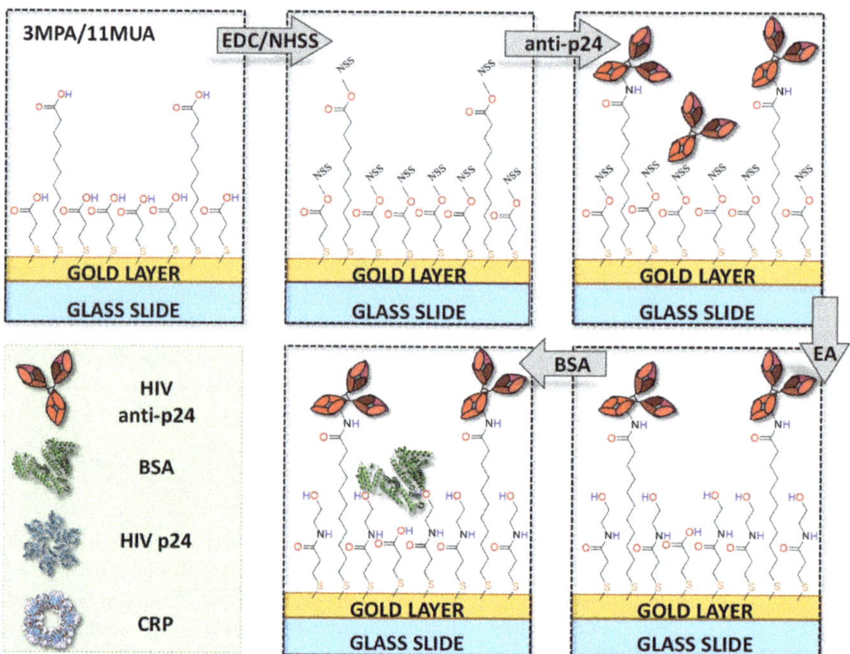

Figure 3. Schematic representation of the anti-p24 covalent bonding on the mixed SAM (3MPA/11MUA) on gold (not in scale). The inset on the left side reports a legend of all the biorecognition elements and analytes involved in the sensing and negative control experiments.

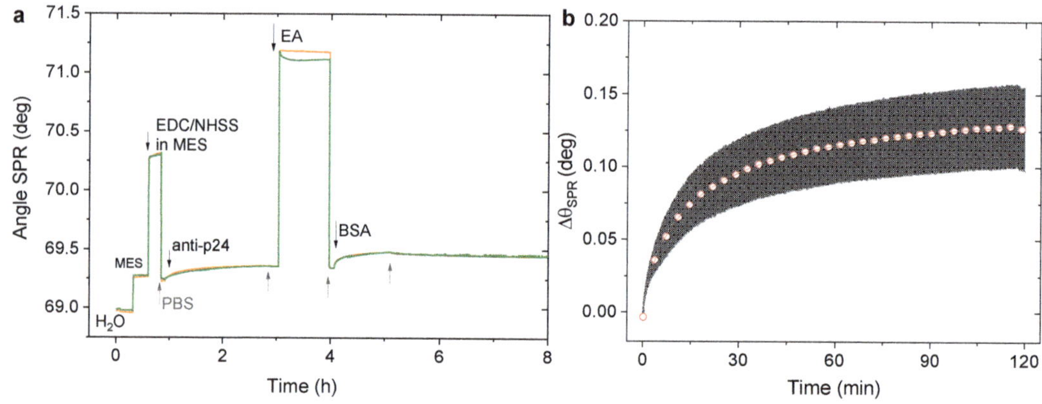

Figure 4. (**a**) SPR sensogram of the anti-p24 covalent immobilization through mixed-SAM on the gold SPR slide. Two area were sampled on the surface in each experiment, shown as green and orange curves. (**b**) Detail of the anchoring of anti-p24, reported as an average signal over four replicate experiments (red circle) along with their standard deviation (grey shadow).

Table 1. Differences in performance between multi-parameter SPR and ELISA gold standard for the detection of HIV-1 p24 proteins.

p24 Detection Method	Detection-Type	Limit of Detection	Assay Steps	Label-Needing	Assay Time
ELISA [22]	quantitative	40 nM	5	yes	at least 5 h
MP-SPR	quantitative	4 nM	2	no	<1 h

Table 2. Experimental condition used for the modification of the activated mixed-SAM in the SPR apparatus. The reagent composition, time of exposure, and SPR response is reported for each biofunctionalization step.

	Reagent	Time	$\Delta\theta_{SPR}$ (deg)
Antibodies conjugation	anti-p24 (50 µg/mL) in PBS	2 h	0.13 ± 0.02
Bond-saturation	EA (1 M) in PBS	45 min	[1] 0.11 ± 0.02
Blocking	BSA (100 µg/mL) in PBS	1 h	0.08 ± 0.01

[1] The angular shift is relevant to the anti-p24 bond after EA deactivation.

Once the bio-conjugation is completed, the surface is exposed to the ethanolamine solution to deactivate the unreacted sites on the SAM. A decrease of 15% in the angular shift was measured after EA had been registered, with the $\Delta\theta_{SPR}$ equal to 0.11 ± 0.02 deg. This is ascribable to the removal of unreacted anti-p24 antibodies after EA injection.

This experimental value can be used to determine the surface coverage of anti-p24 achieved on the SAM. Hence, by using Equation (3), a coverage of 61 ± 16 ng/cm^2 was calculated, or equivalently a value of $(2.4 \pm 0.6) \times 10^{11}$ molecule/cm^2, being 68% lower than the one achieved with the anti-p24 physisorption.

The following conclusions can be drawn from the comparison of the surface coverage accomplished with the two bio-modification methods. The physisorption of antibodies directly on a gold surface determines the formation of a thick layer of bio-recognition elements, as proven by the surface coverage evaluation. On the other hand, the grafting of antibodies on a chemical SAM gave more control on the number of sites on which the antibodies could be attached, in agreement with previous studies on the modification of gold surfaces using amine coupling on mixed SAMs reported elsewhere [31,32,34,36].

3.2. SPR Binding of p24 Proteins to the Anti-p24 Modified Gold Slides

The binding efficacy of the bio-recognition elements was evaluated against HIV-1 p24 capsid proteins (p24) for both physisorption and covalent binding immobilization strategies [37,38].

The analysis was focused on the kinetics of the SPR response registered upon p24 exposure of the physisorbed antibodies or the covalently bound ones. SPR binding of p24 proteins was registered according to the following protocol. The assay was carried out by recording the baseline in PBS and injecting p24 solutions in PBS at different concentrations ranging from 5×10^{-10} M to 1×10^{-6} M (the detailed sensograms are reported in Figures S1 and S2, respectively). Each solution was let to interact with the functionalized interface for 40 min. Upon equilibrium, the protein excess was removed, by rinsing the cell with the PBS buffer solution.

The angular shift, $\Delta\theta_{SPR}$, was calculated for each concentration as the difference between the equilibrium value after rinsing with PBS and the initial baseline. Thus, in Figure 5, the dose-curves for the assayed protein are reported as $\Delta\theta_{SPR}$ vs. [p-24] nominal concentrations (semi-log scale). Here, the response measured for the binding of p24 on physisorbed anti-p24 is shown as blue circles, while the binding with anti-p24 conjugated with the chemical-SAM is reported as red squares. Remarkably, an enhancement in the SPR signal of 65% was registered for the p24 detection occurring on the SAM modified surface, as shown in Figure 5. Indeed, although physical adsorption endows the detecting interface with a higher number of deposited capturing antibodies, it likely reduces the availability of antibody active sites, thus resulting in the lowering of the SPR response upon exposure to the antigen solution [39].

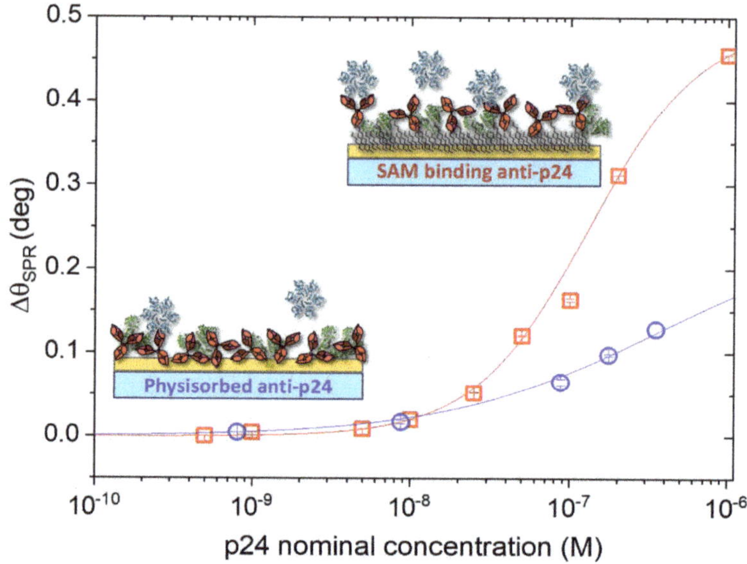

Figure 5. SPR response for the p24 binding on anti-p24 covalently bound on the chemical SAM (red squares) and for the p24 binding on physisorbed anti-p24 (blue circles). The Hill fitting model is shown as red and blue solid lines, respectively.

The kinetic analysis of the experimental data was performed according to Hill's binding model for both assays [40]. The solid lines in Figure 5 are the result of the fitting against the Hill equation:

$$Y = V_{max} \cdot \frac{X^n}{k^n + X^n} \tag{4}$$

which describes the dependence of the assay response at equilibrium, $Y = \Delta\theta_{SPR}$, from the analyte concentration, $X = [\mathrm{p24}]$.

The equation returns the Hill parameter, n, and the apparent dissociation constant, k, (i.e., the analyte concentration corresponding at half of the maximum response (V_{max}) or, equivalently, at half occupied binding sites) [40,41]. In Equation (4), the term k^n represents the equilibrium dissociation constant (K_D) of the binding pairs, which estimates the analyte/antibody binding affinity [41]. Moreover, the Hill parameter, n, reflects the degree of cooperativeness of the target molecules interacting with the available binding sites: $n = 1$ holds for a non-cooperative binding while, $n > 1$ and $n < 1$ apply for positive and negative cooperativity, respectively [42,43]. The fitting parameters found for the two assays are reported in Table 3, along with the calculated K_D.

Table 3. The parameters obtained from the Hill fitting are reported along with their standard error for both immobilization methods.

Hill Fit	Vmax	k	n	R^2	$k^n = K_D$ (M)
SAM-binding	0.492 ± 0.004	$(1.27 \pm 0.03) \cdot 10^{-7}$	1.2 ± 0.03	0.999	$5.30 \cdot 10^{-9}$
Phys-anti-p24	0.25 ± 0.01	$(3.6 \pm 0.2) \cdot 10^{-7}$	0.64 ± 0.01	0.998	$7.46 \cdot 10^{-5}$

By applying Hill's model, a value of $K_D = 5.3 \times 10^{-9}$ M was estimated when the p24 proteins bind to the chemically grafted anti-p24, which is in excellent agreement with previously reported results [44]. Indeed, the affinity binding constant of anti-p24 antibodies vs. p24 showed a $K_A = 1/K_D$ value falling in the 10^8–10^9 M^{-1} range [45]. On the other hand, the K_D calculated for the p24 dose-curve on physisorbed anti-p24 (blue trace of Figure 5) gave a value of 7.46×10^{-5} M. This result evidences a lower binding affinity between the physisorbed bio-recognition element and the analyte. By applying Hill's model, it is worth mentioning that if incompletely bound species accumulate at the detecting interface, the equation could fail to provide physicochemically correct equilibrium concentrations and/or interaction parameters [40,42]. This holds true, especially when the Hill coefficient diverges from $n = 1$, and there is no extreme positive cooperative effect among the binding pairs, leading to an overestimated K_D value [40]. Thus, because of the lower Hill coefficient, $n = 0.64 \pm 0.01$ (Table 3), obtained for the physisorbed system, some effects of the incomplete binding pairs at the sensor surface cannot be ruled out [40].

The biosensing assay, comprising the covalently bound antibodies, was further characterized in terms of selectivity and sensitivity. Indeed, the anti-p24 modified surface through chemical SAM was tested against the exposure to a non-binding protein, the human C-reactive protein (CRP). In Figure 6, the SPR responses of the antigen p24 (in red) and the non-binding CRP (in black) are shown. Both assays were performed in the same experimental conditions (vide infra). The CPR solutions in PBS at increasing concentration in the range from 5×10^{-10} M to 1×10^{-6} M were kept in contact with the modified surface for 40 min. Then, upon equilibrium, the SPR cell was rinsed with PBS and the relevant angle-shift ($\Delta\theta_{CRP}$) was measured.

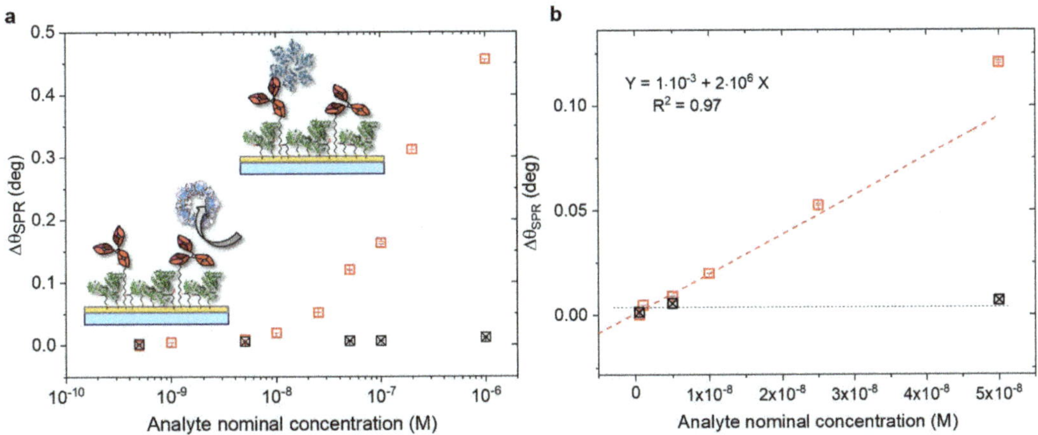

Figure 6. (a) SPR angular shift vs nominal concentration of HIV-1 p24 (red squares) and CRP (black squares) in the cross-reactivity test, performed on the SAM modified surface (semi-log scale). (b) Linear plot of SPR response performed on modified anti-p24 SAM, upon the p24 (red squares) and CRP (black squares) binding vs. analyte nominal concentration. The regression of the linear portion of p24 response is shown as the red dotted line; the average signal of the negative control is depicted as the black dotted line. The average value of three replicate analysis and their standard deviation are reported.

As observed in Figure 6a, the selectivity of the biosensing platform was successfully demonstrated. Indeed, the negative control experiment showed a maximum angle-shift below 0.01 deg, being only 3% of the signal registered for the p24 assay. Accordingly, the selectivity of the assay was estimated as the ratio between the angle-shift measured for CRP and p24 binding, respectively [46]. The resulting value was as low as $\Delta\theta_{CRP}/\Delta\theta_{p24} = 0.01/0.46 = 0.02$, which demonstrated extremely high selectivity performances [47].

The limit of detection of the assay was also evaluated. To this aim, the linear portion of the calibration curve of p24 in linear scale was considered (Figure 6b). In the same figure, the black squares are the data of the negative control experiment involving CRP. Thus, the LOD was calculated as the average signal of the negative control experiment (s_{CRP}) plus three times its standard deviation (σ_{CRP}). This signal was as high as $y = s_{CRP} + 3\sigma_{CRP} = 8 \times 10^{-3}$ deg. Hence, the comparison of this level with the interpolating linear regression of Figure 6b resulted in a LOD of (4.1 ± 0.5) nM. The LOD found in this study was one order of magnitude lower than direct SPR detection methods, which reached limits of detection most at 40 nM, depending on several factors such as the experimental configuration, the sample's optical property, and binding affinity of target molecules [20,22,47,48].

Additionally, to compare the performances of the two biofunctionalization protocols, the gold surface with physisorbed anti-p24 was further characterized, with the same method used for the SAM modified surface. In Figure 7, the SPR responses of the antigen p24 (in blue) and the non-binding CRP (in black) are shown. The physisorbed antibodies were exposed to the non-binding protein, CRP, in the range of concentration between 5×10^{-9} M and 6×10^{-7} M. The contact of the solutions with the surface was kept for 40 min, after which PBS was used to rinse the protein excess. The relevant angle-shift ($\Delta\theta_{CRP}$) was measured and compared with the response of the p24 protein. As observed in Figure 7a, the selectivity of the biosensing platform could be demonstrated. The maximum angle shift measured for the cross-reaction was below 0.02 deg, with 4% of the signal coming from the analyte. Nevertheless, the ratio between the angle-shift measured for CRP and p24 binding, respectively, resulted in a value of $\Delta\theta_{CRP}/\Delta\theta_{p24} = 0.01/0.13 = 0.08$, which implies a slightly lower selectivity compared to the chemically modified surface.

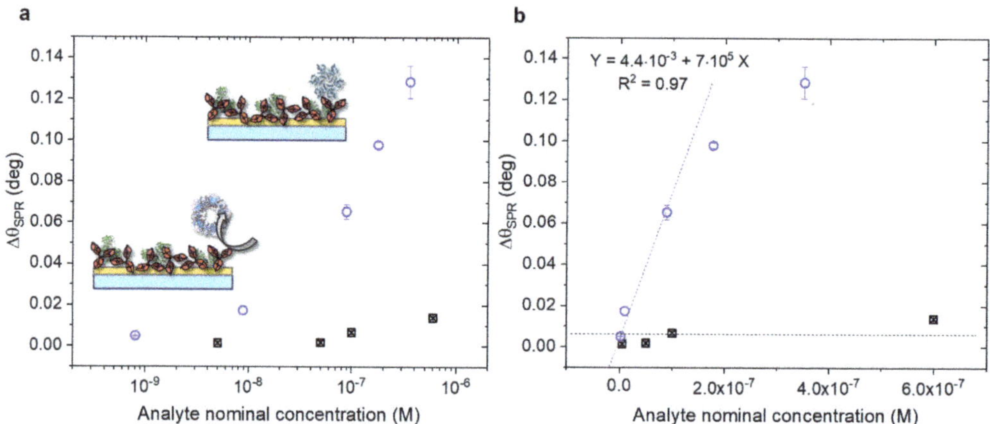

Figure 7. (**a**) SPR angular shift vs nominal concentration of HIV-1 p24 (blue circles) and CRP (black squares) in the cross-reactivity test, performed on physisorbed antibodies (semi-log scale). (**b**) Linear plot of SPR response performed on the modified surface, upon the p24 (blue circles) and CRP (black squares) binding vs, analyte nominal concentration. The regression of the linear portion of p24 response is shown as a blue dotted line; the average signal of the negative control is depicted as a black dotted line. The average value of three replicate analysis and their standard deviation are reported.

Then, to evaluate the LOD of the assay performed on the physisorbed anti-p24, the linear portion of the calibration curve of p24 in linear scale was considered (Figure 7b, blue circles). In the same figure, the value measured for the CRP assay is depicted as black squares and their average value (s_{CRP}) as a black dotted line. Thus, the LOD was calculated as the average signal of the negative control experiment (s_{CRP}) plus three times its standard deviation (σ_{CRP}). This signal was as high as $y = s_{CRP} + 3\sigma_{CRP} = 2.3 \times 10^{-2}$ deg. Hence, the comparison of this level with the interpolating linear regression of Figure 7b resulted in a LOD of (27 ± 1) nM. Relevantly, the LOD found for the SAM modified surface was one order of magnitude lower than that calculated for the physisorbed anti-p24.

The main analytical figures of merit of the two biofunctionalization methods are summarized in Table 4. The percentage relative standard deviation (RSD%) was estimated over three independent experiments on nominally identical experimental conditions. The LOD of the assays was calculated from the linear plot of the SPR response of p24 over the SAM-binding and the physisorbed antibody, respectively, as shown in Figures 6b and 7b (vide infra). The sensitivity was determined from the same plot as the slope of the linear fit. The selectivity is expressed as the ratio between the angular shift measured for the control experiment with CRP and the response of the analyte, p24, at the highest concentration assayed.

Table 4. Summary of the analytical figures of merit of the two biomodification methods with anti-p24 for the detection of the p24 protein.

	RSD (%), n = 3	LOD (nM)	Sensitivity (deg·M^{-1})	Selectivity ($\Delta\theta_{CRP}/\Delta\theta_{p24}$)
SAM-binding	1.0	4.1 ± 0.5	(1.9 ± 0.2)·10^6	0.02
Phys-anti-p24	6.2	27 ± 1	(7 ± 2)·10^5	0.08

4. Conclusions

In conclusion, the characterization of two biofunctionalization strategies for gold surfaces was performed through a multi-parameter SPR assay. The physisorption of antibodies against HIV-1 p24 (anti-p24) directly on the bare gold detecting surface led to the immobilization of $(7.5 \pm 0.3) \times 10^{11}$ molecule/cm^2. The covalent binding of anti-p24 on a mixed SAM of alkanethiols brings a decreased surface coverage of $(2.4 \pm 0.6) \times 10^{11}$ molecule/cm^2, thus being 68% lower than the one registered with physisorbed capturing antibodies.

However, the chemical immobilization endows the detecting interface with a reduced steric hindrance between the closest neighbor biorecognition elements, providing enhanced capturing efficacy toward the target analyte. Indeed, a doubled response was recorded for the latter assay. In addition, compared to the physisorbed antibodies, the covalently bound anti-p24 also resulted in a lower dissociation constant. In fact, K_D values of 7.46×10^{-5} M and 5.30×10^{-9} M were measured, respectively, highlighting a better analyte/antibody binding affinity for the assay on anti-p24 modified SAM.

The biosensing assay of both covalently bound and physisorbed anti-p24 were also characterized in terms of selectivity and sensitivity. The modified surfaces were tested against the exposure to a non-binding protein, the human C-reactive protein (CRP), for the first time and the response was compared to the p24 signal in the same range of concentrations. This allowed for the estimation of a selectivity ratio as low as 0.02, and a limit of detection of (4.1 ± 0.5) nM for the covalently bound antibodies, being one order of magnitude lower than the state-of-the-art limit of detection of 40 nM reported for the direct SPR assays and comparable to the label needing ELISA gold standard. This study thus represents a proof of principle of the early detection of HIV infection. Meanwhile, a selectivity ratio of 0.08 and a limit of detection of (27 ± 1) nM were found for the physisorbed anti-p24 assay. Moreover, this SPR characterization could pave the way toward developing reliable bio-electronic platforms in which the gold sensing electrode can be modified following the biofunctionalization strategy assessed in the present study.

Supplementary Materials: The following are available online at https://www.mdpi.com/article/10.3390/bios11060180/s1, Figure S1: SPR sensogram recorded for the real-time exposure of the HIV-1 anti-p24 modified SAM to the target protein p24. The protein concentration ranged from 500 pM to 1uM. The baseline level was established in PBS; Figure S2: SPR sensogram of the assay of HIV-1 p24 protein at increasing concentrations (0.8 nM–350 nM), performed on the physisorbed anti-p24 antibodies on gold.

Author Contributions: Conceptualization, E.M. and L.T.; methodology, E.M., Z.G. and L.S.; validation, L.S. and E.M.; formal analysis, L.S.; investigation, R.Ö., P.B., G.F.M., and F.T.; data curation, L.S., E.M. and P.B.; writing—original draft preparation, E.M. and L.S.; writing—review and editing, L.T., R.Ö., F.T. and P.B.; supervision, E.M.; funding acquisition, E.M. and L.T. All authors have read and agreed to the published version of the manuscript.

Funding: Academy of Finland projects #316881, #316883 "Spatiotemporal control of Cell Functions", #332106 "ProSiT—Protein Detection at the Single-Molecule Limit with a Self-powered Organic Transistor for HIV early diagnosis"; Biosensori analitici usa-e getta a base di transistori organici auto-alimentati per la rivelazione di biomarcatori proteomici alla singola molecola per la diagnostica decentrata dell'HIV (6CDD3786); Research for Innovation REFIN—Regione Puglia POR PUGLIA FESR-FSE 2014/2020; Dottorati innovativi con caratterizzazione industrial—PON R&I 2014–2020; "Sensore bio-elettronico usa-e-getta per l'HIV autoalimentato da una cella a combustibile biologica" (BioElSens&Fuel); SiMBiT—Single molecule bio-electronic smart system array for clinical testing (Grant agreement ID: 824946); PMGB—Sviluppo di piattaforme meccatroniche, genomiche e bioinformatiche per l'oncologia di precisione—ARS01_01195-PON "RICERCA E INNOVAZIONE" 2014–2020; Åbo Akademi University CoE "Bioelectronic activation of cell functions"; and CSGI are acknowledged for partial financial support.

Institutional Review Board Statement: Not applicable.

Informed Consent Statement: Not applicable.

Data Availability Statement: The data presented in this study are openly available in https://www.fairdata.fi/en/ (accessed on 2 June 2021). Repository Research data storage service IDA (ida.fairdata.fi).

Acknowledgments: Rosaria Anna Picca and Kyriaki Manoli are acknowledged for their useful discussions.

Conflicts of Interest: The authors declare no conflict of interest. The funders had no role in the design of the study; in the collection, analyses, or interpretation of data; in the writing of the manuscript, or in the decision to publish the results.

References

1. Njagi, J.; Kagwanja, S.M. The Interface in Biosensing: Improving Selectivity and Sensitivity. *ACS Symp. Ser.* **2011**, *1062*, 225–247. [CrossRef]
2. Yu, C.; Irudayaraj, J. Quantitative evaluation of sensitivity and selectivity of multiplex nanoSPR biosensor assays. *Biophys. J.* **2007**, *93*, 3684–3692. [CrossRef] [PubMed]
3. Makaraviciute, A.; Ramanaviciene, A. Site-directed antibody immobilization techniques for immunosensors. *Biosens. Bioelectron.* **2013**, *50*, 460–471. [CrossRef] [PubMed]
4. Vörös, J. The density and refractive index of adsorbing protein layers. *Biophys. J.* **2004**, *87*, 553–561. [CrossRef] [PubMed]
5. Zhou, C.; De Keersmaecker, K.; Braeken, D.; Reekmans, G.; Bartic, C.; De Smedt, H.; Engelborghs, Y.; Borghs, G. Construction of high-performance biosensor interface through solvent controlled self-assembly of PEG grafted polymer. In Proceedings of the 2006 NSTI Nanotechnology Conference and Trade Show—NSTI Nanotech 2006 Technical Proceedings, Washington, DC, USA, 18–20 October 2006; pp. 750–753.
6. Kyprianou, D.; Chianella, I.; Guerreiro, A.; Piletska, E.V.; Piletsky, S.A. Development of optical immunosensors for detection of proteins in serum. *Talanta* **2013**, *103*, 260–266. [CrossRef]
7. Sarcina, L.; Torsi, L.; Picca, R.A.; Manoli, K.; Macchia, E. Assessment of gold bio-functionalization for wide-interface biosensing platforms. *Sensors* **2020**, *20*, 3678. [CrossRef] [PubMed]
8. Feller, L.M.; Cerritelli, S.; Textor, M.; Hubbell, J.A.; Tosatti, S.G.P. Influence of poly(propylene sulfide-block-ethylene glycol) di- And triblock copolymer architecture on the formation of molecular adlayers on gold surfaces and their effect on protein resistance: A candidate for surface modification in biosensor research. *Macromolecules* **2005**, *38*, 10503–10510. [CrossRef]
9. Sauer-Eriksson, A.E.; Kleywegt, G.J.; Uhlén, M.; Jones, T.A. Crystal structure of the C2 fragment of streptococcal protein G in complex with the Fc domain of human IgG. *Structure* **1995**, *3*, 265–278. [CrossRef]
10. Bergström, G.; Mandenius, C.-F. Orientation and capturing of antibody affinity ligands: Applications to surface plasmon resonance biochips. *Sens. Actuators B* **2011**, *158*, 265–270. [CrossRef]
11. Wiseman, M.E.; Frank, C.W. Antibody Adsorption and Orientation on Hydrophobic Surfaces. *Langmuir* **2012**, *28*, 1765–1774. [CrossRef]
12. Branson, B.M.; Stekler, J.D. Detection of acute HIV infection: We can't close the window. *J. Infect. Dis.* **2012**, *205*, 521–524. [CrossRef]
13. Faraoni, S.; Rocchetti, A.; Gotta, F.; Ruggiero, T.; Orofino, G.; Bonora, S.; Ghisetti, V. Evaluation of a rapid antigen and antibody combination test in acute HIV infection. *J. Clin. Virol. Off. Publ. Pan Am. Soc. Clin. Virol.* **2013**, *57*, 84–87. [CrossRef] [PubMed]
14. Ly, T.D.; Ebel, A.; Faucher, V.; Fihman, V.; Laperche, S. Could the new HIV combined p24 antigen and antibody assays replace p24 antigen specific assays? *J. Virol. Methods* **2007**, *143*, 86–94. [CrossRef] [PubMed]
15. Rissin, D.M.; Kan, C.W.; Campbell, T.G.; Howes, S.C.; Fournier, D.R.; Song, L.; Piech, T.; Patel, P.P.; Chang, L.; Rivnak, A.J.; et al. Single-molecule enzyme-linked immunosorbent assay detects serum proteins at subfemtomolar concentrations. *Nat. Biotechnol.* **2010**, *28*, 595–599. [CrossRef]
16. Hurt, C.B.; Nelson, J.A.E.; Hightow-Weidman, L.B.; Miller, W.C. Selecting an HIV Test: A Narrative Review for Clinicians and Researchers. *Sex. Transm. Dis.* **2017**, *44*, 739–746. [CrossRef]

17. Zhan, L.; Granade, T.; Liu, Y.; Wei, X.; Youngpairoj, A.; Sullivan, V.; Johnson, J.; Bischof, J. Development and optimization of thermal contrast amplification lateral flow immunoassays for ultrasensitive HIV p24 protein detection. *Microsyst. Nanoeng.* **2020**, *6*. [CrossRef]

18. Sailapu, S.K.; Macchia, E.; Merino-Jimenez, I.; Esquivel, J.P.; Sarcina, L.; Scamarcio, G.; Minteer, S.D.; Torsi, L.; Sabaté, N. Standalone operation of an EGOFET for ultra-sensitive detection of HIV. *Biosens. Bioelectron.* **2020**, *156*, 1–7. [CrossRef]

19. Macchia, E.; Sarcina, L.; Picca, R.A.; Manoli, K.; Di Franco, C.; Scamarcio, G.; Torsi, L. Ultra-low HIV-1 p24 detection limits with a bioelectronic sensor. *Anal. Bioanal. Chem.* **2020**, *412*, 811–818. [CrossRef]

20. Nguyen, H.H.; Park, J.; Kang, S.; Kim, M. Surface Plasmon Resonance: A Versatile Technique for Biosensor Applications. *Sensors* **2015**, *15*, 10481–10510. [CrossRef] [PubMed]

21. Jönsson, U.; Fägerstam, L.; Ivarsson, B.; Johnsson, B.; Karlsson, R.; Lundh, K.; Löfås, S.; Persson, B.; Roos, H.; Rönnberg, I. Real-time biospecific interaction analysis using surface plasmon resonance and a sensor chip technology. *Biotechniques* **1991**, *11*, 620–627.

22. Hifumi, E.; Kubota, N.; Niimi, Y.; Shimizu, K.; Egashira, N.; Uda, T. Elimination of ingredients effect to improve the detection of anti HIV-1 p24 antibody in human serum using SPR apparatus. *Anal. Sci.* **2002**, *18*, 863–867. [CrossRef]

23. Kretschmann, E.; Raether, H. Radiative Decay of Non Radiative Surface Plasmons Excited by Light. *Z. Nat. Sect. A J. Phys. Sci.* **1968**, *23*, 2135–2136. [CrossRef]

24. Miyazaki, C.M.; Shimizu, F.M.; Ferreira, M. Surface Plasmon Resonance (SPR) for Sensors and Biosensors. *Nanocharacterization Tech.* **2017**, 183–200. [CrossRef]

25. Homola, J. Present and future of surface plasmon resonance biosensors. *Anal. Bioanal. Chem.* **2003**, *377*, 528–539. [CrossRef] [PubMed]

26. Schasfoort, R.B.M. *Handbook of Surface Plasmon Resonance*, 2nd ed.; Royal Society of Chemistry: London, UK, 2017; ISBN 9781788010283.

27. De Feijter, J.A.; Benjamins, J.; Veer, F.A. Ellipsometry as a tool to study the adsorption behavior of synthetic and biopolymers at the air–water interface. *Biopolymers* **1978**, *17*, 1759–1772. [CrossRef]

28. BioNavis MP-SPR Navi LayerSolver User Manual. Available online: https://www.bionavis.com/en/publications/biosensors/structural-and-viscoelastic-properties-layer-layer-extracellular-matrix-ecm-nanofilms-and-their-interactions-living-cells-2/ (accessed on 2 June 2019).

29. Ball, V.; Ramsden, J.J. Buffer dependence of refractive index increments of protein solutions. *Biopolymers* **1998**, *46*, 489–492. [CrossRef]

30. Holzer, B.; Manoli, K.; Ditaranto, N.; Macchia, E.; Tiwari, A.; Di Franco, C.; Scamarcio, G.; Palazzo, G.; Torsi, L. Characterization of Covalently Bound Anti-Human Immunoglobulins on Self-Assembled Monolayer Modified Gold Electrodes. *Adv. Biosyst.* **2017**, *1*, 1700055. [CrossRef]

31. Ferretti, S.; Paynter, S.; Russell, D.A.; Sapsford, K.E.; Richardson, D.J. Self-assembled monolayers: A versatile tool for the formulation of bio- surfaces. *TrAC Trends Anal. Chem.* **2000**, *19*, 530–540. [CrossRef]

32. Lee, J.W.; Sim, S.J.; Cho, S.M.; Lee, J. Characterization of a self-assembled monolayer of thiol on a gold surface and the fabrication of a biosensor chip based on surface plasmon resonance for detecting anti-GAD antibody. *Biosens. Bioelectron.* **2005**, *20*, 1422–1427. [CrossRef]

33. Sam, S.; Touahir, L.; Salvador Andresa, J.; Allongue, P.; Chazalviel, J.N.; Gouget-Laemmel, A.C.; De Villeneuve, C.H.; Moraillon, A.; Ozanam, F.; Gabouze, N.; et al. Semiquantitative study of the EDC/NHS activation of acid terminal groups at modified porous silicon surfaces. *Langmuir* **2010**, *26*, 809–814. [CrossRef]

34. Fischer, M.J.E. Amine coupling through EDC/NHS: A practical approach. *Methods Mol. Biol.* **2010**, *627*, 55–73.

35. Lichtenberg, J.Y.; Ling, Y.; Kim, S. Non-specific adsorption reduction methods in biosensing. *Sensors* **2019**, *19*, 2488. [CrossRef]

36. Frederix, F.; Bonroy, K.; Laureyn, W.; Reekmans, G.; Campitelli, A.; Dehaen, W.; Maes, G. Enhanced performance of an affinity biosensor interface based on mixed self-assembled monolayers of thiols on gold. *Langmuir* **2003**, *19*, 4351–4357. [CrossRef]

37. Vashist, S.K.; Dixit, C.K.; MacCraith, B.D.; O'Kennedy, R. Effect of antibody immobilization strategies on the analytical performance of a surface plasmon resonance-based immunoassay. *Analyst* **2011**, *136*, 4431–4436. [CrossRef]

38. Baniukevic, J.; Kirlyte, J.; Ramanavicius, A.; Ramanaviciene, A. Comparison of oriented and random antibody immobilization techniques on the efficiency of immunosensor. *Procedia Eng.* **2012**, *47*, 837–840. [CrossRef]

39. Kausaite-Minkstimiene, A.; Ramanaviciene, A.; Kirlyte, J.; Ramanavicius, A. Comparative Study of Random and Oriented Antibody Immobilization Techniques on the Binding Capacity of Immunosensor. *Anal. Chem.* **2010**, *82*, 6401–6408. [CrossRef] [PubMed]

40. Gesztelyi, R.; Zsuga, J.; Kemeny-Beke, A.; Varga, B.; Juhasz, B.; Tosaki, A.; Laubichler Gesztelyi, M.R.; Varga, B.; Juhasz, B.; Tosaki, A.; et al. The Hill equation and the origin of quantitative pharmacology. *Arch. Hist. Exact Sci* **2012**, *66*, 427–438. [CrossRef]

41. Barlow, R.; Blake, J.F. Hill coefficients and the logistic equation. *Trends Pharmacol. Sci.* **1989**, *10*, 440–441. [CrossRef]

42. Weiss, J.N. The Hill equation revisited: Uses and misuses. *FASEB J.* **1997**, *11*, 835–841. [CrossRef] [PubMed]

43. Blasi, D.; Sarcina, L.; Tricase, A.; Stefanachi, A.; Leonetti, F.; Alberga, D.; Mangiatordi, G.F.; Manoli, K.; Scamarcio, G.; Picca, R.A.; et al. Enhancing the Sensitivity of Biotinylated Surfaces by Tailoring the Design of the Mixed Self-Assembled Monolayer Synthesis. *ACS Omega* **2020**, *5*, 16762–16771. [CrossRef]

44. Rich, R.L.; Myszka, D.G. Spying on HIV with SPR. *Trends Microbiol.* **2003**, *11*, 124–133. [CrossRef]

45. Karlsson, R.; Michaelsson, A.; Mattsson, L. Kinetic analysis of monoclonal antibody-antigen interactions with a new biosensor based analytical system. *J. Immunol. Methods* **1991**, *145*, 229–240. [CrossRef]

46. Vessman, J.; Stefan, R.I.; Van Staden, J.F.; Danzer, K.; Lindner, W.; Burns, D.T.; Fajgelj, A.; Müller, H. Selectivity in analytical chemistry (IUPAC Recommendations 2001). *Pure Appl. Chem.* **2001**, *73*, 1381–1386. [CrossRef]
47. Lee, J.H.; Kim, B.C.; Oh, B.K.; Choi, J.W. Highly sensitive localized surface plasmon resonance immunosensor for label-free detection of HIV-1. *Nanomed. Nanotechnol. Biol. Med.* **2013**, *9*, 1018–1026. [CrossRef] [PubMed]
48. Valizadeh, A. Nanomaterials and Optical Diagnosis of HIV. *Artif. Cells Nanomed. Biotechnol.* **2016**, *44*, 1383–1390. [CrossRef]

 *biosensors*

Article

"Green" Prussian Blue Analogues as Peroxidase Mimetics for Amperometric Sensing and Biosensing [†]

Galina Z. Gayda [1,*], Olha M. Demkiv [1,2], Yanna Gurianov [3], Roman Ya. Serkiz [1], Halyna M. Klepach [4], Mykhailo V. Gonchar [1,4] and Marina Nisnevitch [3,*]

1   Institute of Cell Biology, National Academy of Sciences of Ukraine, 79005 Lviv, Ukraine; demkivo@nas.gov.ua (O.M.D.); roman.serkiz@lnu.edu.ua (R.Y.S.); gonchar@cellbiol.lviv.ua (M.V.G.)
2   Faculty of Veterinary Hygiene, Ecology and Law, Stepan Gzhytskyi National University of Veterinary Medicine and Biotechnologies, 79000 Lviv, Ukraine
3   Department of Chemical Engineering, Ariel University, Kyriat-ha-Mada, Ariel 4070000, Israel; yannag@ariel.ac.il
4   Faculty of Biology and Natural Sciences, Drohobych Ivan Franko State Pedagogical University, 82100 Drohobych, Ukraine; pavlishko@yahoo.com
*   Correspondence: galina.gayda@nas.gov.ua (G.Z.G.); marinan@ariel.ac.il (M.N.); Tel.: +380-322-226-2144 (G.Z.G.); +972-3-914-3042 (M.N.)
†   This paper is an extended version of our paper published in: Gayda, G.; Demkiv, O.; Gurianov, Y.; Serkiz, R.; Gonchar, M.; Nisnevitch, M. "Green" nanozymes: synthesis, characterization and application in am perometric (bio)sensors. In Proceedings of the 1st International Electronic Conference on Biosensors, 2–17 November 2020.

Abstract: Prussian blue analogs (PBAs) are well-known artificial enzymes with peroxidase (PO)-like activity. PBAs have a high potential for applications in scientific investigations, industry, ecology and medicine. Being stable and both catalytically and electrochemically active, PBAs are promising in the construction of biosensors and biofuel cells. The "green" synthesis of PO-like PBAs using oxido-reductase flavocytochrome b_2 is described in this study. When immobilized on graphite electrodes (GEs), the obtained green-synthesized PBAs or hexacyanoferrates (gHCFs) of transition and noble metals produced amperometric signals in response to H_2O_2. HCFs of copper, iron, palladium and other metals were synthesized and characterized by structure, size, catalytic properties and electro-mediator activities. The gCuHCF, as the most effective PO mimetic with a flower-like micro/nano superstructure, was used as an H_2O_2-sensitive platform for the development of a glucose oxidase (GO)-based biosensor. The GO/gCuHCF/GE biosensor exhibited high sensitivity (710 A $M^{-1}m^{-2}$), a broad linear range and good selectivity when tested on real samples of fruit juices. We propose that the gCuHCF and other gHCFs synthesized via enzymes may be used as artificial POs in amperometric oxidase-based (bio)sensors.

Keywords: artificial enzymes; green synthesis; hexacyanoferrates of transition and noble metals; peroxidase mimetic; amperometric (bio)sensor; glucose oxidase; glucose analysis

Citation: Gayda, G.Z.; Demkiv, O.M.; Gurianov, Y.; Serkiz, R.Y..; Klepach, H.M.; Gonchar, M.V.; Nisnevitch, M. "Green" Prussian Blue Analogues as Peroxidase Mimetics for Amperometric Sensing and Biosensing . *Biosensors* **2021**, *11*, 193. https://doi.org/10.3390/bios11060193

Received: 30 April 2021
Accepted: 9 June 2021
Published: 10 June 2021

1. Introduction

Artificial enzymes are stable and low-cost mimetics of natural enzymes. The search for effective novel artificial enzymes, especially nanozymes, and the development of simple methods for their synthesis and characterization, as well as the selection of novel branches for their application, are currently challenging problems in different fields of biotechnology, industry, and medicine [1–9].

Peroxidase (PO) mimetics are the most frequently investigated artificial enzymes [10–12]. One of the well-known effective PO-like artificial enzymes is Prussian blue (PB) or iron(III) hexacyanoferrate (FeHCF). PB is a member of a well-documented family of synthetized coordination compounds with an extensive 300-year history [13–16]. PB and its analogs (PBAs) are cheap and easy to synthesize, environmentally friendly, and have potential applications for basic research and industrial purposes [12–18] in a large variety of fields,

267

particularly in medicine [13,19–23]. Despite their multifunctionality, PBAs have complicated compositions, which are largely dependent on the synthesis methods and storage conditions [14–16]. Insoluble PB can be described by the formula $Fe_4[Fe(CN)_6]_3$, while $KFe[Fe(CN)_6]$ corresponds to a colloidal solution of PB. The general formula of hexacyanoferrate (HCF) is $M_k[Fe(CN)_6] \times H_2O$, where M is a transition metal [14,15].

Due to their capability to insert various ions as counter-ions during the redox process, PB and PBAs have attracted increasing interest as electrode materials for energy storage in fuel cells [15,24,25]. Having remarkable super-magnetic properties, redox and PO-like activities, PBAs are widely applied in bioreactors for detoxification of dangerous chemicals [13,17,26], in molecular magnets, and in optical and electrochemical biosensors [12–16,24,27].

The first report of electrochemical reduction of H_2O_2 on PB-modified electrodes was published by Itaya in 1984 [28]. In 2000, Karyakin named PB as an "artificial PO" and published numerous reports concerning PB-based amperometric biosensors (ABSs) [24,29–33]. Numerous other scientific groups, especially from China, have also worked diligently on this problem [18–23,34–37].

PBAs are usually obtained via various techniques, including chemical [12–16] and biological methods [38–40]. The biosynthesis of materials using plants, microorganisms and their metabolites as biosurfactants can be related to "green synthesis (GS) [41–44]. Purified enzymes were also shown to be capable of reducing metal ions to obtain metallic nanoparticles [40,45,46].

The application of green-synthesized PBAs (gPBAs or gHCFs) for the construction of ABSs is not yet well documented. The main advantages of green-synthesized nanomaterials (gNMs) are the low energy cost of their synthesis, lack of toxic chemicals, simplicity of procedure, high adaptability of the synthesized gNM, and the presence of functional groups on their surface. The latter is promising for simple immobilization of bioorganic molecules, including enzymes, during biosensor construction [40,47]. If the gNM has additional catalytic properties, it plays a dual role in biosensors, simultaneously serving as the carrier of bio-elements and as the enzyme mimetic (nanozyme).

In our previous research, we reported obtaining gHCFs of transition metals using the purified yeast enzyme flavocytochrome b_2 (Fcb_2; L-lactate: ferricytochrome c oxidoreductase, EC 1.1.2.3). The structure, size, composition, electro-catalytic properties, electro-mediator activity, and PO-like properties of the obtained gHCFs, which were synthesized via an enzyme and incorporated with it, were characterized. A more detailed study was performed on copper hexacyanoferrate (gCuHCF or gCuPBA), which was found to be the most effective PO mimetic. When immobilized on a GE, the gCuHCF under special pH conditions and working potential gave the intrinsic amperometric response to hydrogen peroxide. We demonstrated that the synthesized gCuHCF may be successfully used as an artificial PO for sensor analysis of hydrogen peroxide in a real disinfectant sample [40].

In the current work, we describe in more detail the synthesis and characteristics of new gHCFs of transition and noble metals with PO-like activity, an additional structural study of the most effective gCuHCF, development of an improved and highly sensitive ABS using glucose oxidase (GO) and gCuHCF, and testing of the constructed GO/gCuHCF ABS for glucose analysis in real samples of fruit juices.

2. Materials and Methods

2.1. Reagents

Potassium ferricyanide ($K_3Fe(CN)_6$), iron(III) chloride ($FeCl_3 \times 4H_2O$), copper(II) sulfate ($CuSO_4$), Cerium(IV) sulfate tetrahydrate $Ce(SO_4)_2 \times 4H_2O$, palladium chloride ($PdCl_3$), cobalt(II) chloride ($CoCl_2 \times 6H_2O$), zinc(II) sulfate ($ZnSO_4$), manganese(II) chloride ($MnCl_2 \times 4H_2O$), cadmium(II) chloride ($CdCl_2$), neodymium(III) chloride ($NdCl_3$), 2,2′-azinobis (3-ethylbenzothiazoline-6-sulfonate) diammonium salt (ABTS), o-dianisidine, hydrogen peroxide (H_2O_2, 30%), sodium ethylenediaminetetraacetate (EDTA), sodium

L-lactate, Nafion (5% solution in 90% low-chain aliphatic alcohols) and all other reagents and solvents used in this work were purchased from Sigma-Aldrich (Steinheim, Germany). All reagents were analytical grade and were used without further purification. All solutions used ultra-pure water prepared with the Milli-Q® IQ 7000 Water Purification system (Merck KGaA, Darmstadt, Germany).

2.2. Enzymes

Flavocytochrome b_2 (Fcb_2) was isolated from the yeast *Ogataea (Hansenula) polymorpha* 356 and purified, as described earlier [48,49]. The Fcb_2 (20 U·mg^{-1}) was stored at -10 °C in a suspension of 70% ammonium sulfate, prepared with 50 mM phosphate buffer, pH 7.5, containing 1 mM EDTA and 0.1 mM dithiothreitol. To prepare a fresh solution, the enzyme was precipitated from the suspension by centrifugation (10,000 rpm, 10 min, 4 °C) and dissolved in 50 mM phosphate buffer, pH 7.5, up to 50 U·mL^{-1}. An assay of Fcb_2 activity in solution was performed as described earlier [48,49]. One unit of the enzyme activity was defined as the amount of enzyme that oxidizes 1 μmol of L-lactate in 1 min under standard assay conditions (20 °C; 30 mM phosphate buffer, pH 7.5; 0.33 M L-lactate; 0.83 mM $K_3Fe(CN)_6$; 1 mM EDTA).

A commercial lyophilized horseradish peroxidase (PO or HRP, EC 1.11.1.7) from *Armoracia rusticana* (Aster, Lviv, Ukraine) with 600 U·mg^{-1} activity was dissolved in 20 mM phosphate buffer, pH 6.0, up to 400 U·mL^{-1}.

A commercial lyophilized glucose oxidase (GO, EC 1.1.3.4) from *Asperigillus niger* (Sigma, St. Louis, MO, USA) with an activity of 100,000 U·g^{-1} in a solid form was dissolved in 20 mM phosphate buffer, pH 6.0, up to a concentration of 0.1 mg·mL^{-1}. GO activity was assayed in a reaction mixture containing 0.16 mM *o*-dianisidine, 1.61% (*w/v*) glucose and 2 U mL^{-1} of PO in 50 mM sodium acetate buffer (NaOAc), pH 5.0, as described earlier [50].

2.3. Synthesis of Hexacyanoferrates

Synthesis of gHCF was carried out according to the scheme presented in Figure 1 [40]. A reaction mixture containing 6 mM $K_3[Fe(CN)_6]$, 20 mM sodium lactate, 0.03–0.15 U mL^{-1} Fcb_2 in 50 mM phosphate buffer, pH 8.0, was prepared and incubated at 37 °C for 30 min. Formation of gHCF was initiated by the addition of salt to a final concentration of 10–100 mM.

$$\text{L-Lactate} \xrightarrow{\ \textit{Fcb}_2\ } \text{Pyruvate} + 2e + 2H^+$$

$$2[Fe(CN)_6]^{3-} + 2e \longrightarrow 2[Fe(CN)_6]^{4-} \qquad (1)$$

$$4M^{n+} + n[Fe(CN)_6]^{4-} \longrightarrow M_4[Fe(CN)_6]_n \qquad (2)$$

Figure 1. Scheme of green hexacyanoferrate synthesis using flavocytochrome b_2 (Fcb_2) in enzymatic (**1**) and chemical (**2**) reactions; M—metal.

To obtain chemically synthesized HCFs (chHCFs), a solution of 6 mM $K_3Fe(CN)_6$ and 60 mM transition metal salt in 50 mM phosphate buffer, pH 8.0, was mixed with H_2O_2, added dropwise up to 100 mM. After 0.5–10 min incubation, the resulting mixture was fractionated by centrifugation at 13,000 rpm for 1 min, and the precipitate was resuspended in water. The centrifugation–redispersion procedure was repeated 2–4 times. The obtained HCFs were resuspended in water and kept at +4 °C until used.

2.4. Characterization of the Synthesized HCFs

2.4.1. Optical Properties

The optical properties of the synthesized HCFs, their concentrations and PO-like activities were characterized using a Shimadzu UV1650 PC spectrophotometer (Kyoto, Japan).

2.4.2. Scanning Electron Microscopy (SEM)

Morphological analyses of the samples were performed using a SEM microanalyzer REMMA-102-02 (Sumy, Ukraine). The samples of different dilutions (2 µL) were dropped onto the surface of a silicon wafer and dried at room temperature. The distance from the last lens of the microscope to the sample (WD) ranged from 17.1 to 21.7 mm. The accelerator voltage was in the range of 20 to 40 eV.

2.4.3. FTIR Analysis

The infrared spectra were prepared using the KBr pellet technique, by thoroughly mixing 3 µL of a particle suspension with 0.2 g of KBr and pressing at 5 ton$_f$ using a hydraulic press (Carver$^®$ Inc., Wabash, IN, USA). The samples were dried in a desiccator overnight and analyzed by the Spectrum^{TM} One FTIR Spectrometer (Perkin Elmer, Waltham, MA, USA) at room temperature in the 4000–400 cm^{-1} range at an operation number of 20 scans, a resolution of 4.0 cm^{-1}, and a scanning interval of 1 cm^{-1}.

2.4.4. Particle Counter Analysis

Particle concentration was measured using a particle counter (Spectrex Corp., Redwood, CA, USA) in a round-shaped 150 mL transparent glass bottle with a wall thickness of 2 mm. A total of 10 µL of the sample was added to the bottle with 99 mL of water (HPLC grade, Bio-Lab Ltd., Jerusalem, Israel) under continuous stirring. Particle counting was performed with a laser diode at a wavelength of 650 nm.

2.4.5. Dynamic Light Scattering (DLS) Analysis and Zeta-Potential Measurements

The DLS analysis and zeta-potential measurements were performed using a Litesizer 500 type BM10 instrument (Anton Paar GmbH, Graz, Austria) at 25 °C. For measurement of hydrodynamic diameters, the samples were diluted to 1:150, 1:300, and 1:600 with HPLC-grade water, placed into a semi-micro quartz cell, and analyzed using a laser at a wavelength of 660 nm and a side scatter of 90°. Zeta-potential was measured in diluted colloidal solutions at a particle concentration of 1.33×10^4 mL^{-1}, which was determined as described in Section 2.4.4. The solutions were injected into an omega-shaped cuvette and analyzed at an operating voltage of 200 V.

2.4.6. X-ray Diffraction (XRD) Analysis

The phase composition of synthesized particles was studied by XRD analysis using a Rigaku SmartLab SE X-ray powder diffractometer with Cu Kα radiation (λ = 0.154 nm) for phase identification. Full-pattern identification was carried out by a SmartLab Studio II software package, version 4.2.44.0 from the Rigaku Corporation (Tokyo, Japan). Materials identification and analysis were performed by the ICDD base PDF-2 Release 2019 (Powder Diffraction File, ver. 2.1901). XRD patterns were obtained using 40 kV, 30 mA by $\Theta/2\Theta$ (Bragg-Brentano geometry) in the 2Θ range of 10–90° (step size 0.03° and speed 4°/min). The crystallite size was calculated using quantitative analysis based on the Halder–Wagner method, with the help of the program Powder XRD plugin of SmartLab Studio II x64 v4.2.44.0.

2.5. Assay of Enzyme-Like Activities of the Synthesized HCFs in Solution

PO-like activity of the HCFs was measured by the colorimetric method, with *o*-dianisidine and ABTS as chromogenic substrates in the presence of H_2O_2. One unit (U) of PO-like activity was defined as the amount of HCF releasing 1 µmol H_2O_2 per 1 min at 30 °C under standard assay conditions. To estimate special enzyme-like activity (U/mg), the HCFs were dried. The tested solution/suspension was prepared by weighing the solid substance and adding water until the needed concentration was obtained.

The assay of PO-like activity with *o*-dianisidine: 10 µL of the aqueous suspension of HCF (1 mg mL^{-1}) was incubated in a glass tube with 1 mL of 0.17 mM *o*-dianisidine in water (as a control), and with the same substrate in the presence of 8.8 mM H_2O_2 (as a substrate for PO). The addition of NPs to the substrate stimulated the development of an orange color over time, indicating an enzymatic reaction. The enzyme-mimetic activity could be assessed qualitatively with the naked eye and was measured quantitatively with a spectrophotometer. After incubation for an exact time (1–10 min) at 30 °C, and upon the appearance of the orange color, the reaction was stopped by the addition of 0.26 mL 12 M HCl. The generated color was determined at 525 nm using a spectrophotometer. The millimolar extinction coefficient (ε) of the resulting pink dye in the acidic solution was 13.38 mM^{-1}·cm^{-1}.

The assay of PO-like activity with ABTS: 10 µL of the aqueous suspension of HCF was incubated in a 1 mL quartz cuvette with 1 mM ABTS in water (as a chromogenic substrate for oxidase), and with the same substrate in the presence of 12 mM H$_2$O$_2$ (as a substrate for PO-like HCF). The addition of HCF to the corresponding substrate (ABTS for oxidase-like HCF, ABTS with H$_2$O$_2$ for PO-like HCF) stimulated the development of a green color over time, indicating an enzymatic reaction. The enzyme-mimetic activity could be assessed with the naked eye and was measured quantitatively with a spectrophotometer. The speed of appearance of a green color was monitored at 420 nm over time using a spectrophotometer, thus enabling calculation of the enzyme-like activity. The coefficient ε of the resulting green dye was 36.0 mM^{-1}·cm^{-1}.

2.6. Sensor Evaluation

2.6.1. Apparatus and Measurements

The amperometric sensors were evaluated using constant–potential amperometry in a three-electrode configuration with an Ag/AgCl/KCl (3 M) reference electrode, a Pt-wire counter electrode, and a working graphite electrode. Graphite rods (type RW001, 3.05 mm diameter) from Ringsdorff Werke (Bonn, Germany) were sealed in glass tubes using epoxy glue for disk electrode formation. Before sensor preparation, the graphite electrode (GE) was polished on emery paper and on a polishing cloth using decreasing sizes of alumina paste (Leco, Germany). The polished electrodes were rinsed with water in an ultrasonic bath.

Amperometric measurements were carried out using a potentiostat CHI 1200 A (IJ Cambria Scientific, Burry Port, UK) connected to a personal computer, performed in a batch mode under continuous stirring in an electrochemical cell with a 20 mL volume at 25 °C.

All experiments were carried out in triplicate trials. Analytical characteristics of the proposed electrodes were statistically processed using the OriginPro 8.5 software. Error bars represent the standard error derived from three independent measurements. Calculation of the apparent Michaelis–Menten constants (K_M^{app}) was performed automatically by this program according to the Lineweaver–Burk equation.

2.6.2. Immobilization of HCFs and the Enzyme onto Electrodes

The HCFs and enzymes were immobilized on the GEs using the physical adsorption method.

For the development of the HCF or PO-based electrode, 5 µL of HCF or 5 µL of enzyme solution was dropped onto the surface of bulk GEs. After drying for 10 min at room temperature, the layer of HCF or enzyme on the electrode was covered with 10 µL of Nafion. The modified electrodes were rinsed with corresponding buffers and kept in these buffers at 4 °C until used.

To fabricate the glucose oxidase (GO)-based biosensor, 8 µL of GO solution (5 U/mL) was dropped onto the dried surface of the gCuHCF-modified GE. The dried composite was covered by a Nafion membrane. The coated bioelectrode was rinsed with water and stored in 50 mM phosphate buffer, pH 6.0, until used.

3. Results and Discussion

3.1. gHCFs-Modified Electrodes for Hydrogen Peroxide Sensing

According to the literature, chemically synthesized HCFs (chHCFs) of Fe (III), Mn (II) and Cu (II) demonstrate significant PO-like activity in solution and on electrodes [13–16,29,31]. In the current work, several gHCFs were obtained via Fcb_2 from the corresponding salts (Fe, Cu, Pd, Ce, Mn, et al.) and from $K_4Fe(CN)_6$, a product of $K_3Fe(CN)_6$ reduction by L-lactate in the presence of an enzyme (Figure 1). Our first task was to screen the obtained gPBAs for their sensitivity to H_2O_2 on amperometric graphite electrodes (GEs) and to select the best compounds as PO mimetics. For this purpose, the optimal conditions for the amperometric experiments were investigated. The amperometric characteristics of the control GE (not modified with gHCF) as a chemosensor for H_2O_2 were tested using cyclic voltammetry (CV) analysis. Selection of the optimal pH, working potential and scan rate was carried out according to the CV results (data not shown).

Under the experimentally chosen optimal conditions (50 mM NaOAc buffer, pH 4.5 and -50 mV as the working potential), numerous electrodes modified with the synthesized HCFs were screened for their ability to decompose hydrogen peroxide. A low working potential is necessary in order to avoid the effect of possible interfering substances on the electrode's response in the presence of oxygen. This requirement is relevant for the construction of biosensors and their exploitation for the analysis of real samples (food products, biological liquids, and others).

The electrocatalytic activities of the synthesized HCFs immobilized on the surface of GEs were tested by CV and chronoamperometry, as described in Section 2.6.1. The amperometric responses of different HCF/GEs to the added H_2O_2 were compared. Following the chronoamperograms, calibration curves were plotted for H_2O_2 determination by the developed electrodes (Figure 2 and Figure S1). The linear ranges and sensitivities of the electrodes modified with HCF were calculated. The analytical characteristics of the developed HCF/GEs, as deduced from the graphs (Figure 2 and Figure S1), are summarized in Table 1.

Modification of GEs with the gHCFs improved the efficiency of electron transfer due to the increase in the electrochemically accessible electrode surface area. It is worth mentioning that in comparison to native PO, several gHCF/GEs displayed higher current responses (I_{max}) to H_2O_2 at substrate saturation and higher sensitivities (Table 1). The enhancement of current outputs and sensitivities of the electrodes modified with other gHCFs were insignificant. Thus, gCuHCF, gFeHCF, gPdHCF and gCeHCF, when immobilized on graphite electrodes, demonstrated higher PO-like activities in comparison with other gHCFs, as well as with native PO and chemically synthesized chCuHCF (Table 1, Figure 2 and Figure S1). For the most effective electrode (gCuHCF/GE), the current response (I_{max}) to H_2O_2 at substrate saturation was five-fold higher, and the sensitivity was 29-fold higher than those of the PO/GE (Table 1).

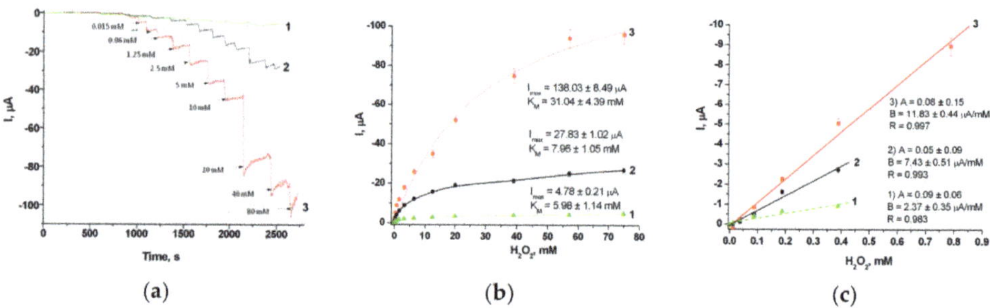

Figure 2. Amperometric characteristics of the modified electrodes: chronoamperograms (**a**), dependences of the response on increasing concentrations of H_2O_2 (**b**), and calibration graphs (**c**) for PO/GE (1), gFeHCF/GE (2), and gCuHCF/GE (3). Conditions: working potential -50 mV versus Ag/AgCl (reference electrode), 50 mM NaOAc buffer, pH 4.5 at 23 °C.

Table 1. Comparative analytical characteristics of HCFs as artificial peroxidases on graphite electrodes.

Sensitive Film	$K_M{}^{app}$, mM	I_{max}, μA	Linear Range, Up to, mM	Sensitivity, A M^{-1}m^{-2}
gCuHCF	31.0 ± 4.4	138.0 ± 8.5	0.8	1620
gPB	8.0 ± 1.1	27.8 ± 1.0	0.4	1090
gPdHCF	33.1 ± 3.9	62.4 ± 3.0	0.8	697
gCeHCF	3.5 ± 0.4	27.3 ± 0.8	3.2	560
PO	4.9 ± 1.1	5.0 ± 0.2	0.4	352
gYHCF	10.1 ± 0.9	21.6 ± 1.1	3.1	214
gCoHCF	9.3 ± 0.9	17.2 ± 1.0	0.8	159
chCuHCF	20.0 ± 3.5	6.5 ± 0.4	0.8	110
gMnHCF	92.3 ± 15.2	21.1 ± 2.1	0.8	98
gZnHCF	25.5 ± 2.2	4.0 ± 0.2	6.5	22
gNdHCF	21.3 ± 1.7	3.1 ± 0.1	6.5	16
gCdHCF	40.0 ± 5.4	2.6 ± 0.2	1.5	15

As seen, the results presented in Table 1 supported the gCuHCF/GE as the most effective PO mimetic. It was therefore studied in more detail.

Many of the reported H_2O_2-sensitive PBA-based sensors have sensitivities similar to the developed gCuHCF/GE sensor (1620 A M^{-1}m^{-2}) [40]. For example, a PB-modified glassy carbon electrode (GCE) demonstrated sensitivity of 2000 A M^{-1}m^{-2} [51], MnPBA/GCE—1472 A M^{-1}m^{-2} [37]. Graphite-paste electrodes, modified with Ni-FePBA and Cu-FePBA, showed sensitivities of 1130 and 2030 A M^{-1}m^{-2}, respectively [52]. Diamond-boron doped (DBD) electrodes, modified with PB and Ni-FePBA, demonstrated sensitivities of 2100 and 1500 A M^{-1}m^{-2}, respectively [53].

Other H_2O_2-sensitive sensors that contain PBA, coupled with other nanomaterials (carbon, graphene, natural polysaccharides, or synthetic polymers), demonstrated significantly higher sensitivities (from 3–5-fold [16,27,29,32–34] up to 300-fold [54]) compared with the gCuHCF/GE. The main peculiarities of the described sensors were high stability, sensitivity, and selectivity towards H_2O_2 in extra-wide linear ranges. These properties led to the successful use of the PBAs in oxidase-based biosensors [29–33,35,36,40,51,54–56].

The results obtained by us indicated that the gCuHCF and other gHCFs may have a potential for use as PO-like composites for the construction of amperometric oxidase-based biosensors.

3.2. Study of Structure, Morphology, and Size of the gCuHCF Composite

The size, morphology, and composition of any materials, especially of NPs, are considered as their basic parameters. A number of noninvasive label-free methods were developed for the characterization of different materials: scanning electron microscopy (SEM), transmission electron microscopy (TEM), dynamic light scattering (DLS), Fourier transform infrared spectroscopy (FTIR), X-ray diffraction (XRD) analysis, Raman spectroscopy, atomic force microscopy (AFM) and other approaches. FTIR spectroscopy allows rapid acquisition of a biochemical fingerprint of the sample under investigation, giving information on its main biomolecule content. DLS allows the rapid determination of diffusion coefficients and also provides information on relaxation time distribution for the macromolecular components of complex systems and their hydrodynamic diameters. XRD provides information regarding the crystallographic structure of a material based on incident X-ray irradiation of the material and measurement of scattering angles and intensities of X-rays leaving the sample. SEM produces images of a sample by scanning the surface with a focused beam of electrons and gives information about the surface topography and composition of the sample. The diversity and ambiguity of green-synthesized materials necessitate the use of multiple techniques for valid characterizations. In our study, the synthesized catalytically active organic-inorganic composite gCuHCF was examined using FTIR, DLS, XRD and SEM (see Section 2.4).

3.2.1. FTIR Characterization

The FTIR spectrum of the sample is presented in Figure S2. The FTIR spectrum was described in detail in our previous work [40]; it demonstrates the presence of the following groups: O-H, N-H, C≡N, C-H, C-O, C-N, Fe-C≡N and H_2O-Cu-CN. Hydroxyl groups were identified by the bands at 3456 and 3050 cm^{-1}, which are related to O-H stretching vibrations; and at 1398 cm^{-1}, which corresponds to O-H bending [57]. Amine groups were determined by the bands at 3437 and 2994 cm^{-1} (primary amine stretching) and at 1638 cm^{-1} (assigned to N-H bending) [57]. The 2875 cm^{-1} band was attributed to C-H stretching; the 1476, 790 and 719 cm^{-1} bands corresponded to C-H bending [57]. The signals at 1122 and 1109 cm^{-1} can be explained by stretching vibrations of C-O and C-N groups, respectively [57]. The presence of C≡N groups was confirmed by the band at 2105 cm^{-1}, reflecting stretching vibrations of this group [58]. The bands in the fingerprint region in the 509–667 cm^{-1} range can be related to Fe-CN linear bending, and the band at 468 cm^{-1} to Fe-C stretching [58]. The 2010 cm^{-1} band indicates the presence of a H_2O-Cu-CN moiety [58]. The results of FTIR showed the presence of copper cyanoferrate particles enveloped by an organic layer with hydroxyl and amine groups, probably of protein origin.

3.2.2. DLS Studies

The main results of the DLS measurements were described in detail in our previous work [40]. The DLS demonstrated heterogeneous mean hydrodynamic diameters of the particles in a tested gCuHCF. It is worth mentioning that very large differences in hydrodynamic diameters were found for various dilutions of the sample. In the most concentrated sample, only one size fraction was detected. There were probably larger agglomerates of particles in the concentrated suspensions that could not be measured by the designated instrument since the upper limit of measurement was 10,000 nm. After dilutions under gentle agitation, large aggregates disintegrated, and two fractions of particles were obtained.

In concentrated suspensions, the hydrodynamic diameter in the smaller particle fraction was 445 nm, whereas after dilution of the sample, two fractions were detected. The polydispersity index exceeded 10% for all dilutions. This result proved that the tested sample was not monodispersed. The zeta-potential was negative, estimated as −20.9 mV. This value characterizes the suspension state of gCuHCF as the threshold of delicate dispersion.

Particle concentration and mean size of the gCuHCF fraction, estimated by the particle counter, were 2.00×10^6 mL^{-1} and 3.04 ± 1.98 μm, respectively.

3.2.3. X-ray Diffraction (XRD) Analysis

The XRD pattern of the particles is shown in Figure S3. Diffraction peak positions and their relative intensities reflect the cubic crystalline structure of gCuHCF. Parameters of the crystal cell were calculated from the XRD pattern data (Table 2). The crystal cell belongs to a cubic type with the parameter a = 7.071 Å. Crystallite size was estimated as 156 ± 13 Å.

Table 2. Crystal cell parameters of gCuHCF.

Characteristics	Data
Crystal System	Cubic
Space group	Fm-3m (225)
Parameter of cell	a = b = c = 7.071 Å V = 250.00 Å^3
Crystal	Centrosymmetric
Pearson Symbol	cF 60.02
ANX	AB2C6X6
Molecular Weight	226.08 g/mol
Structural Density	2.25 g/cm^3

3.2.4. SEM

gCuHCF was characterized by SEM coupled with X-ray microanalysis (SEM-XRM). It was found in our previous work [40] that SEM can supply information on the size, distribution, and shape of the tested sample. Figure 3a–d presents the overall morphology of the flower-like particles formed in the process. The XRM images of the gCuHCF film show the characteristic peaks for Cu and Fe (Figure 3e). According to the SEM results, the synthesized gCuHCFs are not nano-sized but rather microparticles.

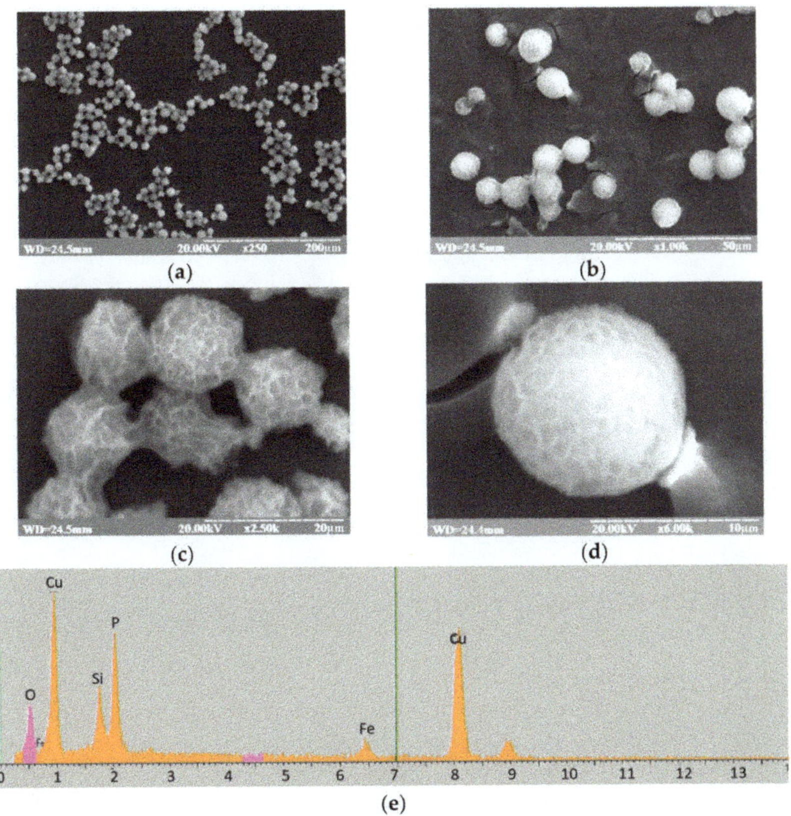

Figure 3. The results of gCuHCF study using SEM with RSM: (**a–d**)—SEM images at different magnifications; (**e**)—X-ray spectral characteristics.

Likewise, the different analytical approaches demonstrated that the synthesized gCuHCF is a suspension of micro-sized particles. These observations were confirmed by different methods: means of particle counting, dynamic light scattering, zeta-potential analysis, and SEM.

Based on the gCuHCF images presented in Figure 3, the studied catalytically active composite material may be described as "organic-inorganic micro/nanoflowers" (hNFs).

hNFs belong to a class of flower-like hybrid materials that self-assemble from metal ions and organic components, such as enzymes, DNA, and amino acids, into flower-like micro/nano superstructures [59,60]. hNFs are widely used for the development of stable, robust, reusable, efficient and cost-effective systems for the immobilization of biomolecules. Some hNFs were shown to exhibit an intrinsic PO-like activity [61,62]. Due to their remarkable performance—the simplicity of their synthesis; their high surface area; excellent thermal, storage, and pH stability; and catalytic activity—hNFs have various potential applications in bioremediation, bioassays, biomedicine, industrial biocatalysis and wastewater treatment [60]. Promising results were reported for hNFs in biosensing, including electrochemical biosensors, colorimetric biosensors and point-of-care diagnostic devices [60–62].

3.3. Application of the gCuHCF as a PO Mimetic in Amperometric (Bio)sensors

The applicability of the gCuHCF as a chemo-sensor for H_2O_2 detection was demonstrated in our previous work [40]. Quantitative analysis of a real sample of commercial disinfectant was carried out. The average H_2O_2 concentration determined by the gCuHCF-based chemo-sensor was shown to be well correlated with the manufacturer's data, with an error of less than 10%.

3.3.1. Properties of gCuHCF

Selectivity of the ABS towards the target analyte is of great importance, especially for the analysis of real samples. In this paper, to study the selectivity of the gCuHCF, a modified GE was tested for its ability to respond to a number of analytes: glucose, alcohols, organic acids, and ammonium ions, etc. The selectivity of the constructed chemo-sensor was estimated for the individual natural substrates (Figure 4a) as well as for their mixture with hydrogen peroxide (Figure 4b). The results presented in Figure 4b demonstrate that the presence of various compounds in the analyzed mixture does not interfere with H_2O_2 determination.

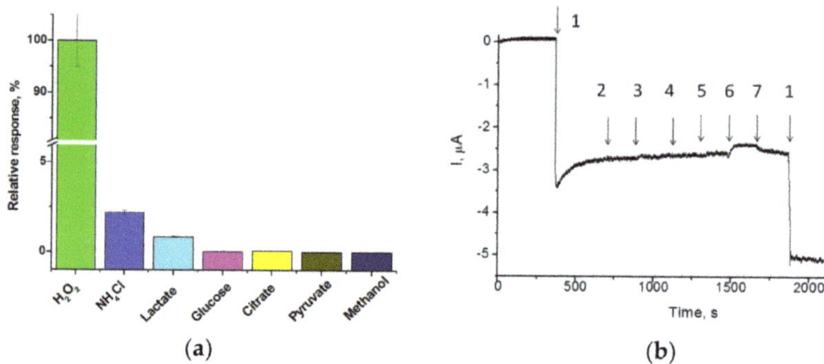

Figure 4. The selectivity tests for gCuHCF/GE: (**a**)—current responses in relative units (%), on the added analytes up to 2 mM concentration, as a ratio of the detected signals to the value of the highest current response; (**b**)—chronoamperograms as outputs on the added analytes (1–7) up to 0.5 mM concentration: (1)—H_2O_2, (2)—glucose, (3)—glycerol, (4)—methanol, (5)—sodium citrate, (6)—sodium lactate, (7)—ammonium chloride. Conditions: working potential −50 mV vs. Ag/AgCl (reference electrode), 50 mM NaOAc buffer, pH 4.5 at 23 °C.

The amperometric analysis was performed using CV and chronoamperometry at different potentials (−50 and +150–200 mV) in different buffer solutions, with a pH from 4.0 to 8.0 (data not shown). It was demonstrated that neither methanol, glycerol, organic acids, nor glucose elicited any signals, while hydrogen peroxide (at −50 mV), ammonium ions and L-lactate (both at +200 mV) were found to elicit significant current responses on the gCuHCF/GE under the tested conditions. Current responses to L-lactate and ammonium under the potential −50 mV were insignificant (Figure 4).

Moreover, we demonstrated that in gCuHCF formation, Fcb_2 was concentrated from the diluted solutions due to co-precipitation with the gCuHCF-based hNFs. When immobilized on a GE, the gCuHCF may become an ABS for L-lactate. CV analysis showed that the current output due to the L-lactate addition correlated with Fcb_2 activity in the sensing layer (data not shown). Thus, the proposed method of hNF formation, using oxido-reductase in the presence of its substrate, may be a promising platform for the concentration and stabilization of any enzyme.

Additionally, using a laccase as a model oxidase, we demonstrated that the gCuHCF not only displayed enzymatic (PO) activity but also an electro-mediator ability (data not shown).

Preliminary experiments for the development of biosensors for primary alcohols and L-amino acids (based on alcohol oxidase and L-amino acid oxidase, respectively) were carried out (data not shown). The obtained results indicated that the gCuHCF and other gHCFs have a potential for use as PO-like composites for the construction of amperometric biosensors with any oxidase.

We conclude that the gCuHCF that was obtained with Fcb_2 assistance, forming a flower-like micro-superstructure, is a prospective organic-inorganic composite material for biosensor construction. It is a stable, catalytically and electrochemically active carrier for enzyme concentration, immobilization and stabilization.

3.3.2. Optimization of H_2O_2 Sensing

To improve the conditions for exploiting the biosensor, the optimal buffer, pH and working potential were estimated. For optimization of the chemo-sensor and further biosensor construction, the quantity of gCuHCF material on the surface of the GE, as well as the enzyme/gCuHCF ratio, were determined experimentally.

We analyzed the correlation of PO-mimetic activity with the effectiveness of H_2O_2 sensing, using the gCuHCF/GE under different conditions of pH and working potential.

The dependence of the chemo-sensor's analytical characteristics on the quantities of gCuHCF placed on the GE surface was studied under the working potential -50 mV in 50 mM NaOAc, pH 4.5. The results are presented in Figure S4 and are summarized in Table 3. Based on the data, the optimal PO-like activity of the gCuHCF for achieving the highest sensitivity under the described conditions is 2–5 mU.

Table 3. Effect of gCuHCF PO-mimetic activity on the analytical characteristics of the modified GEs at pH 4.5.

Number	gCuHCF Volume, µL	Placed on GE Activity, mU	Sensitivity, $A\,M^{-1}m^{-2}$	I_{max}, µA	$K_M{}^{app}$, mM
1	0.5	1	261	59.0 ± 3.6	33.3 ± 4.5
2	1	2	1065	162.3 ± 20.7	54.8 ± 13.4
3	2.5	5	747	114.6 ± 12.7	22.4 ± 5.17
4	5	10	139	66.6 ± 13.9	22.4 ± 9.69

The optimal working potential for H_2O_2 sensing was determined using a CV study (Figure 5), followed by chronoamperometry experiments at pH 6.0 (data not shown). The decision to change the conditions of the experiments, and work under a pH range of 6–8, was necessitated by our plans to develop biosensors using different oxido-reductases. Many microbial enzymes have shown optimal activity near these pH values.

As seen in Figure 5, the optimal working potentials for H_2O_2 sensing under pH 6.0 were lower than -100 mV. To select the best conditions for achieving the highest gCuHCF/GE sensitivity, we determined its analytical parameters under different potentials, namely, -50 and -200 mV (Figure S5). According to the data, the chemo-sensor sensitivity under -200 mV was 2.7-fold higher than under -50 mV.

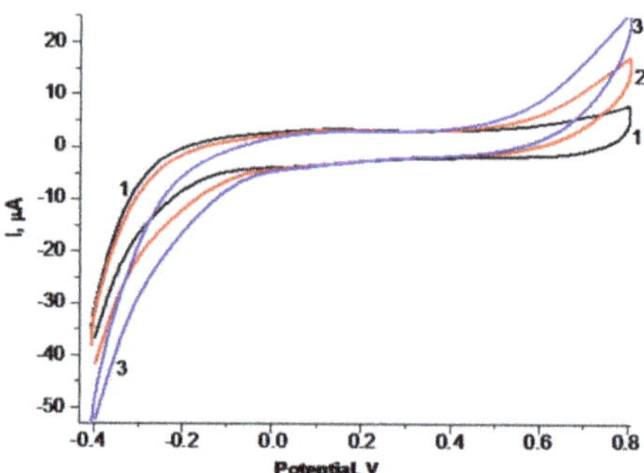

Figure 5. Cyclic voltammograms (CV) of the gCuHCF/GE. CV profiles (1–3) as outputs upon addition of H_2O_2 up to concentrations: (1)—0 mM (black); (2)—0.17 mM (red); (3)—0.5 mM (blue) mM. Conditions: scan rate 50 mV·s^{-1}; Ag/AgCl (reference electrode) in 50 mM PB, pH 6.0. The sensing layer contains 0.35 mU of PO-like activity.

3.3.3. Development of an Amperometric Biosensor for Glucose Determination

In our previous work [40], we reported on the construction of a mono-enzyme amperometric biosensor (ABS) for glucose, using gCuHCF as the PO mimetic and commercial glucose oxidase (GO). It is worth mentioning that the control gCuHCF/GE did not show any amperometric output in response to glucose. The sensitivity of the developed GO/gCuHCF/GE was rather low (76 A·M^{-1}·m^{-2}). In the current study, we set a goal to develop an improved GO/gCuHCF/GE with elevated/optimized analytical characteristics. We carried out the investigation of the gCuHCF as an artificial PO in more detail by studying the influence of various experimental stages on the effectiveness of H_2O_2 sensing; we describe these results in Section 3.3.2. The next task was the optimization of glucose biosensing.

According to Figure 6, the optimal working potential for glucose sensing determined via CV measurement was −450 mV. However, to avoid a possible interference of various substances on the electrode response in the presence of oxygen at high voltage, we chose a lower working potential, namely, −250 mV. This requirement is relevant for the application of the biosensor for the analysis of real samples, e.g., food products.

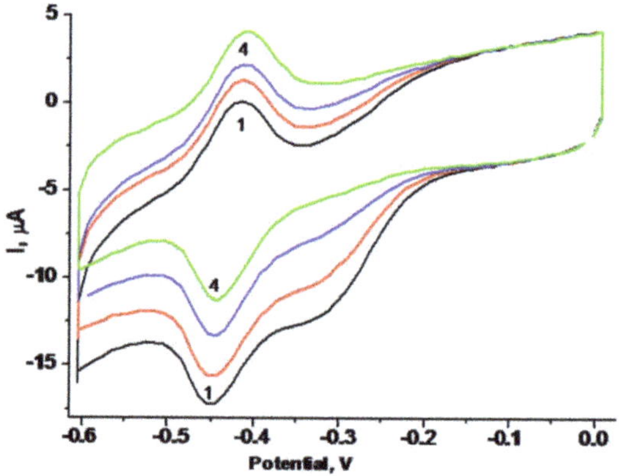

Figure 6. Cyclic voltammograms (CV) of the GO/gCuHCF/GE. CV profiles (1–4) as outputs upon addition of glucose up to concentrations: (1) 0, (2) 0.17, (3) 0.5, (4) 1.3 mM. Conditions: scan rate 50 mV·s^{-1}; Ag/AgCl (reference electrode) in 50 mM PB, pH 6.5. The sensing layer of the biosensor contains 0.5 mU of PO-like gCuHCF and 40 mU of GO.

For optimization of the biosensor composition, the enzyme/gCuHCF ratio on the GE surface was determined experimentally (data not shown). It was found that the optimal ratio, calculated from total activities (GO and PO-like gCuHCF), was 80. Activities of the GO and gCuHCF were estimated with o-dianisidine, as described in Sections 2.2 and 2.5, respectively.

Figure 7 demonstrates the best results obtained from the constructed GO-ABSs. To select the optimal working potential for GO-ABS exploitation, we estimated its analytical parameters under two potentials, at −250 and at −300 mV (Figure 7). Taking into account the parameters (b) from the linear regression graphs (Figure 7b,d) and the square of the electrode surface (7.3 mm^2), we calculated the sensitivities of the GO-ABS to glucose. These and other analytical characteristics of the developed GO/gCuHCF/GEs are summarized in Table 4. According to Table 4, the sensitivity (A M^{-1}m^{-2}) at the potential −250 mV was 2.2-fold higher than at −300 mV, and 9.4-fold higher than at −50 mV. Thus, −250 mV was chosen as the optimal working potential for the exploitation of a GO/gCuHCF-based ABS.

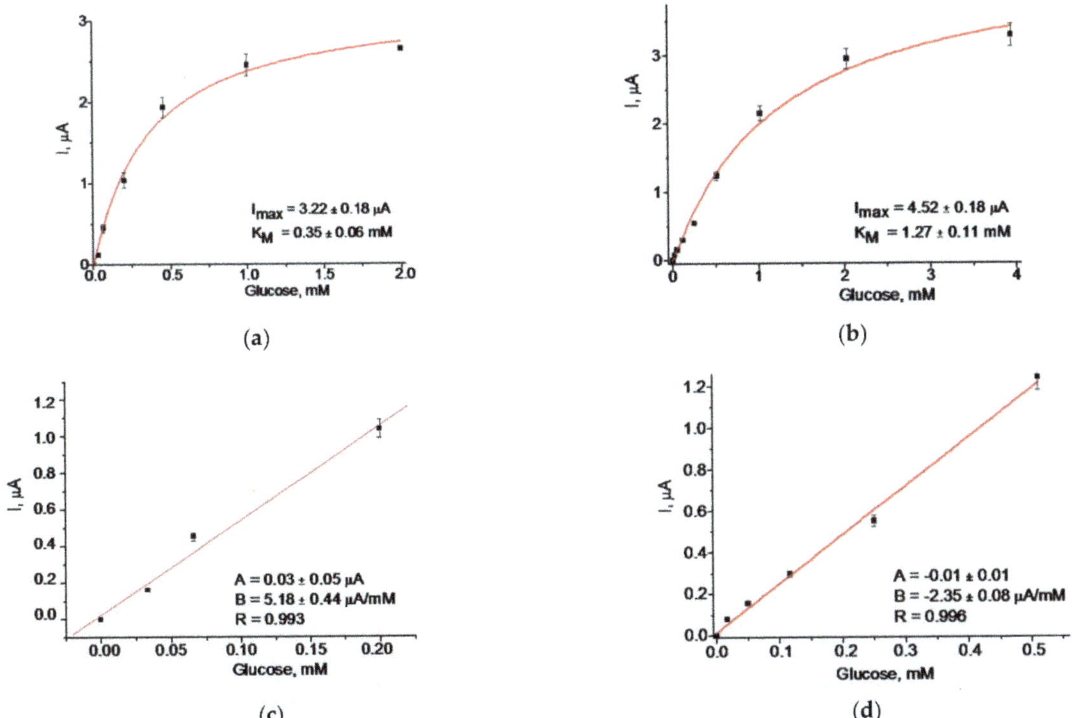

Figure 7. Characteristics of the GO/gCuHCF/GE under different working potentials: (**a,b**)—dependences of the current response on increasing concentrations for glucose determination; (**c,d**)—calibration graphs. Conditions: working potentials −250 (**a,c**) and −300 mV (**b,d**) vs. Ag/AgCl (reference electrode), 50 mM phosphate buffer, pH 6.0 at 23 °C. The GE contains 0.5 mU of PO-like activity and 40 mU GO.

Table 4. Analytical characteristics of the developed GO/gCuHCF/GEs.

Number	Composition of Sensing Film		Voltage, mV	Sensitivity, A M^{-1}m^{-2}	I_{max}, µA	Linear Range, Up to µM	K_M^{app}, mM
	GO, mU	PO Mimic, mU					
1	300	20	−50	76	1.15	3000	1.8
2	40	0.5	−250	710	3.22	200	0.35
3	40	0.5	−300	322	4.52	500	1.3

Thus, we determined the optimal conditions for construction and exploitation of the most effective and highly sensitive GO-based ABS: the ratio of GO activity to PO-like activity of gCuHCF was shown to be 80 under conditions of −250 mV working potential, 50 mM phosphate buffer, and pH 6.0.

3.3.4. Testing of GO/gCuHCF/GE Biosensor for Glucose Analysis in Juice Samples

In order to demonstrate the practical feasibility of the constructed ABS, the developed biosensor was used for glucose analysis in three fruit juice samples using the graphical method known as the standard addition test (SAT). Graphical SAT is a type of quantitative analysis often used in analytical chemistry when a standard is added directly to the aliquots of the analyzed sample. SAT is used in situations where sample components also may contribute to the analytical signal, which makes it impossible to use routine calibration methods. Estimation of glucose concentration in the initial sample was performed using the equation C = AN/B, where A and B are parameters of a linear regression and N is the dilution factor.

Figure 8 demonstrates in detail the algorithm of glucose estimation using two juices as the examples. The results of glucose determination in the juices sampled by the proposed biosensor and by a commercial enzymatic kit are presented in Table 5. The average glucose concentrations determined from the data in Figure 8 differ by less than 10% from the data obtained using the reference method (Table 5).

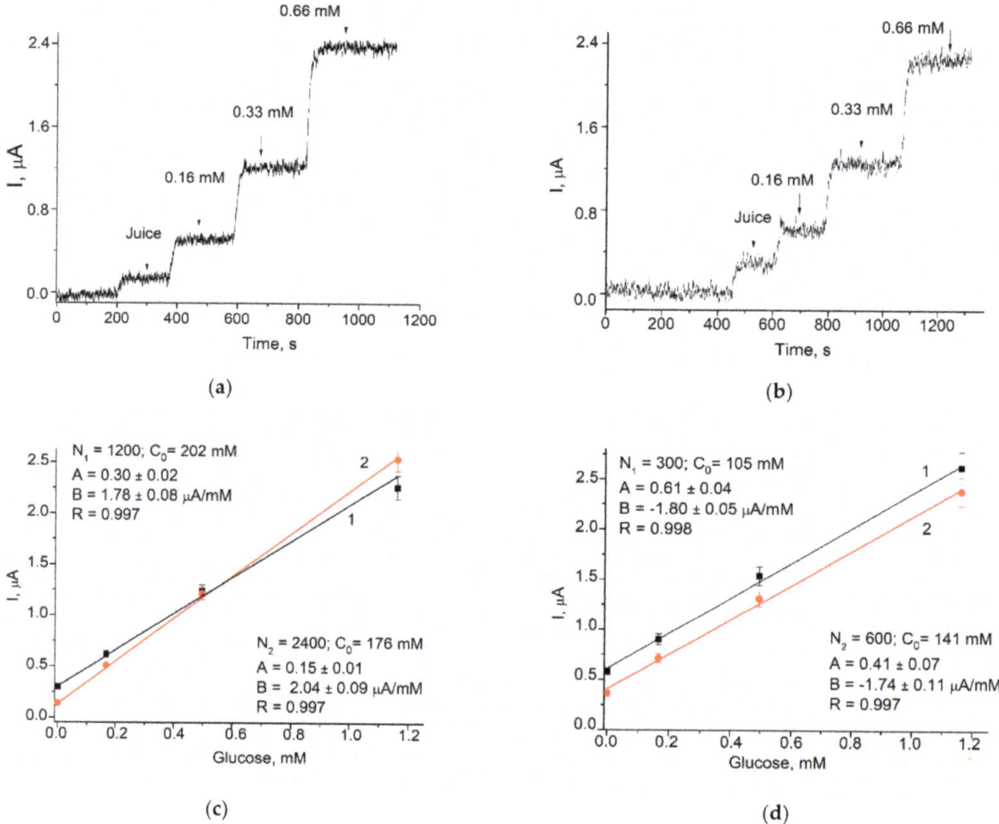

Figure 8. The example of glucose analysis using the biosensor in samples of juices: Multivitamin "Sadochok" (**a–c**), and apple–pear "Galicia" (**d**), in two dilutions; chronoamperograms (**a,b**), and corresponding linear graphs (**c,d**). Conditions: working potential −250 mV vs. Ag/AgCl (reference electrode), 50 mM phosphate buffer, pH 6.0 at 23 °C.

Table 5. Results of glucose estimation in the samples of fruit juices.

Juice	Glucose, mM		
	Biosensor	**Reference**	**Difference, %**
Multi vitamin, "Sadochok"	189 ± 17	206 ± 15	8.6
Apple–pear, "Galicia"	123 ± 10	131 ± 12	6.3
Apple fresh	186 ± 16	202 ± 18	8.2

4. Conclusions

In the current research, we report the development of reagentless amperometric H_2O_2-sensitive sensors with artificial peroxidases (PO). As PO mimetics, "green" hexacyanoferrates (gHCFs) of transition and noble metals were used, which were synthesized via the oxidoreductase F*cb*$_2$. The gCuHCF was identified as the most effective PO mimetic and was characterized in detail concerning its structural, catalytic and electrochemical properties.

SEM analysis demonstrated that the gCuHCF formed a flower-like micro/nano superstructure. Thus, it may be used not only as a H_2O_2-sensitive platform for the development of oxidase-based biosensors but also as a carrier for enzyme concentration, immobilization and stabilization.

An amperometric glucose-oxidase-based biosensor with gCuHCF as the PO mimetic was developed. It exhibited high sensitivity (710 A $M^{-1}m^{-2}$), a broad linear range and good selectivity. The practical feasibility of the constructed biosensor was demonstrated on samples of fruit juices.

The obtained results indicated that the gCuHCF and other gHCFs may have a potential for use as PO-like composites for the construction of amperometric biosensors with any oxidase.

Supplementary Materials: The following are available online at https://www.mdpi.com/article/10.3390/bios11060193/s1, Figure S1. Amperometric characteristics of the several modified electrodes: chronoamperograms (left), dependence of the current response on increasing concentrations of H_2O_2 (middle), and calibration graphs (right); H_2O_2-sensing films are the following: gPdHCF (a); gCeHCF (b); gYHCF (c); gCoHCF (d); gMnHCF (e); gZnHCF (f); gNdHCF (g); gCdHCF (h) and chCuHCF (i). Conditions: working potential −50 mV vs. Ag/AgCl (reference electrode), 50 mM NaOAc buffer, pH 4.5 at 23 °C. Figure S2. FTIR spectrum of the gCuHCF. Figure S3. DLS plots of particle hydrodynamic diameter of the sample at various concentrations: the green line represents a 1.3×10^8 mL^{-1}, the yellow line 6.6×10^7 mL^{-1}, and the red line 3.3×10^7 mL^{-1} particle concentration. Figure S4. X-ray diffraction analysis of the gCuHCF's synthesized particles. Figure S5. Effect of the PO-mimetic activity on the efficiency of H_2O_2 sensing: current response to increasing concentrations of H_2O_2 (a–e); and calibration graphs (f) for the GEs modified with different quantities of gCuHCF: (a) 1 mU, (b) 2 mU, (c) 5 mU, (d) 10 mU, (e,f); combined graph lines (1–4) correspond to graphs (a–d), respectively. Conditions: working potential −50 mV, Ag/AgCl (reference electrode) in 50 mM NaOAc, pH 4.5. Figure S6. Effect of PO-mimetic activity and working potential on analytical characteristics of the gCuHCF/GE: current responses to increasing concentrations of H_2O_2 (a,c); and the corresponding calibration graphs (b,d) for the GE modified with different quantities of gCuHCF: (1)—0.07 mU, (2)—0.15 mU, (3)—0.40 mU. Conditions: working potential −50 mV (a,b) and −200 mV (c,d), Ag/AgCl (reference electrode) in 50 mM phosphate buffer, pH 6.0.

Author Contributions: Conceptualization: G.Z.G. and M.N.; methodology: O.M.D., G.Z.G., Y.G.; investigation: O.M.D., R.Y.S., Y.G. and H.M.K.; resources: M.V.G.; data curation: G.Z.G. and M.N.; writing—original draft preparation: G.Z.G. and O.M.D.; writing—review and editing: G.Z.G. and M.N.; supervision: M.V.G.; project administration: G.Z.G.; funding acquisition: M.V.G. and M.N. All authors have read and agreed to the published version of the manuscript.

Funding: This work was partially funded by NAS of Ukraine (the program "Smart sensor devices of a new generation based on modern materials and technologies", project 0118U006260), by the Ministry of Education and Science of Ukraine (Ukrainian-Lithuanian R&D, project 0120U103398), by the National Research Foundation of Ukraine (project 0100/02.2020 "Development of new nanozymes as catalytic elements for enzymatic kits and chemo/biosensors") and by the Research Authority of the Ariel University, Israel.

Institutional Review Board Statement: Not applicable.

Informed Consent Statement: Not applicable.

Data Availability Statement: Data is contained within the article: Gayda, G.; Demkiv, O.; Gurianov, Y.; Serkiz, R.; Gonchar, M.; Nisnevitch, M. 2020. "Green" nanozymes: synthesis, characterization and application in amperometric (bio)sensors. Proceedings 60(1), 58; doi.org/10.3390/IECB2020-07072 and Supplementary Material section.

Acknowledgments: We acknowledge Oksana M. Zakalska (Institute of Cell Biology, Lviv, Ukraine) for technical support and experimental assistance. The authors would like to thank Alexey Kossenko (Ariel University) for his help with X-ray crystallography analyses.

Conflicts of Interest: The authors declare no conflict of interest.

Abbreviations

ABTS	2,2′-azinobis (3-ethylbenzothiazoline-6-sulfonate) diammonium salt
chHCF	Chemically synthesized HCF of a transition metal
CV	Cyclic voltammetry
DLS	Dynamic light scattering
Fcb_2	Flavocytochrome b_2
FTIR	Fourier transform infrared spectroscopy
gHCF	Green-synthesized hexacyanoferrate of a transitional or noble metal
gHCF/GE	Green-synthesized hexacyanoferrate immobilized on GE
gNPs	Green-synthesized NPs
GE	Graphite electrode
GO	Glucose oxidase
gPBA	Green synthesized Prussian blue analog
HCF	Hexacyanoferrate of a transitional or noble metal
hNFs	Organic-inorganic hybrid nanoflowers
I_{max}	Maximal current response on tested analyte at substrate saturation
$K_M{}^{app}$	Apparent Michaelis–Menten constant
NaOAc	Sodium acetate buffer
NZ	Nanozyme
NP	Nanoparticle
PAAG	Polyacrylamide gel
PB	Prussian blue
PBA	PB analog
PO	Peroxidase
SAT	Standard addition test
SEM-XRM	Scanning electron microscopy coupled with X-ray microanalysis
XRD	X-ray diffraction analysis

References

1. Palomo, J.M. Artificial enzymes with multiple active sites. *Curr. Opin. Green Sustain. Chem.* **2021**, *29*, 100452. [CrossRef]
2. Castillo, N.E.; Melchor-Martínez, E.M.; Ochoa Sierra, J.S.; Ramírez-Torres, N.M.; Sosa-Hernández, J.E.; Iqbal, H.M.N.; Parra-Saldívar, R. Enzyme mimics in-focus: Redefining the catalytic attributes of artificial enzymes for renewable energy production. *Int. J. Biol. Macromol.* **2021**, *179*, 80–89. [CrossRef] [PubMed]
3. Liang, M.; Yan, X. Nanozymes: From New Concepts, Mechanisms, and Standards to Applications. *Acc. Chem. Res.* **2019**, *5*, 2190–2200. [CrossRef] [PubMed]
4. Wu, J.; Wang, X.; Wang, Q.; Lou, Z.; Li, S.; Zhu, Y.; Qin, L.; Wei, H. Nanomaterials with enzyme-like characteristics (nanozymes): Next-generation artificial enzymes (II). *Chem. Soc. Rev.* **2019**, *48*, 1004–1076. [CrossRef] [PubMed]
5. Huang, Y.; Ren, J.; Qu, X. Nanozymes: Classification, Catalytic Mechanisms, Activity Regulation, and Applications. *Chem. Rev.* **2019**, *119*, 4357–4412. [CrossRef]
6. Wang, W.; Gunasekaran, S. Nanozymes-based biosensors for food quality and safety. *TrAC Trends Anal. Chem.* **2020**, *126*, 115841. [CrossRef]

7. Wang, P.; Wang, T.; Hong, J.; Yan, X.; Liang, M. Nanozymes: A New Disease Imaging Strategy. *Front. Bioeng. Biotechnol.* **2020**, *8*, 1–10. [CrossRef] [PubMed]
8. Nayl, A.A.; Abd-Elhamid, A.I.; El-Moghazy, A.Y.; Hussin, M.; Abu-Saied, M.A.; El-Shanshory, A.A.; Soliman, H.M.A. The nanomaterials and recent progress in biosensing systems: A review. *Trends Environ. Anal. Chem.* **2020**, *26*, e00087. [CrossRef]
9. Mahmudunnabi, R.; Farhana, F.Z.; Kashaninejad, N.; Firoz, S.H.; Shim, Y.B.; Shiddiky, M.J.A. Nanozyme-based electrochemical biosensors for disease biomarker detection. *Analyst* **2020**, *145*, 4398–4420. [CrossRef]
10. Attar, F.; Shahpar, M.G.; Rasti, B.; Sharifi, M.; Saboury, A.A.; Rezayat, S.M.; Falahati, M. Nanozymes with intrinsic peroxidase-like activities. *J. Mol. Liq.* **2019**, *278*, 130–144. [CrossRef]
11. Neumann, B.; Wollenberger, U. Electrochemical Biosensors Employing Natural and Artificial Heme Peroxidases on Semiconductors. *Sensors* **2020**, *20*, 3692. [CrossRef] [PubMed]
12. Stasyuk, N.; Smutok, O.; Demkiv, O.; Prokopiv, T.; Gayda, G.; Nisnevitch, M.; Gonchar, M. Synthesis, Catalytic Properties and Application in Biosensorics of Nanozymes and Electronanocatalysts: A Review. *Sensors* **2020**, *20*, 4509. [CrossRef] [PubMed]
13. Guari, Y.; Larionova, J. (Eds.) *Prussian Blue-Type Nanoparticles and Nanocomposites: Synthesis, Devices, and Applications: Synthesis, Devices, and Applications*; Pan Stanford Publishing Pte Ltd.: Singapore, 2019; 314p, ISBN 978-981-4800-05-1.
14. Ivanov, V.D. Four decades of electrochemical investigation of Prussian blue. *Ionics* **2020**, *26*, 531–547. [CrossRef]
15. Ojwang, D.O. Prussian Blue Analogue Copper Hexacyanoferrate: Synthesis, Structure Characterization and Its Applications as Battery Electrode and CO_2 Adsorbent. Ph.D. Thesis, Stockholm University, Stockholm, Sweden, 13 October 2017. Available online: http://www.diva-portal.org/smash/record.jsf?pid=diva2%3A1136799amp;dswid=8693 (accessed on 10 July 2020).
16. Matos-Peralta, Y.; Antuch, M. Review—Prussian Blue and its analogs as appealing materials for electrochemical sensing and biosensing. *J. Electrochem. Soc.* **2020**, *167*, 037510. [CrossRef]
17. Rauwel, P.; Rauwel, E. Towards the Extraction of Radioactive Cesium-137 from Water via Graphene/CNT and Nanostructured Prussian Blue Hybrid Nanocomposites: A Review. *Nanomaterials* **2019**, *9*, 682. [CrossRef]
18. Cheng, L.; Ding, H.; Wu, C.; Wang, S.; Zhan, X. Synthesis of a new Ag+-decorated Prussian Blue analog with high peroxidase-like activity and its application in measuring the content of the antioxidant substances in *Lycium ruthenicum* Murr. *RSC Adv.* **2021**, *11*, 7913. [CrossRef]
19. Jia, Q.; Li, Z.; Guo, C.; Huang, X.; Kang, M.; Song, Y.; He, L.; Zhou, N.; Wang, M.; Zhang, Z.; et al. PEGMA-modified bimetallic NiCo Prussian blue analogue doped with Tb(III) ions: Efficiently pH-responsive and controlled release system for anticancer drug. *Chem. Eng.* **2020**, *389*, 124468. [CrossRef]
20. Wang, X.; Li, H.; Li, F.; Han, X.; Chen, G. Prussian blue-coated lanthanide-doped core/shell/shell nanocrystals for NIR-II image-guided photothermal therapy. *Nanoscale* **2019**, *11*, 22079–22088. [CrossRef]
21. He, L.; Li, Z.; Guo, C.; Hu, B.; Wang, M.; Zhang, Z.; Du, M. Bifunctional bioplatform based on NiCo Prussian blue analogue: Label-free impedimetric aptasensor for the early detection of carcino-embryonic antigen and living cancer cells. *Sens. Actuators B Chem.* **2019**, *298*, 126852. [CrossRef]
22. Tian, M.; Xie, W.; Zhang, T.; Liu, Y.; Lu, Z.; Li, C.M.; Liu, Y. Sensitive lateral flow immunochromatographic strip with Prussian blue nanoparticles mediated signal generation and cascade amplification. *Sens. Actuators B Chem.* **2020**, *309*, 127728. [CrossRef]
23. Chen, W.; Gao, G.; Jin, Y.; Deng, C. A facile biosensor for Aβ 40 O based on fluorescence quenching of Prussian blue nanoparticles. *Talanta* **2020**, *216*, 120390. [CrossRef]
24. Komkova, M.A.; Andreev, E.A.; Ibragimova, O.A.; Karyakin, A.A. Prussian Blue based flow-through (bio)sensors in power generation mode: New horizons for electrochemical analyzers. *Sens. Actuators B Chem.* **2019**, *292*, 284–288. [CrossRef]
25. Nai, J.; Lou, X.W.D. Hollow Structures Based on Prussian Blue and Its Analogs for Electrochemical Energy Storage and Conversion. *Adv. Mater.* **2019**, *31*, 1706825. [CrossRef] [PubMed]
26. Lee, I.; Kang, S.M.; Jang, S.C.; Lee, G.W.; Shim, H.E.; Rethinasabapathy, M.; Roh, C.; Huh, Y.S. One-pot gamma ray-induced green synthesis of a Prussian blue-laden polyvinylpyrrolidone/reduced graphene oxide aerogel for the removal of hazardous pollutants. *J. Mater. Chem. A* **2019**, *7*, 1737–1748. [CrossRef]
27. Keihan, A.H.; Karimi, R.R.; Sajjadi, S. Wide dynamic range and ultrasensitive detection of hydrogen peroxide based on beneficial role of gold nanoparticles on the electrochemical properties of Prussian blue. *J. Electroanal. Chem.* **2020**, *862*, 114001. [CrossRef]
28. Itaya, K.; Shoji, N.; Uchida, I. Catalysis of the reduction of molecular oxygen to water at Prussian blue modified electrodes. *J. Am. Chem. Soc.* **1984**, *106*, 3423–3429. [CrossRef]
29. Karyakin, A.A.; Karyakina, E.E.; Gorton, L. Amperometric Biosensor for Glutamate Using Prussian Blue-Based "Artificial Peroxidase" as a Transducer for Hydrogen Peroxide. *Anal. Chem.* **2000**, *72*, 1720–1723. [CrossRef] [PubMed]
30. Komkova, M.A.; Karyakin, A.A.; Andreev, E.A. Power output of Prussian Blue based (bio)sensors as a function of analyte concentration: Towards wake-up signaling systems. *J. Electroanal. Chem.* **2019**, *847*, 113263. [CrossRef]
31. Karyakin, A.A. Advances of Prussian blue and its analogues in (bio)sensors. *Curr. Opin. Electrochem.* **2017**, *5*, 92–98. [CrossRef]
32. Karyakin, A.A.; Gitelmacher, O.V.; Karyakina, E.E. Prussian Blue-Based First-Generation Biosensor. A Sensitive Amperometric Electrode for Glucose. *Anal. Chem.* **1995**, *67*, 2419–2423. [CrossRef]
33. Vokhmyanina, D.V.; Andreeva, K.D.; Komkova, M.A.; Karyakina, E.E.; Karyakin, A.A. "Artificial peroxidase" nanozyme—Enzyme based lactate biosensor. *Talanta* **2020**, *208*, 120393. [CrossRef]
34. Huang, J.; Lu, S.; Fang, X.; Yang, Z.; Liu, X.; Li, S.; Feng, X. Optimized deposition time boosts the performance of Prussian blue modified nanoporous gold electrodes for hydrogen peroxide monitoring. *Nanotechnology* **2020**, *31*, 045501. [CrossRef]

35. Chen, J.; Yu, Q.; Fu, W.; Chen, X.; Quan Zhang, Q.; Dong, S.; Chen, H.; Zhang, S. A Highly Sensitive Amperometric Glutamate Oxidase Microbiosensor Based on a Reduced Graphene Oxide/Prussian BlueNanocube/Gold Nanoparticle Composite Film-Modified Pt Electrode. *Sensors* **2020**, *20*, 2924. [CrossRef]
36. Niu, Q.; Bao, C.; Cao, X.; Liu, C.; Wang, H.; Lu, W. Ni–Fe PBA hollow nanocubes as efficient electrode materials for highly sensitive detection of guanine and hydrogen peroxide in human whole saliva. *Biosens. Bioelectron.* **2019**, *141*, 111445. [CrossRef]
37. Pang, H.; Zhang, Y.; Cheng, T.; Lai, W.Y.; Huang, W. Uniform manganese hexacyanoferrate hydrate nanocubes featuring superior performance for low-cost supercapacitors and nonenzymatic electrochemical sensors. *Nanoscale* **2015**, *7*, 16012–16019. [CrossRef] [PubMed]
38. Jassal, V.; Shanker, U.; Kaith, B.S. *Aegle marmelos* mediated green synthesis of different nanostructured metal hexacyanoferrates: Activity against photodegradation of harmful organic dyes. *Scientifica* **2016**, *2016*, 2715026. [CrossRef]
39. Jassal, V.; Shanker, U.; Kaith, B.S.; Shankar, S. Green synthesis of potassium zinc hexacyanoferrate nanocubes and their potential application in photocatalytic degradation of organic dyes. *RSC Adv.* **2015**, *5*, 26141–26149. [CrossRef]
40. Gayda, G.Z.; Demkiv, O.M.; Gurianov, Y.; Serkiz, R.Y.; Gonchar, M.V.; Nisnevitch, M. "Green" Nanozymes: Synthesis, Characterization, and Application in Amperometric (Bio)sensors. *Proceedings* **2020**, *60*, 58. [CrossRef]
41. Drummer, S.; Madzimbamuto, T.N.; Chowdhury, M. Green Synthesis of Transition Metals Nanoparticle and Their Oxides: A Review. *Materials* **2021**, *14*, 2700. [CrossRef]
42. Gour, A.; Jain, N.K. Advances in green synthesis of nanoparticles. *Artif. Cells Nanomed. Biotechnol.* **2019**, *47*, 844–851. [CrossRef] [PubMed]
43. Stasyuk, N.Y.; Gayda, G.Z.; Serkiz, R.Y.; Gonchar, M.V. The "green" synthesis of gold nanoparticles by the yeast *Hansenula polymorpha*. *Visnyk Lviv Univ. Ser. Biol.* **2016**, *73*, 96–102.
44. Kharissova, O.V.; Kharisov, B.I.; Oliva González, C.M.; Méndez, Y.P.; López, I. Greener synthesis of chemical compounds and materials. *R. Soc. Open Sci.* **2019**, *6*, 191378. [CrossRef]
45. Chinnadayyala, S.R.; Santhosh, M.; Singh, N.K.; Goswami, P. Alcohol oxidase protein mediated in-situ synthesized and stabilized gold nanoparticles for developing amperometric alcohol biosensor. *Biosens. Bioelectron.* **2015**, *69*, 155–161. [CrossRef]
46. Vetchinkina, E.P.; Loshchinina, E.A.; Vodolazov, I.R.; Kursky, V.F.; Dykman, L.A.; Nikitina, V.E. Biosynthesis of nanoparticles of metals and metalloids by basidiomycetes. Preparation of gold nanoparticles by using purified fungal phenol oxidases. *Appl. Microbiol. Biotechnol.* **2017**, *101*, 1047–1062. [CrossRef] [PubMed]
47. Gayda, G.Z.; Demkiv, O.M.; Stasyuk, N.Y.; Serkiz, R.Y.; Lootsik, M.D.; Errachid, A.; Gonchar, M.V.; Nisnevitch, M. Metallic nanoparticles obtained via "green" synthesis as a platform for biosensor construction. *Appl. Sci.* **2019**, *9*, 720. [CrossRef]
48. Gaida, G.Z.; Stel'mashchuk, S.Y.; Smutok, O.V.; Gonchar, M.V. A new method of visualization of the enzymatic activity of flavocytochrome b_2 in electrophoretograms. *Appl. Biochem. Microbiol.* **2003**, *39*, 221–223. [CrossRef]
49. Gonchar, M.; Smutok, O.; Os'mak, H. Flavocytochrome b2-Based Enzymatic Composition, Method and Kit for L-Lactate. Available online: http://www.wipo.int/pctdb/en/wo.jsp?WO=2009009656 (accessed on 14 August 2020).
50. Synenka, M.M.; Stasyuk, N.Y.; Semashko, T.V.; Gayda, G.Z.; Mikhailova, R.V.; Gonchar, M.V. Immobilization of oxidoreductases at/on gold and silver nanoparticles. *Stud. Biol.* **2014**, *8*, 5–16. [CrossRef]
51. Puganova, E.A.; Karyakin, A.A. New materials based on nanostructured Prussian blue for development of hydrogen peroxide sensors. *Sens. Actuators B Chem.* **2005**, *109*, 167–170. [CrossRef]
52. Pandey, P.C.; Panday, D.; Pandey, A.K. Polyethylenimine mediated synthesis of copper-iron and nickel-iron hexacyanoferrate nanoparticles and their electroanalytical applications. *J. Electroanal. Chem.* **2016**, *780*, 90–102. [CrossRef]
53. Komkova, M.A.; Pasquarelli, A.; Andreev, E.A.; Galushin, A.A.; Karyakin, A.A. Prussian Blue modified boron-doped diamond interfaces for advanced H_2O_2 electrochemical sensors. *Electrochim. Acta* **2020**, *339*, 135924. [CrossRef]
54. Clausmeyer, J.; Actis, P.; Córdoba, A.L.; Korchev, Y.; Schuhmann, W. Nanosensors for the detection of hydrogen peroxide. *Electrochem. Commun.* **2014**, *40*, 28–30. [CrossRef]
55. Valiūnienė, A.; Virbickas, P.; Rekertaitė, A.; Ramanavičius, A. Amperometric Glucose Biosensor Based on Titanium Electrode Modified with Prussian Blue Layer and Immobilized Glucose Oxidase. *J. Electrochem. Soc.* **2017**, *164*, B781–B784. [CrossRef]
56. Virbickas, P.; Valiūnienė, A.; Kavaliauskaitė, G.; Ramanavicius, A. Prussian White-Based Optical Glucose Biosensor. *J. Electrochem. Soc.* **2019**, *166*, B927–B932. [CrossRef]
57. IR Spectrum Table & Chart. Available online: https://www.sigmaaldrich.com/technical-documents/articles/biology/ir-spectrum-table.html (accessed on 7 May 2020).
58. Mink, J.; Stirling, A.; Ojwang, D.O.; Svensson, G.; Mihály, J.; Németh, C.; Drees, M.; Hajba, L. Vibrational properties and bonding analysis of copper hexacyanoferrate complexes in solid state. *Appl. Spectrosc. Rev.* **2018**, *54*, 369–424. [CrossRef]
59. Ge, J.; Lei, J.; Zare, R.N. Protein–inorganic hybrid nanoflowers. *Nat. Nanotechnol.* **2012**, *7*, 428–432. [CrossRef]
60. Cui, J.; Jia, S. Organic–inorganic hybrid nanoflowers: A novel host platform for immobilizing biomolecules. Review. *Coord. Chem. Rev.* **2017**, *352*, 249–263. [CrossRef]
61. Zhu, J.; Wen, M.; Wen, W.; Du, D.; Zhang, X.; Wang, S.; Lin, Y. Recent progress in biosensors based on organic-inorganic hybrid nanoflowers. *Biosens. Bioelectron.* **2018**, *120*, 175–187. [CrossRef] [PubMed]
62. Dong, W.; Chen, G.; Hu, X.; Zhang, X.; Shi, W.; Fu, Z. Molybdenum disulfides nanoflowers anchoring iron-based metal organic framework: A synergetic catalyst with superior peroxidase-mimicking activity for biosensing. *Sens. Actuators B Chem.* **2020**, *305*, 127530. [CrossRef]

biosensors

Article

The Scavenging Effect of Myoglobin from Meat Extracts toward Peroxynitrite Studied with a Flow Injection System Based on Electrochemical Reduction over a Screen-Printed Carbon Electrode Modified with Cobalt Phthalocyanine: Quantification and Kinetics [†]

Ioana Silvia Hosu [1,*], Diana Constantinescu-Aruxandei [1], Florin Oancea [1] and Mihaela Doni [2,*]

[1] Bioproducts Department, National Institute for Research & Development in Chemistry and Petrochemistry—ICECHIM, 202 Spl. Independentei, Bioproducts, Sector 6, 060021 Bucharest, Romania; diana.constantinescu@icechim.ro (D.C.-A.); florin.oancea@icechim.ro (F.O.)

[2] Biotechnology and Bioanalysis Department, National Institute for Research & Development in Chemistry and Petrochemistry—ICECHIM, 202 Spl. Independentei, Sector 6, 060021 Bucharest, Romania

[*] Correspondence: ioana.hosu@icechim.ro (I.S.H.); mihaela.badea@icechim.ro (M.D.); Tel.: +40-213-163-063 (M.D.)

[†] This article is the extended version of our paper: Hosu, I.S.; Constantinescu-Aruxandei, D.; Oancea, F.; Doni, M. Studying Modified Screen-Printed Carbon Electrodes and a Flow Injection Analysis System. In Proceedings of the 1st International Electronic Conference on Biosensors, 2–17 November 2020.

Citation: Hosu, I.S.; Constantinescu-Aruxandei, D.; Oancea, F.; Doni, M. The Scavenging Effect of Myoglobin from Meat Extracts toward Peroxynitrite Studied with a Flow Injection System Based on Electrochemical Reduction over a Screen-Printed Carbon Electrode Modified with Cobalt Phthalocyanine: Quantification and Kinetics. *Biosensors* **2021**, *11*, 220. https://doi.org/10.3390/bios11070220

Received: 13 June 2021
Accepted: 29 June 2021
Published: 2 July 2021

Publisher's Note: MDPI stays neutral with regard to jurisdictional claims in published maps and institutional affiliations.

Abstract: The scavenging activity of myoglobin toward peroxynitrite (PON) was studied in meat extracts, using a new developed electrochemical method (based on cobalt phthalocyanine-modified screen-printed carbon electrode, SPCE/CoPc) and calculating kinetic parameters of PON decay (such as half-time and apparent rate constants). As reactive oxygen/nitrogen species (ROS/RNS) affect the food quality, the consumers can be negatively influenced. The discoloration, rancidity, and flavor of meat are altered in the presence of these species, such as PON. Our new highly thermically stable, cost-effective, rapid, and simple electrocatalytical method was combined with a flow injection analysis system to achieve high sensitivity (10.843 nA μM^{-1}) at a nanomolar level LoD (400 nM), within a linear range of 3–180 μM. The proposed biosensor was fully characterized using SEM, FTIR, Raman spectroscopy, Cyclic Voltammetry (CV), Differential Pulse Voltammetry (DPV), and Linear Sweep Voltammetry (LSV). These achievements were obtained due to the CoPc-mediated reduction of PON at very low potentials (around 0.1 V vs. Ag/AgCl pseudoreference). We also proposed a redox mechanism involving two electrons in the reduction of peroxynitrite to nitrite and studied some important interfering species (nitrite, nitrate, hydrogen peroxide, dopamine, ascorbic acid), which showed that our method is highly selective. These features make our work relevant, as it could be further applied to study the kinetics of important oxidative processes in vivo or in vitro, as PON is usually present in the nanomolar or micromolar range in physiological conditions, and our method is sensitive enough to be applied.

Keywords: electrocatalysis; peroxynitrite; flow injection analysis; meat extracts; myoglobin; cobalt phthalocyanine; electrochemical reduction; screen-printed carbon electrode; amperometric detection; decay kinetics

1. Introduction

For the food industry and for the consumers, it is very important to monitor the quality and freshness of raw meat. Different factors are a sign of meat alteration (e.g., discoloration, rancidity, alteration of flavor) [1–3]. One pathway of alteration is the scavenging activity of myoglobin toward nitro-oxidative species (such as peroxynitrite, PON). For example, the formation of metmyoglobin can alter the flavor due to lipid and protein oxidation [4]. The lack of metmyoglobin (MbFe^{3+}OH$_2$ or metMb) reducing enzymatic systems in meat

285

after slaughter determines the irreversibility of the oxidation processes of myoglobin [5]. The color changes are a sign of these processes, and some possible oxidation pathways are described in Figure 1 [6]. The aspect of meat, by itself, has a great impact on consumers, and the impact on the food industry is huge. Adding nitrites to the raw meat helps keeping the pink color of the meat, as NO (nitric oxide) can bind to the iron ion in a similar way as oxygen molecule does. Nitrites and nitrates are also two of the decomposition compounds of peroxynitrite. Distinguishing between these species is important for meat quality.

The detection of peroxynitrite, being a short-living ROS in biological samples, is a big challenge that scientists still try to solve nowadays. Even if there are different methods of detection presented in the literature, most of them rely on indirect methods after the formation of secondary species. Forming different species, such as 3-nitrotyrosine, that can be detected by immunochemical or chromatographic techniques, or oxidizing different probes with peroxynitrite, further to be detected with fluorescent and chemiluminescent methods, are the usual methods [7]. The problem is that the selectivity toward peroxynitrie is not assured using these methods, as other ROS/RNS species could give the same response. Other usually used methods are high-performance liquid chromatography, UV-Vis absorbance spectroscopy, electron spin resonance, and electrochemistry [8]. These methods usually use antioxidants such as resveratrol, polyphenols, or catechins [3], especially in batch analysis, but also using flow injection analysis, for example by injecting antioxidants that can quench the peroxynitrite [9]. The microfluidic injection analysis presents different advantages such as the presence of laminar flow with no dilution effects, necessity of low volume of analyte (lower than 150 µL), miniaturization, the possibility of real-time continuous monitoring (process control), faster and more sensitive response, or the possibility of automated processes [10].

Electrochemistry uses usually low-cost instrumentation, has fast response time, and can be coupled with online analysis. Electrochemical methods are a better alternative than the usually used methods as they can assure direct, label-free, specific, real-time measurements. Different electrochemically active matrices are described in the literature, such as polymeric films (based on porphyrins, metal phthalocyanine, and/or conducting polymers) hybridized or not with graphene [7,11–16]. Only very few reports present the batch reduction (or oxidation at low potential) of peroxynitrite using a chemically modified electrode (presented in Table 1). Ligands based on extended π conjugated systems can create coordinative chemical bonds with different metals and act as good electrochemical mediators for different redox processes, even nowadays. Phthalocyanines (PCs) are part of this class, and due to different oxidation states of various metallic centers and high conductivity, they are a good platform for the detection of oxygen/nitrogen reactive species [11] or other molecules [17]. PCs are not toxic and have high thermal resistance and are quite stable at room temperature, assuring the stability of the biosensors in time. Except for the metallic centers, the ring-based redox processes may also influence the catalytical activity.

Table 1. Literature study of the developed sensors used to detect PON via electroreduction.

Biosensor	Potential (V)	Sensitivity (nA mM^{-1})	LOD (nM)	pH	Ref.
Microelectrode Pt/Mn-pDPB (manganese-[poly-2,5-di-(2-thienyl)-1H-pyrrole)-1-(p-benzoicacid)]) coated with PEI (polyethyleneimine)	0.2	157.0	1.9	7.4	[14]
Nanoelectrode carbon fibers/manganese(III)-[2]paracyclophenylporphyrin	−0.35	1	50	-	[15]
Microelectrode Pt/MnTPAc (manganese tetraaminophthalocyanine)	−0.45	14.6	5000	10.2	[12]
Electrode SPCE/2,6-dihydroxynaphthalene	0.15	4.12	200	9–12	[16]
Electrode SPCE/cobalt phthalocyanine	0.1	10.84	400	9	this work

The sensitivity of chemically modified electrodes is greatly improved using a flow analysis system (FIA) compared to batch [18]. FIA is easier to use in comparison with batch; it increases the reproducibility and simplifies quantification. In addition, the optimization of a method is more rapid, and testing electrodes is more efficient [10]. Understanding the mechanism of reaction is very important, and FIA provides several advantages in electrochemistry that could be useful for this purpose. For example, by using microreactors, one can narrow the diffusion layers of the electrodes that could also "overlap", which helps to optimize the reaction conditions and facilitates the determination of the mechanism. Channon et al. achieved pharmaceutical detection limits with an FIA electrochemical method for hydrazine detection [19]. They described that convective mass transport enhances the electrochemical signal by comparison with diffusive stationary experiments (batch). Recently, nanomolar levels were achieved for the detection of acetaminophen and codeine, using an FIA system combined with multiple pulse amperometry, and the analytes were also quantified in urine and human serum with excellent recoveries [20].

Herein, we describe the electrochemical reduction of peroxynitrite at 0.1 V, using a commercial cobalt (II) phthalocyanine complex (CoPc) and screen-printed carbon electrodes (SPCE). This electrochemical sensor is able to select between the most important interfering species of peroxynitrite (nitrite, nitrate, and hydrogen peroxide) and other molecules (e.g., ascorbic acid and dopamine), due to a specific, but simple design, combined with the advantages of a micro-fluidic system.

In the last part, we show that our proposed method could be used to further study the decay kinetics of PON in the absence and presence of myoglobin. Our method is both a detection and quantification method and a further tool for kinetic studies, as the RSDs values between the classical static UV-Vis method and our method are low (less than 10%).

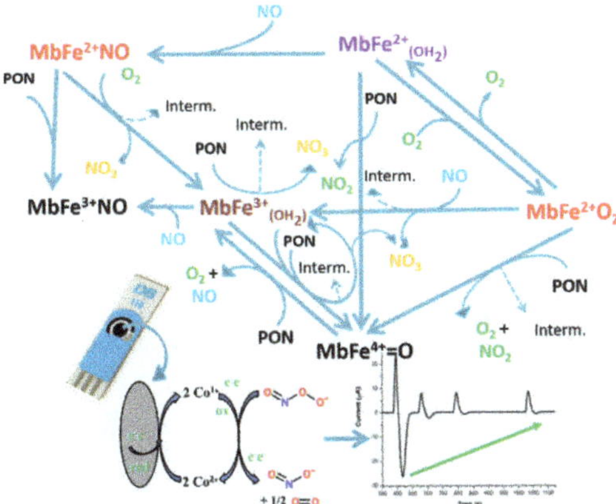

Figure 1. Graphical abstract. ferrylmyoglobin: MbFe^{4+} = O, oxymyoglobin: MbFe^{2+}O$_2$, deoxymyoglobin: MbFe^{2+}(OH$_2$), metmyoglobin: MbFe^{3+}(OH$_2$), nitrosylmetmyoglobin: MbFe^{3+}NO, nitrosylmyoglobin: MbFe^{2+}NO. This is a schematic representation of the chemical reactions of different forms of myoglobin with peroxynitrite and other interfering species/decomposition products. This scheme is not exhaustive and was inspired from data from different literature references [2,21–26].

2. Materials and Methods

Sodium nitrite, hydrogen peroxide (30%), manganese dioxide (MnO$_2$), myoglobin from equine skeletal muscle, sodium hydroxide (NaOH), sodium phosphate dibasic dihydrate (Na$_2$HPO$_4$ 2H$_2$O), cobalt (II) phthalocyanine (CoPc), phthalocyanine (H$_2$Pc), DMF (dimethylformamide), and TBATBF$_4$ (tetrabutylammonium tetrafluoroborate 99%), hydrochloric acid (HCl), hydrogen peroxide 30% (H$_2$O$_2$), sodium nitrite (NaNO$_2$), sodium nitrate (NaNO$_3$), ascorbic acid, and dopamine were acquired from Sigma-Aldrich. Screen-printed carbon electrodes were acquired from DropSens, Spain.

2.1. Peroxynitrite Synthesis

Peroxynitrite (PON) was synthesized following a slightly modified procedure [27]. Briefly, a solution of 0.7 M HCl + 0.6 M H_2O_2 was added over an ice-cooled stirring solution of 0.6 M $NaNO_2$, and almost simultaneously, a solution of 3M NaOH was added over the mixture to quench the decomposition of peroxynitrite (a yellow solution). After several reaction minutes, a few grams of MnO_2 (0.1 g/mL) were added to the mixture, to catalyze the decomposition of hydrogen peroxide. After the gas liberation was finished (approximatively 15 min), the MnO_2 was filtered under vacuum, and the solution was divided into small aliquots (1 mL) and stored in the freezer ($-20\,^{\circ}$C).

2.2. Electrode Chemical Modification

The SPCE electrodes were modified by drop casting 2 μL of a (cobalt phthalocyanine) CoPc solution (1 mg/mL in DMF). The CoPc solution was prepared by dissolving CoPc in DMF (1 mg/mL), using ultrasonication over 1 h (power 100%, frequency 37 Hz). After the drop-cast, the electrodes were dried at 60 $^{\circ}$C in the oven (for 15 min). Before different drop-casting steps, the electrodes were rinsed with DMF and dried with nitrogen. The process was repeated 3 times, without rinsing. The electrode was stabilized by cycling between -0.6 and 0.6 V (in PBS pH 12). Two reduction pre-treatments were proposed for the optimization of the SPCE/CoPc for PON detection: amperometry at -0.3 V for different time periods and chemical reduction with 25 mM sodium borohydride, during 20 min, followed by rinsing with deionized water.

2.3. Meat Extracts and Myoglobin Solutions

Manz meat (veal under the age of 2) was achieved from a local store. Yellow filtering paper (Filtrak n. 389), ROTI$^{®}$Spin MINI-3 25 units CL12.1 (gel ultrafiltration, 1.5 Eppendorf tubes, for 3 kDa), and Sephadex G-25 in PD-10 Desalting Columns were used to remove the strong reducer (sodium borohydride) from the metmyoglobin (metMb) reduced system (redMb). The separation systems were bought from Sigma Aldrich. For oxymyoglobin (redMb) synthesis, a solution of 75 mM of sodium borohydride ($NaBH_4$) in PBS pH 9 was added to a solution of 25 μM metMb.

The meat extraction was done according to the procedure from [28]. Briefly, 200 g of meat were cut in small pieces and blended with 100 mL of PBS pH 9. In addition, to the mixture, 400 mL of 0.1 M PBS (pH 9) were added, and the solution was stirred during 30 min, on an ice bath. After stirring, the mixtures were centrifuged 20 min, at 15 $^{\circ}$C, at 9000 RPM, and the supernatant was centrifuged in the same conditions. Filtration on yellow filter paper was performed under vacuum and the pH was adjusted to 9, using sodium hydroxide. The desired pH values of the solutions were 12, 9, and 7.4, and the concentration of the phosphate was 100 mM.

For PBS pH 9, $Na_2HPO_4·2H_2O$ (0.1 M) and KCl (0.1 M) were dissolved in 500 mL of ultra-pure water. The solution of pH 12 was prepared in the same way, but NaOH was added: 450 mL solution was titrated with NaOH (approximately 35 mL of 1.4 M NaOH) until pH 12 and brought to 500 mL at the end. For PBS, pH 7.4 prepared tablets were used. As we designed the synthesis of PON to obtain high-concentration stock solutions and only added very small amounts of alkaline PON solutions to PBS pH 9 buffer, the pH 9 was practically constant [29].

2.4. Electrochemistry

A single-line flow system was coupled with potentiostat using a flow cell provided by DropSens, for the SPCE electrodes (DRP110). The Electrochemical Flow Injection Analysis system (FIA-EC) used is composed of a four-channel Peristaltic Pump—MINIPULS$^{®}$ 3 (Gilson, Villiers-le-Bel, France), Injection Valve 77521 Rheodyne (with a 100 μL sample loop), Flow Cell from DropSens (model DRP-FLWCL), and Potentiostat (Autolab PGSTAT 101). A boxed connector for screen-printed electrodes from Dropsens was used to connect the SPCEs. Each measurement was performed in triplicate.

2.5. Determination of Apparent Rate Constants and Half-Lives

The kinetics of PON decay and the scavenging effect of myoglobin on peroxynitrite at pH 9 were assessed using the calculation of half-lives and apparent rate constants of peroxynitrite decay. The static method was used for the measurements of concentrations over time, as pH 9 offers the possibility of having slower decays and changes dynamics toward specific chemical decay reactions. The terms "peroxynitrite" and "PON" are widely accepted for a mixture of $ONOO^-$ and $ONOOH$, depending on the pH. At pH 9, $ONOO^-$ is assumed to be in excess over the protonated form $ONOOH$, based on the pKa of PON 6.8. The term "peroxynitrate" refers to a mixture of O_2NOO^- and O_2NOOH, depending on the pH. Trivial names: $ONOO^-$, peroxynitrite; $ONOOH$, peroxynitrous acid; O_2NOO^-, peroxynitrate; O_2NOOH, peroxynitric acid.

For the determination of the kinetic parameters, we took in consideration the previously proposed model by which the bimolecular decomposition of PON is dominant at pH 9, according to the following reaction [30,31]:

$$HOONO + ONOO^- -> 2\,NO_2^- + O_2 + H^+ + 2e^- \tag{1}$$

According to this model, the rate law should follow pseudo first-order kinetics at pH 9, due to the excess of $ONOO^-$. Our aim was to validate the FIA-EC method as a tool for the determination of rate order and kinetic parameters by comparing the results with the classical UV-Vis method. In this particular case, the first purpose was to establish by both UV-Vis and FIA methods if the reaction follows indeed pseudo first-order kinetics or second/higher-order kinetics at pH 9.

Two approaches were used for the confirmation of rate order and the calculation of apparent rate constants and half-lives: the first method (namely called from now on "Method A") uses the plotting of all the integrated rate law data, according to the assumed rate order. Briefly, this was done as follows: the linearity (from the value of R^2) of the graphs *ln(concentration)* (for (pseudo) first-order) or *1/concentration* (for second-order) vs. *time* and the correlation of the observed half-life (extrapolation from the graph $t_{1/2obs}$) with the calculated half-life ($t_{1/2calc}$). The half-lives for (pseudo) first-order and second-order reactions were calculated with the following formulas, respectively: $ln2/k$ and $1/k \cdot C_{0PON}$. C_{0PON} is the initial concentration of PON and k is the (apparent) rate constant. For the linearity, we considered the R^2 values. If R^2 approaches 1, is significantly higher than the R^2 of the other model, *and* there is good similarity between $t_{1/2calc}$ and $t_{1/2obs}$, the corresponding apparent order of the reaction is attributed to the detriment of the other apparent order. Each decay rate constant determination was plotted for a total of 180 s.

The second method (namely called from now on "Method B") was the "half-life method" described by Ira Levine, in the "Physical Chemistry" book, chapter 16, "Reaction kinetics" [32]. This method can be applied when the rate law has the form $r = k[A]^n$. Based on Equation (1) and considering the excess of $ONOO^-$, this method would follow the equation $r = k[ONOOH]^n$, therefore determining the rate order in ONOOH. According to this method adapted to our particular case, one first plots the *concentration* vs. *time* and then should be able to fit the equation with a single-exponential decay function if n = 1 (based on Equation (2), where parameter k (the apparent pseudo first-order rate constant in our case, k_{obs}) is solved after compilation. Secondly, one performs the extrapolation of $t_{1/2}$ for various concentrations and then plots the *logt_{1/2}* vs. *logC_0* [32]. The slope of the logarithmic graph (that should have a linear fit) will establish the order of the reaction in reactant A, *n*, where *n = 1 − slope*. The two methods, A and B, were compared at the end.

$$C_{PON} = C_{0PON} \times e^{-kt} \tag{2}$$

where C_{PON} is the PON concentration at a specific moment in time.

Both A and B kinetic methods were used for the UV-Vis method and for our proposed FIA-EC method. The molar extinction coefficient of 1670 $M^{-1}cm^{-1}$ was used to calculate the concentration of $ONOO^-$ at 302 nm [33,34].

2.6. Surface Characterization

Fourier transform infrared (FTIR) spectra were recorded using a ThermoScientific FTIR instrument (Nicolet 8700, Pleasantville, NY, USA) equipped with a VariGATR accessory (Harrick Scientific Products, Inc., New York, NY, USA). KBr pellets were used as reference, and the powders were ground with KBr to create a solid pellet. For the ATR measurement technique, the Perkin Elmer GX FTIR spectrometer equipped with a Pike MIRacle having a 1.8 mm round diamond crystal was used. Diamond has an intrinsic absorption from approximately 2300 to 1800 cm^{-1}, which limits is usefulness in this region.

Micro-Raman spectroscopy measurements were performed on a Horiba Jobin Yvon LabRam HRMicro-Raman system combined with a 473 nm laser diode as an excitation source. Visible light was focused using a $100\times$ objective. The scattered light was collected by the same objective in backscattering configuration, dispersed by an 1800 mm focal length monochromator, and detected using a CCD.

SEM images were obtained using an electron microscope FEI-QUANTA 200 equipped with wolfram filament (W), at the ICECHIM laboratory (Bucharest, Romania). The SEM images were taken at an accelerating voltage of 25–30 kV using a gaseous secondary electron detector (GSED).

Absorption spectra were recorded using Ocean Optics UV-VIS-NIR spectrometer, in the 200–1100 nm range.

The Limit of Detection and the Limit of Quantification were calculated with the following formulas: $LOD = 3\,\sigma_b/m$, and $LOQ = 10\,\sigma_b/m$, in which σ_b represents the standard deviation of the background and m is the slope of the calibration graph. The sensitivity of the sensor was calculated from the slope of the calibration curve (plot *concentration* vs. *current*).

3. Results and Discussion

Detecting reactive oxygen and nitrogen species is of great importance for many domains. Peroxynitrite, despite being a short half-life oxidative species, induces powerful oxidative stress effects on cells. Scavengers are important tools to eliminate this oxidative stress. Myoglobin is one of the scavengers [35], as it is an oxygen-binding protein from the heme group. Cobalt phthalocyanine was already used in the literature as a bio-mimetic material for biosensors [36]. It has a similar scavenging role when it comes to PON. As the metal cobalt center of the heme, similar to iron, has multiple oxidation states, the reduction of PON seems to be catalyzed by CoPc through redox reactions. We demonstrate here that the Co^{2+}/Co^{1+} redox couple is more effective than the high potential electrochemical methods reported in the literature for the electrochemical detection of PON, as it offers better selectivity. Cyclic voltammetry as well as other techniques (linear sweep voltammetry or differential pulse voltammetry) were used to determine that this redox reaction of PON is apparently an irreversible process.

Before presenting and discussing the results, several other important factors regarding PON are to be mentioned, which are factors that are also important challenges to overcome in developing a selective and sensitive sensor for PON for meat extracts. (i) PON is most stable in cold alkaline solutions, without any metals and carbonyl compounds [30] and even so, it slowly decomposes, mainly, to nitrite. (ii) Below pH 12, the decomposition rate increases, and at pH 7, the half-life of PON is below 1 s. (iii) Acidic pH favors the decomposition to nitrate. (iv) Temperature, buffers, or several scavengers (e.g., myoglobin in meat as in Figure 1) influence the stability of PON [30,31]. Proteins tend to precipitate at pH values around 12, so such alkaline environment is not suitable for our purpose. As the decay of PON is slower at pH 9 than at neutral pH values, pH 9 is an optimal compromise for the detection of PON. At this pH value, the decay to nitrite is the predominant pathway, and it depends on peroxynitrite concentration. Due to these factors, we needed a fast response technique to help us distinguish between PON, nitrite, and nitrate, and to combat batch problems (mixing of the aliquots added for detection would increase the noise of the measurement, the concentration of the analyte will not be homogenous, etc.).

3.1. Batch Determination of Peroxynitrite Using SPCE/CoPc Electrodes

3.1.1. Characterization of the Deposited CoPc Films on the SPCE

SEM analysis of the morphology revealed that in accordance with the literature, the CoPc molecules tend to form random agglomerates [37] of various sizes (between 0.4 and 10 µm in length and a few hundred nm in thickness and width) depending on the surface used for the deposition (Figure 2a,b). The thin films were obtained by drop casting CoPc, due to π stacking (as the macrocycle has 18 delocalized π electrons), similar to other materials, such as graphene. This method is deposition is cost-effective and rapid.

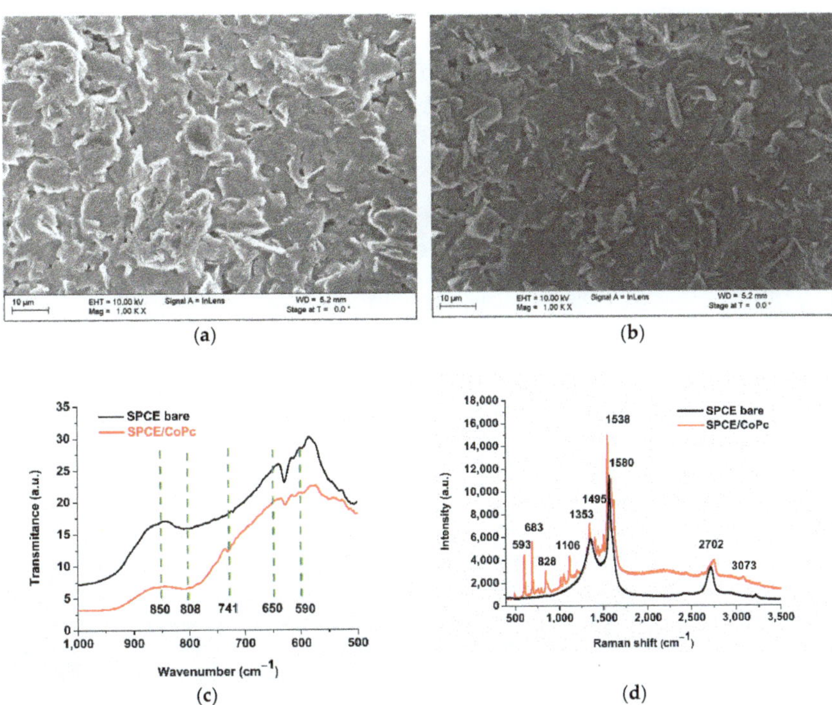

Figure 2. SEM characterization of the (**a**) unmodified SPCE and (**b**) SPCE/CoPc, (**c**) FTIR spectra of unmodified SPCE and modified SPCE with CoPc (SPCE/CoPc) [1], (**d**) Raman spectra of unmodified and CoPc modified SPCE.

The FTIR analysis (Figure 2c) proves the presence of CoPc on the surface of the electrode, especially due to the 741 cm^{-1} band in the fingerprint region, corresponding to phthalocyanine in plane vibrations [38]. Other vibrations belonging to the graphitic structure of SPCE are also observed: deformation of C-C out of aromatic plane (650 cm^{-1}), vibrations of C-H in the aromatic plane (850 and 808 cm^{-1}).

In the Raman spectrum of CoPc (Figure 2d), there are active modes of the symmetry A1g at 592 cm^{-1} (benzene radial), 834 cm^{-1}, B1g at 684 cm^{-1} (macro breathing and benzene radial), 1542 cm^{-1} (C=N stretching mode), and B2g at 1498 cm^{-1} (pyrrole stretch) [39,40]. The HUMO–LUMO gap energy of CoPc is 1.9 eV.

After the morphological characterization of the surface of the electrode, we also used electrochemistry to extensively characterize the electrode. Cyclic voltammetry gave rise to two anodic and two cathodic peaks (corresponding to Co^{3+}/Co^{2+} and Co^{2+}/Co^{1+} redox processes), as previously reported [11]. In this paper, we targeted the exploitation of the Co^{2+}/Co^{1+} redox couple (Figure 3a), due to the occurrence at more negative potentials ($E_0 \approx 0.1$ V), than the couple Co^{3+}/Co^{2+} ($E_0 \approx 0.65$ V), and because, besides good sensitivity, there is the possibility to reach a remarkable selectivity for peroxynitrite. It can be observed that upon several scans (five cycles), the electrode reaches a steady state. Similar peaks were observed in the literature [41].

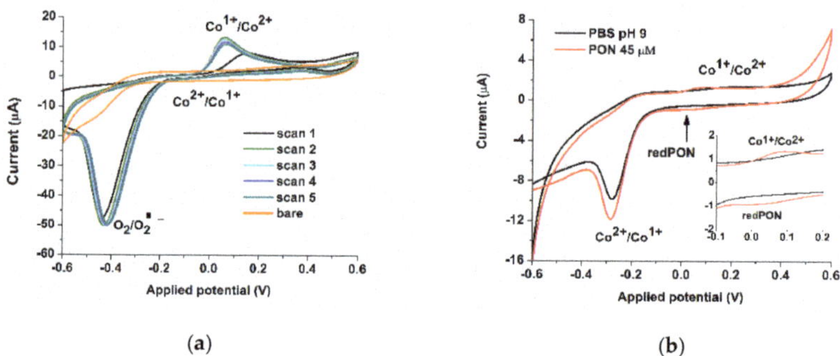

Figure 3. Cyclic Voltammetry (CV) of (**a**) the screen-printed carbon electrode (SPCE)/cobalt phthalocyanine (CoPc) electrode upon different scans, in PBS pH 12 [1]. The oxido-reduction processes are presented for the cobalt metallic center. (**b**) Cyclic voltammetry was registered using the SPCE/CoPc electrode in the absence (black) and in the presence (red) of 45 µM PON (scan rate 9 mV/s), PBS pH 9. Zoom in of the redox peaks for PON from the same plot.

Cyclic voltammetry (CV), in the −0.6 to 0.6 V range, was used to determine the redox process of PON, using the SPCE/CoPc electrode, at pH 9 and pH 12. A higher pH value (12) presented enhanced anodic and cathodic peaks of the biosensor (Figure 3a) than a lower pH value (pH 9, Figure 3b, both in the presence and absence of PON), but high alkalinity interferes with the integrity of proteins (the final goal of our work) and may interfere with the integrity of the film, so it was not used further for analytical purposes, only for biosensor characterization. The redox process involving peroxynitrite at pH 9 takes place at around 0.1 V, with Ec = 0.047 and Ea = 0.072 ($\Delta E = 25$ mV $= E_{cathode} - E_{anode}$), but there is a small current in the reduction, suggesting that the oxidation might be irreversible. The calculated formal potential (the redox standard potential, E^0) is 0.06 V vs. Ag/AgCl pseudoreference electrode. Following the Nernst equation (Equation (3), where a_{red} represents the concentration of reduced species and a_{ox} represents the concentration of oxidized species, [32]) described below, we can conclude that two electrons are involved in the redox process (transferred in the cell reaction) at 25 °C. This conclusion is consistent with the nature of the peroxynitrite oxidant (usually described as a two-electron oxidant).

$$E_{cell} = E_{cathode} - E_{anode} = E^0 + 0.059/n \cdot \lg a_{red}/a_{ox} \tag{3}$$

Furthermore, under another probable mechanism, an irreversible reduction takes place around −0.3 V, probably involving also the chemical oxidative reaction of peroxynitrite over the metallic center: cobalt being chemically oxidized by PON, which is electro-reduced, with a higher current, depending on the concentration of PON (Figure 3b).

If the anodic/cathodic current is proportional to the scan rate, the process is an adsorption-controlled process, and if the plot I vs. $v^{1/2}$ is linear, the redox processes are more likely diffusion-controlled ones [42]. For this purpose, cyclic voltammetry was used to determine the correlation between the cathodic current and the scan rate for scan rates in the range of 16–800 mV/s (Figure 4a). At lower scan rates, a linear correlation was obtained only for the square root of the scan rate, suggesting a diffusion-controlled process. Deviation from linearity usually means that other processes are involved, processes related to the surface, such as adsorption, ligand–species interaction, etc. At higher scan rates (above 256 mV/s), the correlation between Ipc vs $v^{1/2}$ is not linear anymore, suggesting more a mixture of surface and diffusion-controlled processes (Figure 4b).

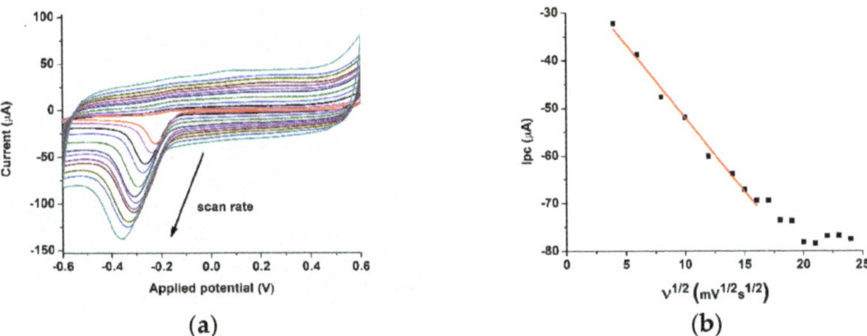

Figure 4. (**a**) Cyclic voltammetry of the SPCE/CoPc electrode for 50 µM PON, PBS pH 12. (**b**) The dependence of the cathodic current (around −0.3 V) with the square root of the scan rate.

The SPCE is a three-electrode electrochemical system. The screen-printed area of the carbon-based working electrode (WE, black circle) of the SPCE is 0.126 cm^2, which is surrounded by the Ag/AgCl pseudoreference electrode (silver) and counter/auxiliary electrode (CE, black). The chemical modification should not cover the CE and the reference electrode, but the surface coverage should be optimal. Starting from a solution of classical concentration for CoPc (1 mg/mL), we drop-casted 1 and 2 µL of the solution on the electrodes and calibrated them by DPV, from −0.4 to 0.3 V, step potential 5 mV, amplitude 25 mV, modulation time 50 ms, scan rate 10 mV/s (Figure 5a). The sensitivity for the 2 µL drop-casted CoPc was 0.083 nA µM^{-1}, in comparison to 0.057 nA µM^{-1} for the 1 µL drop-casted CoPc. More than 2 µL is difficult to deposit without covering the other electrodes. These sensitivities are very low, but further optimization helps us improve them, especially using chronoamperometry at the optimized potential. The efficiency of different cobalt phthalocyanine layers was studied using Cyclic Voltammetry measurements. By drop-casting different layers of solution of CoPc, the best sensitivity for PON was achieved for three layers (drop-casting 2µL in three successive steps, that also included drying steps). The sensitivity for one layer of 2 µL CoPc was 3.3 nA µM^{-1}, and for three layers of 2 µL CoPc, it was 8.5 nA µM^{-1}, which is already one order and respectively two orders of magnitude improvements from the DPV method, but the LODs remained almost the same (5.45 µM and 5.14 µM, respectively).

Chronoamperometry was used to study the preliminary potential for an improved quantitative detection of PON. Two different potentials were used, and both worked on the catalytical oxidation of PON with sensitivities of 8.75 nA µM^{-1} (0.10 V) and 0.91 nA µM^{-1} (0.15 V).

Based on the data described above, we propose a simplified mechanism of the catalytic process, involving two electrons as calculated from Nernst equation:

$$Co^{II}PC + e^- \rightarrow Co^{I}PC \tag{4}$$

(reduction)

$$2\,Co^{I}PC + ON^{V}OO^- \rightarrow \text{intermediary complex} \rightarrow 2\,Co^{II}PC + N^{III}O_2{}^- + \frac{1}{2}\,O_2 + 2e^- \tag{5}$$

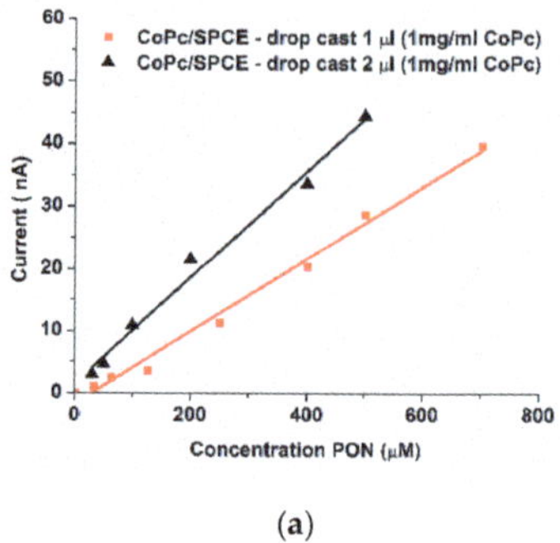

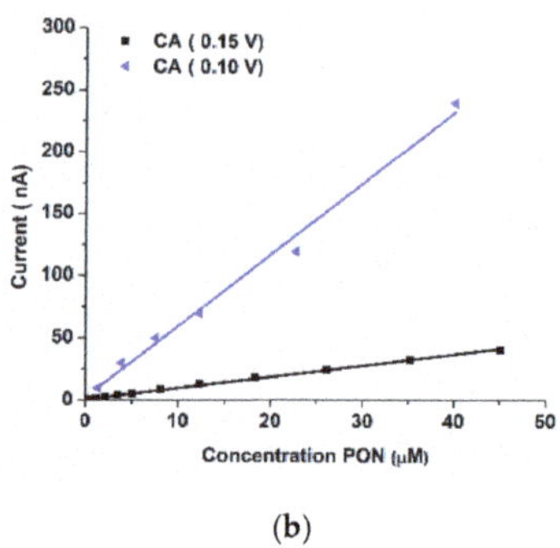

(a) (b)

Figure 5. (**a**) Calibration curves for 1 µL and 2 µL drop-casted CoPc solution (1 mg/mL in DMF) on the SPCE (from DPV measurements) (**b**) Calibration curves from the amperometric response at +0.15 V (black) and +0.10 V (blue), using a GCE/CoPc electrode.

(oxidation/reduction)

In order to prove that the proposed mechanism involves the catalytic site of the metal phthalocyanine, we compared our electrochemical signals with the metal-free phthalocyanine (the exclusion of the catalytic site). Experiments were performed with H_2Pc (dihydrogen phthalocyanine), which is the phthalocyanine ring without any metal coordinated to the nitrogen ligands. The H_2Pc was not responsive to peroxynitrite, underlining the importance of the metallic center in this catalytic process (data not shown).

The surface coverage was calculated using the equation Equation (6):

$$\Gamma = Q/nFA \tag{6}$$

where Γ is the coverage of CoPC immobilized upon the desired electrode surface (mol cm^{-2}), Q is the charge taken from the integration of the oxidation wave resulting from the Co$^{1+/2+}$ couple recorded in a pH 7.4 phosphate buffer solution (PBS) at slow scan rates, n is the number of electrons taking place in the electrochemical process, F is the Faraday constant, and A is the geometrical electrode area (without recourse to any surface roughness corrections, and it was calculated to be 0.126 cm^{-2}, as the diameter of the WE is 4 mm) [43]. Using this calculation method, there was a surface coverage of 7.0558×10^{-9} mol cm^{-2} for the drop-casted CoPC SPEs, while the commercial (DRP 410, Drop Sense electrode) had a similar surface coverage (8.9643×10^{-9} mol cm^{-2}). These kinds of commercial electrodes are recommended for the detection of hydrogen peroxide at low potentials (0.4 V).

3.1.2. Batch Optimization of the CoPc-Modified Electrodes for PON Detection

Bedoui et al. [44] mention that species such as ascorbic acid, nitrite and nitrate, uric acid, hydrogen peroxide, and others could interfere in blood or other biological samples with the signal that one may obtain for peroxynitrite. For this purpose, we studied a series of interfering species using Cyclic Voltammetry and a GCE-modified CoPc and compared the signal with the ones for PON at the same concentration, 100 µM (Figure 6). The biological concentrations of these interfering species are low, but we used the same concentration as for PON measurements, which usually is produced at a rate of up to 50–100 µM/min, but the steady-state reaction is in the nanomolar range, for hours [45]. Ascorbic acid gave rise to an oxidation peak around 0.4 V, hydrogen peroxide around 0.6 V, and PON was electro-catalyzed around 0.1 V (as already described above). Nitrate is not electrochemically active, and nitrite was not responsive within the chosen potential windows, meaning that using potentials around 0.1 V gives a very good selectivity toward PON.

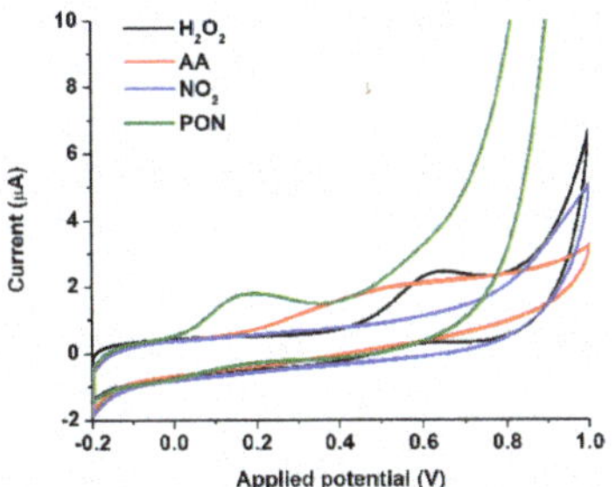

Figure 6. Evaluation of the interfering species using CV for 100 μM peroxynitrite (PON—green): 100 μM ascorbic acid (AA—red), 100 μM hydrogen peroxide (H_2O_2—black), 100 μM nitrite (NO_2—blue), using a glassy carbon electrode (GCE)/CoPc electrode. Scan rate 100 mV/s, electrolyte: PBS pH 9 0.1 M + 0.1 M KCl [1].

More sensitive techniques than CV (Cyclic Voltammetry) such as LSV (Linear Sweep Voltammetry) and DPV (Differential Pulse Voltammetry) were used for further characterization of the electrodes. LSV voltammograms (Figure 7a) in the potential window of −0.4 to 0.3 V were used to observe the electro-oxidation of PON over the SPCE/CoPc electrode at different PON concentrations. The oxidation potential shifts from 0.075 V (for 72 μM PON) to higher potentials (around 0.1 V for 145 μM PON).

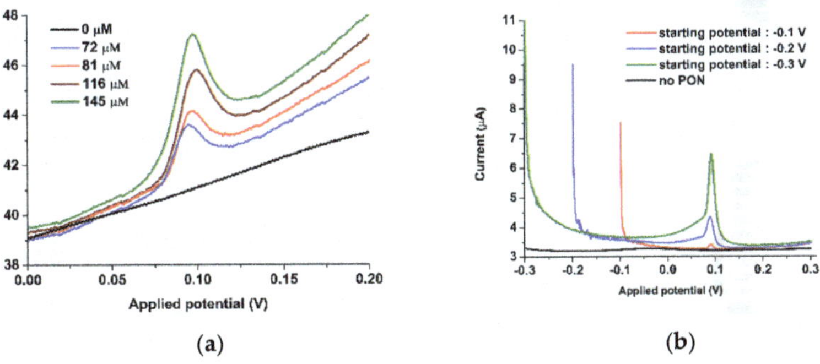

(a)   (b)

Figure 7. (a) LSV using SPCE/CoPc for the detection of PON, using droplets of analyte over the electrode, −0.4 to 0.3 V (electrolyte: PBS pH 9 0.1M + 0.1 M KCl), using different PON concentrations. (b) DPV of the SPCE/CoPc in the presence (red, blue, green) and absence (black) of 125 μM of PON, using different negative starting potentials (−0.1 V, −0.2 V, and −0.3 V).

Differential Pulse Voltammetry (DPV) revealed that upon reduction of the CoPc films by starting the scans from lower potentials, the cumulative current (both cathodic and anodic) increases (Figure 7b). This suggests that the pre-treatment of the electrode could improve the response of the electrode for PON because of the reduction of Co^{2+} to Co^{1+}. In addition, the shape of the DPV peak suggests that, besides diffusion, other processes occur (e.g., adsorption of product or reactant molecules on the surface of the electrode or even the coordination of the $ONOO^-$ molecule to the metallic catalytic center).

3.2. FIA Optimization of the SPCE/CoPc Electrodes for PON Detection

Initially, we did hypothesize that while the oxidation of Co^{1+} to Co^{2+} takes place, it can also be part of the redox process involving the reduction of peroxynitrite to nitrite. Starting from this hypothesis, an important issue was understanding how to overcome the apparent irreversible oxidation of PON from batch electrochemical analysis. The ability of flow injection analysis in understanding reaction mechanisms and complicated electrode kinetics [46] served as an important tool to select between the oxidation potential of peroxynitrite and of other species, such as hydrogen peroxide, and the electrocatalytic reduction potential of peroxynitrite was unraveled and exploited.

FIA coupled with chronoamperometry gave us the opportunity to develop a very selective sensor for PON, using the gathered information from the batch electrochemistry. First, we wanted to establish the optimal potential, so we used chronoamperometry and changed the applied potentials (0.00, 0.10, and 0.25 V) on a SPCE/CoPc electrode. This study helped us to determine that the reduction of PON occurs below 0.1 V, as opposed to the oxidation of PON that occurs above 0.1 V (Figure 8a). Due to the several other advantages already described, a single line flow injection system was employed for further experiments using the revealed reduction potential for further optimization of the PON sensor.

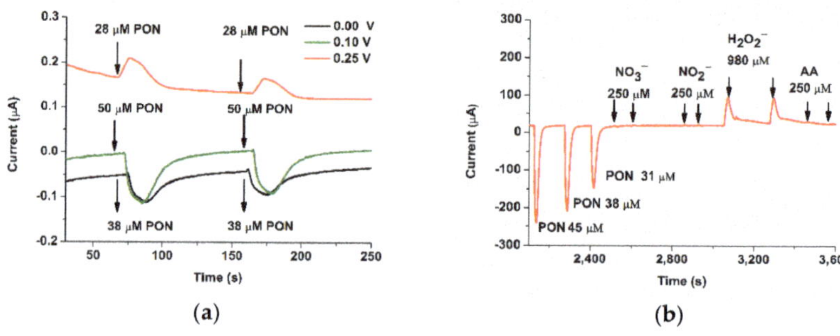

Figure 8. (**a**) Chronoamperometry (CA) spectra of the flow injection analysis (FIA) using different potentials: 0.00 V (black), 0.10 V (green), and 0.25 V (red) [1]. (**b**) Interfering species study using the FIA equipment and chronoamperometry at 0.1 V, in PBS pH 9: 250µM nitrate, 250 µM nitrite, hydrogen peroxide 980 µM and 250 µM ascorbic acid (AA), as compared to 31, 38, and 45 µM PON signals.

Determination of the optimum flow rate was done using injections of 200 µM of PON (PBS pH 12) and amperometry at 0.1 V. The curve *flow rate* vs. *current* was fitted with $R^2 = 0.9195$ with a fourth-grade polynomial function. The optimal flow rate of 0.4 mL/min and a voltage of 0.1 V were used to further optimize the detection process, as the response time of this flow rate is fast, around 10 s. Interfering species were also evaluated using chronoamperometry at 20-fold higher concentration than PON for H_2O_2 and 5-fold higher concentration for the other species. As determined also by batch Cyclic Voltammetry, hydrogen peroxide gave rise to oxidation signals, but with very low sensitivity (Figure 8b) in contrast to PON, which gave rise to sensitive reduction signals. SPCE/CoPc electrodes are known to be used for the oxidation of hydrogen peroxide at 0.4 V.

As we have already shown for in batch electrochemistry (Figure 8b), the pre-treatment of electrodes with a reduction potential might be very important before the quantification of PON. We optimized the amount of time needed to reduce the CoPc film to obtain the best CA signal for PON. We used FIA amperometry for different time periods, 0, 20, 30, 60, 120, and 180 s and the reducing potential -0.3 V, which is the redox potential for Co^{2+}/Co^{1+}. We determined that applying a potential of -0.3 V for 60 s was the optimal procedure (data not shown).

We have performed the calibration (Figure 9a) using the flow injection system, at 0.1 V with the chronoamperometric method (Figure 9b), and we obtained a sensitivity of 6.31 nA μM^{-1} ($R^2 = 0.9938$), after the pre-treatment at -0.3 V for 60 s. The calculated LOQ = 2.41 µM, the calculated LOD = 0.72 µM, and the linear range is 3–180 µM. The reproducibility varied from 95% to 99% (50 µM PON) and the RSD for each calibration concentration (in triplicates) did not exceed 10%. The analytical parameters are very good if we consider several facts: (i) our unstable oxidative anion species are hard to detect, and (ii) screen-printed carbon electrodes are disposable electrodes.

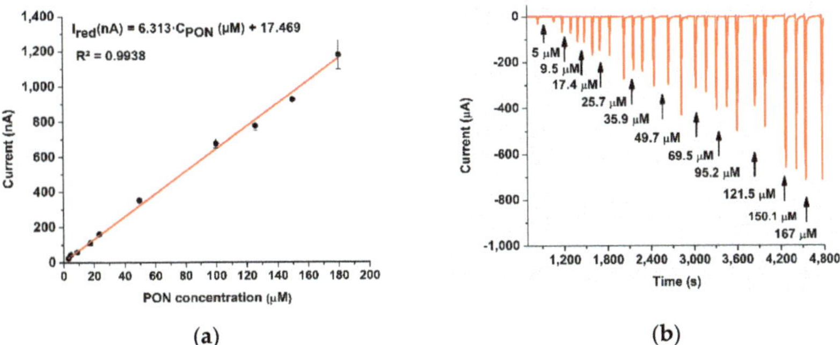

Figure 9. (**a**) Calibration curve of the SPCE/CoPc electrode for PON, PBS pH 9. (**b**) Chronoamperogram measured using the FIA system and the SPCE/CoPc electrode, E = 0.1 V, flow rate = 0.4 mL/min [1].

By replacing the electrochemical polarization of the SPCE/CoPc electrode during the 60 s, at −0.3 V with the chemical oxidation of the CoPc film using sodium borohidrate (25 mM), for 20 min, the calibration was improved to Ired (nA) = 10.843·C_{PON} (μM) − 36.484 (R^2 = 0.9925). The new LOD was equal to 0.42 μM, and the LOQ was 1.4 μM.

The LODs of our method reached nanomolar level, the same level of PON under physiological conditions. Even though we did not study the interaction of PON in the absence or in the presence of myoglobin at physiological pH, our method has physiological relevance because it could be further used for this purpose by miniaturization of the electrodes with the same electrocatalytic bio-mimetic film. The micro-dimension of the surface-active area of the electrode offers enough sensitivity to study, for example, the formation of PON by cells (the cells being also in the micrometer range) involved in the oxidative burst (micromolar range of PON), in the redox signaling, or even in the steady state of PON (nanomolar range), as the literature suggests [47,48].

3.3. UV-Vis and Determination of Synthesized PON for Kinetic Studies

Molina et al. [30] describe the influence of buffer and pH over the stability of peroxynitrite solutions ("peroxynitrite" being the term widely accepted for peroxynitrite and peroxynitrate). The rate constants depend on pH, ionic strength, temperature, scavengers, and other parameters. Several mechanisms of PON decay have been already proposed in the literature. In acidic conditions, the isomerization of peroxynitrite occurs (mainly present as ONOOH, the form that decays rapidly to nitrate), independent of total peroxynitrite concentration. If pH is ≥7, nitrite is the main decomposition product of PON (the higher the pH value and concentration of PON, the higher the conversion yield to nitrite) [31]. As mentioned in Section 2.5, the Equation (1) describing the bimolecular decomposition of PON (as opposite to the mononuclear isomerization, as termed by IUPAC) is supposed to be predominant at pH = 9, especially when PON concentration is higher than 0.1 mM [32,49].

The reaction from Equation (1) was proposed to follow second-order or pseudo first-order kinetics, depending on pH [31].

Molina et al. [30] proposed that the decomposition of PON to nitrite has more intermediate steps than the one in Equation (1) (disproportionation reaction followed by the formation of intermediary/adduct species and then followed by decomposition to nitrite). The formation of the adduct is very rapid in the range of 10^4 M^{-1} at alkaline pH. The last direct decomposition step to nitrite is much slower than the other elementary steps [31]. So, this could be the rate-determining step in these series of proposed reactions involved. In addition, the disproportionation of PON at pH 9 seems to favor the equilibrium toward ONOO⁻, as knowing the pKa of ONOOH (6.8), one can calculate the concentration of ONOOH at pH 9. For a concentration of 150 μM ONOO⁻ in PBS at pH 9, there are only 0.946 μM ONOOH, and for 50 μM ONOO⁻ in PBS pH 9, there are 0.3154 μM

ONOOH. At pH 9, one can say that ONOOH concentration is insignificant (less than 1% of the ONOO$^-$ concentration) if pKa 6.8 is taken in consideration. Nevertheless, the pKa of ONOOH depends on the ionic strength and pH, so these calculated values could be further refined [49].

If ONOOH is present in small amounts at pH 9, one can assume that the isomerization is insignificant (as the only form to isomerize is ONOOH) and PON decay should follow the pseudo first-order reaction at this pH.

Although the mechanism of PON decay is not fully understood, we took in consideration the generally agreed model depicted in Equation (1) for pH 9, as the aim was to determine the accuracy of the FIA-EC method compared to the classical UV-Vis one in determining the kinetic parameters. We checked both possibilities of reaction order for Equation (1), i.e., the pseudo first-order and the second-order.

We calculated the decay (apparent) rate constant and half-lives of PON using two methods: Method A involves the assumption that both (the (pseudo) first and second order) integrated rate laws could be possible. One finds the rate constant value from the slope of the two graphs: $ln(C_{PON})$ vs. *time*, where $t_{1/2} = ln2/k$ (for the (pseudo) first-order reaction) and $1/C_{PON}$ vs. *time*, where $t_{1/2} = 1/(C_{0PON} \cdot k)$ (for the second-order reaction). Using the slope of the most linear plot, one calculates the corresponding half-time.

The second method, Method B, also named the "half-life method", is based on the several half-time values from the *concentration* vs. *time* plot of the data, and it plots them after as $log(t_{1/2})$ vs. $log(C_{PON})$, where $n = 1$-slope (n being the (apparent) rate order). This method is more precise, as it is difficult to assess the linearity of a plot (as in Method A).

The idea was to compare the UV-Vis results with our proposed FIA-EC method$^-$ using SPCE/CoPc electrodes. The UV-Vis spectrum of synthesized genuine PON is less complicated than other synthetic methods (such as using the nitric oxide and superoxide donor-based synthesis, SIN-1). The molar extinction coefficient of 1670 M^{-1} cm^{-1} can be used to calculate the concentration of genuine ONOO$^-$ at 302 nm (Figure 10a) [33]. We evaluated the stability of our synthesized PON solution using UV-Vis spectrometry. As we have already described [1], because we performed our measurements at alkaline pH values, we had a significant amount of nitrite in the genuine PON solutions. The amount of nitrite is correlated with the absorbance at 355 nm. Calibration was performed at this wavelength using a Griess reagent-based protocol, ($y = 0.0488 \cdot C_{NO2}(\mu M) + 0.0076$, $R^2 = 0.9971$, data not shown), the amount of nitrite was also assessed to be 72 ± 5 mM for a 117 mM PON solution (improved PON synthesis), at pH 9, in PBS 0.1 M. This amount of nitrite is confirmed in the literature [24]. Nitrite reaches a plateau between pH 9 and 10 [50].

The kinetic plots at 302 nm (*concentration* vs. *time*, Figure 10b for UV-Vis data and Figure 10d for FIA-EC data) were fitted with a single exponential curve in Origin 8.5 software, as suggested in the literature [2,24,30,34,51]. The $log(t_{1/2})$ vs. $log(C_{PON})$ was plotted, and the slope was equal to *slope* $= 1 - n$, where n is the reaction order (Figure 10c,e). As it can be observed, Method B can be applied for both UV-Vis, as well as for the FIA-EC described in Section 3.2, and it will be further descried and compared with Method A (Figure 10f).

Table 2. Calculated pseudo first-order decay rates constants for PON, in PBS pH 9 (0.1M), at 25 °C, determined using classical UV-Vis method at 302 nm and the FIA-EC method, using the SPCE/CoPc (Method A).

Pseudo First-Order Decay Rates	k (s^{-1})			
Method / Samples	SPCE/CoPc	R^2	UV-Vis	R^2
PON 50 µM	0.0084 ± 0.0011	0.9830	0.0089 ± 0.0010	0.9985
PON 150 µM	0.0028 ± 0.0012	0.9975	0.0025 ± 0.0004	0.9877
Mb 15 µM + PON 50 µM	0.0134 ± 0.0010	0.9815	Not possible	-
Mb 15 µM + PON 150 µM	0.0086 ± 0.0020	0.9842	Not possible	-
Meat diluted 10 + PON 50 µM	0.0200 ± 0.0048	0.9020	Not possible	-

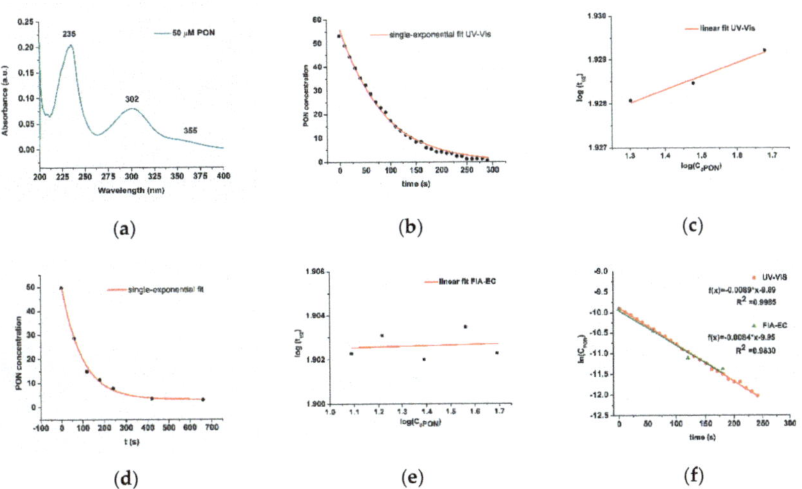

Figure 10. (**a**) UV-Vis spectra of 50 µM PON, pH 9, PBS 0.1 M. The illustration of the second method "half-lives method": plot of C_{PON} vs. *time*, fitted with a single-exponential equation (y = y$_0$·e^{-kx}) for the data obtained using (**b**) the UV-Vis method and (**d**) our proposed FIA-EC method. Plot of $log(C_{0PON})$ vs. $log(t_{1/2})$ from (**c**) UV-Vis and from (**e**) FIA-EC, fitted linearly for the determination of apparent reaction order from the slope of the equation (Method B). The illustration of the first method (Method A): (**f**) plot of $ln(C_{PON})$ vs. *time* according to pseudo first order, fitted linearly for both UV-Vis and our proposed FIA-EC method (linear correlation function, R^2 linear correlation coefficient and apparent kinetic constant is determined from these plots, data presented in Table 2). All spectra represent the data for 50 µM PON, pH 9, PBS 0.1 M.

3.4. UV-Vis Determination of Different Forms of Myoglobin

In meat extracts, different oxidation forms of myoglobin are present, as we depicted in Figure 1. The conversion of reduced myoglobin (MbFe^{2+}OH$_2$ or MbFe^{2+}O$_2$) to met-myoglobin (MbFe^{3+}(OH$_2$)) can be followed using UV-Vis due to PON scavenging activity (Figure 11): the concentration of the redMb solutions can be verified by measuring the absorbance at 417, 542, and/or 580 nm (ε_{417} = 128 mM^{-1} cm^{-1}, ε_{542} = 13.9 mM^{-1} cm^{-1}, and ε_{580} = 14.4 mM^{-1} cm^{-1}) [21], and the spectrum of metMb has a maximum of absorbance at 502 [ε_{502} = 10.2 mM^{-1}cm^{-1}] and 610 nm at pH 6.4, and the Soret band absorbance maximum is at 408 nM at pH 7.4 [52]. So, the scavenging effect could be identified using UV-Vis, but no quantification of PON decay can be done in a direct, rapid, sensitive, and selective manner.

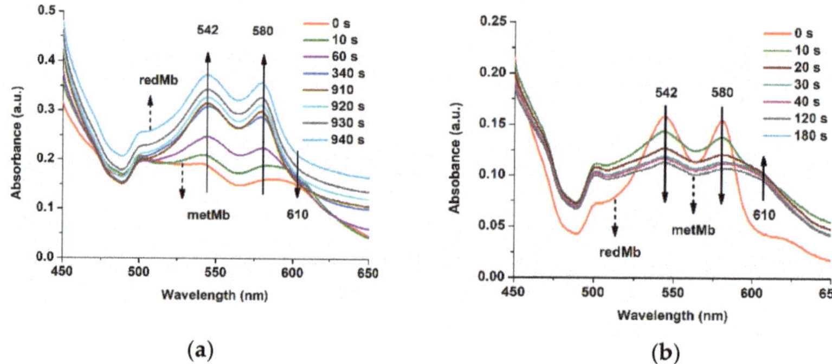

Figure 11. (**a**) UV-Vis spectra for different reaction times of 75 mM NaBH$_4$ and 25 µM metMb (PBS pH 9). (**b**) UV-Vis spectra for different incubation times (0, 10, 20, 30, 40, 120, and 180 s) of 100 µM PON with 10 µM redMb (PBS pH 9).

The quantification of the concentration of redMb was realized with measurements at 580 nm. The concentration of the stock meat extract was determined to be 480 µM, which corresponds to ca. 20 mg of myoglobin for 1 g of meat. Moreover, if we use the absorption at 525 nm (representing the isosbestic point for the absorption in visible range for the 3 forms of myoglobin) and a molar extinction coefficient of 7.6 mM^{-1}cm^{-1}, we obtained a value of 485 µM of myoglobin for the same stock solution. A more complex method for determining the content of myoglobin was described by Krzywicki, and improved by Tang, in 2004 [53].

The reactions of PON with myoglobin or meat were studied both with the electrochemical and spectrophotometric methods. Figure 12 describes the evolution of the Mb absorption peaks during Mb incubation with 50 and 150 µM PON. For PON 50 µM, almost no change was observed after 11 min of incubation, but the same incubation of metMb with 150 µM PON induced a more significant change, with decrease in absorbance at 542 and 580 nm, and the appearance of a band at 700 nm, corresponding to a qualitative evaluation of the catalytic reaction. The same incubation was studied with our FIA-EC method, also to obtain quantitative information regarding the catalyzed PON decay.

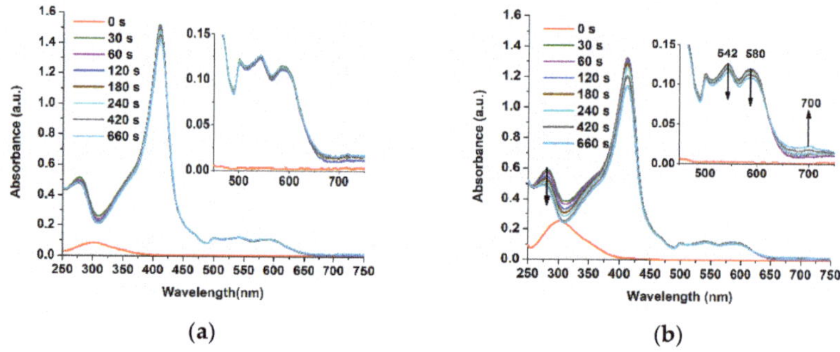

Figure 12. UV-Vis spectra of (**a**) 50 µM PON and (**b**) 150 µM PON incubated with 15 µM metMb, at different incubation periods.

3.5. Studying the Reaction of Myoglobin with Peroxynitrite with FIA-EC

The incubated solutions of 50 µM and 150 µM PON with 15 µM Mb were also investigated, in parallel, with our FIA-EC optimized method (Figure 13). It was expected that the kinetics between the two concentrations of PON to be different due to the ratio between PON and Mb, although the difference could be overcome by the kinetics of spontaneous PON decay independent of Mb, which presents opposite behavior. More precisely, increasing PON concentrations favor pseudo first-order reaction in relation to Mb and deviation to second-order reaction in the case of spontaneous PON decay. Moreover, approximatively at least eight equivalents of PON are necessary for the complete oxidation of myoglobin (depending on experimental conditions and pseudo first-order kinetics) [21]. The recovered current was converted into PON concentration with the calibration curve, and in the following sections, we discuss and compare the decay of PON using these data.

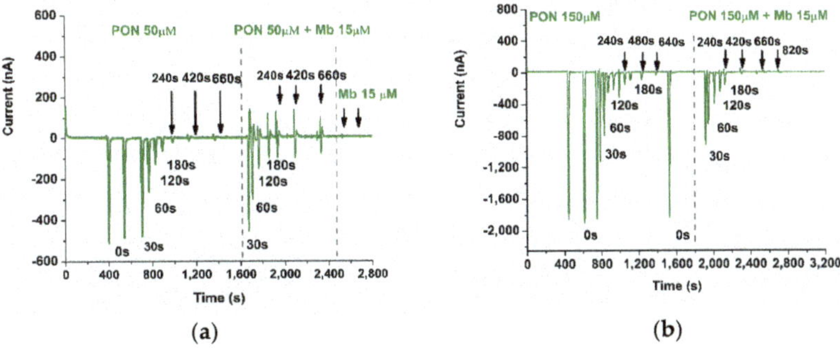

Figure 13. Chronoamperogram using the FIA-EC system for (**a**) 50 µM and (**b**) 150 µM PON, in the presence and in the absence of 15 µM metMb, at different incubation periods (0, 30, 60, 120, 240, 420, 660, 820 s). The decomposition of PON alone was studied at the same incubation periods to prove that the recovery of the current is due to the presence of PON in the metMb solution and not to the metMb itself (PBS pH 9, 0.1M, E = 0.1 V and flow rate 0.4 mL/min). The values obtained in these studies were further used for kinetic calculations.

3.6. Studying the Reaction between Myoglobin from Meat Extracts and Peroxynitrite Using FIA-EC

Meat extracts were first diluted 10-fold and analyzed with UV-Vis to determine the quantity of myoglobin. Using the isosbestic point at 525 nm, as described above, we determined 15 µM of myoglobin (independent of the oxidation form, Figure 14a). The incubation of the meat extracts with PON was studied both with UV-Vis and FIA-EC. Using the optimized calibration curve of the electrochemical method, we developed (Ired (nA) = 10.843·C_{PON} (µM)–36.484, R^2 = 0.9925), quantified PON during its incubation with both Mb 15 µM and meat extracts diluted 10 times in PBS pH 9, and compared it to a normal decomposition rate of PON, at pH 9, without any scavenger (Figure 14b).

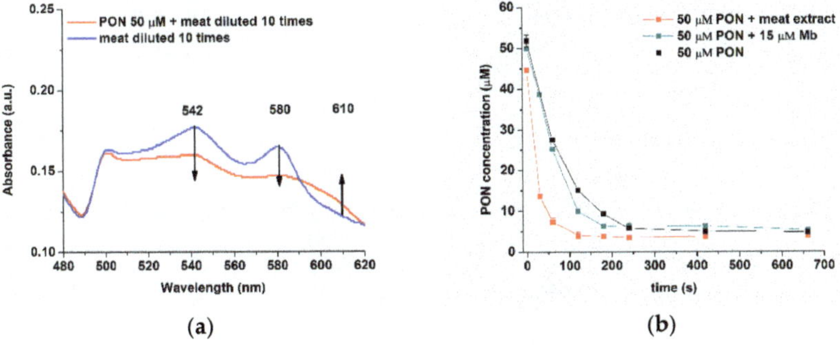

Figure 14. (**a**) UV-Vis spectra of meat diluted 10 times in PBS pH 9 in the presence of 50 µM PON and in absence, during 12 min. (**b**) The graphs of concentration (from FIA-EC) over time for 50 µM PON decay in the absence (black) or in the presence of 15 µM Mb (cyan) or meat extract (red).

It can be clearly observed that the decomposition of PON takes place at a faster rate for the PON samples incubated with both Mb and meat extracts.

Calculated apparent reaction orders, apparent rate constants, and half-lives determined using UV-Vis spectrophotometry and our FIA-EC method are presented in Tables 2–5 and further discussed in the following two sections. All measurements and calculations were done for PBS pH 9, 0.1 M, at 25 °C.

Table 3. The "half-life" method: Determined values of half-life for PON and calculation of rate orders. PBS pH 9 (0.1 M), at 25 °C, classical UV-Vis method at 302 nm and the FIA-EC method, using the SPCE/CoPc (Method B).

Half-Lives and Rate Orders	$t_{1/2}$ (s)		Calculated Values for Rate Order *	
	SPCE/CoPc	UV-Vis	SPCE/CoPc	UV-Vis
PON 50 μM	81.33 ± 8.69	84.73 ± 3.53	1.0000 ± 0.0014	1.0030 ± 0.0016
PON 150 μM	252.12 ± 2.97	230.75 ± 5.13	1.0001 ± 0.0001	1.0054 ± 0.0009
Mb 15 μM + PON 50 μM	64.83 ± 1.64	Not possible	1.0004 ± 0.0008	Not possible
Mb 15 μM + PON 150 μM	88.12 ± 0.03	Not possible	1.0001 ± 0.0001	Not possible
Meat diluted 10 + PON 50 μM	19.77 ± 0.10	Not possible	1.9442 ± 0.0587	Not possible

* Rate order = n = 1- slope of the $log(C_0)$ vs. $log(t_{1/2})$, the data were fitted with a single exponential function based on equation $y = y_0 \cdot e^{-kx}$. Estimations of the rate orders were done in the case of values lower than 10^{-4}.

Table 4. Observed first-order decay rates constants for PON, in PBS pH 9 (0.1M), at 25 °C, determined using the classical UV-Vis method at 302 nm and the FIA-EC method, using the SPCE/CoPc (Method B).

Apparent First-Order Rate Constants	k (s^{-1})	
	SPCE/CoPc	UV-Vis
PON 50 μM	0.00862 ± 0.0007	0.0080 ± 0.0018
PON 150 μM	0.00275 ± 0.0012	0.0030 ± 0.0004
Mb 15 μM + PON 50 μM	0.01690 ± 0.0010	Not possible
Mb 15 μM + PON 150 μM	0.00780 ± 0.0020	Not possible

Table 5. Calculated second-order decay rates constants for PON, in PBS pH 9 (0.1M), at 25 °C, determined using classical UV-Vis method at 302 nm and the FIA-EC method, using the SPCE/CoPc (Method B).

Apparent Second-Order Rate Constants	k (M^{-1} s^{-1})	
	SPCE/CoPc	UV-Vis
Mb 15 μM + PON 50 μM	311.87 ± 7.9600	Not possible
Meat diluted 10 + PON 50 μM	891.76 ± 220.54	Not possible

3.6.1. Estimation of the Apparent Rate Decay Orders of PON in the Absence and Presence of Myoglobin

Using Method A (Table 2), for 50 μM PON, the R^2 values for the plot according to a first-order apparent kinetics (0.9985) were higher than for a second-order one (0.9246); as for 150 μM PON, the results were not conclusive (0.9872 for pseudo first order and 0.9877 for second order). Taking in consideration Equation (1) and the excess of ONOO$^-$ comparing to ONOOH, these results from Method A are relevant and indicate pseudo first-order kinetics.

Using Method B, the apparent decay of PON is also a pseudo first-order reaction for both 50 μM and 150 μM (see Table 3), with calculated order values of 1.0030 and 1.0054 for the UV-Vis method, in comparison with 1.0000 and 1.0001 for the FIA-EC method. The differences between these two methods were less than 3% (acceptable error values).

The fitting curves for both 50 μM PON (Chi-Sqr = 3.9770, R^2 = 0.9865) and 150 μM PON Chi-Sqr = 1.9293, R^2 = 0.9990) were proper, as tolerance criteria were satisfied.

Our results are in accordance with the literature, describing that especially concentrations above 100 μM present deviations from (pseudo) first-order reactions as second-order in total peroxynitrite concentration at pH 9 [30,31]. These deviations are most pronounced at pH 9 [50].

Method B helped us to assess numerically the reaction order out of our data acquired with the FIA-EC method for the interaction of 50 and 150 μM PON with 15 μM Mb, and the calculated reaction orders were 1.0222 and 1.0001, suggesting also a pseudo first-order kinetics in both cases. Deviations in apparent second order from the single-exponential curve were higher in the case of 50 μM PON + 15 μM Mb, as expected (meaning that the model could be adjusted, as pseudo first-order kinetics were not fulfilled for PON:Mb, Chi-Sqr = 15.7763, R^2 = 0.9446). These deviations were smaller for 150 μM PON + 15 μM Mb (Chi-Sqr = 17.9047, R^2 = 0.9941), as pseudo first-order kinetics were fulfilled: PON: Mb was 10:1.

Using both methods (A and B), the only studied interaction that gave clear second-order decay for PON was the meat diluted 10 times with 50 μM (R^2 = 0.9020 vs. R^2 = 0.9968 for (pseudo) first order and second order, respectively, using method A). The reaction order assessed with Method B is 1.94, which is very close to a second order. One explanation can be that in meat extracts, the interactions are more complicated/complex than with standard Mb, as other scavengers could be present, so an apparent second-order decay is easier to determine even using Method A (which is less precise than Method B). Method B will be used further on. Nevertheless, the fitting should be replaced with a more significant equation in this case (Chi-Sqr = 11.10, R^2 = 0.9452), because this high deviation from a single-exponential equation proves once again that a second-order decay might be involved. A double-exponential equation could be helpful, as suggested in reference [54].

Nevertheless, the single-exponential fitting of pseudo first-order situations (C_{PON} = $C_{0PON} \cdot e^{-kx}$) determines k_{obs}, which allows us to detect the (apparent) second-order rate constants (k_{cat}), for a larger concentration range of scavengers or other reactants involved (especially the concentration of the catalyzer, myoglobin, or other) [24,25]. The k_{cat} obtained from a linear fit of k_{obs} vs. *catalyst concentration* will refine our findings on second-order kinetics and improve the investigation of the scavenging effect of myoglobin over PON in meat extracts or other biological samples.

3.6.2. Determination of Apparent Rate Constants and Half-Lives for the Decay of PON

Molina et al. [30] describe the observed half-life $t_{1/2}$ = 9.4 $\pm$ 0.1 s for PBS 0.07 M, at pH 8, for PON 250 μM. We have determined that apparent k = 0.0086 s^{-1} with Method A and k = 0.0078 s^{-1} with Method B (the apparent first-order decomposition rate constant), for PBS 0.1 M, at pH 9, for a concentration of 150 μM ($t_{1/2}$ = 252.12 $\pm$ 2.97). As it can be seen, in the chosen conditions, PON is more stable than in the conditions described by Molina et al., mainly because of more alkaline pH, more ionic strength in the buffer, and smaller concentrations of PON. Kissner et al. described this kind of behavior [31].

The decomposition of PON occurs faster at lower concentrations at pH 9, as it can be observed from the UV-Vis data in Table 3. The same conclusion can be drawn from our SPCE/CoPc developed method: PON 50 μM will decay faster (reaction rate of 0.426 $\times$ 10^{-6} s^{-1}, using $t_{1/2}$ = 81.33 s) than PON 150 μM (reaction rate of 0.412 $\times$ 10^{-6} s^{-1}, using $t_{1/2}$ = 252.12 s), where k = 0.693/$t_{1/2}$ and r = k[C_{PON}]. Reaction rates in the range of 10^{-6} for 50 μM PON at pH 9 are described by Kissner et al. [31], and this is in very good accordance with the data we obtained.

The myoglobin-mediated decay of PON is described as a second-order rate interaction. Most probably, the scavenging effect gives rise to ferrylmyoglobin (MbFe^{4+}=O) with a rate constant of 4.6 $\pm$ 0.2 $\times$ 10^4 M^{-1} s^{-1}, at pH 8.3, in the absence of CO$_2$, that will further react with PON with a rate constant of 1.2 $\pm$ 0.2 $\times$ 10^4 M^{-1} s^{-1} to form metMb [21]. The oxidation of redMb to metMb is not a stoichiometric reaction, as eight to 25 equivalents of PON are required for the oxidation reaction to be completed (depending on the absence or presence of CO_2), as the natural decay of PON takes place at the same time. A 10-fold excess of PON is necessary for pseudo first-order conditions to be fulfilled. As the concentration of nitrite does not influence the interaction of myoglobin with peroxynitrite in a direct manner [21], our PON synthesized using the alkaline method is suitable for kinetic calculations (or otherwise saying, the decay of PON in the presence of myoglobin is a zero-order reaction in nitrite [2,21]).

The metMb scavenges PON at a lower rate than redMb or even than ferrylMb. MetMb catalyzes the isomerization of PON to nitrate, at pH 7, at 20 °C, with a rate of $29,000 \pm 100$ $M^{-1}s^{-1}$, and an iron Fe^{3+} core of metMb is involved in this process. The k values decrease with increasing pH; thus, the decay rate is expected to be smaller for pH 9 [24]. So, scavenging PON with metMb is less effective at pH 8 (or above) than at biological pH, with a PON decay rate constant k of 2700 ± 30 $M^{-1}s^{-1}$, for pH 8, 0.1 M PBS, for 100 µM of PON, in the absence of CO_2, at 20 °C [24]. In our case, at pH 9, PBS 0.1 M, 25 °C, the apparent k value of the reaction of 15 µM myoglobin and 50 µM PON was calculated to be 311.87 ± 7.96 $M^{-1}s^{-1}$ using Method B (where deviations from pseudo first order are second order). This value is lower than the value that Herold et al. obtained, which is probably due to the pH values and the temperature difference. As far as our knowledge goes, the kinetics of this exact decay conditions were not described in the literature before. Nevertheless, more PON or target molecule (myoglobin) concentrations are to be varied for refining better apparent second orders.

The scavenging was evaluated by comparing the decomposition rate of PON in the absence and in the presence of myoglobin (from both standard and meat extracts). The standard Mb decreased the half-life on PON from 81 to 65 s (for 50 µM) and from 250 to 88 s (for 150 µM), thus increasing the rate of decomposition in both cases. As in accordance with the UV-Vis spectra (Figure 12a) of metMb incubated with 50 µM of PON, the scavenging effect of 15 µM metMb with 50 µM PON is not very effective, and the calculated $t_{1/2}$ values are in accordance with the UV-Vis data (with less 10% error values between the two methods). This is because the number of equivalents of PON (here around 3) were not sufficient to oxidize myoglobin. When we increase the number of equivalents to 10 (150 µM PON), the scavenging effect can be observed with both UV-Vis and FIA-EC.

The scavenging effect of the myoglobin from meat was stronger than we initially thought (as we expected the concentration of Mb to be similar to the standard Mb), with a $t_{1/2}$ of 19.77 ± 0.10 s. Even if we estimated the concentration of myoglobin to be around 15 µM in the extracted meat (diluted 10 times in PBS pH 9), when we compare the decay constants of PON, we can observe that the standard Mb has a lower apparent k constant $(311.87 \pm 7.96$ $M^{-1}s^{-1})$ than the meat extract $(891.76 \pm 220.54$ $M^{-1}s^{-1})$; thus, it has a higher half-life. Nevertheless, this may come from different oxidation states of the myoglobin in meat extract, and the presence of other possible scavengers in meat is not excluded. Other studies including varying concentrations of PON, myoglobin, and/or other catalysts from meat or other biological samples are still to be performed further.

4. Conclusions

Our label-free electrochemical method proposes a cobalt phthalocyanine deposition on the screen-printed carbon electrode (SPCE/CoPc) for the direct detection of peroxynitrite via electrocatalyzed reduction. This method is simple, sensitive, highly selective, rapid, and cost-effective. A simple flow injection system based on amperometric detection at potentials near +0.1 V was designed for PON determination, bringing the possibility for the automation and fast-response measurements. CA (chronoamperometry) was used to establish that changing the applied potential, the sensing mechanism of the electrode is changing/becomes easier to distinguish: the oxidation of PON occurs near 0.07 V and the reduction occurs near 0.05 V. At those low redox potentials (60 mV), the most important interfering species are less likely to appear, especially for the reduction potential.

We optimized different parameters (flow rate, reduction potential, the quantity of deposited CoPc on the surface of the electrode, determined surface coverage etc.), and we suggested a mechanism that described the electro-reduction of PON, involving two electrons, based on the electrochemical characterization of the electrodes. We also determined that the electrochemical reactions taking place at the surface of the electrode are not simply diffusion-controlled ones, and other processes such as adsorption, ligand-based processes may take place. Interfering species were studied at a 5-fold concentration compared to PON, and only hydrogen peroxide responded with a very low sensitivity and under an electro-oxidation catalyzed mechanism, not influencing our PON detection.

Compared to the literature, this is one of the few articles based on the electro-catalyzed reduction of PON. Even though we obtained good sensitivity (with an LOD of 0.4 μM compared to 1.9 nM [14] and 1.0 nM [15], but higher than 5 μM [16]), we achieved a much better selectivity against all the important interfering species (ascorbic acid, hydrogen peroxide, nitrate, and nitrite) at biological concentrations.

The optimized SPCE/CoPc calibration helped us monitor and quantify PON and the reaction between peroxynitrite and myoglobin. The literature presents the fact that the PON to redMb ratio must be 1:10 for efficient scavenging and that metMb is less efficient in scavenging PON (as the metal center is already oxidized) [24,52]. The scavenging of PON with myoglobin (in the form of metMb and redMb) decreases the amount of PON, and this decrease was measured/quantified with our SPCE/CoPc electrode. As the UV-Vis technique cannot be used to study the interaction with myoglobin (as the protein peak at 280 nm interferes with the PON peak at 302 nm), our FIA-EC method is an alternative technique that is able to study the scavenging effect of myoglobin from meat extracts toward peroxynitrite. The similarity of absorption spectra, especially in the UV zone, corresponding to redMb and meat extracts after incubation with PON, proving once again that PON is decomposed during the irreversible oxidation of redMb to metMb.

We propose a simple method that has great impact on the PON sensor choice, as it can be used to quantify PON in complex media (such as meat samples) but also to study the kinetics of PON decay. We have also demonstrated that one can study the interaction of PON with different forms of myoglobin. Kinetic studies were also performed and correlated with the literature to study the scavenging effect. Meat extracts scavenged better PON, which was probably due to different forms of Mb and the possibility of another scavenger. The scavenging effect is a second-order decay rate (when PON is not in excess, when pseudo first orders apply), and the rate constants were determined.

Our study for the kinetics should be regarded more as an alternative possible method ("proof of concept") for studying the interaction between PON and myoglobin using the FIA-EC method than a finalized kinetic study. The kinetic parameters obtained with both proposed kinetic methods (A and B) were in good correlation with each other. We can conclude that our FIA-EC method is precise enough for further studying the interaction of PON with scavengers. Evaluation of decay order for the interactions is impossible with the classical static UV-Vis method (that also has a low sensitivity toward PON, as an absorbance of 0.08 a.u. corresponds to 50 μM PON).

A further research direction is to use the already described hybrid materials (CoPc-graphene) to increase the solubility of the hybrid material in aqueous solvents and electron transfer at the surface of the electrode. The electroactive film could be also deposited on micro-electrodes to be used for the detection of PON in physiological conditions.

Author Contributions: Conceptualization, M.D. and I.S.H.; methodology I.S.H., investigation, I.S.H. and D.C.-A.; project administration, M.D. and F.O.; supervision, M.D. and F.O.; validation, M.D., visualization, M.D. and I.S.H., writing—original draft preparation, I.S.H.; writing—review and editing, M.D., I.S.H., D.C.-A. and F.O. All authors have read and agreed to the published version of the manuscript.

Funding: This work was supported by Romanian National Authority for Scientific Research and Innovation, CCDI-UEFISCDI, Project PN-II-ID-PCE-2011-3-1076 and project ERANET-MANUNET-NITRISENS, contract no. 216/2020 within PNCDI III. The authors also acknowledge the financial support to Ministry of Research and Innovation, Nucleu Programme ChemEmergent, Project PN.19.23.01.01—Smart-Bi.

Institutional Review Board Statement: Not applicable.

Informed Consent Statement: Not applicable.

Acknowledgments: The authors are grateful to Sanda Doncea for FTIR measurements and Alina Vasilescu for all the advices and the cobalt phthalocyanine.

Conflicts of Interest: The authors declare no conflict of interest. The funders had no role in the design of the study; in the collection, analyses, or interpretation of data; in the writing of the manuscript, or in the decision to publish the results.

References

1. Hosu, I.S.; Constantinescu-Aruxandei, D.; Oancea, F.; Doni, M. Studying the Reaction of Peroxynitrite with Myoglobin for Meat Extract Samples Using Cobalt Phthalocyanine-Modified Screen-Printed Carbon Electrodes and a Flow Injection Analysis System. *Proceedings* **2020**, *60*, 46. [CrossRef]
2. Exner, M.; Herold, S. Kinetic and mechanistic studies of the peroxynitrite-mediated oxidation of oxymyoglobin and oxyhemoglobin. *Chem. Res. Toxicol.* **2000**, *13*, 287–293. [CrossRef]
3. Vasilescu, A.; Vezeanu, A.; Liu, Y.; Hosu, I.S.; Worden, R.M.; Peteu, S.F. Meat Freshness: Peroxynitrite's Oxidative Role, Its moltural Scavengers, and New Measuring Tools. In *Instrumental Methods for the Analysis and Identification of Bioactive Molecules*; Jayprakasha, G.K., Patil, B.S., Pellati, F., Eds.; American Chemical Society: Washington, DC, USA, 2014; pp. 303–332.
4. Brannan, R.G.; Decker, E.A. Peroxynitrite-induced oxidation of lipids: Implications for muscle foods. *J. Agric. Food Chem.* **2001**, *49*, 3074–3079. [CrossRef] [PubMed]
5. Jeong, J.Y.; Kim, G.D.; Yang, H.S.; Joo, S.T. *Pigments and Color of Muscle Foods*; CRC Press: Boca Raton, FL, USA, 2014; pp. 44–61.
6. Møller, J.K.S.; Skibsted, L.H. Myoglobins: The link between discoloration and lipid oxidation in muscle and meat. *Quim. Nova* **2006**, *29*, 1270–1278. [CrossRef]
7. Szunerits, S.; Peteu, S.; Boukherroub, R. CHAPTER 4 Peroxynitrite-Sensitive Electrochemically Active Matrices. In *Peroxynitrite Detection in Biological Media: Challenges and Advances*; The Royal Society of Chemistry: Cambridge, UK, 2016; pp. 63–77.
8. Li, M.; Gong, X.; Li, H.-W.; Han, H.; Shuang, S.; Song, S.; Dong, C. A fast detection of peroxynitrite in living cells. *Anal. Chim. Acta* **2020**, *1106*, 96–102. [CrossRef] [PubMed]
9. Wada, M.; Kira, M.; Kido, H.; Ikeda, R.; Kuroda, N.; Nishigaki, T.; Nakashima, K. Semi-micro flow injection analysis method for evaluation of quenching effect of health foods or food additive antioxidants on peroxynitrite. *Luminescence* **2011**, *26*, 191–195. [CrossRef] [PubMed]
10. Ruzicka, J.; Ruzicka, J. Flow injection analysis—A survey of its potential as solution handling and data gathering technique in chemical research and industry. *Z. Anal. Chem.* **1988**, *329*, 653–655. [CrossRef]
11. Hosu, I.S.; Wang, Q.; Vasilescu, A.; Peteu, S.F.; Raditoiu, V.; Railian, S.; Zaitsev, V.; Turcheniuk, K.; Wang, Q.; Li, M.; et al. Cobalt phthalocyanine tetracarboxylic acid modified reduced graphene oxide: A sensitive matrix for the electrocatalytic detection of peroxynitrite and hydrogen peroxide. *RSC Adv.* **2015**, *5*, 1474–1484. [CrossRef]
12. Cortes, J.S.; Granados, S.G.; Ordaz, A.A.; Jimenez, J.A.L.; Griveau, S.; Bedioui, F. Electropolymerized manganese tetraaminophthalocyanine thin films onto platinum ultramicroelectrode for the electrochemical detection of peroxynitrite in solution. *Electroanalysis* **2007**, *19*, 61–64. [CrossRef]
13. Oprea, R.; Peteu, S.F.; Subramanian, P.; Qi, W.; Pichonat, E.; Happy, H.; Bayachou, M.; Boukherroub, R.; Szunerits, S. Peroxynitrite activity of hemin-functionalized reduced graphene oxide. *Analyst* **2013**, *138*, 4345–4352. [CrossRef]
14. Koh, W.C.A.; Son, J.I.; Choe, E.S.; Shim, Y.-B. Electrochemical Detection of Peroxynitrite Using a Biosensor Based on a Conducting Polymer–Manganese Ion Complex. *Anal. Chem.* **2010**, *82*, 10075–10082. [CrossRef]
15. Mason, R.; Jacob, R.; Corbalan, J.; Szczesny, D.; Matysiak, K.; Malinski, T. The favorable kinetics and balance of nebivolol-stimulated nitric oxide and peroxynitrite release in human endothelial cells. *BMC Pharmacol. Toxicol.* **2013**, *14*, 48. [CrossRef] [PubMed]
16. Hosu, I.S.; Constantinescu-Aruxandei, D.; Jecu, M.-L.; Oancea, F.; Badea Doni, M. Peroxynitrite Sensor Based on a Screen Printed Carbon Electrode Modified with a Poly(2,6-dihydroxynaphthalene) Film. *Sensors* **2016**, *16*, 1975. [CrossRef] [PubMed]
17. Boni, A.C.; Wong, A.; Dutra, R.A.F.; Sotomayor, M.D.T. Cobalt phthalocyanine as a biomimetic catalyst in the amperometric quantification of dipyrone using FIA. *Talanta* **2011**, *85*, 2067–2073. [CrossRef] [PubMed]
18. Angnes, L.; Azevedo, C.M.N.; Araki, K.; Toma, H.E. Electrochemical detection of NADH and dopamine in flow analysis based on tetraruthenated porphyrin modified electrodes. *Anal. Chim. Acta.* **1996**, *329*, 91–95. [CrossRef]
19. Channon, R.B.; Joseph, M.B.; Bitziou, E.; Bristow, A.W.T.; Ray, A.D.; Macpherson, J.V. Electrochemical Flow Injection Analysis of Hydrazine in an Excess of an Active Pharmaceutical Ingredient: Achieving Pharmaceutical Detection Limits Electrochemically. *Anal. Chem.* **2015**, *87*, 10064–10071. [CrossRef]

20. Santos, A.M.; Silva, T.A.; Vicentini, F.C.; Fatibello-Filho, O. Flow injection analysis system with electrochemical detection for the simultaneous determination of nanomolar levels of acetaminophen and codeine. *Arab. J. Chem.* **2020**, *13*, 335–345. [CrossRef]

21. Herold, S.; Exner, M.; Boccini, F. The mechanism of the peroxynitrite-mediated oxidation of myoglobin in the absence and presence of carbon dioxide. *Chem. Res. Toxicol.* **2003**, *16*, 390–402. [CrossRef]

22. Ascenzi, P.; Marinis, E.D.; Masi, A.d.; Ciaccio, C.; Coletta, M. Peroxynitrite scavenging by ferryl sperm whale myoglobin and human hemoglobin. *Biochem. Biophys. Res. Commun.* **2009**, *390*, 27–31. [CrossRef]

23. Connolly, B.J.; Brannan, R.G.; Decker, E.A. Potential of peroxynitrite to alter the color of myoglobin in muscle foods. *J. Agric. Food Chem.* **2002**, *50*, 5220–5223. [CrossRef] [PubMed]

24. Herold, S.; Shivashankar, K. Metmyoglobin and methemoglobin catalyze the isomerization of peroxynitrite to nitrate. *Biochemistry* **2003**, *42*, 14036–14046. [CrossRef] [PubMed]

25. Kalinga, S. Peroxynitrite-Induced Modifications of Myoglobin: A Kinetic and Mechanistic Study. Ph.D. Thesis, ETH Zurich, Zürich, Switzerland, 2005.

26. Connolly, B.J.; Decker, E.A. Peroxynitrite induced discoloration of muscle foods. *Meat Sci.* **2004**, *66*, 499–505. [CrossRef]

27. Robinson, K.M.; Beckman, J.S. Synthesis of Peroxynitrite from Nitrite and Hydrogen Peroxide. In *Methods in Enzymology*; Academic Press: Cambridge, MA, USA, 2005; Volume 396, pp. 207–214. [CrossRef]

28. Badea, M.; Amine, A.; Benzine, M.; Curulli, A.; Moscone, D.; Lupu, A.; Volpe, G.; Palleschi, G. Rapid and Selective Electrochemical Determination of Nitrite in Cured Meat in the Presence of Ascorbic Acid. *Microchim. Acta* **2004**, *147*, 51–58. [CrossRef]

29. Herold, S. Kinetic and spectroscopic characterization of an intermediate peroxynitrite complex in the nitrogen monoxide induced oxidation of oxyhemoglobin. *FEBS Lett.* **1998**, *439*, 85–88. [CrossRef]

30. Molina, C.; Kissner, R.; Koppenol, W.H. Decomposition kinetics of peroxynitrite: Influence of pH and buffer. *Dalton Trans.* **2013**, *42*, 9898–9905. [CrossRef] [PubMed]

31. Kissner, R.; Koppenol, W.H. Product Distribution of Peroxynitrite Decay as a Function of pH, Temperature, and Concentration. *J. Am. Chem. Soc.* **2002**, *124*, 234–239. [CrossRef]

32. Levine, I. *Physical Chemistry*; McGraw-Hill Education: New York, NY, USA, 2009.

33. Ischiropoulos, H.; Al-Mehdi, A.B. Peroxynitrite-mediated oxidative protein modifications. *FEBS Lett.* **1995**, *364*, 279–282. [CrossRef]

34. Carballal, S.; Bartesaghi, S.; Radi, R. Kinetic and mechanistic considerations to assess the biological fate of peroxynitrite. *Biochim. Biophys. Acta Gen. Subj.* **2014**, *1840*, 768–780. [CrossRef] [PubMed]

35. Herold, S.; Shivashankar, K.; Mehl, M. Myoglobin Scavenges Peroxynitrite without Being Significantly Nitrated. *Biochemistry* **2002**, *41*, 13460–13472. [CrossRef]

36. Dashteh, M.; Safaiee, M.; Baghery, S.; Zolfigol, M.A. Application of cobalt phthalocyanine as a nanostructured catalyst in the synthesis of biological henna-based compounds. *Appl. Organomet. Chem.* **2019**, *33*, e4690. [CrossRef]

37. Kumar, P.; Kumar, A.; Sreedhar, B.; Sain, B.; Ray, S.S.; Jain, S.L. Cobalt Phthalocyanine Immobilized on Graphene Oxide: An Efficient Visible–Active Catalyst for the Photoreduction of Carbon Dioxide. *Chem. A Eur. J.* **2014**, *20*, 6154–6161. [CrossRef]

38. Tackley, D.R.; Dent, G.; Ewen Smith, W. IR and Raman assignments for zinc phthalocyanine from DFT calculations. *Phys. Chem. Chem. Phys.* **2000**, *2*, 3949–3955. [CrossRef]

39. Szybowicz, M.; Bała, W.; Dümecke, S.; Fabisiak, K.; Paprocki, K.; Drozdowski, M. Temperature and orientation study of cobalt phthalocyanine CoPc thin films deposited on silicon substrate as studied by micro-Raman scattering spectroscopy. *Thin Solid Film.* **2011**, *520*, 623–627. [CrossRef]

40. Aroca, R.; Pieczonka, N.; Kam, A.P. Surface-enhanced Raman scattering and SERRS imaging of phthalocyanine mixed films. *J. Porphyr. Phthalocyanines* **2001**, *5*, 25–32. [CrossRef]

41. Shih, Y.; Zen, J.M.; Kumar, A.S.; Chen, P.Y. Flow injection analysis of zinc pyrithione in hair care products on a cobalt phthalocyanine modified screen-printed carbon electrode. *Talanta* **2004**, *62*, 912–917. [CrossRef]

42. Fotouhi, L.; Hashkavayi, A.B.; Heravi, M.M. Electrochemical behaviour and voltammetric determination of sulphadiazine using a multi-walled carbon nanotube composite film-glassy carbon electrode. *J. Exp. Nanosci.* **2013**, *8*, 947–956. [CrossRef]

43. Foster, C.W.; Pillay, J.; Metters, J.P.; Banks, C.E. Cobalt phthalocyanine modified electrodes utilised in electroanalysis: Nanostructured modified electrodes vs. bulk modified screen-printed electrodes. *Sensors* **2014**, *14*, 21905–21922. [CrossRef] [PubMed]

44. Bedioui, F.; Quinton, D.; Griveau, S.; Nyokong, T. Designing molecular materials and strategies for the electrochemical detection of nitric oxide, superoxide and peroxynitrite in biological systems. *Phys. Chem. Chem. Phys.* **2010**, *12*, 9976–9988. [CrossRef]

45. Storkey, C.; Pattison, D.I.; Ignasiak, M.T.; Schiesser, C.H.; Davies, M.J. Kinetics of reaction of peroxynitrite with selenium- and sulfur-containing compounds: Absolute rate constants and assessment of biological significance. *Free Radic. Biol. Med.* **2015**, *89*, 1049–1056. [CrossRef]

46. Møinichen, C. Microfluidic Flow Cells for Studies of Electrochemical Reactions. Master's Thesis, Institutt for Materialteknologi, Oslo, Norway, 2012.

47. Ai, F.; Chen, H.; Zhang, S.H.; Liu, S.Y.; Wei, F.; Dong, X.Y.; Cheng, J.K.; Huang, W.H. Real-time monitoring of oxidative burst from single plant protoplasts using microelectrochemical sensors modified by platinum nanoparticles. *Anal. Chem.* **2009**, *81*, 8453–8458. [CrossRef]

48. Amatore, C.; Arbault, S.; Bruce, D.; de Oliveira, P.; Erard, M.; Vuillaume, M. Characterization of the electrochemical oxidation of peroxynitrite: Relevance to oxidative stress bursts measured at the single cell level. *Chem. Eur. J.* **2001**, *7*, 4171–4179. [CrossRef]

49. Kissner, R.; Nauser, T.; Bugnon, P.; Lye, P.G.; Koppenol, W.H. Formation and properties of peroxynitrite as studied by laser flash photolysis, high-pressure stopped-flow technique, and pulse radiolysis. *Chem. Res. Toxicol.* **1998**, *11*, 1285–1292. [CrossRef]
50. Kirsch, M.; Korth, H.-G.; Wensing, A.; Sustmann, R.; de Groot, H. Product formation and kinetic simulations in the pH range 1–14 account for a free-radical mechanism of peroxynitrite decomposition. *Arch. Biochem. Biophys.* **2003**, *418*, 133–150. [CrossRef] [PubMed]
51. Ascenzi, P.; De Simone, G.; Tundo, G.R.; Platas-Iglesias, C.; Coletta, M. Ferric nitrosylated myoglobin catalyzes peroxynitrite scavenging. *J. Biol. Inorg. Chem.* **2020**, *25*, 361–370. [CrossRef]
52. Rehmann, F.-J. Mechanistic Studies of the Nitrogen Monoxide-and Nitrite-Mediated Reduction of Ferryl Myoglobin and Ferryl Hemoglobin. Ph.D. Thesis, ETH Zürich, Zürich, Switzerland, 2002.
53. Tang, J.; Faustman, C.; Hoagland, T.A. Krzywicki Revisited: Equations for Spectrophotometric Determination of Myoglobin Redox Forms in Aqueous Meat Extracts. *J. Food Sci.* **2004**, *69*, C717–C720. [CrossRef]
54. Carballal, S.; Cuevasanta, E.; Yadav, P.K.; Gherasim, C.; Ballou, D.P.; Alvarez, B.; Banerjee, R. Kinetics of Nitrite Reduction and Peroxynitrite Formation by Ferrous Heme in Human Cystathionine β-Synthase*. *J. Biol. Chem.* **2016**, *291*, 8004–8013. [CrossRef]

Article

Market Perspectives and Future Fields of Application of Odor Detection Biosensors within the Biological Transformation—A Systematic Analysis [†]

Johannes Full [1,*], Yannick Baumgarten [1], Lukas Delbrück [1], Alexander Sauer [1,2] and Robert Miehe [1]

[1]   Fraunhofer Institute of Manufacturing Engineering and Automation IPA, 70569 Stuttgart, Germany; yannick.baumgarten@ipa.fraunhofer.de (Y.B.); lukas.delbrueck@ipa.fraunhofer.de (L.D.); alexander.sauer@ipa.fraunhofer.de (A.S.); robert.miehe@ipa.fraunhofer.de (R.M.)

[2]   Institute for Energy Efficiency in Production (EEP), University of Stuttgart, 70569 Stuttgart, Germany

[*]   Correspondence: johannes.full@ipa.fraunhofer.de; Tel.: +49-711-970-1434

[†]   This paper is an extended version of our paper published in: Full, J.; Delbrück, L.; Sauer, A.; Miehe, R. Market Perspectives and Future Field of Application of Odor Detection Biosensors—Systematic Analysis. In Proceedings of the 1st International Electronic Conference on Biosensors, 2–17 November 2020.

Abstract: The technological advantages that biosensors have over conventional technical sensors for odor detection and the role they play in the biological transformation have not yet been comprehensively analyzed. However, this is necessary for assessing their suitability for specific fields of application as well as their improvement and development goals. An overview of biological basics of olfactory systems is given and different odor sensor technologies are described and classified in this paper. Specific market potentials of biosensors for odor detection are identified by applying a tailored methodology that enables the derivation and systematic comparison of both the performance profiles of biosensors as well as the requirement profiles for various application fields. Therefore, the fulfillment of defined requirements is evaluated for biosensors by means of 16 selected technical criteria in order to determine a specific performance profile. Further, a selection of application fields, namely healthcare, food industry, agriculture, cosmetics, safety applications, environmental monitoring for odor detection sensors is derived to compare the importance of the criteria for each of the fields, leading to market-specific requirement profiles. The analysis reveals that the requirement criteria considered to be the most important ones across all application fields are high specificity, high selectivity, high repeat accuracy, high resolution, high accuracy, and high sensitivity. All these criteria, except for the repeat accuracy, can potentially be better met by biosensors than by technical sensors, according to the results obtained. Therefore, biosensor technology in general has a high application potential for all the areas of application under consideration. Health and safety applications especially are considered to have high potential for biosensors due to their correspondence between requirement and performance profiles. Special attention is paid to new areas of application that require multi-sensing capability. Application scenarios for multi-sensing biosensors are therefore derived. Moreover, the role of biosensors within the biological transformation is discussed.

Keywords: odor sensor; market analysis; technology assessment; application field; performance profile; requirement profile; biointelligence; biological transformation

Citation: Full, J.; Baumgarten, Y.; Delbrück, L.; Sauer, A.; Miehe, R. Market Perspectives and Future Fields of Application of Odor Detection Biosensors within the Biological Transformation—A Systematic Analysis . *Biosensors* **2021**, *11*, 93. https://doi.org/10.3390/bios11030093

Received: 12 February 2021
Accepted: 20 March 2021
Published: 23 March 2021

Publisher's Note: MDPI stays neutral with regard to jurisdictional claims in published maps and institutional affiliations.

1. Introduction

This paper is the extended version of the proceedings paper presented at the 1st International Electronic Conference on Biosensors, 2–17 November 2020 [1].

The biological transformation is viewed by many as the next technological leap [2–4]. As such, it describes an interdisciplinary innovation pathway that seeks to increase the use of biological materials, structures, processes and organisms in technical systems through intensive collaboration and combination of production, information and biotechnology [2,5].

According to Miehe et al. [6], the development of the biological transformation can be divided into three modes. The first mode, bioinspiration, describes the transfer of evolutionarily optimized biological principles and processes to technology. In this context, the well-established research field of biomimicry (the transfer of phenomena, systems, elements or processes from nature to technology) is often mentioned, which has already led to numerous developments, for instance in lightweight construction or information technology (artificial intelligence, swarm intelligence, etc.). In addition to inspiration, first steps towards an integration of biological systems in production and products can be seen today. This second development mode includes the biotechnological production of vaccines or enzymes, for example as well as other innovative technologies, some of which show great potential, such as additive manufacturing of biobased materials [7]. The latter also include tissue engineering technologies that enable direct cell growth in such a way that specific shapes are achieved (e.g., bioprinting of functional tissues) [8,9]. The third development mode involves the interaction between technical, biological and information systems. This form of systemic interaction enables the creation of self-regulating value-added systems and products, which are summarized under the term biointelligence [6]. A system or product is considered to be biointelligent if there is a real time exchange of information between biological and technical components [10]. As part of a comprehensive pilot survey commissioned by the German government with regard to the biological transformation, biosensors were identified as one of the key enabling technologies. In this course, among others, odor-detecting biosensors were presented as promising examples [6,11].

Research has been conducted on odor sensors since the 1980s [12]. Over the years, various methods and technologies have been developed, ranging from analytical instruments such as mass spectrometry to smaller sensor arrays collectively known as "electronic noses". Although analytical instruments are considered the "gold standard" in medical diagnostics, for example, the current trend is towards smaller and less expensive sensors for use in the field [13]. Numerous sensor arrays such as metal-oxide (MOX) sensors or carbon nanotubes [14,15], e.g., for the detection of chemical substances and gases (NO, O_2, CO, H_2, etc.) have demonstrated high sensitivity at low cost [13,16,17]. However, most of these sensors face challenges in terms of specificity, dependence on external conditions, and ability to discriminate between analytes, to name a few [13]. While advances are being made in electronic noses [18,19], biosensors appear to be a promising alternative. New developments in biotechnology, such as gene editing methods (CRISPR, etc.), make it possible to develop odor sensors that are able to identify gases and volatile organic compounds (VOCs) in low concentrations with high selectivity by specifically modifying the biological receptor components [20,21]. Because of their specific properties, these new technologies manage to open up new application possibilities. These include, for example, new forms of environmental monitoring or reliable testing of food quality [12]. It is essential to specify these application possibilities at an early stage in order to enable a more targeted market entry. To this date a comprehensive analysis of the requirements for application fields of odor detection sensors has not been presented nor has a meta-analysis showing the specific advantages of biosensors in this context. This paper thus addresses the following research questions:

1. What are the specific market potentials of odor detection biosensors?
2. What are therefore the most promising application fields for odor detection biosensors?
3. What new fields of application can arise for odor detection biosensors from their specific properties?
4. What role do odorant detection biosensors play in the biological transformation of industrial value creation?

This paper presents an evaluation method oriented towards the technology potential analysis of Spath et al. [22]. In this case, it is adapted to develop requirement profiles for different application areas and performance profiles for bio-based odor sensor technologies. By comparing the performance and requirement profiles, these questions can be answered and fundamental statements about application-specific market potentials for biosensors can be made. An overview of biological basics of olfactory systems is given in the following chapter and different odor sensor technologies are described and classified in order to create a common and consistent understanding.

2. Basics

All odor detecting biosensors are based on the fundamental principles of odor sensing in biological system. Therefore, the following section first gives an overview of the biological basics of the olfactory system before addressing the different odor sensor technologies and their classifications. Furthermore, the technical criteria used to create the performance profile are briefly described. At the end of this section, potential markets and applications for odor sensors are discussed.

2.1. Biological Foundations of the Olfactory System

Before addressing the available odor sensing technologies, the basic principles of odor sensing in biological systems are outlined. The system responsible for the reception of the stimulus initiated by odorants and the transmission and processing of the signals is referred to as the olfactory system. Stimulus perception occurs inside the nasal cavity on a sensory area of a few square centimeters, called the olfactory epithelium. Air reaches the olfactory epithelium via two access points, the orthonasal access during inhalation and the retronasal access during exhalation [23]. Figure 1 illustrates the structure of the olfactory epithelium which mainly comprises three cell types: olfactory sensory neurons, sustentacular cells, and basal cells. The olfactory sensory neurons are primary sensory cells and are constantly renewed as they lie unprotected on the surface and come into direct contact with hazardous substances in the breath [24]. They form dendrites, at the ends of which many small hairs, so-called cilia, are located which serve to enlarge the surface [25]. The cilia protrude into the mucus layer, which protects the skin from drying out and plays an important function in trapping VOCs [26]. The mucus contains water-soluble odorant-binding proteins that form a pocket for binding the VOCs and are therefore considered to play a major role in the transport the VOCs to the olfactory sensory neurons. These neurons contain receptors in the cell membrane of the cilia to which the VOCs bind and thus trigger transduction [23].

The transduction process of sensory olfactory neurons shown in Figure 1 begins upon binding of a VOC to a receptor. This G-protein-coupled receptor is a transmembrane protein in the olfactory cilia of the olfactory sensory cell. Extracellular binding of an odorant results in activation of the receptor, triggering an intracellular signaling cascade that is spatially separated from the receptor. The complex transduction process results in the influx of positive ions, namely calcium and sodium, and the efflux of chlorine ions.

After transduction, the intensity of the stimulus is mirrored in the amplitude of the triggered membrane potential. The potential is conducted to the axon hillock where, after reaching a threshold value, it is converted into electrical impulses of equal amplitude, called action potentials. In this way, the intensity of the stimulus is encoded in the frequency of the action potentials [27].

The process and structure for transmission and preprocessing of the action potentials before they enter the brain is substantially determined by the genetically encoded type of receptors. In the human genome, coding sequences for olfactory receptors are present in approximately 950–1000 genes, although more than 50% of these are nonfunctional due to mutations [25]. Expression of the active genes leads to the formation of a receptor. Only one type of receptor is formed per olfactory sensory neuron [23]. A receptor shows specificity for an odorant or a class of odorants of similar molecular structure. However, activation of the receptor is also possible by other odorants for which the receptor shows a lower affinity. At sufficiently high concentrations, the probability of binding substances for which the receptor shows a lower affinity increases, so that a signal may well result [25]. In addition, the receptor also determines where the signal from the olfactory sensory neuron is directed and processed. Thus, the axons of all olfactory sensory neurons of one receptor type lead to a collecting point, the so-called glomeruli, in the olfactory bulb. This is a significant first step of signal preprocessing since, therefore, subsequent processing structures only have to process signals from 4000 glomeruli instead of the signals from 100 million olfactory sensory neurons [26]. Accordingly, the principle that one odorant matches one specific receptor type, i.e., one olfactory sensory neuron, whose depolarization leads directly to odor sensation does not apply. Instead, several odorants can bind to several receptor types, i.e., various olfactory sensory neurons, to different degrees, which generates a characteristic pattern of activity in the olfactory epithelium that is transmitted via the olfactory signaling pathway to the brain for pattern analysis [23]. The ability to encode odorants combinatorically results in an almost unlimited differentiability of the very high variability of odorant molecules [25].

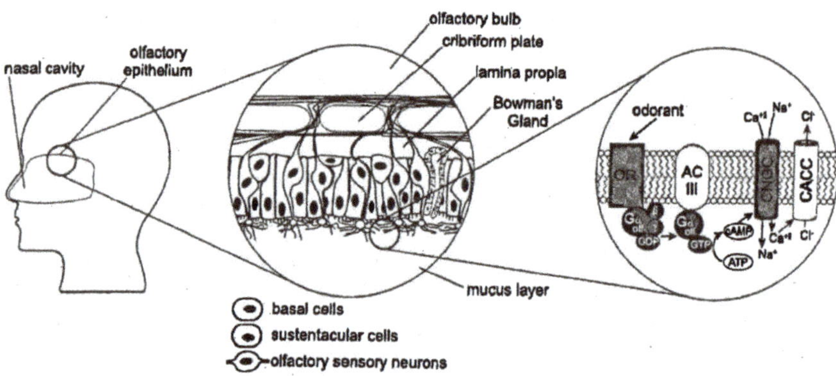

Figure 1. Structure of the human olfactory system according to Cave et al. [28]. The olfactory epithelium, located in the nasal cavity, mainly consists of basal and sustentacular cells as well as olfactory sensory neurons. On the upper side, neurons run through the cribriform plate of the skull towards the olfactory bulb of the brain. On the lower side, their thin cilia protrude in the mucus layer secreted by the Bowman's gland. Odorant receptor proteins embedded in the ciliary membrane allow the binding of volatile organic compounds (VOCs) solubilized in mucus. Binding of a VOC to the receptor triggers a molecular transduction cascade, as shown on the right, resulting in an influx of mainly calcium ions. Additional efflux of chloride ions is thought to further increase the depolarization.

2.2. Types of Odor Sensing Technologies

Odor detection technologies can be divided into biosensors and technical sensors, each with several subgroups. Figure 2 illustrates a classification scheme developed in accordance with [20,28–31]. Biosensors contain integrated biological elements, such as cells, cell tissue, proteins, or nanovesicles, which are fundamental for their functionality and rely in part on the structures and processes outlined in the previous section. Technical sensors consist exclusively of technical components and can be divided into so-called electronic noses and conventional instrumental analysis. In the following, these technologies are briefly described.

An electronic nose is a technical system consisting of several chemical sensors that are connected to form a sensor array. Among the various sensor types shown in Figure 2, metal oxide (MOX) and conducting polymer sensors are the most commonly used [32]. These sensors form the detection element of the electronic nose and convert the chemical information into an analytical signal. Depending on the number and type of sensors used, the result of a measurement may be a complex signal pattern [33]. This pattern has to be compared with a reference pattern derived primarily from previous knowledge acquired from an existing data set. Only by matching the signal pattern with the reference pattern, a result with analytical significance is obtained [32]. Conventional analytical methods include, for example, gas chromatography and mass spectrometry. Here, the mixtures of substances in a liquid or gaseous state are examined for their chemical composition using different physical measuring principles, such as the detection of mass-to-charge ratio in mass spectrometry or polarity in gas chromatography [20]. Biosensors, rather than conventional sensors, use bio-elements such as proteins, nanovesicles, individual cells, or entire cell tissues as recognition elements. In a biosensor, the analyte to be measured docks to a bioreceptor. This creates a specific compound leading to a biochemical reaction that can be technically recorded and evaluated. For example, the reaction may involve a change in the thickness of the bioreceptor layer, the refractive index, light absorption, or electrical charge. These changes are detected by means of a transducer and converted into a signal, which is usually amplified and processed by an electronic system. Thus, a specific signal is generated for each specific substance [34].

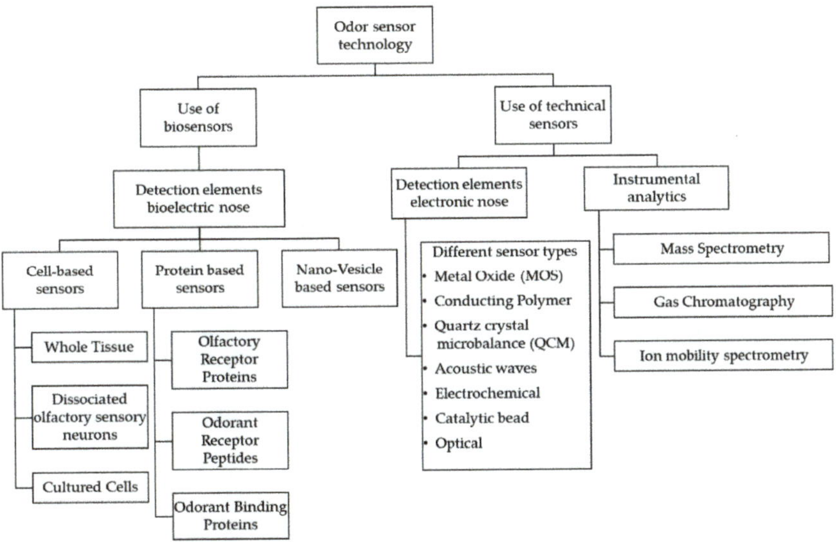

Figure 2. Overview of different technologies for odor detection. Own illustration, based on [20,28–31].

As depicted in Figure 2, bioelectronic nose technologies comprise of three categories: cell-based, protein-based, and nanovesicle-based. Cave et al. [28] provides a comprehensive overview of these categories, which are briefly summarized below. A number of protein-based approaches aim to utilize olfactory receptor proteins embedded in membrane fragments, nanodiscs, or nanovesicles as the receiving element [35–38]. Membrane fragments of bacterial or yeast cells that express olfactory receptor proteins are solubilized with detergents in order to implement them in the sensor. The detergent acts as solubilizer and mimics the cell membrane environment, so that the olfactory receptor protein maintains its structure and function. This method is limited by missing other proteins, either in cytosol or membrane, relevant for e.g., signal transduction. Moreover, approaches in which odorant receptor proteins are isolated from cell membrane fragments and inserted into nanodiscs, synthetic membranes, including transmembrane proteins that help establishing an environment very similar to the original one, still separate the olfactory receptor proteins. Through sonication and centrifugation or chemical treatment of heterologous cells that express odorant receptor proteins along with proteins responsible for transduction, so-called nanovesicles can be formed. These are small spherical membrane fragments that contain all proteins for complete transduction cascade. Other authors see these approaches as a separate category of bioelectronic noses [29]. They may argue that spherical nanovesicles represent an intermediate form between cells and membrane fragments or nanodiscs. Irrespective of the applied introduced approaches, quartz microbalance electrode, electrochemical impedance spectroscopy, field effect transistors are often used as detectors. Challenges remain the intensive labor, reproducibility of performance, due to varying amounts of olfactory receptor proteins and variability in performance depending on temperature and humidity. Instead of monitoring the level of interaction of complex odorant receptor proteins with VOCs, simpler peptides derived from odorant receptor proteins can also be used offering the advantage of higher stability. Usually, these peptides are coupled with field-effect transistors and achieve a high sensitivity and reproducibility [39,40]. However, the number of available peptides is currently very limited [28]. As described in Section 2.1, odorant binding proteins are thought to facilitate the transport of VOCs through the mucus to the cilia of the olfactory sensory neuron. Due to this level of interaction with VOCs, they are interesting for an implementation in odorant sensors [41–43]. In comparison to sensors utilizing surface plasmon resonance, electrochemical impedance spectroscopy, sonic acoustic wave resonators, quartz microbalance electrodes those using field-effect transistors achieve a higher sensitivity. Advantages over odorant receptor protein include easier expression in heterologous cells and a higher stability towards environmental conditions. A major disadvantage lies in the limited amount of molecules being recognized due to very low specificity [28,44].

However, protein-based electronic noses have a limited lifetime, as their functionality decreases over time, due to the degradation and denaturation of receptor proteins and peptides. Another disadvantage is that necessary cofactors to restore the initial state after a transduction cascade must be actively provided from the outside, otherwise failure is imminent. Against this background, cell-based bioelectronic noses represent an attractive alternative through their ability to restore defective proteins and to produce the essential cofactors [28]. In approaches focusing on whole tissues, microelectrode sensors are inserted into the olfactory bulb of living animals to measure their neuronal activity when exposed to different odors [45,46]. Analyzing the activity patterns by computer programs to draw conclusions about specific odorants remains a complex challenge. Even approaches that involve existing olfactory epithelial tissue and placing it on microelectrode arrays, for example, encounter similar issues [47]. In addition, the preparation effort is very high and the comparability is low, since different signals are obtained depending on the recording area on the epithelium. The use of olfactory receptor organs from insects allow an easier analysis of distinct activity patterns, but the organs show stability issues [48,49]. Instead of using whole tissues, approaches using dissociated olfactory sensory neurons measure only single neurons that are either suspended, immobilized on a microelectrode array, or trapped in microfluidic chambers. The proof-of-concepts available so far, encounter the problem that there is no method to isolate or organize neurons of a specific receptor type, resulting in the readout signal being arbitrary [28,50–52].

Considering the advantages of cell-based sensor methods over peptide or protein-based ones as well as the outlined drawbacks of strategies utilizing whole tissues and dissociated olfactory sensory neurons, approaches using cultured cells are considered one of the most promising methods [11,28]. In cultured cells approaches, specific DNA segments encoding olfactory receptor proteins from olfactory neurons are transfected into a heterologous cell line so that it expresses olfactory receptors. Alternatively, there are strategies to extract olfactory receptor proteins and implant them into heterologous cells. The cells require a suitable culture environment (nutrient medium) on the sensor platform to remain functional and viable [28]. Veithen et al. [53] present a number of cell lines that are typically considered for in vitro expression and implantation of olfactory receptor proteins initially, regardless of the selection technology used. These include heterologous mammalian cells, such as the human embryonic kidney 293 cell line (HEK293) and Hela cells, as well as insect cells, Xenopus oocytes and yeast cultures [54–60]. To enhance the delivery of the expressed olfactory receptor molecule into the cell membrane, additional DNA segments of receptor transport proteins (RTP) are transcribed so that the cell line stably expresses chaperones such as the RTP1, RTP1s, RTP2, in addition to the expression of the olfactory receptor protein [56]. If the display of olfactory receptor proteins on the membrane succeeds, the binding of a VOC to this receptor triggers a reaction process that is comparable to the transduction cascade explained in Section 2.1 and also leads to the influx of calcium ions. One method to visualize this influx is by calcium imaging based functional assays. Calcium-sensitive indicators are used to detect deviations in the intracellular calcium concentration. For example, GCaMP is a genetically encoded calcium indicator that changes structure when calcium ions bind to it, activating its attached green fluorescent protein (GFP). Other commonly used calcium-sensitive dyes include Fura-2 and Fluo-4 [53,61]. Cell-based approaches are not limited to optical readout techniques, for instance, the surface plasmon resonance method or microelectrode arrays can also serve as transducers [28].

2.3. Technical Performance Criteria of Odor Sensing Technologies

In order to describe the performance of an odor sensor, both static parameters, such as selectivity and sensitivity, and dynamic parameters, such as service life, can be considered. In the following, the individual criteria considered in this paper are described. The selection and definitions were developed in the course of an expert workshop of the authors based on Fraden et al. and verified by the review and supplementation of external experts in different fields of sensor technology [62].

Measurement quality:

- Sensitivity: describes the degree to which the output signal (measured value) changes in relation to the change of the input signal (measuring signal);
- Accuracy: describes the deviations of the sensor's predicted measurement values from the real (ideal) value (typically 2 or 3 sigma of the error fluctuations);
- Selectivity: describes the response of the sensor to a certain group of analytes or one specific analyte;
- Specificity: indicates the probability that the measured value is falsely positive or falsely negative;
- Resolution: describes the smallest measurable change the sensor is able to register;
- Repeatability: indicates the error that occurs with repeated measurements, under the same situation.

Handling:

- Reliability: describes the performance of the sensor that must be maintained over a defined period;
- Resistance to environmental influences or stability: describes the accuracy of the measurement results in case of changing environmental influences, such as temperature, humidity, radiation or magnetism;
- Maintenance effort: describes the overall effort of measures that keep the system in a functional state;
- Multi-sensing capability: describes the ability to measure several different substances in parallel;
- Operability: describes the simplicity of use;
- Measurement duration: the time required to complete a measurement process.

Technical construction and production:

- Durability: describes the period of time during which the sensor remains functional, i.e., the performance remains within certain predefined specifications (e.g., a maximum drop of the measurable signal below 50% of its original value).
- Dimensions: describes the flexibility of relevant, characteristic geometric dimensions of the sensor shape.
- Weight: mass of the body in kg per measuring unit or sensor;
- Cost: describes the monetary costs of the manufacturing process for materials and the production process.

2.4. Markets and Application Fields for Biosensors

The annual turnover of all suppliers in the biosensor market was USD $11.5 billion in 2014 and is expected to grow to USD $28.78 billion by 2021. This corresponds to a growth rate of 12.2% per year [63]. In the following, the fields of application for the use of odor sensors are listed and described regarding their market volumes in Germany. Figure 3 gives an overview of the findings.

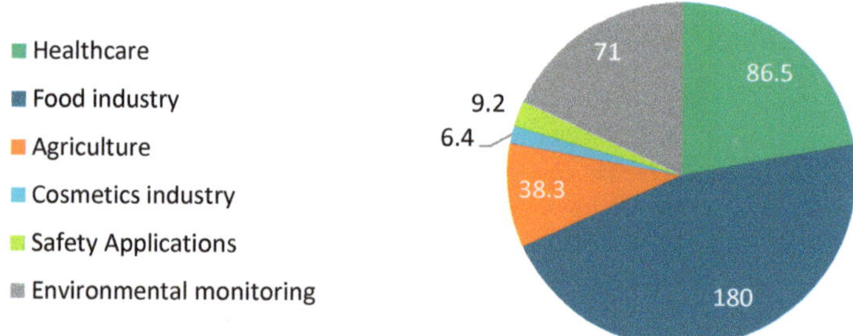

Figure 3. Overview of market volumes in billion euro (EUR) of different application fields in Germany [64–69].

- Healthcare: the healthcare market includes the ambulatory and stationary achievement contribution by established physicians, dentists, and hospitals, as well as other service providers [70]. In 2019, the health care system in Germany had a turnover of EUR 86.5 billion [64]. In 2018, 48,346 companies in Germany were active in the healthcare sector [71]. One example of a future field of application is diagnostics. Compared to healthy people, people with diseases excrete either different concentrations of certain VOCs or different types of VOCs. These VOCs are used as biomarkers and can be identified by breath, urine, and other body fluids. A diagnosis based solely on a patient's odor requires very accurate diagnostic equipment [72]. Odor sensors prove to be a suitable diagnostic tool when it comes to diagnosing diseases. There is a great demand for non-invasive diagnostic methods in the healthcare sector. These sensor devices should be able to perform real-time monitoring, and they should be portable and inexpensive [73].
- Food industry: the food industry comprises food and feed manufacturers together with the beverage industry. Altogether, there are about 6000 companies with more than 20 employees in the German food industry [74]. In 2018, these companies employed more than half a million people. With an annual turnover of almost EUR 180 billion, the food industry is one of the largest industries in Germany [65]. The odor sensors in this industry should enable fast detection of quality changes during production. During quality control, impurities and pathogens are identified. Furthermore, the correct composition of the produced food and its smell and taste can be analyzed [75].
- Agriculture: agriculture is the economic activity where soil, livestock, labor, and know-how produce agricultural products that ensure the supply of plant and animal food to the people [76]. In 2018, there were 266,600 active companies in Germany [77]. They had a turnover of EUR 38.3 billion in 2018 [66]. Odor sensors can be used in agriculture to determine the quality of products and stocks based on odors or VOCs, or to detect pests and other negative influences already in the field [20]. Another application example is the monitoring of livestock odors [78].
- Cosmetics industry: cosmetics include all products that have a healing effect but are also used for beauty care. The industry is mainly determined by the large consumer goods groups. In 2018, there were 137 companies in the German industry for the production of cosmetics [79], generating sales of approximately EUR 6.4 billion [67]. Fields of application for odor sensors in the cosmetics industry are mainly quality control of production goods. Odor sensors can also be used in production to check the

correct composition of the products, in order to be able to analyze odors and develop them more specifically, for example [20].

- Safety applications: safety applications are all applications that aim to detect hazardous substances. Smells contain important information about the environment and activities relevant to military and safety-oriented applications. This includes the detection of explosive materials or hazardous chemicals. However, an odor sensor can also be used for crime prevention tasks, such as security checks at airports or drug detection [80]. In 2021, the security industry in Germany is forecast to generate sales of EUR 9.2 billion [68].

- Environmental monitoring: in environmental monitoring, indoor and outdoor air is analyzed in order to detect air quality issues caused by harmful VOCs. These issues occur, for example, during the manufacturing of furniture [81]. The detection of harmful and toxic substances is also one of the areas of application for odor sensors. Furthermore, air quality and factory emissions can be monitored as well as the quality of ground and surface water. Sensors can be either installed stationary or mounted on drones [82]. Because of increased environmental awareness and pollution, the market for technological solutions for environmental monitoring applications is growing [75]. The turnover of the German environmental protection industry in 2018 amounted to EUR 71 billion [69].

3. Methodology

In each field of application (see Section 2.3) for odor detection sensors, there are different requirements for the technology used, which can be reflected in assessments of technical performance criteria. Sixteen technical performance criteria (see Section 2.2) that can be used especially for the description of the requirements of odor detection applications were established within this study through expert workshops and literature research. In order to specifically assess the importance of these criteria for the fields of application, a comprehensive expert survey was conducted with 11 experts from renowned research institutes and companies active in the fields of olfactory sensing electronic noses. Each of the participating experts had extensive experience in the research and development of odor sensors. The quality of the survey was, therefore, ensured by the targeted selection of experts who were able to classify the complex relationships between the product characteristics and their respective importance in the application fields and markets. The experts were asked to answer questions about which criteria were more or less important for each application field. For this purpose a scoring model was introduced to quantify the qualitative estimates for visualization, as follows: 0 = not important, 1 = rather unimportant, 2 = important, 3 = very important. In summary, the results of this survey were visualized in specific requirement profiles for each of the fields of application considered by forming the mean values of the scoring points. Additionally, all individual criteria were assigned to three related classes or categories. The first category combined all criteria related to measurement quality. This included the resolution and sensitivity of the sensors. The second category included the handling and the operability of the sensors. For example, measuring duration, maintenance effort, and multi-sensing-capability were assigned in this category. The third category combined production parameters such as manufacturing costs, weight, durability, and dimensions. Similarly, a performance profile was drawn up for individual technologies, showing the degree to which the performance criteria were fulfilled by the respective technology. The performance profile for bioelectronics noses was derived to enable statements about the fulfillments of the criteria in order to compare them to the competing technologies of technical sensor, as shown in Figure 2. To evaluate the performance of biosensors and instrumental analysis, numerous existing studies and research results were analyzed in a comprehensive meta-analysis regarding the performance perspectives of biosensors in comparison to those competing technologies. For the evaluation, the properties of biosensors were rated with the scale: 0 = is fulfilled worse by comparison; 1 = is fulfilled equally well or no clear statement can be made; 2 = is fulfilled comparatively better. In conclusion, by comparing the performance criteria with the requirement criteria, conclusions can be drawn about specific market potentials for the individual fields of application. In addition, an empirical investigation regarding the statistical variances of the survey results is analyzed for the assessments of importance for the respective criteria determined in the survey. The purpose of this investigation is to be able to conduct more in-depth research on criteria with high variances between the expert assessments in order to reduce the associated uncertainties. Criteria with high variances will consequently be examined and discussed in more detail to provide a better understanding of future market potentials. In particular, special attention will be paid to criteria that can potentially be better fulfilled by means of biosensors. This complementary analysis will serve as a basis for the goal of specific development targets for innovative biosensor concepts and business models. Furthermore, the analysis is supported by outlining the important role of biosensors against the background of the biological transformation.

4. Results

In the following subsections, the generated performance profile for biosensors (Section 4.1) and the requirement profiles of different application fields (Section 4.2) are presented. In addition, the market potential of biosensors is outlined for specific application scenarios within the application fields is outlined (Section 4.3).

4.1. Performance Profile of Biosensors for Odor Detection

The evaluation results show the performance of biosensors in comparison to technical sensors (electronic noses or instrumental analytics). All references and statements are summarized in Table 1 and the performance profile is graphically illustrated in Figure 4.

Table 1. Evaluation of the fulfillments of performance criteria by bioelectronic odor sensors; fields: 0 = is fulfilled worse by comparison; 1 = is fulfilled equally well or no clear statement to be made; 2 = is fulfilled comparatively better.

Properties	Fulfillment	Rating	References
High sensitivity	Because of the natural binding of olfactory receptors (ORs) with the specific ligand, the sensor can react even to very small amounts of analyte.	2	[12,83–85]
High accuracy	High accuracy due to natural binding of OR with specific ligand.	2	[12,83]
High resolution	Substances can be detected in very high resolutions at a level of nanomoles (or lower).	2	[12,86–88]
High repeat accuracy	Currently there are still problems with the stability of the results. No high repeat accuracy can be guaranteed yet.	0	[86,89,90]
High selectivity	It can be tested very specifically for certain substances.	2	[12,20,84,85]
High specificity	Good results for falsely positive and falsely negative measurements.	2	[12,20]
Low weight	A compact and light design for biosensors in comparison to analytical instruments allows online monitoring. Portable devices (sensors on chip) are currently in testing phases. No advantages. Probably no significant advantages over electronic noses to be expected.	1	[75,85]
Small dimensions	Analytical instruments are large benchtop systems permanently installed in laboratories. There are electronic noses with a diameter of a few cm. The same is possible for biosensors. Probably no significant advantages over electronic noses to be expected.	1	[17,85,89,91]
Low cost	The manufacturing costs for biological odor sensors are not yet finally known. Because of high research and development costs and complex production processes, a high sales price can be expected. For comparison, analytical instruments can cost up to USD 30,000. Electronic noses are available from USD 200.	1	[13,17,28,85,86,92,93]

Table 1. *Cont.*

Properties	Fulfillment	Rating	References
High durability	Sensors, which use cells as bioreceptors, currently have a lifetime of just about a few weeks. The durability of these systems, especially for use as industrial sensors, are not reported.	0	[12,20,84,86,92]
Low maintenance effort	Bioreceptors must be replaced regularly. Replacement receptors must be stored correctly.	0	[28,86,94]
Short measuring duration	Measuring times for biosensors are reported from 5–30 s. Total measuring process takes 5 min due to sample preparation and pauses between measurements. This is comparatively faster than analytical instruments but in the same range as electronic noses.	1	[12,28,85]
Operability	Usability cannot be conclusively evaluated yet. However, odor sensors allow a non-invasive measuring method that does not require the extraction of sample material.	1	[33,73,85]
Resistant to environmental influences	Sensors must be protected against environmental influences. Susceptible to humidity and temperature fluctuations.	0	[20]
Multi-sensing capability	Biosensors are able to measure several different substances simultaneously. By multiplexing/multi-channeling various naturally occurring or synthetically optimized biological detection elements (olfactory receptors, olfactory receptor derived proteins, odorant binding protein), the bioelectronic nose can detect a variety of combinations of different VOCs. Although only a few multiplexed systems have been presented so far, multi-sensing is considered to be a decisive advantage over technical odor sensors in terms of mimicking and digitizing the sense of smell.	2	[29,33,83–85,95,96]

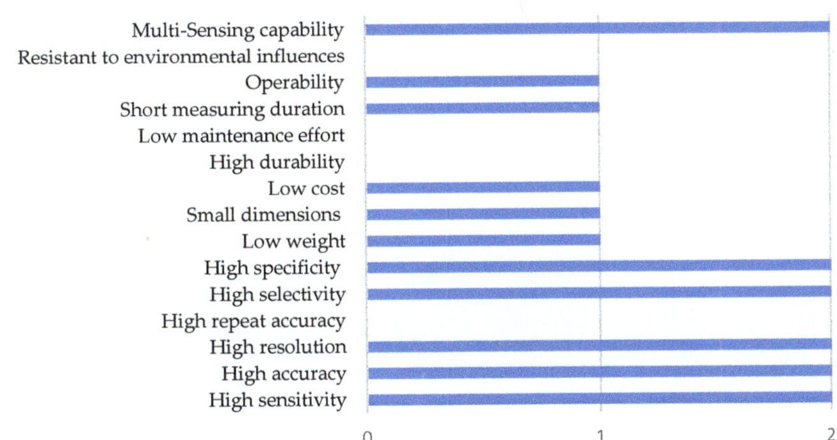

Figure 4. Performance profile of biosensors for odor detection in comparison to competitive technical odor sensors, based on Table 1; fields: 0 = is fulfilled worse by comparison; 1 = is fulfilled equally well or no clear statement to be made; 2 = is fulfilled comparatively better.

As illustrated in Figure 4, biosensors have advantages in terms of sensitivity, selectivity, specificity, accuracy, and resolution. This is due to physical bindings of the olfactory receptors with specific ligands. Therefore, the sensor can react to even very small amounts of analyte or single molecules within gas mixtures [83]. While there are advantages in terms of weight and dimensions compared to analytical instruments, this cannot be assumed for comparisons with electronic noses. The design of biosensors can be smaller than most technical analysis devices, such as mass spectrometers [75]. Disadvantages compared to technical sensors can be seen in terms of durability, maintenance effort, repeat accuracy, and resistance to environmental influences. The main reason for this is the limited lifetime and fragility of the used biomolecules. Users are forced to change the biomolecules after a certain time. This requires an enormous maintenance effort, which many users are not prepared to bear. Furthermore, the low resistance to environmental influences such as humidity, radioactive radiation, or high temperatures is a problem of biosensors that limits the application possibilities. All biosensors used, for example, for medical applications must meet the demanding and specific requirements of the medical industry. In addition, improvements and developments of other medical devices create more competitors for biological sensors [63]. A further disadvantage compared to technical sensors is cost. The long development cycles of biological sensors, which can only adapt to the new competitors with difficulty, play a key role here, according to the biosensor manufacturer Koniku Inc. Regarding the operability and the measuring duration, there are neither clearly defined advantages nor disadvantages for biosensors.

4.2. Requirement Profiles for Different Application Fields of Odor Sensing Technologies

In order to derive application-specific requirement profiles, each defined requirement criterion was evaluated with regard to its importance for a successful product in the respective fields of application. The evaluation was carried out in a survey, leading to the results shown in Figure 5, assigned to three related categories. The first category (a) combined all criteria related to measurement quality. The second category (b) included the handling and the operability of the sensors, and the third category (c) combined production parameters. The following sections describe the results grouped by these categories.

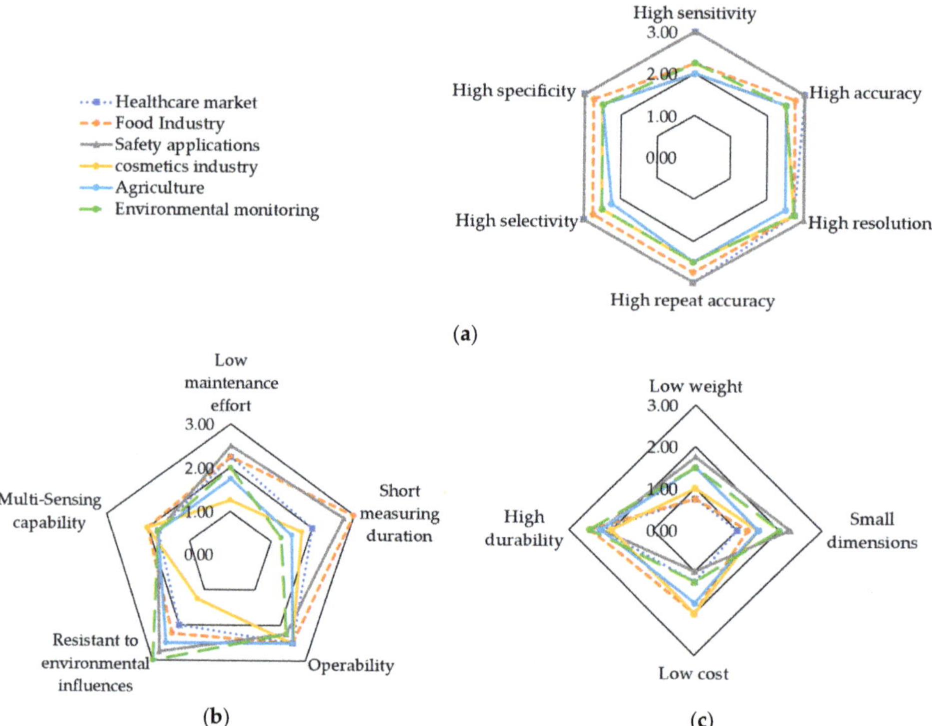

Figure 5. Evaluation of the requirement criteria on their importance for the categories of (**a**) measurement quality; (**b**) handling; (**c**) technical construction and production; 0 = not important, 1 = rather unimportant 2 = important, 3 = very important; sample size: 11.

In Figure 5a, criteria are shown concerning the measuring quality of odor sensors. Overall, all the measurement quality criteria shown were assessed as "important" or "very important" across all application fields. In a comparison of the application fields, it can be seen that all quality criteria shown were of even higher importance for the application fields in the healthcare market and for safety applications, compared to the other fields. According to the experts, all quality-related criteria shown in Figure 5 were very important for these fields of application. Since critical safety and sometimes vital data are to be collected in these industries, high measurement quality is essential. For example, for the detection of explosives and medical diagnostics, it must be possible to detect even small trace elements and individual molecules with high specificity, sensitivity, accuracy, resolution, and selectivity. For safety applications, all criteria were rated 3, thus, as "very important." The only exception for healthcare applications was high resolution, which was rated 2.75. High resolution was very important for all other fields of application with a rating of 2.75, as well, except agriculture. However, with a rating of 2.5, the criterion was still considered very important for agricultural applications.

In the category of handling and operation, shown in Figure 5b, there were stronger differences in the importance ratings for the considered fields of application compared to the criteria of measurement quality shown in Figure 5a. It is illustrated that short measuring times were very important for the food industry, due to the tendency of high throughputs of units to be measured coupled to large production numbers in this field of application. Because of the high risk of time delays, short measurement times were also very important for safety applications. According to the survey, the multi-sensing capability was particularly interesting and rated as important for the food industry, where taste analyses are performed. Tastes are usually defined by compositions of a large number of individual odorous substances. The multi-sensing capability was also evaluated as important for the cosmetics industry, since the composition of many different scents is also relevant for fragrances. The resistance to environmental influences was very important for applications in environmental monitoring, according to the experts, as these have to be used in changing environmental conditions outside the laboratory. This circumstance must not lead to any deviation of the measurement results. Resistance to environmental influences was also very important for safety applications and agriculture. In the cosmetics industry, however, this criterion was not very important, since the measuring systems can be used in a sterile and defined environment and fewer environmental influences are expected to affect the measurement results. Ease of operation or operability played an important role in all industries, since the measuring systems should be operable by ordinary employees who have no special training in the operation of these systems. For companies this was a decisive cost-saving factor, if no major training of the employees for the operation of the measuring system was necessary.

The criteria related to the construction and production of the sensors are summarized in Figure 5c. The geometric dimensions of the sensor tended to play a more important role in safety applications, since mobile applications, such as explosives' detection or people searches are potentially more common there. This could also be the case for environmental monitoring, which is why the criterion for this field of application was also rated important. The weight of the sensors was also considered important for safety applications due to mobile applications. Rather unimportant ratings were, however, given to this criterion for environmental and agricultural applications. Weight tended to play a smaller role for mobile applications than dimensions. Due to the large areas to be monitored by sensors, drone applications can play a central role in agricultural applications in the future, which was the reason for the relatively higher importance of this field. Weight would be a decisive factor here. The durability rates varied in their importance for all application fields between a narrow range of 2 for the cosmetics industry and 2.5 for the environmental monitoring and food industry. Therefore, this criterion is important for all application fields. According to the experts, low cost production tended to play a more important role in the food and cosmetics industries than in the other fields of application rated as rather important. This could be due to the high competitive situation in this market, where manufacturing costs play a major role in gaining a competitive advantage over the competition.

The statistical variances of the survey results are summarized in Figure 6. It can be seen that in some cases there was a high degree of uncertainty regarding the assessment of the importance of technical performance criteria in certain fields of application. The measurement time in the healthcare market and the cosmetics industry as well as the multi-sensing capability for the health care market, safety applications, agriculture, and environmental monitoring should be emphasized, with variances higher than 2.

	Healthcare Market	Food Industry	Safety Applications	Cosmetics Industry	Agriculture	Environmental Monitoring
High sensitivity	0.00	0.92	0.00	0.67	0.67	0.92
High accuracy	0.00	0.25	0.00	0.33	0.33	0.33
High resolution	0.25	0.25	0.00	0.25	0.33	0.25
High repeat accuracy	0.00	0.25	0.00	1.00	1.00	1.00
High selectivity	0.00	0.25	0.00	0.33	0.92	0.33
High specificity	0.00	0.25	0.00	0.33	0.33	0.33
Low weight	0.92	0.92	1.58	0.67	1.00	1.67
Small dimensions	0.67	0.92	0.25	0.33	1.00	0.67
Low cost	0.25	0.67	0.67	0.00	0.25	0.92
High durability	0.92	0.33	0.92	0.00	0.92	0.33
Low maintenance effort	0.25	0.25	0.33	0.92	0.25	0.67
Short measuring duration	2.00	0.00	0.25	2.25	1.67	1.58
Operability	1.00	1.00	0.92	1.00	1.00	0.92
Resistant to environmental influences	0.67	0.92	0.25	0.92	0.33	0.00
Multi-Sensing capability	2.25	1.33	2.25	1.33	2.25	2.25

Figure 6. Statistical variances of the survey results shown in Figure 5 (red = higher values, green = lower values); sample size: 11.

4.3. Market Potentials of Biosensors in Different Application Scenarios

To evaluate the findings, the experts in the Delphi survey were invited to assess the market potential of the applications known to date within the fields of application. In addition, the experts were asked in which future applications they see the use of odorant detecting biosensors. The economic potential of these new applications was also assessed in the second phase of the Delphi survey. The results are depicted in Figure 7. The economic potential was indicated on a scale from 1 = very low potential to 5 = very high potential.

The results of the requirement profiles of the application fields presented in Section 4.3 show that the requirements considered to be the most important were high specificity, high selectivity, high repeat accuracy, high resolution, high accuracy, and high sensitivity. These criteria describing the measurement quality were classified as "important" or "very important" in every considered field of application. All these criteria except for the repeat accuracy are potentially better met by biosensors than by technical sensors. It can be concluded that biosensor technology has a high potential for application in the considered fields and will play a decisive role in the market for odor sensors. Specific fields of application that can be covered specifically with biosensors, resulting from the high correspondence between requirement and performance profiles, are healthcare and security applications. These findings were confirmed by the results of the assessments of market potentials by the experts. In Figure 7a it can be seen that medicinal diagnostics, early cancer detection as well as detection of drugs or persons and detection of hazardous substances achieve high values from 4.0 and above.

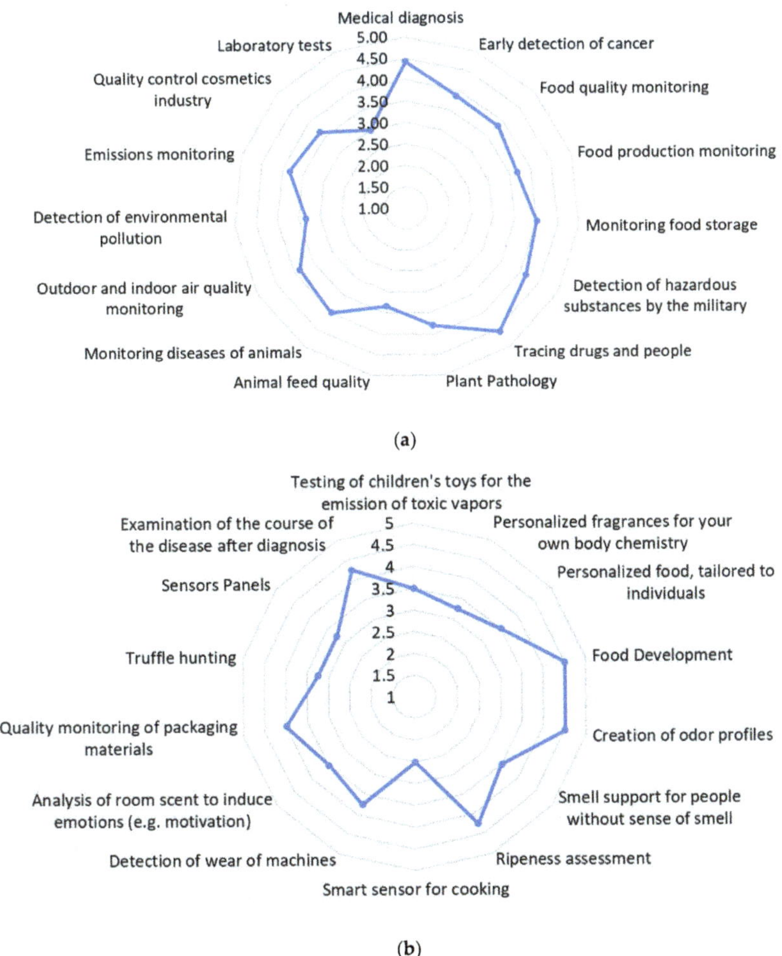

Figure 7. Results of the expert survey on the assessment of the market potential of (**a**) existing applications and (**b**) newly identified applications; 1 = very low potential, 2 = low potential, 3 = neutral potential, 4 = high potential, 5 = very high potential; sample size: 11.

Despite the high statistical deviations in multi-sensing capability, as shown in Figure 6, it can be considered as one of the key potentials, scoring 2.0 in importance among all application fields, as shown in Figure 5b. This finding can also be verified by the ranking of the newly identified application scenarios in Figure 7b. Food development, odor profiling, and ripeness assessment were ranked as having the highest market potential. All three application scenarios rely heavily on multi-sensing capability. Further elaboration and discussion on this topic is covered in the following chapter.

5. Discussion

In the following, the empirical investigation regarding the statistical variances of the survey results (Section 5.1) and the role of biosensors against the background of the biological transformation (Section 5.2) are discussed.

5.1. Discussion of the Empirical Investigation

An analysis of the empirical investigation regarding the statistical variances for the assessments of importance for respective criteria reveals a need for discussion of controversial statements and their importance in the markets under consideration. The experts agree on the importance of biosensors for the fields of application under consideration for most of the criteria that biosensors potentially fulfill better than conventional sensors, namely sensitivity, accuracy, resolution, selectivity and specificity (see Figure 5). However, the high variance within the expert responses regarding the multi-sensing capability indicates a high degree of uncertainty. Thus, it can be seen in Figure 6 that a comparatively high uncertainty across all fields of application occurred for the criterion of multi-sensing capability. Hence, a special focus is placed on the specific market potential associated with this criterion. To this end, specific application scenarios are described in the following for all fields of application that are directly related to the multi-sensing capability of the sensors. The identified and classified application scenarios provide approaches for specific development goals and are thus intended to reduce the identified uncertainties by providing ideas for specific application scenarios for multi-sensing capable sensors.

For healthcare market, technology leaps in diagnostics can be achieved. Diseases are often manifest through odorous substances secreted by the body before they can be detected with today's analytical methods. Often, complex compositions of different molecules can be decisive in order to make a specific diagnosis. The fact that this is possible is known, for example, from studies in which dogs were trained to detect those substances. In this context the function that multi-sensing biosensors could fulfil can be seen. It was proven that through the odor substances, for example, tumor diseases in early stages but also mental disorders and even moods can be recognized purely on odor substances that are secreted by the body [65–68]. Imaging this capability with multi-sensing biosensors could thus represent a technological leap in medical diagnostics. Even after diagnosis, these sensors could be used to continuously monitor disease progression, providing a better basis for decisions regarding further treatment (see Figure 7b).

Tastes and smells play a major role in the food industry. In all cases, it is odor compositions and not individual analytes that play a decisive role. In order to be able to detect these compositions in a targeted and collected manner, multi-sensing sensors may play a decisive role in the future. Especially in quality assurance through the targeted in-line detection of fermentation or digestion processes, there may be an extremely high market potential for multi-sensing. As shown in Figure 7b, the development of new food products with predefined flavor profiles is another promising application enabled by multi-sensing. As in the food industry, odor compositions also play a decisive role in cosmetics industry. In addition to quality testing and assurance, multi-sensing can also provide a technological leap forward in research and development for odorants. The digital recording and visualization of odor profiles using multi-sensing biosensors for the targeted demand-oriented development of cosmetic products can become a game-changer in the cosmetics industry (see Figure 7b).

Further, the odor detection of safety applications can be continuously developed by multi-sensing. This means that a wide variety of hazards, such as pollutants or even traces of explosives or drugs in security-relevant areas such as airports, can be detected together in a single device. Clean air and water are very important for our health and key business cases for environmental monitoring. External influences, for example nitrogen oxide and particulate matter pollution in many cities with high traffic volumes, endanger it. But the health of our natural ecosystems is also affected by changes in the smallest particles in the air. To collect them in a multi-sensing device for the respective areas showing the indicators of health hazards or natural pollution as completely as possible would be a great step forward for environmental monitoring.

In addition, potentials through multi-sensing could be exploited in agricultural applications. Digitalization already plays a major role in the optimization of agricultural processes under the term "smart agriculture". All related applications depend on suitable sensor technology. The integration of individual applications, such as pest detection, the degree of ripeness (see Figure 7b) and nutrition can be detected via various messengers. In order to be able to collect them, multi-sensing sensors could be used in the future. A more targeted nitrification, irrigation, and pest control and thus a reduction of the resources used and environmental impact can be achieved.

5.2. The Role of Odorant Sensing Biosensors within the Biological Transformation

As outlined in chapter 1 the biological transformation is progressing in three modes: bioinspiration, biointegration and biointelligence. The latter is characterized by the interaction between technical, biological and information systems. Consequently, a biointelligent system requires the implementation of an interface between biological and technical components. In addition to the identification of biosensors as key enabling technologies, so-called biology–technology interfaces (BTI) were identified as one of the core areas of future research [6]. The generic concept of a BTI comprises the recording and processing of information as well as control actions derived from it. The main interface components are corresponding sensors and actuators. They are based on electrical, chemical, mechanical or optical principles of action and realize a communication between the biological and technical system [10]. In this context, biosensors assume a special position as they can be considered an application of a BTI-based system on the one hand, and can also be seen as an enabler of superordinate BTI-based systems on the other. Odor detection biosensors provide a good example on both scenarios as illustrated in Figure 8. Firstly, the biological components (living cells, proteins, etc.) are in direct contact with the technical system (field-effect transistor, microelectrode array, etc.) and may be stimulated by adding VOCs and read out simultaneously. An information system evaluates the data and connects the biological and technical system, thus forming the basis of a BTI-based system. Secondly, the odorant detection biosensor may be deployed, for instance, in a food production bioreactor for an inline or online control. The reactor with its technical housing, its producing biological cells and the intelligent control system, based on the information of the odor detection biosensor along with other sensors form the superordinate BTI-based system.

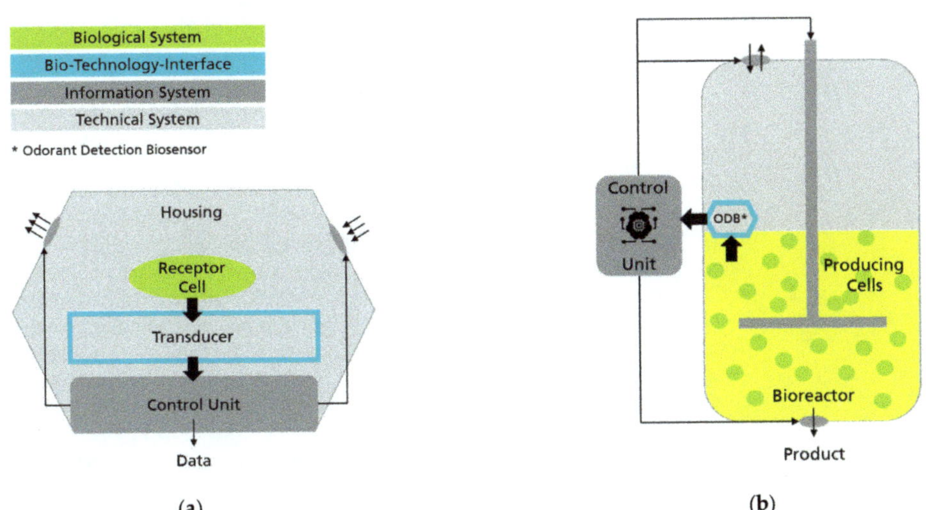

Figure 8. Odor detection biosensors as examples for biology–technology interface (BTI) systems following Miehe et al. [10]. Biosensors assume a special position as they can be considered as (**a**) an application of a BTI-based system, as well as (**b**) an enabler of a superordinate BTI-based system (i.e., bioreactor with cells that are generating a product).

With regard to the categorization of odorant detection biosensors in the context of the three development modes of the biological transformation the state-of-the-art technologies described in Section 2.2 can be classified in the second mode, biointegration. While the integration of the biological components into the technical sensor is feasible, an intelligent readout and control system still leave room for improvement. Nonetheless, bioelectronic noses constitute the foundation of biointelligent systems (third mode of biological transformation), in particular due to their ability to map complex odor patterns (multi-sensing capability). As explained in Section 2.2, the evaluation of odor patterns is associated with intelligent information processing. Gutherie et al. [20] present various biomimetic approaches to the analysis of electronic sensor signals by using frameworks that mimic parts of the biological sense of smell (neural networks, etc.). One of the goals is to reproduce the temporally and spatially resolved information of the human sense of smell. Since these approaches are inspired by nature, they can in turn be assigned to the first mode of biological transformation, bioinspiration. However, as indicated earlier, they require further development to transform biosensors into fully biointelligent systems.

6. Summary and Outlook

In this paper, the specific market requirements for odor sensors were empirically assessed on the basis of 16 technical properties for various fields of application. The properties were classified into criteria concerning measurement quality, handling as well as construction and production-related criteria. The fulfillment of these criteria by biosensors compared to technical sensors was evaluated in order to derive specific market potentials. Biosensors were found to have advantages in measurement quality criteria (sensitivity, selectivity, specificity, accuracy, resolution), which are important for all application fields, especially for safety and healthcare applications. It can, therefore, be predicted that biosensors have comparatively high potential in these markets, with possible applications in the odor-based diagnosis and monitoring of various diseases or the detection of traces of drugs or explosives in security-relevant facilities. However, compared to technical sensors, disadvantages are seen in terms of durability, maintenance effort, repeat accuracy, cost, and resistance to environmental influences. Durability is rated as important to very important for all fields of application considered and should therefore be one of the focal points in the further development of biosensors. For applications in the cosmetics, food, and agricultural sectors, cost optimizations are necessary, since these markets are very price-sensitive due to either the high number of throughput and measurement cycles or high competition. For outdoor applications (environmental monitoring, safety, agriculture), resistance to environmental influences must also be improved. In addition, the analysis of the empirical investigation regarding the statistical variances of the survey results for the assessments of importance for the respective criteria determined in the survey has shown a high degree of uncertainty concerning the multi-sensing capability. Since this criterion also appeared to be an advantage of biosensors over technical sensors with moderate to high importance for all application fields, this uncertainty was addressed by identifying specific application scenarios in all application fields and providing approaches for specific development goals. Furthermore, in the context of the biological transformation, odorant detecting biosensors assume a special position as they are not only considered an application of a biology–technology interface (BTI) based system, but can also be seen as enablers of superordinate BTI-based systems (e.g., deployed in a bioreactor). However, for odorant detecting biosensors to make the step from biointegrated to truly biointelligent systems, further development in the area of pattern recognition (multi-sensing capability) is still necessary.

All in all, with the results obtained, market specific application potential and development goals can be discussed more clearly on the basis of qualitative assessments shown in this paper. However, the authors would like to point out that each application should be further regarded separately and can sometimes differ considerably from the requirement profiles of the respective application field. In addition, for investigations based on this results, weightings can be established for the criteria, the values of which can be determined from existing or future market volumes and the requirement profiles, for example. Furthermore, additional performance criteria, such as limit of detection, power consumption, and response time, as considered, e.g., by Burgués et al. [97,98], may be evaluated with respect to the performance profile of the biosensors as well as the application field requirements profiles. Moreover, there are strong dependencies between the specificity and selectivity criteria that make differentiation difficult and should therefore be discussed further. This paper can be referred to as a basis for further examinations.

Author Contributions: Writing, J.F., Y.B. and L.D.; supervision, R.M. and A.S. All authors have read and agreed to the published version of the manuscript.

Funding: This research was funded by the Ministry of Economic Affairs, Labor, and Housing of Baden-Wuerttemberg and the Fraunhofer Gesellschaft.

Institutional Review Board Statement: Not applicable.

Informed Consent Statement: Not applicable.

Data Availability Statement: Not applicable.

Acknowledgments: The results shown in this paper were conducted within a research project supported by the Ministry of Economic Affairs, Labor, and Housing of Baden-Wuerttemberg and Fraunhofer Gesellschaft. The authors gratefully acknowledge the financial support.

Conflicts of Interest: The authors declare no conflict of interest.

References

1. Full, J.; Delbrück, L.; Sauer, A.; Miehe, R. Market Perspectives and Future Fields of Application of Odor Detection Biosensors—A Systematic Analysis. *Proceedings* **2020**, *60*, 7029. [CrossRef]
2. Byrne, G.; Dimitrov, D.; Monostori, L.; Teti, R.; van Houten, F.; Wertheim, R. Biologicalisation: Biological transformation in manufacturing. *CIRP J. Manuf. Sci. Technol.* **2018**, *21*, 1–32. [CrossRef]
3. Drossel, W.; Dani, I.; Wertheim, R. Biological transformation and technologies used for manufacturing of multifunctional metal-based parts. *Procedia Manuf.* **2019**, *33*, 115–122. [CrossRef]
4. Herles, B. Industry 5.0: The Next Industrial Revolution is Biological (German). Available online: https://www.capital.de/wirtschaft-politik/industrie-5-0-die-naechste-industrielle-revolution-ist-biologisch (accessed on 31 January 2021).
5. Miehe, R.; Bauernhansl, T.; Schwarz, O.; Traube, A.; Lorenzoni, A.; Waltersmann, L.; Full, J.; Horbelt, J.; Sauer, A. The biological transformation of the manufacturing industry—Envisioning biointelligent value adding. *Procedia CIRP* **2018**, *72*, 739–743. [CrossRef]
6. Miehe, R.; Bauernhansl, T.; Beckett, M.; Brecher, C.; Demmer, A.; Drossel, W.-G.; Elfert, P.; Full, J.; Hellmich, A.; Hinxlage, J.; et al. The biological transformation of industrial manufacturing—Technologies, status and scenarios for a sustainable future of the German manufacturing industry. *J. Manuf. Sys.* **2020**, *54*, 50–61. [CrossRef]
7. Guvendiren, M.; Molde, J.; Soares, R.M.D.; Kohn, J. Designing Biomaterials for 3D Printing. *ACS Biomater. Sci. Eng.* **2016**, *2*, 1679–1693. [CrossRef]
8. Gungor-Ozkerim, P.S.; Inci, I.; Zhang, Y.S.; Khademhosseini, A.; Dokmeci, M.R. Bioinks for 3D bioprinting: An overview. *Biomater. Sci.* **2018**, *6*, 915–946. [CrossRef] [PubMed]
9. Vijayavenkataraman, S.; Yan, W.-C.; Lu, W.F.; Wang, C.-H.; Fuh, J.Y.H. 3D bioprinting of tissues and organs for regenerative medicine. *Adv. Drug Deliv. Rev.* **2018**, *132*, 296–332. [CrossRef]
10. Miehe, R.; Fischer, E.; Berndt, D.; Herzog, A.; Horbelt, J.; Full, J.; Bauernhansl, T.; Schenk, M. Enabling bidirectional real time interaction between biological and technical systems: Structural basics of a control oriented modeling of biology-technology-interfaces. *Procedia CIRP* **2019**, *81*, 63–68. [CrossRef]
11. Fraunhofer-Gesellschaft zur Förderung der Angewandten Forschung e. V. Identified Research and Design Fields in the Field of Action Further and New Development of Biology-Technology Interfaces (German). 2018. Available online: https://www.biotrain.info/projektveroeffentlichungen/ (accessed on 2 February 2021).
12. Dung, T.T.; Oh, Y.; Choi, S.-J.; Kim, I.-D.; Oh, M.-K.; Kim, M. Applications and Advances in Bioelectronic Noses for Odour Sensing. *Sensors* **2018**, *18*, 103. [CrossRef] [PubMed]
13. Bohbot, J.D.; Vernick, S. The Emergence of Insect Odorant Receptor-Based Biosensors. *Biosensors* **2020**, *10*, 26. [CrossRef]
14. Haghnegahdar, N.; Abbasi Tarighat, M.; Dastan, D. Curcumin-functionalized nanocomposite AgNPs/SDS/MWCNTs for electrocatalytic simultaneous determination of dopamine, uric acid, and guanine in co-existence of ascorbic acid by glassy carbon electrode. *J. Mater. Sci. Mater. Electron.* **2021**. [CrossRef]
15. Shan, K.; Yi, Z.-Z.; Yin, X.-T.; Dastan, D.; Altaf, F.; Garmestani, H.; Alamgir, F.M. Mixed conductivity evaluation and sensing characteristics of limiting current oxygen sensors. *Surf. Interfaces* **2020**, *21*, 100762. [CrossRef]

16. Zhou, X.; Wang, Y.; Wang, Z.; Yang, L.; Wu, X.; Han, N.; Chen, Y. Synergetic p+n Field-Effect Transistor Circuits for ppb-Level Xylene Detection. *IEEE Sens. J.* **2018**, *18*, 3875–3882. [CrossRef]
17. Wu, Z.; Zhang, H.; Sun, W.; Lu, N.; Yan, M.; Wu, Y.; Hua, Z.; Fan, S. Development of a Low-Cost Portable Electronic Nose for Cigarette Brands Identification. *Sensors* **2020**, *20*, 4239. [CrossRef]
18. Yin, X.-T.; Li, J.; Dastan, D.; Zhou, W.-D.; Garmestani, H.; Alamgir, F.M. Ultra-high selectivity of H2 over CO with a p-n nanojunction based gas sensors and its mechanism. *Sens. Actuators B Chem.* **2020**, *319*, 128330. [CrossRef]
19. Shan, K.; Yi, Z.-Z.; Yin, X.-T.; Dastan, D.; Dadkhah, S.; Coates, B.T.; Garmestani, H. Mixed conductivities of A-site deficient Y, Cr-doubly doped SrTiO3 as novel dense diffusion barrier and temperature-independent limiting current oxygen sensors. *Adv. Powder Technol.* **2020**, *31*, 4657–4664. [CrossRef]
20. Guthrie, B. Machine Olfaction. In *Springer Handbook of Odor*; Büttner, A., Ed.; Springer International Publishing: Cham, Switzerland, 2017; pp. 459–504, ISBN 9783319269306.
21. Cali, K.; Persaud, K.C. Modification of an Anopheles gambiae odorant binding protein to create an array of chemical sensors for detection of drugs. *Sci. Rep.* **2020**, *10*, 3890. [CrossRef] [PubMed]
22. Spath, D.; Ardilio, A.; Laib, S. The potential of emerging technologies: Strategy-planning for technology-providers throughout an application-radar. In Proceedings of the PICMET '09—2009 Portland International Conference on Management of Engineering & Technology, Portland, OR, USA, 2–6 August 2009; pp. 462–477.
23. Müller, W.A.; Frings, S.; Möhrlen, F. *Animal and Human Physiology (German)*; Springer: Berlin/Heidelberg, Germany, 2019; ISBN 978-3-662-58461-3.
24. Breer, H.; Fleischer, J.; Strotmann, J. Odorant Sensing. In *Springer Handbook of Odor*; Büttner, A., Ed.; Springer International Publishing: Cham, Switzerland, 2017; pp. 585–603, ISBN 9783319269306.
25. Feigenspan, A. *Principles of Physiology (German)*; Springer: Berlin/Heidelberg, Germany, 2017; ISBN 978-3-662-54116-6.
26. Frings, S.; Müller, F. *Biology of the Senses (German)*; Springer: Berlin/Heidelberg, Germany, 2019; ISBN 978-3-662-58349-4.
27. Breer, H.; Pfannkuche, H.; Sann, H.; Deeg, C.A. Sensory Physiology (German). In *Physiology of Pets (German)*; von Engelhardt, W., Breves, G., Diener, M., Gäbel, G., Eds.; Georg Thieme Verlag: Stuttgart, Germany, 2015; ISBN 9783830412595.
28. Cave, J.W.; Wickiser, J.K.; Mitropoulos, A.N. Progress in the development of olfactory-based bioelectronic chemosensors. *Biosens. Bioelectron.* **2019**, *123*, 211–222. [CrossRef] [PubMed]
29. Son, M.; Park, T.H. The bioelectronic nose and tongue using olfactory and taste receptors: Analytical tools for food quality and safety assessment. *Biotechnol. Adv.* **2018**, *36*, 371–379. [CrossRef]
30. Karakaya, D.; Ulucan, O.; Turkan, M. Electronic Nose and Its Applications: A Survey. *Int. J. Autom. Comput.* **2020**, *17*, 179–209. [CrossRef]
31. Scheider-Häder, B.; Müller, M.; Hambacher, E.; Wiech, H.; Wortelmann, T. Instrumental Sensory Analysis in the Food Industry (German): Teil 1: Elektronische Nasen. Available online: https://www.dlg.org/de/lebensmittel/themen/publikationen/experten wissen-sensorik/elektronische-nasen/ (accessed on 3 April 2020).
32. Röck, F.; Weimar, U. Electronic nose and signal extraction (German). In *Information Fusion in Measurement and Sensor Technology (German)*; Beyerer, J., Ed.; Universitätsverlag: Karlsruhe, Germany, 2006; pp. 261–278, ISBN 3-86644-053-7.
33. Wilson, A.D.; Baietto, M. Applications and advances in electronic-nose technologies. *Sensors* **2009**, *9*, 5099–5148. [CrossRef]
34. Karunakaran, C.; Bhargava, K.; Benjamin, R. (Eds.) *Biosensors and Bioelectronics*; Elsevier: Amsterdam, The Netherlands, 2015; ISBN 012803100X.
35. Dacres, H.; Wang, J.; Leitch, V.; Horne, I.; Anderson, A.R.; Trowell, S.C. Greatly enhanced detection of a volatile ligand at femtomolar levels using bioluminescence resonance energy transfer (BRET). *Biosens. Bioelectron.* **2011**, *29*, 119–124. [CrossRef]
36. Lee, S.H.; Jin, H.J.; Song, H.S.; Hong, S.; Park, T.H. Bioelectronic nose with high sensitivity and selectivity using chemically functionalized carbon nanotube combined with human olfactory receptor. *J. Biotechnol.* **2012**, *157*, 467–472. [CrossRef]
37. Park, S.J.; Kwon, O.S.; Lee, S.H.; Song, H.S.; Park, T.H.; Jang, J. Ultrasensitive flexible graphene based field-effect transistor (FET)-type bioelectronic nose. *Nano Lett.* **2012**, *12*, 5082–5090. [CrossRef] [PubMed]
38. Yang, H.; Kim, D.; Kim, J.; Moon, D.; Song, H.S.; Lee, M.; Hong, S.; Park, T.H. Nanodisc-Based Bioelectronic Nose Using Olfactory Receptor Produced in Escherichia coli for the Assessment of the Death-Associated Odor Cadaverine. *ACS Nano* **2017**, *11*, 11847–11855. [CrossRef] [PubMed]
39. Lee, K.M.; Son, M.; Kang, J.H.; Kim, D.; Hong, S.; Park, T.H.; Chun, H.S.; Choi, S.S. A triangle study of human, instrument and bioelectronic nose for non-destructive sensing of seafood freshness. *Sci. Rep.* **2018**, *8*, 547. [CrossRef] [PubMed]
40. Lim, J.H.; Park, J.; Ahn, J.H.; Jin, H.J.; Hong, S.; Park, T.H. A peptide receptor-based bioelectronic nose for the real-time determination of seafood quality. *Biosens. Bioelectron.* **2013**, *39*, 244–249. [CrossRef] [PubMed]
41. Reiner-Rozman, C.; Kotlowski, C.; Knoll, W. Electronic Biosensing with Functionalized rGO FETs. *Biosensors* **2016**, *6*, 17. [CrossRef]
42. Di Pietrantonio, F.; Benetti, M.; Cannatà, D.; Verona, E.; Palla-Papavlu, A.; Fernández-Pradas, J.M.; Serra, P.; Staiano, M.; Varriale, A.; D'Auria, S. A surface acoustic wave bio-electronic nose for detection of volatile odorant molecules. *Biosens. Bioelectron.* **2015**, *67*, 516–523. [CrossRef] [PubMed]
43. Mulla, M.Y.; Tuccori, E.; Magliulo, M.; Lattanzi, G.; Palazzo, G.; Persaud, K.; Torsi, L. Capacitance-modulated transistor detects odorant binding protein chiral interactions. *Nat. Commun.* **2015**, *6*, 6010. [CrossRef] [PubMed]
44. Shiao, M.-S.; Chang, A.Y.-F.; Liao, B.-Y.; Ching, Y.-H.; Lu, M.-Y.J.; Chen, S.M.; Li, W.-H. Transcriptomes of mouse olfactory epithelium reveal sexual differences in odorant detection. *Genome Biol. Evol.* **2012**, *4*, 703–712. [CrossRef] [PubMed]

45. Gao, K.; Li, S.; Zhuang, L.; Qin, Z.; Zhang, B.; Huang, L.; Wang, P. In vivo bioelectronic nose using transgenic mice for specific odor detection. *Biosens. Bioelectron.* **2018**, *102*, 150–156. [CrossRef] [PubMed]
46. Zhuang, L.; Guo, T.; Cao, D.; Ling, L.; Su, K.; Hu, N.; Wang, P. Detection and classification of natural odors with an in vivo bioelectronic nose. *Biosens. Bioelectron.* **2015**, *67*, 694–699. [CrossRef]
47. Dennis, J.C.; Aono, S.; Vodyanoy, V.J.; Morrison, E.E. Development, Morphology, and Functional Anatomy of the Olfactory Epithelium. In *Handbook of Olfaction and Gustation*, 3rd ed.; Doty, R.L., Ed.; John Wiley & Sons Inc.: Hoboken, NJ, USA, 2015; pp. 93–108, ISBN 9781118971758.
48. Strauch, M.; Lüdke, A.; Münch, D.; Laudes, T.; Galizia, C.G.; Martinelli, E.; Lavra, L.; Paolesse, R.; Ulivieri, A.; Catini, A.; et al. More than apples and oranges–detecting cancer with a fruit fly's antenna. *Sci. Rep.* **2014**, *4*, 3576. [CrossRef]
49. Myrick, A.J.; Park, K.-C.; Hetling, J.R.; Baker, T.C. Real-time odor discrimination using a bioelectronic sensor array based on the insect electroantennogram. *Bioinspir. Biomim.* **2008**, *3*, 46006. [CrossRef]
50. Wu, C.; Chen, P.; Yu, H.; Liu, Q.; Zong, X.; Cai, H.; Wang, P. A novel biomimetic olfactory-based biosensor for single olfactory sensory neuron monitoring. *Biosens. Bioelectron.* **2009**, *24*, 1498–1502. [CrossRef]
51. Du, L.; Zou, L.; Wang, Q.; Zhao, L.; Huang, L.; Wang, P.; Wu, C. A novel biomimetic olfactory cell-based biosensor with DNA-directed site-specific immobilization of cells on a microelectrode array. *Sens. Actuators B Chem.* **2015**, *217*, 186–192. [CrossRef]
52. Figueroa, X.A.; Cooksey, G.A.; Votaw, S.V.; Horowitz, L.F.; Folch, A. Large-scale investigation of the olfactory receptor space using a microfluidic microwell array. *Lab Chip* **2010**, *10*, 1120–1127. [CrossRef] [PubMed]
53. Veithen, A.; Philippeau, M.; Chatelain, P. High-Throughput Receptor Screening Assay. In *Springer Handbook of Odor*; Büttner, A., Ed.; Springer International Publishing: Cham, Switzerland, 2017; pp. 505–526, ISBN 9783319269306.
54. Sanz, G.; Schlegel, C.; Pernollet, J.-C.; Briand, L. Comparison of odorant specificity of two human olfactory receptors from different phylogenetic classes and evidence for antagonism. *Chem. Senses* **2005**, *30*, 69–80. [CrossRef]
55. Kajiya, K.; Inaki, K.; Tanaka, M.; Haga, T.; Kataoka, H.; Touhara, K. Molecular Bases of Odor Discrimination: Reconstitution of Olfactory Receptors that Recognize Overlapping Sets of Odorants. *J. Neurosci.* **2001**, *21*, 6018–6025. [CrossRef]
56. Saito, H.; Kubota, M.; Roberts, R.W.; Chi, Q.; Matsunami, H. RTP family members induce functional expression of mammalian odorant receptors. *Cell* **2004**, *119*, 679–691. [CrossRef] [PubMed]
57. Shirokova, E.; Schmiedeberg, K.; Bedner, P.; Niessen, H.; Willecke, K.; Raguse, J.-D.; Meyerhof, W.; Krautwurst, D. Identification of specific ligands for orphan olfactory receptors. G protein-dependent agonism and antagonism of odorants. *J. Biol. Chem.* **2005**, *280*, 11807–11815. [CrossRef] [PubMed]
58. Matarazzo, V.; Zsürger, N.; Guillemot, J.-C.; Clot-Faybesse, O.; Botto, J.-M.; Dal Farra, C.; Crowe, M.; Demaille, J.; Vincent, J.-P.; Mazella, J.; et al. Porcine odorant-binding protein selectively binds to a human olfactory receptor. *Chem. Senses* **2002**, *27*, 691–701. [CrossRef]
59. Abaffy, T.; Malhotra, A.; Luetje, C.W. The molecular basis for ligand specificity in a mouse olfactory receptor: A network of functionally important residues. *J. Biol. Chem.* **2007**, *282*, 1216–1224. [CrossRef] [PubMed]
60. Fukutani, Y.; Nakamura, T.; Yorozu, M.; Ishii, J.; Kondo, A.; Yohda, M. The N-terminal replacement of an olfactory receptor for the development of a yeast-based biomimetic odor sensor. *Biotechnol. Bioeng.* **2012**, *109*, 205–212. [CrossRef]
61. Paredes, R.M.; Etzler, J.C.; Watts, L.T.; Zheng, W.; Lechleiter, J.D. Chemical calcium indicators. *Methods* **2008**, *46*, 143–151. [CrossRef] [PubMed]
62. Fraden, J. Sensor Characteristics. In *Handbook of Modern Sensors*; Fraden, J., Ed.; Springer International Publishing: Cham, Switzerland, 2016; pp. 35–68, ISBN 978-3-319-19302-1.
63. Frost & Sullivan. *Analysis of the Global Biosensors Market: Biosensors Monitoring Stimulates Prevention and Control*; Frost & Sullivan: Mountain View, CA, USA, 2015.
64. Statistisches Bundesamt. Sales Development in the Healthcare Sector in Germany in the Years from 2006 to 2023 (German). Available online: https://de.statista.com/statistik/daten/studie/247979/umfrage/prognose-zum-umsatz-im-gesundheitswesen-in-deutschland/ (accessed on 12 October 2020).
65. Statistisches Bundesamt. Sales of the Food Industry in Germany in the Years 2008 to 2019 (German). Available online: https://de.statista.com/statistik/daten/studie/75611/umfrage/umsatz-der-deutschen-ernaehrungsindustrie-seit-2008/ (accessed on 12 October 2020).
66. Statistisches Bundesamt. Net Sales of Agriculture in Germany in the Years 2002 to 2018 (German). Available online: https://de.statista.com/statistik/daten/studie/323340/umfrage/umsatz-der-landwirtschaft-in-deutschland/ (accessed on 12 October 2020).
67. Statistisches Bundesamt. Sales of the German Personal Care and Fragrance Manufacturing Industry from 2008 to 2019 (German). Available online: https://de.statista.com/statistik/daten/studie/256917/umfrage/umsatz-der-deutschen-kosmetik-und-koerperpflegeindustrie/ (accessed on 12 October 2020).
68. Statistisches Bundesamt. Forecasted Sales Development in the Security Industry in Germany in the Years from 2007 to 2021 (German). Available online: https://de.statista.com/statistik/daten/studie/248225/umfrage/prognose-zum-umsatz-in-der-sicherheitsbranche-in-deutschland/ (accessed on 12 October 2020).
69. Statistisches Bundesamt. Sales of the German Environmental Protection Industry in the Years 2008 to 2018 (German). Available online: https://de.statista.com/statistik/daten/studie/240324/umfrage/umsatz-mit-umweltschutz-klimaschutzguetern-in-deutschland/ (accessed on 12 October 2020).

70. Werding, M. Definition: Healthcare (German). Springer Fachmedien Wiesbaden GmbH [Online]. 19 February 2018. Available online: https://wirtschaftslexikon.gabler.de/definition/gesundheitswesen-34513/version-258015 (accessed on 12 October 2020).

71. Statista. Healthcare 2020 (German): Statista Branchenreport—WZ-Code 86. 2020. Available online: https://de.statista.com/statistik/studie/id/62/dokument/gesundheitswesen/ (accessed on 12 October 2020).

72. Chen, S.; Wang, Y.; Choi, S. Applications and Technology of Electronic Nose for Clinical Diagnosis. *OJAB* **2013**, *2*, 39–50. [CrossRef]

73. Capelli, L.; Taverna, G.; Bellini, A.; Eusebio, L.; Buffi, N.; Lazzeri, M.; Guazzoni, G.; Bozzini, G.; Seveso, M.; Mandressi, A.; et al. Application and Uses of Electronic Noses for Clinical Diagnosis on Urine Samples: A Review. *Sensors* **2016**, *16*, 1708. [CrossRef]

74. Statistisches Bundesamt. Anzahl der Betriebe in der Lebensmittelindustrie in Deutschland in den Jahren 2008 bis 2019 (German). Available online: https://de.statista.com/statistik/daten/studie/321182/umfrage/betriebe-in-der-lebensmittelindustrie-in-deutschland/ (accessed on 12 October 2020).

75. Wasilewski, T.; Gębicki, J.; Kamysz, W. Advances in olfaction-inspired biomaterials applied to bioelectronic noses. *Sens. Actuators B Chem.* **2018**, *257*, 511–537. [CrossRef]

76. Berwanger, J. Definition: Agriculture (German). Available online: https://wirtschaftslexikon.gabler.de/definition/landwirtschaft-41331/version-264696 (accessed on 12 October 2020).

77. Statistisches Bundesamt. Number of Establishments in the Agricultural Sector in Germany in the Years 1975 to 2019. Available online: https://de.statista.com/statistik/daten/studie/36094/umfrage/landwirtschaft---anzahl-der-betriebe-in-deutschland/ (accessed on 12 October 2020).

78. Pan, L.; Yang, S.X. A new intelligent electronic nose system for measuring and analysing livestock and poultry farm odours. *Environ. Monit. Assess.* **2007**, *135*, 399–408. [CrossRef]

79. Statistisches Bundesamt. Number of Establishments in the German Personal Care and Fragrance Manufacturing Industry in the Years 2008 to 2019 (German). Available online: https://de.statista.com/statistik/daten/studie/256938/umfrage/betriebe-in-der-deutschen-kosmetik-und-koerperpflegeindustrie/ (accessed on 12 October 2020).

80. Koniku Inc. Koniku—Intelligence is Natural: Application. Available online: https://koniku.com/applications (accessed on 1 October 2020).

81. Brown, N. Indoor air quality monitoring (German). *Elektronik Industrie* **2017**, *48*, 62–85.

82. Burgués, J.; Marco, S. Environmental chemical sensing using small drones: A review. *Sci. Total Environ.* **2020**, *748*, 141172. [CrossRef]

83. Prickril, B.; Rasooly, A. *Biosensors and Biodetection*; Springer: New York, NY, USA, 2017; ISBN 978-1-4939-6910-4.

84. Hurot, C.; Scaramozzino, N.; Buhot, A.; Hou, Y. Bio-Inspired Strategies for Improving the Selectivity and Sensitivity of Artificial Noses: A Review. *Sensors* **2020**, *20*, 1803. [CrossRef]

85. Son, M.; Lee, J.Y.; Ko, H.J.; Park, T.H. Bioelectronic Nose: An Emerging Tool for Odor Standardization. *Trends Biotechnol.* **2017**, *35*, 301–307. [CrossRef] [PubMed]

86. Wasilewski, T.; Gębicki, J.; Kamysz, W. Bioelectronic nose: Current status and perspectives. *Biosens. Bioelectron.* **2017**, *87*, 480–494. [CrossRef] [PubMed]

87. Beccherelli, R.; Zampetti, E.; Pantalei, S.; Bernabei, M.; Persaud, K.C. Design of a very large chemical sensor system for mimicking biological olfaction. *Sens. Actuators B Chem.* **2010**, *146*, 446–452. [CrossRef]

88. Du, L.; Wu, C.; Liu, Q.; Huang, L.; Wang, P. Recent advances in olfactory receptor-based biosensors. *Biosens. Bioelectron.* **2013**, *42*, 570–580. [CrossRef]

89. Oh, J.; Yang, H.; Jeong, G.E.; Moon, D.; Kwon, O.S.; Phyo, S.; Lee, J.; Song, H.S.; Park, T.H.; Jang, J. Ultrasensitive, Selective, and Highly Stable Bioelectronic Nose That Detects the Liquid and Gaseous Cadaverine. *Anal. Chem.* **2019**, *91*, 12181–12190. [CrossRef]

90. Tan, J.; Xu, J. Applications of electronic nose (e-nose) and electronic tongue (e-tongue) in food quality-related properties determination: A review. *Artif. Intell. Agric.* **2020**, *4*, 104–115. [CrossRef]

91. Arroyo, P.; Meléndez, F.; Suárez, J.I.; Herrero, J.L.; Rodríguez, S.; Lozano, J. Electronic Nose with Digital Gas Sensors Connected via Bluetooth to a Smartphone for Air Quality Measurements. *Sensors* **2020**, *20*, 786. [CrossRef]

92. Pelosi, P.; Zhu, J.; Knoll, W. From Gas Sensors to Biomimetic Artificial Noses. *Chemosensors* **2018**, *6*, 32. [CrossRef]

93. Macías Macías, M.; Agudo, J.E.; García Manso, A.; García Orellana, C.J.; González Velasco, H.M.; Gallardo Caballero, R. A compact and low cost electronic nose for aroma detection. *Sensors* **2013**, *13*, 5528–5541. [CrossRef] [PubMed]

94. Yang, A.; Yan, F. Flexible Electrochemical Biosensors for Health Monitoring. *ACS Appl. Electron. Mater.* **2020**. [CrossRef]

95. Son, M.; Kim, D.; Ko, H.J.; Hong, S.; Park, T.H. A portable and multiplexed bioelectronic sensor using human olfactory and taste receptors. *Biosens. Bioelectron.* **2017**, *87*, 901–907. [CrossRef]

96. Kwon, O.S.; Song, H.S.; Park, S.J.; Lee, S.H.; An, J.H.; Park, J.W.; Yang, H.; Yoon, H.; Bae, J.; Park, T.H.; et al. An Ultrasensitive, Selective, Multiplexed Superbioelectronic Nose That Mimics the Human Sense of Smell. *Nano Lett.* **2015**, *15*, 6559–6567. [CrossRef]

97. Burgués, J.; Jiménez-Soto, J.M.; Marco, S. Estimation of the limit of detection in semiconductor gas sensors through linearized calibration models. *Anal. Chim. Acta* **2018**, *1013*, 13–25. [CrossRef] [PubMed]

98. Burgués, J.; Marco, S. Low Power Operation of Temperature-Modulated Metal Oxide Semiconductor Gas Sensors. *Sensors* **2018**, *18*, 339. [CrossRef]

Article

Detection of Chymotrypsin by Optical and Acoustic Methods

Ivan Piovarci [1], Tibor Hianik [1,*] and Ilia N. Ivanov [2]

[1] Department of Nuclear Physics and Biophysics, Faculty of Mathematics, Physics and Informatics, Comenius University, Mlynska Dolina F1, 842 48 Bratislava, Slovakia; piovarci6@uniba.sk

[2] Center for Nanophase Materials Sciences, Oak Ridge National Laboratory, Oak Ridge, TN 37831, USA; ivanovin@ornl.gov

* Correspondence: tibor.hianik@fmph.uniba.sk

Abstract: Chymotrypsin is an important proteolytic enzyme in the human digestive system that cleaves milk proteins through the hydrolysis reaction, making it an interesting subject to study the activity of milk proteases. In this work, we compared detection of chymotrypsin by spectrophotometric dynamic light scattering (DLS) and quartz crystal microbalance (QCM) methods and determined the limit of chymotrypsin detection (LOD), 0.15 ± 0.01 nM for spectrophotometric, 0.67 ± 0.05 nM for DLS and 1.40 ± 0.30 nM for QCM methods, respectively. The sensors are relatively cheap and are able to detect chymotrypsin in 3035 min. While the optical detection methods are simple to implement, the QCM method is more robust for sample preparation, and allows detection of chymotrypsin in non-transparent samples. We give an overview on methods and instruments for detection of chymotrypsin and other milk proteases.

Keywords: chymotrypsin; β-casein; nanoparticles; UV-vis spectroscopy; dynamic light scattering; quartz crystal microbalance

Citation: Piovarci, I.; Hianik, T.; Ivanov, I.N. Detection of Chymotrypsin by Optical and Acoustic Methods. *Biosensors* **2021**, *11*, 63. https://doi.org/10.3390/bios11030063

Received: 30 January 2021
Accepted: 23 February 2021
Published: 26 February 2021

Publisher's Note: MDPI stays neutral with regard to jurisdictional claims in published maps and institutional affiliations.

1. Introduction

Proteases represent a very wide and important group of enzymes found in a broad range of biological systems [1]. Proteases play an important role in the digestion process and participate in various pathological processes [2,3]. Chymotrypsin is a serine protease present in the human digestive system that participates in protein cleavage in the intestines [4]. Together with trypsin, chymotrypsinogen is ejected into the duodenum, where trypsin cleaves it into the active form [5]. Chymotrypsin activity is closely related to the activity of trypsin, which, along with plasmin, is an important enzyme in milk. Activity of plasmin is correlated to the quality of milk where the protease cleaves the proteins, mainly casein micelles affecting the milk flavor, shelf-life or cheese yield [6]. In pathology and medicine, chymotrypsin also has anti-inflammatory effects and has been successfully used to reduce post-operation complications after cataract surgery [7]. Measuring chymotrypsin activity can also be used for differential diagnosis [8].

Thus, development of sensitive, inexpensive, fast, and easy to use methods for detection of chymotrypsin or other milk proteases would be beneficial to disease diagnostics and control of dairy quality. However, there are no simple and effective assays that can be used for these purposes yet available. Protease detection is currently based on the detection of α-amino groups cleaved from the protein substrate using optical or high-performance liquid chromatography (HPLC) methods. The method that can be used for fast analysis of the protease concentration is based on enzyme-linked immunosorbent assay (ELISA) with a limit of detection (LOD) of about 0.5 nM for chymotrypsin [9,10]. However, the above-mentioned methods do not allow study of the kinetics of substrate digestion.

In this paper we test three methods for chymotrypsin detection: QCM, spectrophotometric, and DLS.

The QCM method is based on measurement of the resonant frequency, f, of shearing oscillations of AT-cut quartz crystal, as well as motional resistance, R_m, and is also known as thickness shear mode method (TSM). The protease substrates, such as β-casein or short specific peptides, are immobilized on thin gold layers sputtered at a QCM transducer. High frequency voltage, typically in the range of 5–20 MHz, induces shearing oscillations of the crystal. The fundamental resonance frequency of the crystal, f_0, depends on the physical properties of the quartz viscosity of the medium to which the crystal surface is exposed, as well as on the molecular interactions at the surface. The R_m value is sensitive to shearing viscosity, which is due to the molecular slip between the protein layer and surrounding water environment. Using Sauerbrey Equation (1) [11], one can link the change in resonant frequency to the mass bound to the surface of the electrode.

$$\Delta f = -2f_o{}^2 \Delta m / A (\mu_q \rho_q)^{1/2}, \tag{1}$$

where f_o is the fundamental resonant frequency (Hz), A is the active crystal area (in our case: 0.2 cm^2), ρ_q is quartz density (2.648 g cm^{-3}), Δm is the mass change (g), ρ_q is the shear modulus of the crystal (2.947 × 10^{11} g cm^{-1} s^{-2}). This Equation is valid only for a rigid layer in vacuum. In a liquid environment and for relatively soft layers, the viscosity contribution can be estimated by measurements of R_m.

We modified the surface of the QCM crystal with a layer of β-casein. The resulting mass added to the sensor leads to the decrease of the resonant frequency, f, and increase of motional resistance, R_m. Chymotrypsin will cleave β-casein, which results in an increase in f and decrease in R_m values. The mass sensitive QCM method was used for the detection of trypsin activity using synthesized peptide chains [12]. Poturnayova et al. used β-casein layers to detect activity of plasmin and trypsin with LOD around 0.65 nM [13]. Incorporation of machine learning algorithm for analysis of multiharmonic QCM response allowed detection of trypsin and plasmin with LOD of 0.2 nM and 0.5 nM, respectively. The applied algorithm in the work of Tatarko et al. allowed us to distinguish these two proteases within 2 min [14].

We also used the spectrophotometric method based on measurement of absorbance of the dispersion of gold nanoparticles (AuNPs) coated by 6-mercapto-1-hexanol (MCH) and β-casein. AuNPs demonstrate a surface plasmon resonance (SPR) effect, which arises from the oscillating electromagnetic field of light rays getting into contact with the free electrons in metallic nanoparticles and induces their coherent oscillation, which have strong optical absorption in the UV-vis part of the spectrum. The SPR absorbance of AuNPs depends on the surrounding medium and on the distance between nanoparticles [15]. In the work by Diouani, AuNPs modified with casein were used to detect Leishmania infantum using amperometric methods [16]. Chen et al. modified AuNPs with a trypsin-specific peptide sequence [17]. After the trypsin cleavage, the gold nanoparticles aggregated, which was detected by monitoring changes in the UV-vis spectrum. The detection limit of this method was estimated to be around 5 nM. Svard et al. modified gold nanoparticles with casein or IgG antibodies for detection trypsin or gingipain activity, by measuring SPR peak shift (blue shift for trypsin and red shift for gingipain) and reporting LOD of less than 4.3 nM for trypsin and gingipain [18]. Goyal et al. developed method of immobilization of gold nanoparticles on a paper membrane [19]. The protease activity then led to aggregation of the gold nanoparticles on the membrane and resulted in a colorimetric response in a visible part of the spectrum detectable by the naked eye. AuNPs modified by gelatin that served as a substrate for proteinase digestion have also been used for detection of other proteases such as trypsin and matrix metalloproteinase-2 [20]. In our work, we modified the gold nanoparticles with β-casein and MCH using protocol from Ref. [20]. The β-casein protects the AuNPs from aggregation. Addition of the chymotrypsin and subsequent cleavage of the β-casein caused nanoparticles aggregation due to loss of the protective shell. This effect was observed by measuring UV-vis spectra of nanoparticle dispersion.

We also used dynamic light scattering (DLS) method which uses Brownian motion and the Rayleigh scattering of the light from particles to assess their size [21]. The intensity of the scattered light (which depends on particle concentration) changes over time because of particle aggregation. The auto-correlation function that correlates the intensity of scattered light with its intensity after an arbitrary time is used to discern the size of the particles. The auto-correlation function also depends on diffusion coefficient of the nanoparticles [22]. In DLS experiments we used AuNPs modified with β-casein. After addition of the chymotrypsin, we were able to observe the cleavage of the casein layer without AuNPs aggregation that resulted in a decrease of the size of nanoparticles.

This report is an extension of a manuscript published in proceeding of the 1st International Electronic Conference on Biosensors [23].

2. Materials and Methods

2.1. Chemicals

Auric acid (HAuCl$_4$), sodium citrate, β-casein (Cat. No. C6905), 6-mercapto-1-hexanol (MCH, Cat. No. 725226), phosphate buffered saline (PBS) tablets (Cat. No. P4417), 11-mercaptoundecanoic acid (MUA, Cat. No. 450561), *N*-(3-Dimetylaminopropyl)-*N'*-etylcarbodiimid (EDC, Cat. No. E6383), *N*-Hydroxisuccinimid (NHS, Cat. No. 130672) and α-chymotrypsin (Cat. No. C3142) were of highest purity and purchased from Sigma-Aldrich (Darmstadt, Germany). Standard chemicals (p.a. grade), NaOH, HCl NaOH, NH$_3$, and H$_2$O$_2$ were from Slavus (Bratislava, Slovakia). Deionized water was prepared by Purelab Classic UV (Elga, High Wycombe, UK).

2.2. Spectrophotometric UV-vis Method

Gold nanoparticles (AuNPs) were prepared by modified citrate method [24]. In short, 100 mL of 0.01% chloroauric acid (HAuCl$_4$) was heated at around 98 °C and then 5 mL of 1% sodium tris-citrate was added. This solution was maintained at the temperature 98 °C and stirred by magnetic stirrer until it turned deep red (for about 15 min). Then the solution of AuNPs was cooled down and stored in the dark. To modify the gold nanoparticles with β-casein, we added 2 mL of 0.1 mg/mL aqueous β-casein into 18 mL of the AuNPs solution. After 2 h of incubation at room temperature without stirring, the gold nanoparticles were further incubated with 200 μL of 1 mM MCH overnight for approximately 18 h. MCH removes the surface charge of nanoparticles and thus facilitates their aggregation [20]. This is reflected by a color change to violet. However, nanoparticles (NPs) are protected from full aggregation due to the presence of a β-casein layer. Addition of chymotrypsin caused cleavage of β-casein, and as a result, the NPs aggregate. This was reflected by changes of the color of the solution to blue and then it became colorless. For the experiments, we prepared 0.95 mL of NPs. Chymotrypsin was dissolved in deionized water and 0.05 mL of chymotrypsin from the stock solution (concentration 100 nM) was added to each cuvette (1 mL standard cuvette, type UV transparent, Sarstedt, Nümbrecht, Germany). The concentration of chymotrypsin in cuvettes was 0.1, 0.3, 0.5, 0.7, 1; 5, and 10 nM at 1 mL of the total volume of solution. We also used a reference cuvette where only 0.05 mL of protease-free water was added to the AuNPs solution (total volume 1 mL). The spectra of the AuNPs were measured before protease addition (t = 0 min), just after protease addition (approximately 30 s) and then every 15 min up to 60 min. The measurements were repeated 3 times. The value of absorbance at time t = 0 has been multiplied by the dilution factor to correct the changes in absorbance intensity caused by the initial protease addition. Absorbance was measured by UV-1700 spectrophotometer (Shimadzu, Kyoto, Japan). The scheme of AuNPs modification and chymotrypsin cleavage is presented in Figure 1.

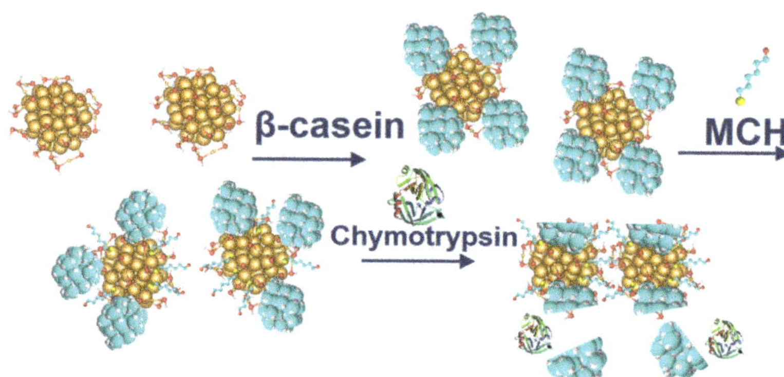

Figure 1. The scheme of gold nanoparticles (AuNPs) modification by 6-mercapto-1-hexanol (MCH) and β-casein and cleavage of β-casein by chymotrypsin.

2.3. DLS Method

The AuNPs prepared as described in Section 2.2. were incubated with 0.1 mg/mL of aqueous β-casein solution overnight in a volume ratio of 1:9 of β-casein to AuNPs (MCH was not used in this case). Addition of the chymotrypsin to the solution of AuNPs modified by β-casein did not lead to any discernible color change; however, it was possible to detect the decreased size of the AuNPs using ZetaSizer Nano ZS (Malvern Instruments, Malvern, UK). First, the size of AuNPs was measured in 1 mL standard cuvette (1 mL standard cuvette, type UV transparent, Sarstedt, Nümbrecht, Germany). Then the 0.1 mL of water solution containing various chymotrypsin concentrations was added to each cuvette. The final concentration of chymotrypsin in cuvettes was 0.1, 0.3, 0.5, 0.7, 1; 5 and 10 nM at 1.1 mL total volume of solution. The size of nanoparticles was measured before addition of chymotrypsin ($t = 0$) right after the addition (approximately 1 min) and after 30 min.

2.4. Quartz Crystal Microbalance (QCM) Method

The acoustic QCM sensor was prepared using an AT-cut quartz piezocrystals ($f_o = 8$ MHz, ICM, Oklahoma, OK, USA) with sputtered thin gold layers of an area $A = 0.2$ cm^2, that served as electrodes. First, the crystal was carefully cleaned as follows. It was exposed to a basic Piranha solution (H_2O_2:NH_3:H_2O = 1:1:5 mL). The crystals were immersed for 25 min in this solution, in beakers in a water bath (temperature was approximately 75 °C). Subsequently, the crystals were withdrawn, rinsed with distilled water, and returned to the beaker with a new dose of Piranha solution on the reverse side of the crystal. This was repeated three times. On the last extraction, the crystals were washed three times with distilled water and then with ethanol and placed in a bottle containing ethanol for storage at room temperature. The clean crystal was incubated overnight for 16–18 h at room temperature with 2 mM MUA dissolved in ethanol. MUA is a carboxylic acid with a sulfide group (SH). The sulfide moiety interacts with the gold on the crystal to form a self-assembled layer. After incubation, the crystal was washed with ethanol, distilled water, and 20 mM EDC and 50 mM NHS were applied for 25 min. These substances react with the carboxyl moiety of MUA and activate them to form a covalent bond with amino acids. Subsequently, the crystal was washed by distilled water, dried with nitrogen, and placed in an acrylic flow cell (JKU Linz, Austria). The cell was filled with PBS buffer using a Genie plus 2011 step pump (Kent Scientific, Torrington, CT, USA) at a flow rate of 200 µL/min. After filling the cell, we switched the flow to the rate of 50 µL/min. Then, 1 mg/mL of β-casein dissolved in PBS was allowed to flow under the crystal modified by MUA layer. After 35 min, only pure PBS was flowed in order to remove the unbound β-casein. All steps of the preparation of β-casein layer were recorded using a research quartz crystal microbalance (RQCM) instrument (Maxtek, East Syracuse, NY, USA).

After binding of β-casein to the electrode surface and stabilizing the resonant frequency (washing out all unbound residues), we applied chymotrypsin to the crystal at concentrations of 1 pM, 10 pM, 100 pM, 1 nM, 10 nM and 20 nM. After 35 min of chymotrypsin application, the PBS was let to flow into the cell until the resonant frequency stabilized. The change in casein coated QCM resonant frequency from application of chymotrypsin to stabilization in PBS corresponds to the amount of casein cleaved from the layer. After lower concentrations (1 pM, 10 pM, 100 pM), we applied a higher concentration of chymotrypsin (at least at a concentration 2 orders of magnitude higher). In such measurements, we analyzed the degree of cleavage as the change in frequency from the initial state to a steady-state value. All measurements were performed at PBS, pH 7.4.

For optical and gravimetric methods, the limit of detection (LOD) was determined using following Equation:

$$LOD = 3.3 \times (SD)/S, \tag{2}$$

where SD is standard deviation of the sample with lowest concentration and S is slope determined from fit of linear part of the calibration curve. The sequence of QCM operation including surface modification and sensing cleavage of β-casein by chymotrypsin using QCM piezocrystal and is presented in Figure 2.

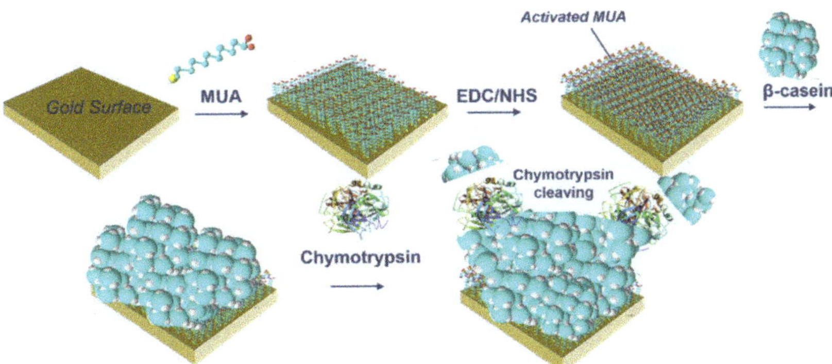

Figure 2. The scheme of modification of the piezocrystal and the cleavage of β-casein by chymotrypsin.

3. Results and Discussion

3.1. Detection of Chymotrypsin by Optical Method

In the first series of experiments, we studied the cleavage of the β-casein at the surface of the AuNPs by UV-vis and DLS methods. AuNPs modified by β-casein and MCH were used in optical detection method. Figure 3 shows the change in the absorption spectra after each step of AuNPs modification. The modification of AuNPs with β-casein resulted in a shift of the maximum of absorbance by around 5 nm toward higher wavelengths and in a slight increase in absorbance. After addition of MCH which replaces the β-casein protective layer leads to broadening of absorption peak, and shifts by 60 nm toward higher wavelengths, indicating increase in size due to aggregation of AuNP. The results agree well with Ref. [20] for AuNPs modified by gelatin and MCH.

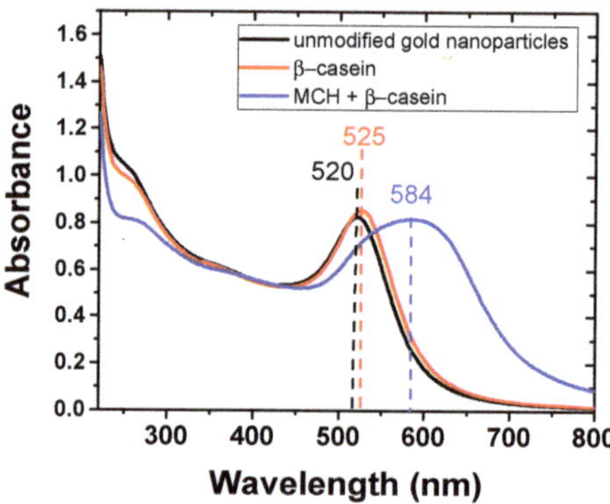

Figure 3. Absorption spectra of unmodified gold nanoparticles (AuNP) (black) and those modified by β-casein (red) and by β-casein + MCH (blue). The numbers at upper part of absorption peaks are wavelengths in nm.

The changes of absorbance spectra of AuNPs suspension have been measured during the chymotrypsin cleavage at 0 min, 0.5 min,15 min, 30 min and 60 min. Changes in spectra over time for two different concentration of chymotrypsin are presented in Figure 4. At a relatively low concentration of chymotrypsin (0.1 nM), we did not observe significant changes of the absorbance (Figure 4a). However, at higher chymotrypsin concentration, around 10 nM, a substantial red shift of the spectra was observed (up to 615 nm). It can be also seen that after maximum shifting substantial decrease of the absorbance with time occurred.

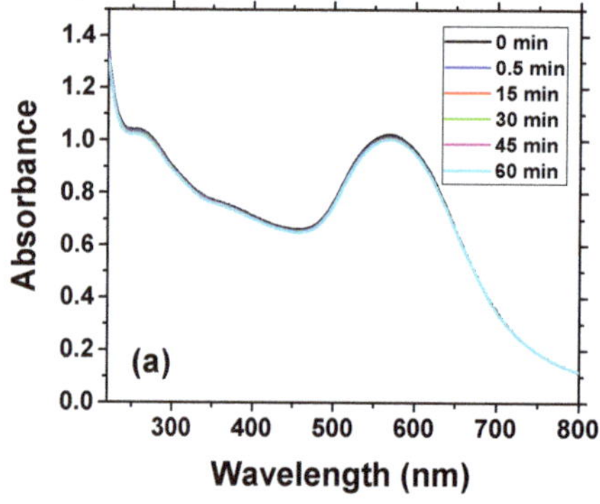
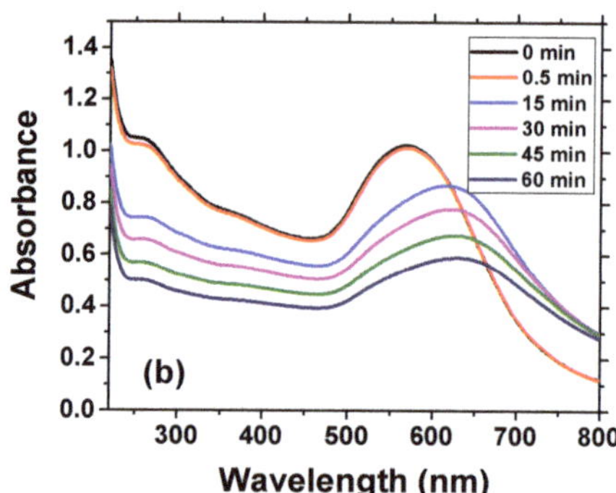

Figure 4. Changes of absorbance spectra of the suspension of AuNPs modified by β-casein and MCH in time for (**a**) 0.1 nM chymotrypsin and (**b**) 10 nM chymotrypsin.

Figure 5 shows the absorbance and change of maximum position of absorbance peak in time for all concentrations of chymotrypsin studied.

The rate of decrease in AuNP absorbance is higher for concentration of chymotrypsin 5 and 10 nM, and at lower concentration of chymotrypsin the rate of change is much slower (Figure 5a). The maximum position of absorbance peak shifted with time substantially at higher chymotrypsin concentrations (5 and 10 nM). For 5 and 10 nM chymotrypsin the maximum position of absorption peak was stabilized at around 615 nm, while for lower concentrations it increased with time almost linearly (Figure 5b). In order to prepare the calibration curve, we fit the change of absorbance vs. time with linear curve and then differentiated this model numerically to obtain the values of dA/dt. The calibration curve is presented on Figure 6a.

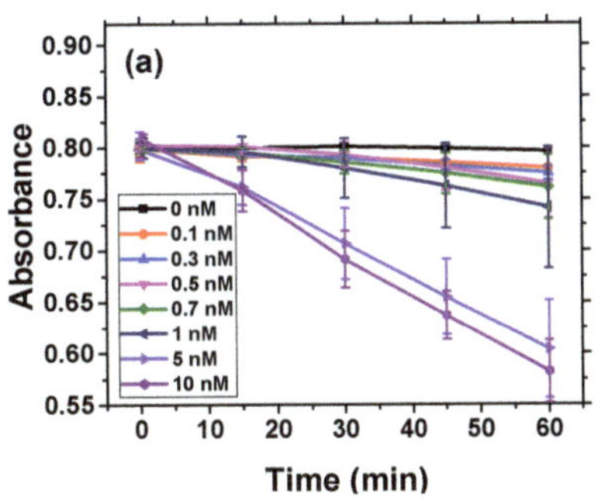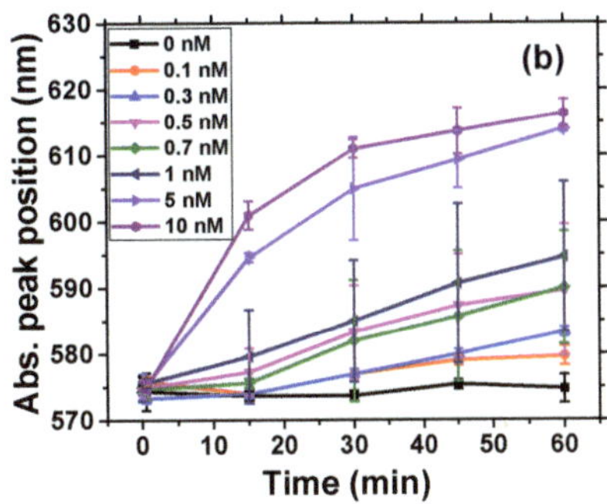

Figure 5. Changes in the absorbance (**a**) and in maximum position of absorbance peak (**b**) vs. time for different chymotrypsin concentrations in a suspension of AuNPs modified by β-casein and MCH. The results represent mean ± SD obtained from 3 independent measurements at each concentration of chymotrypsin.

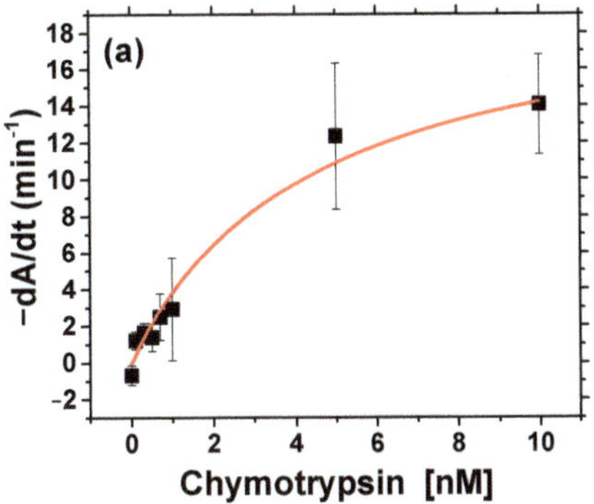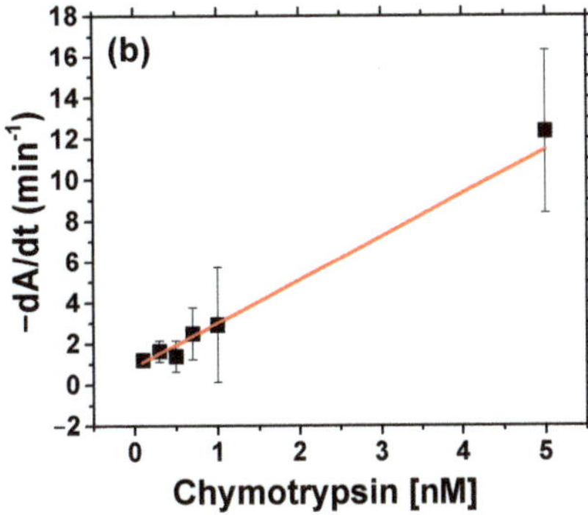

Figure 6. (**a**) Calibration curve for chymotrypsin fitted by reverse Michaelis—Menten model. dA/dt is numerical derivation of linear model of absorbance change at time $t = 0$ and corresponds to the rate of enzyme reaction. (**b**) Linear part of calibration curve $-dA/dt$ vs. concentration of chymotrypsin for calculation of limit of detection (LOD). $-dA/dt = (6.30 \pm 0.23) \times 10^{-4}$ min^{-1} nM^{-1} + $(1.9 \pm 0.4) \times 10^{-4}$ min^{-1}. ($R^2 = 0.993$), LOD = 0.15 ± 0.01 nM.

We were able to use reverse Michaelis—Menten model to analyze the obtained data. However, instead of substrate concentration, the concentration of chymotrypsin, c, has been used in this model: $v = v_{max} [c/(K_M + c)]$, where v and v_{max} are the rate and maximum rate of enzyme reaction, respectively, and $K_M = 3.89 \pm 1.24$ nM is reverse Michaelis—Menten constant obtained from the fit using the Michaelis—Menten model ($R^2 = 0.96$) (Figure 6b). In our case, $v = dA/dt$ and $v_{max} = (5.3 \pm 0.9) \times 10^{-3}$ min^{-1}. However, this model was used only formally because of different restraints. The main assumptions of the excess enzyme and limited substrate concentration was reversed in this case and instead, the concentration of the enzyme changed while the substrate was presented in excess. This implies different meaning of K_M (compared to the Michaelis—Menten model) which now represents the concentration of enzyme at which the rate of reaction is half of the maximum instead of concentration of substrate. A limitation of this approach is the assumption of substrate excess; nevertheless, it can be used for good approximation [25]. To calculate LOD, we used only part of the calibration curve from 0 to 5 nM, where the plot of dA/dt vs. c was almost linear. The results are shown in Figure 6b.

The LOD of the optical method of chymotrypsin detection, 0.15 ± 0.01 nM, was calculated from the Equation (2) using SD = 0.29 min^{-1} and S = 6.3 min^{-1} nM^{-1}. This value is 3.3 times lower than that of the ELISA method reported in the literature, around 0.5 nM [10]. However, in contrast with ELISA which requires specific antibodies, the method based on AuNPs is much easier and faster. Detection time of chymotrypsin using the optical method is about 30 min. The detection of the chymotrypsin with β-casein and MCH modified AuNPs can be done in one step. The disadvantage of this method is that only transparent samples can be used for detection. This restriction can be lifted using the surface sensitive gravimetric method (see Section 3.3).

3.2. Detection of Chymotrypsin by DLS Method

We measured the Z-average size of the non-modified AuNPs, which was found to be around 20 nm. The size is bigger than the assumed size of the prepared AuNPs of around 15 nm [22]. This is explained by the fact that DLS technique tends to overestimate the size of the gold nanoparticles due to the hydration sphere around the AuNPs. The Z-average size of the AuNPs modified with β-casein was around 35 nm. Figure 7. shows the plot of the Z-average size of AuNPs modified by β-casein at time 0 and 30 min at presence of various concentrations of chymotrypsin. Incubation of AuNPs with chymotrypsin resulted in decrease of Z-average size, which is more remarkable at presence of 5 nM and 10 nM protease concentrations. The variation in AuNP size at time 0 is related to the original size of nanoparticles, as well as rather fast cleavage of casein by protease, especially at its higher concentrations. However, even at relatively high chymotrypsin concentrations (10 nM) the average size did not reach those of naked AuNPs. This is evidence that cleavage was not complete and there is still a residual β-casein layer around AuNPs. This also explains why incomplete aggregation was observed in UV-vis experiments.

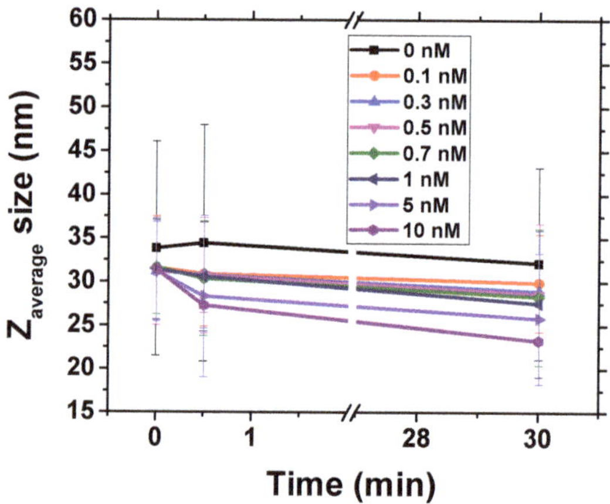

Figure 7. Change in Z-average size of AuNPs modified by β-casein at time 0 and 30 min at presence of various concentrations of chymotrypsin (see the insert). The results represent mean ±SD obtained from 3 independent measurements at each concentration of chymotrypsin.

We also constructed a calibration curve based on the percentual change of Z-average size in 30 min (Figure 8a) and fitted this by reverse Michaelis—Menten model. About 25% of the Z-average size was observed in 30 min. However, we can also see that the data has quite large (5 nm) standard deviation (obtained from 3 independent experiments at each concentration of chymotrypsin) affecting accuracy of concentration measurements, and the LOD of the sensor. Nevertheless, it is still a useful method to detect presence of chymotrypsin in the sample. The standard deviation could be improved by increasing number of measurements of the sample. It is important to note that the enzyme reaction was used without buffering the solution, which could also lead to a large value of standard deviation. The recommended buffer for chymotrypsin is 100 mM Tris-HCl at pH 7.8 (optimum pH) containing 10 mM $CaCl_2$ for stability. However, since we observed AuNPs aggregation in buffer, the experiments were carried in un-buffered solution. The calibration curve seems to be saturated near the 10 nM chymotrypsin.

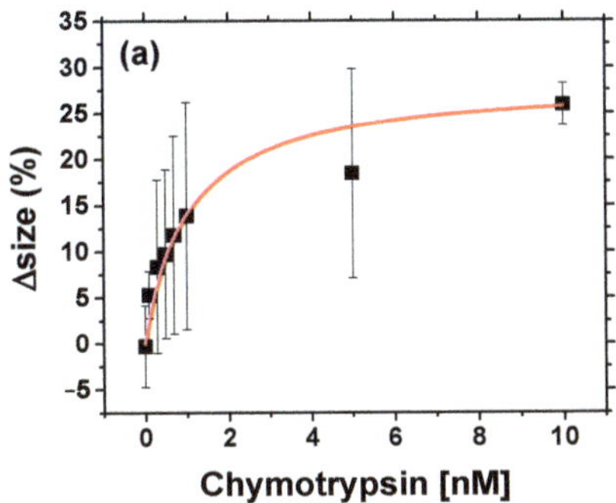
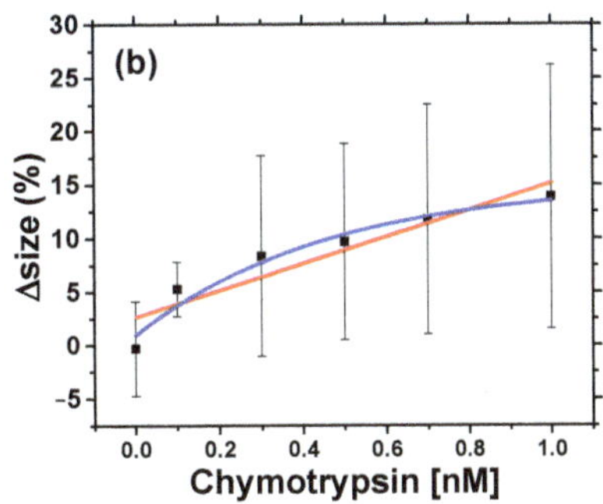

Figure 8. (**a**) The dependence of the changes of Z-average size (Δsize) vs. concentration of chymotrypsin for suspension of AuNPs modified by β-casein measured by DLS method. The curve represents fit according to reverse Michaelis—Menten model (see above). Δsize = $[Z_{average}(30\text{ min}) - Z_{average}(0\text{ min})]/Z_{average}(0\text{ min})$. (**b**) The linear part of calibration curve (red) used for calculation of LOD (Δsize = 12.47% nM^{-1} c + 2.71%, R^2 = 0.87, where c is the concentration of chymotrypsin). For comparison we show first order reaction fit (blue color) (Δsize = 15.09% − 13.83%$e^{-(c-0.006\text{ nM})/0.45nM}$, R^2 = 0.998, where c is the concentration of chymotrypsin). The results represent mean ± SD obtained from 3 independent measurements at each concentration of chymotrypsin.

Using a Michaelis—Menten reverse model from the fit of the results presented on Figure 8a we obtained for K_M = 1.03 ± 0.26 nM (R^2 = 0.998). This value is almost four times less than that obtained via spectrophotometric methods. This can be explained by addition of MCH in the spectrophotometric method, which can interfere with the cleavage of β-casein and decelerate the reaction. Both optical methods, however, should be able to detect the protease activity with similar precision. We took the linear part of the calibration curve (Figure 8b) and calculated LOD = 0.67 ± 0.05 nM. LOD value was calculated from the Equation (2) using SD = 2.54 (%) and S = 12.47 (%) nM^{-1}. This value is 4.5 times higher in comparison with those obtained by spectrophotometric method. The possible reason is less reproducible data in the case of $Z_{average}$ measurement in comparison with the absorbance method. The time of measurement is practically the same for both optical methods. In the case of the DLS method, the preparation of the AuNPs is by one step easier and since there is no MCH in the sample the AuNPs are more stable than those used in the spectrophotometric method.

3.3. Detection of Chymotrypsin by Gravimetric Method

For the gravimetric method, we first modified the surface of the QCM piezocrystal with MUA and then by β-casein. By monitoring resonant frequency, f, and motional resistance, R_m, it was possible to study all steps of preparation of the sensing surface. The value of motional resistance reflects the viscosity contribution caused by non-ideal slip between β-casein-layer and the surrounding water environment [26]. This is presented in Figure 9. The activation of carboxylic groups of MUA with EDC/NHS lead to only a small shift in resonance frequency. The addition of β-casein resulted in a fast drop of resonant frequency of about 170 Hz. After changing the flow with buffer, the frequency increased due to the removal of nonspecifically bound β-casein. The resulting frequency shift after washing of the surface corresponded to 120 Hz. From the Sauerbrey equation, we can calculate the change in mass on QCM biosensor, which corresponded to about 165 ng of mass added. With the knowledge of the molecular weight of β-casein M_w = 24 kDa, we could calculate the surface density of β-casein: Γ = 34.5 pM/cm^2. From the changes in motional resistance R_m, we could estimate the contribution of surface viscosity into resonant frequency. Since the R_m value decreases and increases proportionally to the change of frequency on a rather small value, the change in motional resistance is caused mainly by added weight. Therefore, one can assume that the β-casein layer was relatively rigid, which justifies application of the Sauerbrey equation.

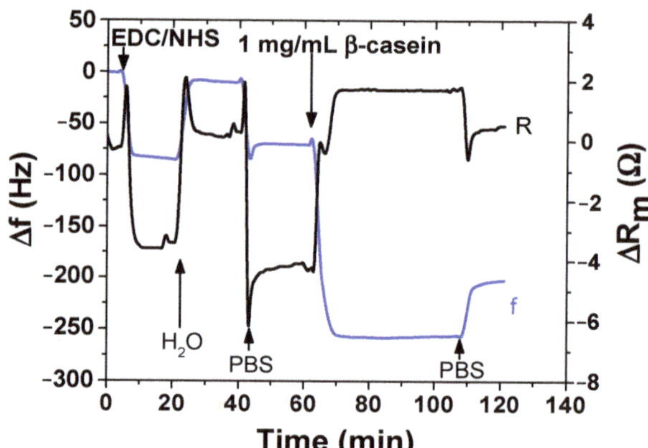

Figure 9. Kinetics of resonant frequency, f (blue), and motional resistance, R_m (black), changes during modification of piezocrystal by β-casein. The carboxylic groups of MUA that were chemisorbed at the crystal were first activated by EDC/NHS. The moments of addition of various compounds as well as washing the surface by water and PBS are shown by arrows.

After β-casein was bound to the surface, we could study its cleavage by different concentrations of chymotrypsin under flow condition for 35 min. An example of the changes of resonant frequency and motional resistance following the addition of 10 nM chymotrypsin are shown in Figure 10. In the presence of chymotrypsin, the resonant frequency increased by 35 Hz, but motional resistance decreased by 1.6 Ω. This is clear evidence of the cleavage of β-casein by chymotrypsin. Decrease of motional resistance can be due to an increase of molecular slip, which can be caused by weaker viscosity contribution.

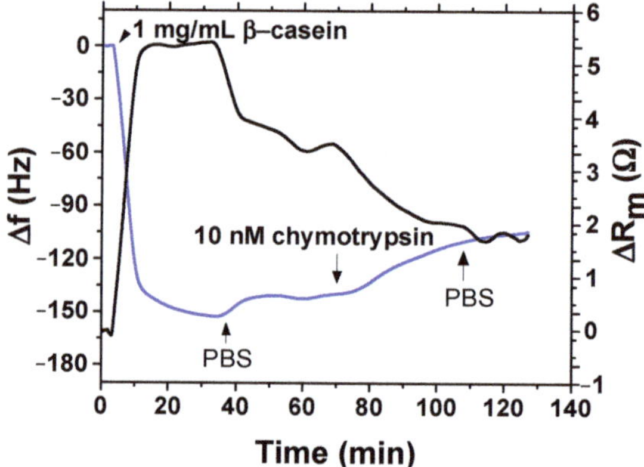

Figure 10. Kinetics of the changes of resonant frequency, f (blue), and motional resistance, R_m (black), following modification of piezocrystal by β-casein and addition of 10 nM of chymotrypsin. Addition of various compounds as well as washing of the surface by PBS is shown by arrows.

Based on the changes of the frequency, we constructed a calibration curve for different concentrations of chymotrypsin (Figure 11a). For determination of LOD, we also prepared calibration curve at a low concentration range where the dependence was almost linear (Figure 11b). The LOD for gravimetric detection of chymotrypsin, 1.40 ± 0.30 nM was calculated from the Equation (2) using SD = 2.20 (%) and S = 5.17 (%) nM^{-1}. This value was 2.8 times higher than that reported by ELISA methods and 9.3 times higher than for optical method of detection reported here. We should mention that gravimetric detection requires careful handling of the cell because even small changes in liquid pressure can affect the measurements. This method is, however, more tolerant to the presence of different components in the sample, such as added fat present in natural milk. The gravimetric method of protease detection requires time similar to optical methods. It is also important to mention the uniformity of the modified materials. The changes of frequency in gravimetric methods and absorbance maximum position in optical methods did not differ significantly from sample to sample, suggesting that it is not a source of significant error (large standard deviation).

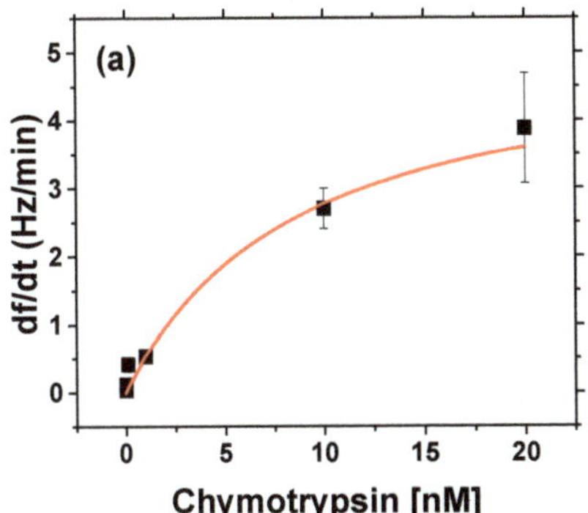
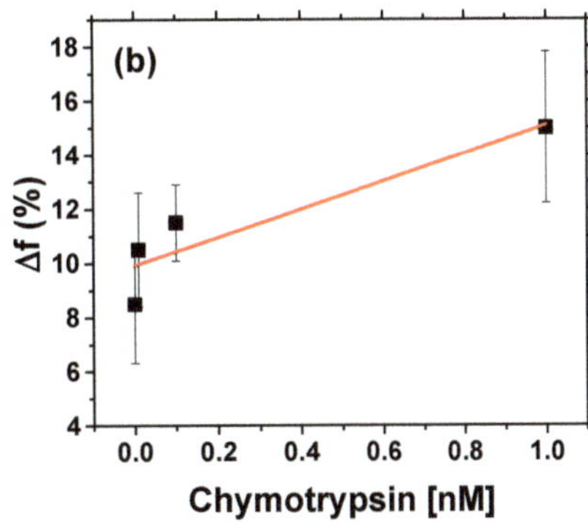

Figure 11. (**a**) Calibration curve for chymotrypsin fitted by reverse Michaelis—Menten model, v = df/dt was the first derivation of frequency obtained from kinetic curve. (**b**) Calibration curve: changes in frequency $\Delta f = (\Delta f_{chymo})/(\Delta f_{casein}) \times 100$ where $\Delta f_{chymo} = f - f_0$ (where f is steady state frequency following addition of chymotrypsin and washing the surface by PBS and f_0 those prior addition of chymotrypsin) is change in frequency after chymotrypsin cleavage and Δf_{casein} is frequency change after formation of β-casein layer. vs. lower chymotrypsin concentration range for determination of LOD. ($\Delta f = 5.17\%nM^{-1}c + 9.94\%$, $R^2 = 0.84$, where c is the concentration of chymotrypsin). The results represent mean $\pm$ SD obtained from 3 independent measurements at each concentration of chymotrypsin.

It is also interesting to compare the K_M values determined in optical and gravimetric experiments. From the results presented above, it can be seen that K_M value in the case of AuNPs-based optical assay (3.89 nM and 1.03 nM for spectrophotometric and DLS methods respectively) were lower in comparison with those based on gravimetric measurements ($K_M = 8.6 \pm 3.6$ nM, $R^2 = 0.999$). This can be evidence of better access of β-casein substrate for chymotrypsin in AuNPs in comparison with those immobilized at the surface of piezo-electric transducer. This effect can probably explain the lower LOD for spectrophotometric method of chymotrypsin detection with those based on gravimetric method. In Table 1 we present comparison of other published methods for detection of chymotrypsin. Results obtained in our work have either comparable or higher LOD, but in most cases are faster in comparison to other methods.

Table 1. The comparison of LOD and detection time of chymotrypsin detection.

Method	LOD	Detection Time	Reference
ELISA	0.5 nM	3.5 h	[10]
Liquid crystals protease assay	4 pM	3 h	[27]
Electrophoresis	20 pM	1 h	[28]
NIR fluorescent probe	0.5 nM	35 min	[29]
Ratiometric fluorescence probe	0.34 nM	30 min	[30]
UV-Vis, AuNPs	0.15 ± 0.01 nM	30 min	This work
DLS, AuMPs	0.67 ± 0.05 nM	30 min	This work
TSM	1.40 ± 0.30 nM	35 min	This work

4. Conclusions

We determined LOD for detection of chymotrypsin by gravimetric (LOD = 1.40 ± 0.30 nM) spectrophotometric (LOD = 0.15 ± 0.01 nM) and DLS method (LOD = 0.67 ± 0.05 nM). Spectrophotometric method showed the best value of LOD, even when compared to commercial ELISA (LOD = 0.5 nM). We also determined a steady-state constant K_M for the different methods with reverse Michaelis—Menten equation. The largest K_M value was found for gravimetric method of chymotrypsin detection (K_M = 8.6 ± 3.6 nM), followed by spectrophotometric method (K_M = 3.89 ± 1.24 nM),) and then DLS method (K_M = 1.03 ± 0.26 nM). We can explain observed differences in K_M values by difference in α-chymotrypsin activity, which is highest and least impeded on gold nanoparticles modified with β-casein. Addition of MCH decelerates the reaction and immobilization of β-casein on the gold surface slows the α-chymotrypsin ability to cleave β-casein. The detection time for methods that we tested was comparable and takes around 30 min for chymotrypsin determination. All methods required preparation of the sensing layers or modification of AuNPs overnight. The AuNPs or gravimetric sensors could be stored for a long time (more than one month) at 4 °C. In terms of difficulty in operation, the optical methods offered the easier way to measure chymotrypsin. With prepared AuNPs modified by β-casein and MCH, the spectrophotometric method required only one step of protease detection based on measurement of absorbance changes after 30 min, which was simpler in comparison with ELISA. DLS method based on AuNPs requires also only one step of measurement of the Z-average. The spectrophotometric method required only 50 µL of sample, the DLS method used 100 µL, while the gravimetric method used around 2 mL. One of the advantages of the gravimetric method is that it is more robust to "impurities" in the sample. The gravimetric method can be used with natural, no-transparent samples containing fat, minerals, or other proteins, just like in milk. Optical assays require a transparent sample; however, DLS method is a little less sensitive to changes in chymotrypsin concentrations. In terms of cost of analysis, the production of gold nanoparticles is relatively inexpensive and can be scaled to industrial amounts. For optical detection of chymotrypsin, gold nanoparticles should be surface-modified using inexpensive chemicals (β-casein and MCH). While gravimetric methods also use inexpensive chemicals for modification, but the cost of quartz crystal would raise the overall cost of the sensor. This cost offset can be reduced by multiple use of the same crystal when the sensing layer is regenerated. All methods have a distinct advantage and disadvantage compared to the currently used ELISA. In contrast with ELISA, the optical and gravimetric assay are not specific to the protease. Non-specificity of response can be addressed by using chymotrypsin-specific peptide substrate [13] or by integration of advanced machine learning algorithms [14]. In conclusion, we demonstrated advantages and disadvantages of spectrophotometric, DSL and gravimetric methods in detecting chymotrypsin. These methods can be applied also for detection of other proteases and can be useful for further application in the food industry and in medicine for real-time monitoring of the protease activity. In future work we plan to explore application of the presented techniques for analysis of natural milk samples (paying particular attention to gravimetric methods). Many new analytical methods use fluorometric or colorimetric molecules for detection of protease activity [31]. Gold nanoparticles seem to be good alternative component for colorimetric detection or for amplification of existing signal (for example increase of Raman signal from a sample using gold nanoparticles). It is clear that efforts furthering the development of new low-cost methods, easily implementable in practice, which would be sensitive, and exhibiting long-term stability, still need to be developed [32].

Author Contributions: Conceptualization, T.H. and I.N.I.; methodology, I.P.; formal analysis, I.P.; investigation, I.P.; resources, T.H. and I.N.I.; writing—original draft preparation, I.P.; writing—review and editing, T.H. and I.N.I.; visualization, I.P.; supervision, T.H. and I.N.I.; project administration, T.H.; funding acquisition, T.H. and I.N.I. All authors have read and agreed to the published version of the manuscript.

Funding: A portion of this research was conducted at the Center for Nanophase Materials Sciences, which is a DOE Office of Science User Facility, project No. CNMS2018-293. This work was funded under the European Union's Horizon 2020 research and innovation program through the Marie Skłodowska-Curie grant agreement No 690898 and by Science Agency VEGA, project No. 1/0419/20.

Institutional Review Board Statement: Not applicable.

Informed Consent Statement: Not applicable.

Data Availability Statement: Not applicable.

Conflicts of Interest: The authors declare no conflict of interest. The funders had no role in the design of the study; in the collection, analyses, or interpretation of data; in the writing of the manuscript; or in the decision to publish the results.

References

1. Bond, J.S. Proteases: History, discovery, and roles in health and disease. *J. Biol. Chem.* **2019**, *294*, 1643–1651. [CrossRef] [PubMed]
2. Lojda, Z. The importance of protease histochemistry in pathology. *Histochem. J.* **1985**, *17*, 1063–1089. [CrossRef] [PubMed]
3. Antalis, T.M.; Shea-Donohue, T.; Vogel, S.N.; Sears, C.; Fasano, A. Mechanisms of disease: Protease functions in intestinal mucosal pathobiology. *Nat. Clin. Pract. Gastroenterol. Hepatol.* **2007**, *4*, 393–402. [CrossRef]
4. Appel, W. Chymotrypsin: Molecular and catalytic properties. *Clin. Biochem.* **1986**, *19*, 317–322. [CrossRef]
5. Tomlinson, G.; Shaw, M.C.; Viswanatha, T. Chymotrypsin(s). *Methods Enzymol.* **1974**, *34*, 415–420. [CrossRef] [PubMed]
6. Ismail, B.; Nielsen, S.S. Invited review: Plasmin protease in milk: Current knowledge and relevance to dairy industry. *J. Dairy Sci.* **2010**, *93*, 4999–5009. [CrossRef] [PubMed]
7. Harris, G.S. Alpha-chymotrypsin in cataract surgery. *Can. Med. Assoc. J.* **1961**, *85*, 186–188. [CrossRef] [PubMed]
8. Skacel, M.; Ormsby, A.H.; Petras, R.E.; McMahon, J.T.; Henricks, W.H. Immunohistochemistry in the differential diagnosis of acinar and endocrine pancreatic neoplasms. *Appl. Immunohistochem. Mol. Morphol.* **2000**, *8*, 203–209. [CrossRef] [PubMed]
9. Chen, L.; Daniel, R.M.; Coolbear, T. Detection and impact of protease and lipase activities in milk and milk powders. *Int. Dairy J.* **2003**, *13*, 255–275. [CrossRef]
10. Chymotrypsin ELISA Kit. Available online: http://img2.creative-diagnostics.com/pdf/DEIA10041,Chymotrypsin.pdf (accessed on 7 October 2020).
11. Kankare, J. Sauerbrey equation of quartz crystal microbalance in liquid medium. *Langmuir* **2002**, *18*, 7092–7094. [CrossRef]
12. Dong, Z.-M.; Cheng, L.; Zhang, P.; Zhao, G.-C. Label-free analytical performances of peptide-based QCM biosensor for trypsin. *Analyst* **2020**, *145*, 3329–3338. [CrossRef] [PubMed]
13. Poturnayova, A.; Castillo, G.; Subjakova, V.; Tatarko, M.; Snejdarkova, M.; Hianik, T. Optimization of the cytochrome c detection by acoustic and electrochemical methods based on aptamer sensors. *Sens. Actuators B Chem.* **2017**, *228*, 817–827. [CrossRef]
14. Tatarko, M.; Muckley, E.; Subjakova, V.; Goswami, M.; Sumpter, B.; Hianik, T.; Ivanov, I. Machine learning enabled acoustic detection of sub-nanomolar concentration of trypsin and plasmin in solution. *Sens. Actuators B Chem.* **2018**, *272*, 282–288. [CrossRef]
15. Huang, X.; El-Sayed, M.A. Gold nanoparticles: Optical properties and implementations in cancer diagnosis and photothermal therapy. *J. Adv. Res.* **2010**, *1*, 13–28. [CrossRef]
16. Diouani, M.F.; Ouerghi, O.; Belgacem, K.; Sayhi, M.; Laouini, D. Casein-conjugated gold nanoparticles for amperometric detection of leishmania infantum. *Biosensors* **2019**, *9*, 68. [CrossRef] [PubMed]
17. Chen, G.; Yusheng, X.; Zhang, H.; Wang, P.; Cheung, H.; Yang, M.; Sun, H. A general colorimetric method for detecting protease activity based on peptide-induced gold nanoparticle aggregation. *RSC Adv.* **2014**, *4*, 6560–6563. [CrossRef]
18. Svard, A.; Neilands, J.; Palm, E.; Svensater, G.; Bengtsson, T.; Aili, D. Protein-Functionalized Gold Nanoparticles as Refractometric Nanoplasmonic Sensors for the Detection of Proteolytic Activity of Porphyromonas gingivalis. *ACS Appl. Nano Mater. Am. Chem. Soc.* **2020**, *3*, 9822–9830. [CrossRef]
19. Goyal, G.; Alagappan, P.; Liedberg, B. Protease Functional Assay on Membrane. *Sens. Actuators B Chem.* **2019**, *305*, 127442. [CrossRef]
20. Chuang, Y.-C.; Li, J.-C.; Chen, S.-H.; Liu, T.-Y.; Kuo, C.-H.; Huang, W.-T.; Lin, C.-S. An optical biosensing platform for proteinase activity using gold nanoparticles. *Biomaterials* **2010**, *31*, 6087–6095. [CrossRef]
21. Goldburg, W.I. Dynamic light scattering. *Am. J. Phys.* **1999**, *67*, 1152–1160. [CrossRef]
22. Naiim, M.; Boualem, A.; Ferre, C.; Jabloun, M.; Jalocha, A.; Ravier, P. Multiangle dynamic light scattering for the improvement of multimodal particle size distribution measurements. *Soft Matter* **2015**, *11*, 28–32. [CrossRef] [PubMed]
23. Piovarci, I.; Hianik, T.; Ivanov, I.N. Comparison of optical and gravimetric methods for detection of chymotrypsin. *Proceedings* **2020**, *60*, 42. [CrossRef]
24. Kimling, J.; Maier, M.; Okenve, B.; Kotaidis, V.; Ballot, H.; Plech, A. Turkevich method for gold nanoparticle synthesis revisited. *J. Phys. Chem. B* **2006**, *110*, 15700–15707. [CrossRef]
25. Románszki, L.; Tatarko, M.; Jiao, M.; Keresztes, S.; Hianik, T.; Thompson, M. Casein probe-based fast plasmin determination in the picomolar range by an ultra-high frequency acoustic wave biosensor. *Sens. Actuators B Chem.* **2018**, *275*, 206–214. [CrossRef]
26. Martin, S.J.; Granstaff, V.E.; Frye, G.C. Characterization of a quartz crystal microbalance with simultaneous mass and liquid loading. *Anal. Chem.* **1991**, *63*, 2272–2281. [CrossRef]

27. Chen, C.-H.; Yang, K.-L. Oligopeptide immobilization strategy for improving stability and sensitivity of liquid-crystal protease assays. *Sens. Actuators B Chem.* **2014**, *204*, 734–740. [CrossRef]
28. Lefkowitz, R.B.; Schmid-Schönbein, G.W.; Heller, M.J. Whole blood assay for elastase, chymotrypsin, matrix metalloproteinase-2, and matrix metalloproteinase-9 activity. *Anal. Chem.* **2010**, *82*, 8251–8258. [CrossRef] [PubMed]
29. Mu, S.; Xu, Y.; Zhang, Y.; Guo, X.; Li, J.; Wang, Y.; Liu, X.; Zhang, H. A non-peptide NIR fluorescent probe for detection of chymotrypsin and its imaging application. *J. Mater. Chem. B.* **2019**, *7*, 2974–2980. [CrossRef]
30. Chen, Y.; Cao, J.; Jiang, X.; Pan, Z.; Fu, N. A sensitive ratiometric fluorescence probe for chymotrypsin activity and inhibitor screening. *Sens. Actuators B Chem.* **2018**, *273*, 204–210. [CrossRef]
31. Magdalena, W.; Lesner, A. Future of Protase Activity Assays. *Curr. Pharm. Des.* **2012**, *19*, 1062–1067. [CrossRef]
32. Shi, H.; Liu, C.; Cui, J.; Cheng, J.; Lin, Y.; Luo, R. Perspective on chymotrypsin detection. *New J. Chem. R. Soc. Chem.* **2020**, *44*, 20921–20929. [CrossRef]

Communication

Microfluidic Impedance Biosensor Chips Using Sensing Layers Based on DNA-Based Self-Assembled Monolayers for Label-Free Detection of Proteins

Khaled Alsabbagh [1], Tim Hornung [1], Achim Voigt [1], Sahba Sadir [2], Taleieh Rajabi [1] and Kerstin Länge [1,*]

[1] Institute of Microstructure Technology, Karlsruhe Institute of Technology, Hermann-von-Helmholtz-Platz 1, 76344 Eggenstein-Leopoldshafen, Germany; khaled.alsabbagh@icloud.com (K.A.); timhornung@posteo.de (T.H.); achim.voigt@kit.edu (A.V.); tala_rajabi@yahoo.de (T.R.)

[2] Institute for Micro Process Engineering, Karlsruhe Institute of Technology, Hermann-von-Helmholtz-Platz 1, 76344 Eggenstein-Leopoldshafen, Germany; sadir.sahba@gmail.com

* Correspondence: kerstin.laenge@kit.edu

Abstract: A microfluidic chip for electrochemical impedance spectroscopy (EIS) is presented as bio-sensor for label-free detection of proteins by using the example of cardiac troponin I. Troponin I is one of the most specific diagnostic serum biomarkers for myocardial infarction. The microfluidic impedance biosensor chip presented here consists of a microscope glass slide serving as base plate, sputtered electrodes, and a polydimethylsiloxane (PDMS) microchannel. Electrode functionalization protocols were developed considering a possible charge transfer through the sensing layer, in addition to analyte-specific binding by corresponding antibodies and reduction of nonspecific protein adsorption to prevent false-positive signals. Reagents tested for self-assembled monolayers (SAMs) on gold electrodes included thiolated hydrocarbons and thiolated oligonucleotides, where SAMs based on the latter showed a better performance. The corresponding antibody was covalently coupled on the SAM using carbodiimide chemistry. Sampling and measurement took only a few minutes. Application of a human serum albumin (HSA) sample, 1000 ng/mL, led to negligible impedance changes, while application of a troponin I sample, 1 ng/mL, led to a significant shift in the Nyquist plot. The results are promising regarding specific detection of clinically relevant concentrations of biomarkers, such as cardiac markers, with the newly developed microfluidic impedance biosensor chip.

Keywords: biosensor; immunosensor; cardiac troponin I; single-strand DNA; electrochemical impedance spectroscopy; label-free; proteins; microfluidic chip; self-assembled monolayers

Citation: Alsabbagh, K.; Hornung, T.; Voigt, A.; Sadir, S.; Rajabi, T.; Länge, K. Microfluidic Impedance Biosensor Chips Using Sensing Layers Based on DNA-Based Self-Assembled Monolayers for Label-Free Detection of Proteins. *Biosensors* **2021**, *11*, 80. https://doi.org/10.3390/bios11030080

Received: 13 January 2021
Accepted: 10 March 2021
Published: 13 March 2021

Publisher's Note: MDPI stays neutral with regard to jurisdictional claims in published maps and institutional affiliations.

1. Introduction

Label-free biosensors allow direct detection of analyte molecules and cells and, hence, offer a tool for fast detection of biomarkers and pathogens. Particularly electrochemical biosensors offer advantages here, as they can be fabricated cost-effectively, array-compatible, and customized in a comparatively easy way. Combining them with microfluidics results in efficient analytical devices, e.g., for biomarker detection in point-of-care applications, where clinically relevant protein concentrations are often in the range of a few ng/mL. In principle, a (micro-)fluidic channel can easily be made separately, e.g., from polydimethylsiloxane (PDMS), and then connected to the sensor unit. The requirements for biosensing layers include both the capability of analyte-specific binding and the minimization of nonspecific binding. The latter is particularly important for label-free bio-sensors to avoid false-positive results [1–7].

Label-free electrochemical biosensors include impedimetric biosensors which measure the impedance, i.e., the opposition presented to a current in an alternating current (AC) circuit when a voltage is applied. The impedance is a complex quantity, and a common graphical representation is the Nyquist plot. This is a frequency response plot, where the values of the real part are plotted on the x-axis and those of the imaginary part on the y-axis. An ideal Nyquist plot shows a semicircle resulting from the dominating, kinetically limited charge transfer through the electric double layer at the electrode. Binding of analyte molecules to the electrode will influence the charge transfer and, hence, result in a shift of the Nyquist plot. As a consequence, the opportunity for charge transfer is an additional requirement for sensing layers of impedance biosensors. At low frequencies, Nyquist plots may show straight lines with a slope of 45°. This is characteristic for diffusion limited processes and described by the Warburg impedance [8–12].

Electrodes of impedance biosensors are typically made of gold. A well-established procedure for the introduction of functional groups on this material is to use suitably substituted thiols forming self-assembled monolayers (SAMs). Thiols with aliphatic hydrocarbon spacers of sufficient chain length lead to well-defined and stable SAMs of high density. The brush-like structure of such layers makes it possible to effectively reduce nonspecific protein adsorption on the underlying gold surface. However, such SAMs may result in insulating layers, hindering the charge transfer required for the transduction principle of impedimetric biosensors. The use of aromatic hydrocarbons featuring delocalized π-electrons would be more beneficial for charge transfer processes, but nonspecific protein adsorption in the subsequent measurements may increase because of a reduced density of the layer. Conductive polymers would offer an alternative, but are often linked with coating procedures more complex than wet chemistry [7,12].

Thiolated single-strand DNA (ssDNA) oligomers, on the other hand, can be packed densely on the gold surface by wet chemistry methods similar to those of thiolated hydrocarbons. Coimmobilization of thiolated ssDNA with thiolated hydrocarbons may be recommended to improve the integrity of the brush-like structure. The negatively charged backbone of the DNA oligomers—resulting from the composition of alternating sugar (deoxyribose) and phosphate groups—promises lower initial impedance values and, hence, the possibility of charge transfer events [13,14]. With impedance sensors, immobilized ssDNA has been used directly as probe for DNA detection (hybridization). Similar to this, immobilized oligonucleotides have been designed as aptamers for protein detection. Furthermore, ssDNA has been hybridized with the complementary ssDNA carrying an analyte-specific capture molecule. A more flexible approach would be the hybridization of surface-bound ssDNA with complementary ssDNA carrying functional groups allowing the covalent coupling of any capture molecule. However, this has rarely been used on impedance electrodes so far [15–18].

As an example, for the applicability of DNA-based SAMs in impedimetric biosensors for protein detection, we used a cardiac biomarker. According to the World Health Organization (WHO), cardiovascular diseases (CVDs) are the leading cause of death worldwide. In 2016, 85% of the people who died of a CVD suffered a stroke or a heart attack. In 2015, 37% of people who died under the age of 70 due to noncommunicable diseases died from a CVD. One way to reduce these numbers is to increase the survival rate by starting the treatment as early as possible, which requires an early diagnosis [19,20]. Diagnosis criteria of a heart attack (also called myocardial infarction) include the detection of biomarkers in blood. Particularly, a concentration increase of cardiac troponins I and T (cTnI, cTnT) indicate myocardial necrosis with normal levels being <0.5–2.0 ng/mL (lab-specific) and <0.1 ng/mL, respectively, and a factor increase of up to 40-fold and 40–60-fold, respectively, in case of an infarction [21,22].

In this work, we introduce an own design for a microfluidic impedance biosensor chip. The base plate is a microscope glass slide carrying the gold electrodes, while the microfluidic channel is added as a PDMS component. Biosensing layers used thiol-SAMs based on both aromatic hydrocarbons and DNA, where the latter showed best performance regarding suppression of nonspecific binding of human serum albumin (HSA) and specific binding of the cardiac marker troponin I via the corresponding antibody.

2. Materials and Methods

2.1. Fabrication of the Microfluidic Impedance Biosensor Chip

2.1.1. Base Plate and Electrode Sputtering

Standard microscope glass slides were used as base plates for the impedance biosensor chips. They were thoroughly cleaned by rinsing with a detergent solution and brushing with a toothbrush. After rinsing with water, they were sonicated with bidistilled water for 5 min using an ultrasonic cleaning bath. Finally, they were rinsed with filtered 2-propanol and blown dry with filtered nitrogen gas; the filter pore size was 0.2 µm in both cases.

A parylene C (poly(2-chloro-p-xylylene)) layer with a thickness of 0.1 µm was applied on the cleaned glass slides as adhesion layer (SCS Labcoter® 1, PDS 2010, Specialty Coating Systems) [23]. Since the parylene C coating did not interfere with the later bonding of the PDMS microfluidic channel (see Section 2.1.2), it could be deposited on the entire surface of the microscope glass slides without the need of a mask or the removal of the coating outside the electrodes (see below). This made the coating process easy and convenient, which is why parylene C was preferred over the metal adhesion layers that were otherwise used for gold electrodes.

Working and counter electrodes made of gold were applied on the parylene C-coated glass slides by using a corresponding mask made of steel and a sputtering system (Balzers MED 010). The gold was sputtered at approximately 0.06 mbar and a current of 30 mA with the time set to 15 min, which led to a thickness in the range 15–35 nm. The diameter of the working electrode was 0.5 mm, corresponding to an area of 0.2 mm². The ratio of the areas of working electrode to counter electrode was approximately 1:100. A microscope glass slide with sputtered electrodes is shown in Figure 1a. Conducting paths led from the electrodes to the edge of the glass slide, where they formed contact pads for the connection to the measurement setup (see Section 2.3.1). To keep things simple, a reference electrode was not included in this work; but it can be included, if required (Figure 1b,c).

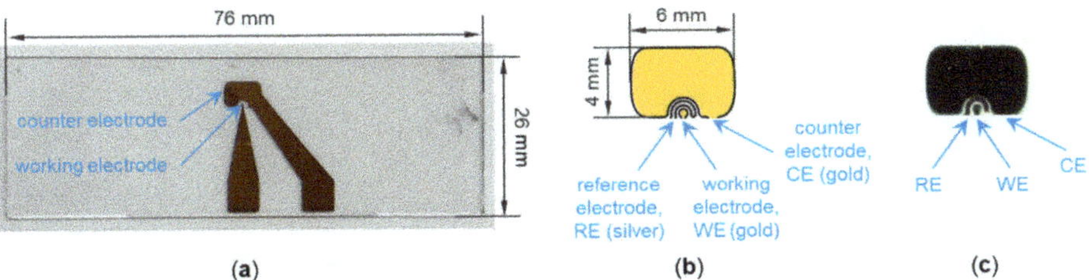

Figure 1. (**a**) Microscope glass slide with sputtered electrodes. (**b**) Detailed view of working and counter electrode, including a potential reference electrode. (**c**) Realized three-electrode setup sputtered on glass.

2.1.2. Microfluidic Channel Fabrication and Connecting

The microfluidic channel was formed from PDMS, the channel design is shown in Figure 2a. PDMS base and curing agent (Sylgard™ 184, Dow) were mixed in a weight ratio of 10:1, following the manufacturer's instructions. Air bubbles introduced by mixing the components were removed by evacuating the mixture in a vacuum. The bubble-free mixture was casted into a milled form made of polymethyl methacrylate (PMMA) and incubated at 70 °C for 2 h. To combine the PDMS channel with the impedance biosensor chip carrying the electrodes, both parts were plasma-activated and assembled with light pressure. During the plasma treatment, the electrodes were covered by impermeable polystyrene pieces, as they had been functionalized with SAM compounds or antibody sensing layer before (see Section 2.2). A picture of the combined parts is shown in Figure 2b.

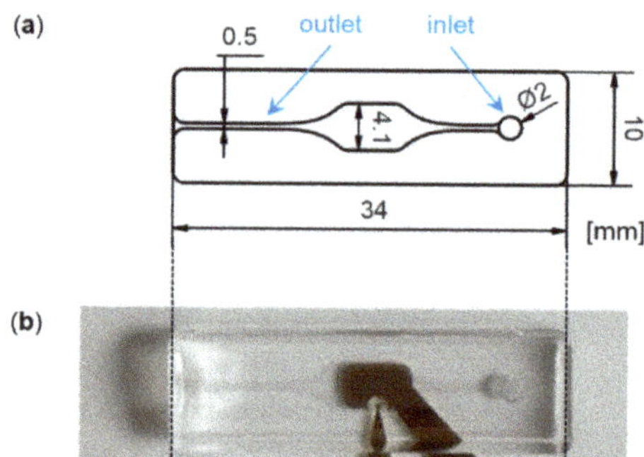

Figure 2. (**a**) Design of the microfluidic channel delivering samples across the electrodes. The channel height was 0.2 mm leading to a channel volume below 10 µL. (**b**) Microfluidic polydimethylsiloxane (PDMS) channel bonded on the impedance biosensor chip.

2.2. Surface Functionalization

The deposition of the 0.05 or 10 µL drops of the reaction solutions described below was carried out manually by using microliter syringes and a magnifying lamp. Rinsing was performed with wash bottles. The antibody used in the following was monoclonal anti-troponin I, clone 1H11L19 (Fisher Scientific, Schwerte, Germany).

2.2.1. Antibody Adsorption

The electrodes were cleaned by plasma treatment. A 0.05 µL drop of an anti-troponin I solution, diluted with phosphate buffered saline (PBS) to 5 µg/mL, was deposited on the working electrode. After 4 h of incubation at 6 °C and rinsing with PBS, the chip was assembled with the PDMS microfluidic channel (see Section 2.1.2).

2.2.2. Application of Thiol-SAMs with Hydrocarbon Spacer, Antibody Immobilization

To test chemicals for use as SAM, the electrodes were cleaned by plasma treatment, and 10 µL of an ethanolic solution containing 50 mM 4-mercaptobenzoic acid, 1,4-benzenedithiol, or 6-mercapto-1-hexanol was deposited on both working and counter electrodes. After incubation overnight at ambient temperature, the electrodes were rinsed with ethanol and the chip assembled with the PDMS channel (see Section 2.1.2).

A schematic representation of the antibody immobilization via SAM with hydrocarbon spacer and subsequent assay is given in Figure S1 in Supplementary Material. The electrodes were cleaned by plasma treatment, and a 0.05 µL drop containing 20 mM 4-mercaptobenzoic acid dissolved in ethanol was deposited on the working electrode. After 24 h of incubation at 6 °C, the chip was rinsed with ethanol. Another plasma cleaning step of the counter electrode was performed, during which the working electrode was covered with a piece of polystyrene. A 10 µL drop containing 20 mM 1,4-benzenedithiol dissolved in ethanol was deposited on the counter electrode and incubated overnight at 6 °C. After rinsing with ethanol and drying, 10 µL of a freshly prepared aqueous solution containing 0.05 M N-hydroxysuccinimide (NHS) and 0.2 M 1-ethyl-3-(3-dimethylaminopropyl)carbodiimide (EDC) was deposited on the electrodes and incubated for 30 min. The mixture was supposed to react only with the carboxyl groups of the 4-mercaptobenzoic acid SAM on the working electrode resulting in an active ester [24]. After rinsing with PBS, 0.05 µL of an anti-troponin I solution, diluted with PBS to 5 µg/mL, was deposited on the working electrode and incubated for 30 min. After rinsing with PBS, 0.05 µL of an aqueous solution of ethanolamine hydrochloride, 1 M, pH = 8.5, was deposited on the working electrode and incubated for 30 min to deactivate potentially available still reactive active ester groups [24]. Finally, the impedance biosensor chip was thoroughly rinsed with PBS, dried, and assembled with the PDMS microfluidic channel (see Section 2.1.2).

2.2.3. Application of Thiol-SAMs with DNA Spacer, Antibody Immobilization

Figure S2 shows the sequences of the ssDNA used below, including the schematic arrangement in the sensing layer. A schematic representation of the antibody immobilization via SAM with DNA spacer is given in Figure S3. Electrodes were cleaned by plasma treatment. SH-ssDNA (5′−thiol-C6-TTT TTT TTTTCC TGC GTC GTT TAA GGA AGT AC-3′, purchased from Metabion, Planegg, Germany) was coimmobilized with thiol compounds with hydrocarbon spacer for stabilization of the DNA-based SAM. The immobilization mixture contained either 0.025 mM SH-ssDNA and 15 mM 6-mercapto-1-hexanol or 0.033 mM SH-ssDNA and 13.3 mM 1,4-benzenedithiol dissolved in ethanol. A 0.05 µL drop of the immobilization mixture was deposited on the working electrode and incubated overnight at 6 °C. After rinsing with ethanol, the counter electrode was again cleaned by plasma treatment, while the working electrode was covered with impermeable polystyrene. A 10 µL drop containing 20 mM 1,4-benzenedithiol dissolved in ethanol was deposited on the counter electrode and incubated overnight at 6 °C. After that, 0.05 µL amino-ssDNA (5′-amino-C6-GTA CTT CCT TAA ACG ACG CAG G-3′, purchased from Metabion, Planegg, Germany), which was diluted with phosphate buffer to a concentration of 0.1 mM, was deposited onto the working electrode and incubated overnight at 6 °C. To convert the amino groups to carboxyl groups for antibody coupling, glutaric anhydride was dissolved in 8 M sodium hydroxide solution at a concentration of 0.5 mg/µL [25]; 10 µL of this solution was applied on the electrodes. The glutaric anhydride was supposed to react only with the amino groups of the functionalized working electrode. After 48 h of incubation at 6 °C, the chip was rinsed with bidistilled water. The following protocol of antibody coupling via EDC/NHS mixture and subsequent bonding of the PDMS channel was the same as described in the section before (Section 2.2.2).

2.3. Measurements with the Microfluidic Impedance Biosensor Chip

2.3.1. Measurement Setup

Impedance measurements were performed with the IMPSPEC device from Meodat (Ilmenau, Germany), which was designed for fast and broadband impedance spectroscopy. The starting frequency was set to 5.895 Hz. The frequency was increased linearly by adding up frequency intervals of 5.895 Hz, until the final frequency of about 10 kHz was reached. The real part and the imaginary part of the impedance were displayed for each frequency. The impedance biosensor chip was connected to the IMPSPEC device cable by means of conductive silver and crocodile clamps. A peristaltic pump delivered the liquid samples through the PDMS channel and across the electrodes.

2.3.2. HSA Adsorption for Testing SAMs

PBS redox was prepared by adding potassium hexacyanoferrate(II) and potassium hexacyanoferrate(III) to PBS to a final concentration of 15 mM each. HSA was dissolved in PBS redox at concentrations of 0/1/10/100/1000 ng/mL, resulting in five HSA samples. Starting with a zero sample containing PBS redox only and corresponding to HSA, 0 ng/mL, the four other HSA samples were subsequently applied with increasing HSA concentration to the same impedance biosensor chip. Each sample was applied for 2 min at a flow rate of 0.04 mL/min. After sample application, the pump was stopped, and after 30 s, the impedance measurement was performed before the next sample was applied. After measuring the last HSA sample (1000 ng/mL), the impedance biosensor chip was disposed, and a new biosensor chip was used.

2.3.3. HSA Adsorption and Troponin I Assay

An impedance biosensor chip with freshly prepared antibody coating (see Section 2.2) and bonded PDMS microfluidic channel (see Section 2.1.2) was rinsed with PBS redox for 2 min at a flow rate of 0.04 mL/min. After switching off the pump and waiting for 30 s, the impedance of the antibody-coated biosensor chip was measured. After that, HSA dissolved at 1000 ng/mL in PBS was applied to the biosensor chip at a flow rate of

0.04 mL/min. Sampling time was 8 min if the antibody coating was performed without SAM (see Section 2.2.1) or 3 min if the antibody coating was performed via SAM (see Sections 2.2.2 and 2.2.3). This was followed by PBS redox rinsing for 2 min at the same flow rate. The pump was stopped, and after 30 s, the effect of HSA on the respective biosensor impedance was measured. Finally, troponin I (Fisher Scientific, Schwerte, Germany) dissolved at 1 ng/mL in PBS was applied to the biosensor chip for 1 min at 0.04 mL/min, followed by PBS redox for 2 min at the same flow rate. The pump was stopped, and after 30 s, the impedance resulting from troponin I binding on the respective biosensor chip was measured. After measuring the troponin I sample, the impedance biosensor chip was disposed, and a new biosensor chip was used.

3. Results and Discussion

3.1. Basic Performance of the Microfluidic Impedance Biosensor Chip

A preliminary experiment to test the performance of the microfluidic impedance bio-sensor chip was carried out by simply adsorbing the antibody on the working electrode, without the use of any SAM. While the working electrode was coated with anti-troponin I, the counter electrode remained uncoated. After application of HSA, 1000 ng/mL, for blocking remaining potentially accessible adsorption sites, troponin I, 1 ng/mL, was applied on the sensor surface. The results are shown in Figure 3.

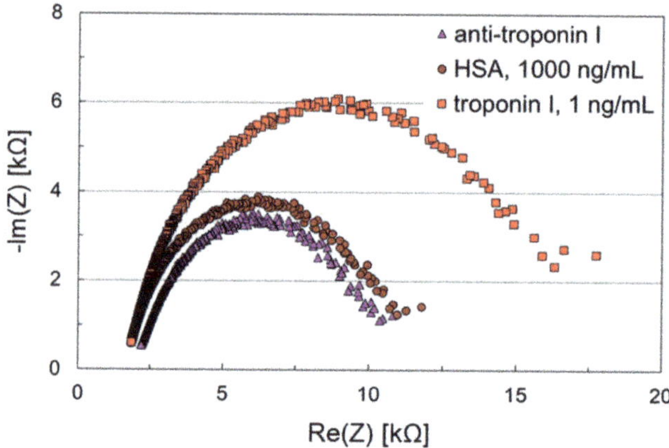

Figure 3. Nyquist plots of an impedance biosensor chip with no intermediary self-assembled mono-layer (SAM) for troponin I detection. Anti-troponin I was adsorbed on the working electrode, while the counter electrode remained uncoated. Samples containing human serum albumin (HSA), 1000 ng/mL, and troponin I, 1 ng/mL, were applied subsequently to the biosensor chip.

The resulting Nyquist plots show almost ideal semicircles, i.e., charge transfer kinetics prevails the diffusion to the layer. Changes in the double layer, such as affinity binding to the surface, can be observed. Hence, the electrode design of the microfluidic impedance biosensor chip together with the measurement protocol with the redox buffer allows basic biosensor measurements. The recording of the impedances requires only a few seconds, the sampling time a few minutes.

The initial impedance after antibody adsorption on the working electrode was 11 kΩ. It increased only slightly to 12 kΩ after HSA blocking, and the change in the complete Nyquist plot was negligible. After troponin I binding, however, the initial impedance increased to 18 kΩ, and a significant shift in the Nyquist plot was observed. This demonstrates that the chosen anti-troponin I is able to bind troponin I, while nonspecific HSA adsorption is blocked. Though this assay even worked with adsorbed antibody, antibody immobilization via SAM was to be performed in the following to show that the newly developed microfluidic impedance biosensor chip is also able to handle standard layer setups generally recommended for impedance biosensors.

3.2. Testing Thiol-SAMs Based on Aliphatic and Aromatic Hydrocarbon Spacer

SAMs on working and counter electrode are recommended to minimize nonspecific protein binding on the electrode surface, as this would lead to false positive signals, particularly in real samples with high concentrations of nonanalyte proteins. Furthermore, a suitable SAM on the working electrode needs to provide functional groups for antibody immobilization. To test the performance of potential SAMs with the microfluidic impedance biosensor chip, both working and counter electrode were coated with the respective thiol compound. After the SAM coating, samples containing increasing concentrations of HSA were successively applied on the electrodes. The results are shown in Figure 4.

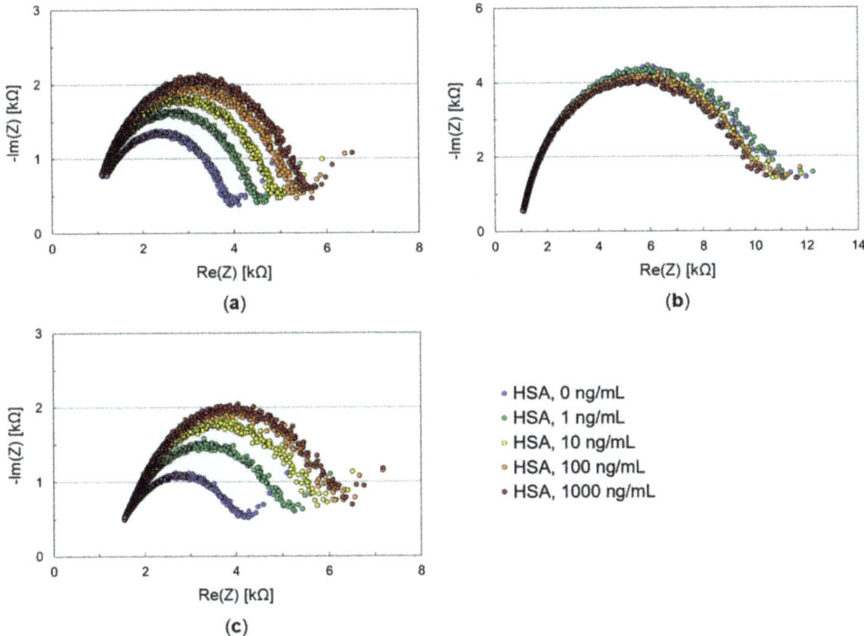

Figure 4. Nyquist plots of impedance biosensor chips for testing SAMs with HSA samples. Both working and counter electrode of such a chip were coated with (**a**) 4-mercaptobenzoic acid, (**b**) 1,4-benzenedithiol, or (**c**) 6-mercapto-1-hexanol. Samples containing HSA in increasing concentrations were applied subsequently to the biosensor chips.

Thiols with aromatic hydrocarbon spacers were chosen as they promise low initial impedance [12] and, hence, are beneficial for the charge transfer through the sensing layer. Both 4-mercaptobenzoic acid (Figure 4a) and 1,4-benzenedithiol (Figure 4b) resulted in SAMs yielding almost ideal semicircles except for a few frequencies at the beginning, confirming the possible charge transfer.

The initial impedance of the 4-mercaptobenzoic acid SAM was 4.9 kΩ, after HSA adsorption it was 6.6 kΩ. i.e., the shielding against nonspecific adsorption was not perfect. However, 4-mercaptobenzoic acid offers a carboxyl group allowing further protein coupling by simple carbodiimide chemistry. The additional immobilization steps could increase the density of the sensing layer in a way that nonspecific HSA adsorption would be reduced to a greater extent. For that reason, experiments with 4-mercaptobenzoic acid as SAM on the working electrode were continued (see Section 3.3).

The initial impedance of the 1,4-benzenedithiol SAM was 12 kΩ. This was higher than that obtained with 4-mercaptobenzoic acid, but it remained in this range after HSA sampling. Both 4-mercaptobenzoic acid and 1,4-benzenedithiol were applied in the same concentration (see Section 2.2.2), and their structure differs only in one functional group. However, 1,4-benzenedithiol has two thiol groups located *para* to each other which promotes the parallel or close to parallel orientation of the molecule on the gold surface [26,27], which is not the case for 4-mercaptobenzoic acid. Hence, a better surface coverage is obtained with 1,4-benzenedithiol, which makes it an ideal protection layer against unwanted protein adsorption.

Except for a few frequencies at the beginning, almost ideal semicircles in the Nyquist plot were also obtained for the 6-mercapto-1-hexanol SAM (Figure 4c), showing that the aliphatic hydrocarbon spacer is short enough to allow charge transfer through the sensing layer. The initial impedance was 5.2 kΩ, after HSA adsorption it was 7.3 kΩ. Hence, the results after HSA sampling were similar to those obtained with 4-mercaptobenzoic acid (Figure 4a) regarding both the initial impedances and the insufficient shielding against nonspecific protein adsorption. Better shielding might be achieved by using a thiol with a longer hydrocarbon chain, but this bears the risk of impeding the charge transfer. As 6-mercapto-1-hexanol in this study was not supposed to act as single SAM but as cocomponent in a DNA-based SAM (see Section 3.4), it is more important that the impedance values obtained with 6-mercapto-1-hexanol are not too high. This requirement is fulfilled, as shown above.

3.3. Troponin I Assay Using Thiol-SAMs Based on Aromatic Hydrocarbon Spacer

The working electrode was functionalized with 4-mercaptobenzoic acid acting as SAM with low impedance and providing functional groups for covalent antibody coupling. The counter electrode was coated with 1,4-benzenedithiol, as this showed the best shielding abilities against nonspecific protein adsorption compared with the other SAMs used here (see Figure 4). After antibody immobilization, samples containing HSA, 1000 ng/mL, and troponin I, 1 ng/mL, were applied successively to the microfluidic impedance biosensor chip. Figure 5 shows the results obtained with this assay.

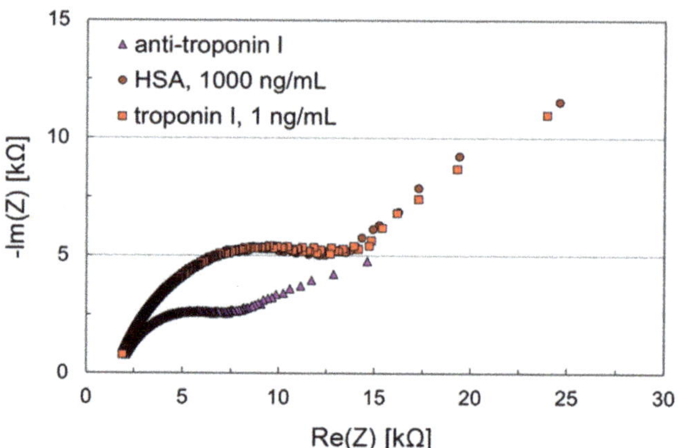

Figure 5. Nyquist plots of an impedance biosensor chip with aromatic hydrocarbon-based SAM for troponin I detection. The working electrode was coated with anti-troponin I, which was immobilized on a SAM consisting of 4-mercaptobenzoic acid. The counter electrode was coated with 1,4-benzenedithiol. Samples containing HSA, 1000 ng/mL, and troponin I, 1 ng/mL, were applied subsequently to the biosensor chip.

The initial impedance after antibody coating was 15 kΩ. Antibody immobilization, however, did not reduce HSA adsorption as anticipated. Instead, after applying

1000 ng/mL HSA, the initial impedance almost doubled (27 kΩ) and remained in this range after troponin I, 1 ng/mL, was applied. Hence, troponin I detection was not possible with this impedance biosensor, at least not at that concentration. The functionality of the anti-troponin I used here was confirmed before (see Figure 3). It is unlikely that the antibody was harmed by the covalent coupling protocol, because this is a standard procedure for protein immobilization. Furthermore, the suitability of this approach is confirmed in the next section (Section 3.4). However, the accessibility of the antibody binding sites may be hindered by the comparatively large amount of nonspecifically adsorbed HSA.

A closer look at the Nyquist plots shows significantly increased linear ranges than obtained before. This is characteristic for the Warburg impedance and indicates that the diffusion of the charge carriers prevails over the charge transfer kinetics. Hence, in order to promote the charge transfer, another SAM was to be applied. SAMs with hydrocarbon spacers of larger chain lengths are favorable regarding prevention of nonspecific protein adsorption, but may result in high initial impedances hindering the charge transfer even more. As wet chemistry methods are preferred due to the less complex coating procedures, conductive polymers are not tested here. Instead, oligonucleotide spacers were tested, as DNA strands represent large chains with negative backbone, promising a lower initial impedance.

3.4. Troponin I Assay Using Thiol-SAMs Based on DNA Spacer

The SAM on the working electrode was formed by ssDNA carrying a thiol group and 6-mercapto-1-hexanol serving as coimmobilization agent. The latter was added to support the alignment of the ssDNA strands perpendicular to the surface so that the subsequent hybridization is not hindered [13,14]. After hybridization of the thiolated ssDNA with the complementary ssDNA carrying an amino group, functional groups for antibody immobilization were available (for details see Section 2.2.3). The counter electrode was coated with 1,4-benzenedithiol as before, because of the excellent shielding abilities. After antibody immobilization on the working electrode, HSA and troponin I samples were applied successively to the microfluidic impedance biosensor chip. The results are shown in Figure 6.

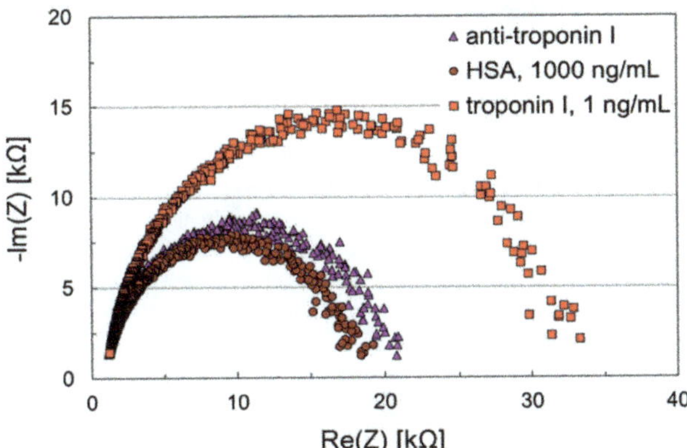

Figure 6. Nyquist plots of an impedance biosensor chip with DNA-based SAM for troponin I detection. The working electrode was coated with anti-troponin I, which was immobilized on a SAM consisting of thiolated DNA and 6-mercapto-1-hexanol. The counter electrode was coated with 1,4-benzenedithiol. Samples containing HSA, 1000 ng/mL, and troponin I, 1 ng/mL, were applied subsequently to the biosensor chip.

In contrast to the sensing layer based on a SAM with aromatic hydrocarbon spacer (Figure 5), the Nyquist plots obtained with the sensing layer using a DNA-based SAM showed ideal semicircles (Figure 6), indicating an enhanced charge transfer. After antibody coupling, the initial impedance was 21 kΩ. Application of the HSA sample, 1000 ng/mL, did not have a significant effect on the impedance values. Sampling with troponin I, 1 ng/mL, however, led to an initial impedance of 33 kΩ and a significant shift in the Nyquist plot. Hence, aside from kinetically controlled charge transfer, the DNA-based SAM allowed both shielding against nonspecific protein adsorption and detection of the cardiac marker corresponding to the antibody in a clinically relevant concentration. Furthermore, it is confirmed that the carbodiimide coupling procedure used here did not affect the integrity of the antibody.

Finally, 1,4-benzenedithiol was used to replace 6-mercapto-1-hexanol as coimmobilization agent. This allowed the addition of potential benefits arising from the aromatic nature of the hydrocarbon spacer. Furthermore, this would simplify the coating procedure with regard to the use of fewer chemicals. The further immobilization and assay procedures remained the same. The results are shown in Figure 7.

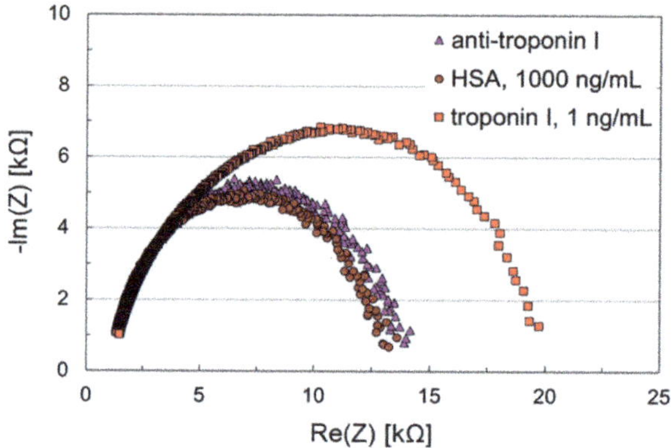

Figure 7. Nyquist plots of an impedance biosensor chip with DNA-based SAM for troponin I detection. The working electrode was coated with anti-troponin I, which was immobilized on a SAM consisting of thiolated DNA and 1,4-benzenedithiol. The counter electrode was coated with 1,4-benzenedithiol. Samples containing HSA, 1000 ng/mL, and troponin I, 1 ng/mL, were applied subsequently to the biosensor chip.

Again, ideal semicircles were obtained in the Nyquist plot, confirming the improved charge-transfer in DNA-based sensing layers. The initial impedances decreased to 14 kΩ and remained in this range after application of the HSA sample, which led only to a negligible shift in the Nyquist plot. An initial impedance of 20 kΩ and a significant shift of the Nyquist plot was obtained by applying the troponin I sample. Though 1,4-benzenedithiol does not have the linear chain structure like 6-mercapto-1-hexanol, it still stabilizes the thiolated ssDNA strands in a way that the accessibility of the binding sites in the sensing layer is not hindered. As a result, the troponin I assay was also performed successfully with this DNA-based sensing layer.

The next step would include tests in real sample media, such as blood and blood derivatives (serum, plasma) or saliva. Those media contain enzymes that may degrade the DNA-based sensing layer. It has to be tested whether this effect is notable already at a sampling time of 3 min (see Section 2.3.3). If so, the sampling time could easily be reduced. Furthermore, the addition of ethylenediaminetetraacetic acid (EDTA) to the sample or the use of EDTA plasma instead of serum could be considered. EDTA complexes Ca^{2+} and Mg^{2+} which are required as cofactors for the DNA-degrading enzymatic reaction. Hence, this complexation would reduce the interfering effects of the DNA-degrading enzymes, while troponin detection is still possible as is the detection of other proteins [28–30].

4. Conclusions

A microfluidic impedance biosensor chip was developed allowing the determination of 1 ng/mL cardiac biomarker troponin I. The use of thiols with DNA spacer showed excellent results regarding both reduction of nonspecific protein adsorption (HSA) and detection of low, clinically relevant concentrations of a biomarker (here: troponin I, 1 ng/mL). Sampling and measurement took only a few minutes, whereby the main part of the time was taken up by the sampling, which can easily be reduced by an optimized fluidic setup. The results were promising regarding a future cost-effective biosensor array chip for the rapid detection of clinically relevant biomarkers in real samples, such as serum and saliva. Future measurements will include measurements in real sample media to investigate the effect of potential DNA-degrading effects by enzymes in the sample media more closely.

Supplementary Materials: The following are available online https://www.mdpi.com/article/10.3 390/bios11030080/s1, Figure S1: Schematic representation of antibody immobilization on a thiol-SAM with aromatic hydrocarbon spacer (4-mercaptobenzoic acid), bonding of the PDMS microfluidic channel and subsequent assay with HSA blocking and troponin I sampling, Figure S2: Schematic representation of the single strand DNAs (ssDNAs) forming the thiol-SAM with DNA spacer for subsequent antibody immobilization (see Figure S3), Figure S3: Schematic representation of antibody immobilization on a thiol-SAM with DNA spacer (co-immobilization compound: 1,4-benzenedithol) and bonding of the PDMS microfluidic channel. The formation of peptide bonds in steps (5) and (6) is not included in this scheme for the sake of clarity. The subsequent assay with HSA blocking and troponin I sampling was performed as shown in Figure S1, steps (6) and (7).

Author Contributions: Conceptualization, T.R., S.S., K.L.; data curation, K.L.; formal analysis, K.A., T.H., K.L.; investigation, K.A., T.H.; methodology, T.R., S.S., A.V., K.L.; project administration, T.R., S.S., K.L.; resources, T.R., K.L.; supervision, T.R., K.L.; validation, K.A., K.L.; visualization, K.A., T.H., K.L.; writing—original draft preparation, K.L.; writing—review and editing, K.L. All authors have read and agreed to the published version of the manuscript.

Funding: This research received no external funding.

Acknowledgments: We acknowledge support by the KIT-Publication Fund of the Karlsruhe Institute of Technology.

Conflicts of Interest: The authors declare no conflict of interest.

References

1. Fernández-la-Villa, A.; Pozo-Ayuso, D.F.; Castaño-Álvarez, M. Microfluidics and electrochemistry: An emerging tandem for next-generation analytical microsystems. *Curr. Opin. Electrochem.* **2019**, *15*, 175–185. [CrossRef]
2. Sassa, F.; Biswas, G.C.; Suzuki, H. Microfabricated electrochemical sensing devices. *Lab Chip* **2020**, *20*, 1358–1389. [CrossRef]
3. Tu, J.B.; Torrente-Rodríguez, R.M.; Wang, M.Q.; Gao, W. The Era of Digital Health: A Review of Portable and Wearable Affinity Biosensors. *Adv. Funct. Mater.* **2020**, *30*, 1906713. [CrossRef]
4. Yang, M.; Lim, C.C.; Liao, R.; Zhang, X. A novel microfluidic impedance assay for monitoring endothelin-induced car-diomyocyte hypertrophy. *Biosens. Bioelectron.* **2007**, *22*, 1688–1693. [CrossRef] [PubMed]
5. Boehm, D.A.; Gottlieb, P.A.; Hua, S.Z. On-chip microfluidic biosensor for bacterial detection and identification. *Sens. Actuators B Chem.* **2007**, *126*, 508–514. [CrossRef]
6. Song, H.; Wang, Y.; Rosano, J.M.; Prabhakarpandian, B.; Garson, C.; Pant, K.; Lai, E. A microfluidic impedance flow cytometer for identification of differentiation state of stem cells. *Lab Chip* **2013**, *13*, 2300. [CrossRef] [PubMed]
7. Gruhl, F.J.; Rapp, B.E.; Länge, K. Biosensors for Diagnostic Applications. *Adv. Biochem. Eng. Biotechnol.* **2013**, *133*, 115–148. [CrossRef]
8. Lisdat, F.; Schäfer, D. The use of electrochemical impedance spectroscopy for biosensing. *Anal. Bioanal. Chem.* **2008**, *391*, 1555–1567. [CrossRef]
9. Kokkinos, C.; Economou, A.; Prodromidis, M.I. Electrochemical immunosensors: Critical survey of different architectures and transduction strategies. *TrAC Trends Anal. Chem.* **2016**, *79*, 88–105. [CrossRef]
10. Bertok, T.; Lorencova, L.; Chocholova, E.; Jane, E.; Vikartovska, A.; Kasak, P.; Tkac, J. Electrochemical Impedance Spectroscopy Based Biosensors: Mechanistic Principles, Analytical Examples and Challenges towards Commercialization for Assays of Protein Cancer Biomarkers. *Chem. Electron. Chem.* **2019**, *6*, 989–1003. [CrossRef]

11. Traynor, S.M.; Pandey, R.; Maclachlan, R.; Hosseini, A.; Didar, T.F.; Li, F.; Soleymani, L. Review-Recent advances in electrochemical detection of prostate specific antigen (PSA) in clinically-relevant samples. *J. Electrochem. Soc.* **2020**, *167*, 037551. [CrossRef]

12. Carneiro, L.P. *Development of An Electrochemical Biosensor Platform and a Suitable Low-Impedance Surface Modi-Fication Stategy*; KIT Scientific Publishing: Karlsruhe, Germany, 2014.

13. Steel, A.B.; Levicky, R.L.; Herne, T.M.; Tarlov, M.J. Immobilization of nucleic acids at solid surfaces: Effect of oligonu-cleotide length on layer assembly. *Biophys. J.* **2000**, *79*, 975–981. [CrossRef]

14. Keighley, S.D.; Li, P.; Estrela, P.; Migliorato, P. Optimization of DNA immobilization on gold electrodes for label-free detection by electrochemical impedance spectroscopy. *Biosens. Bioelectron.* **2008**, *23*, 1291–1297. [CrossRef] [PubMed]

15. Benvidi, A.; Firouzabadi, A.D.; Tezerjani, M.D.; Moshtaghiun, S.M.; Mazloum-Ardakani, M.; Ansarin, A. A highly sensitive and selective electrochemical DNA biosensor to diagnose breast cancer. *J. Electroanal. Chem.* **2015**, *750*, 57–64. [CrossRef]

16. Ribovski, L.; Zucolotto, V.; Janegitz, B.C. A label-free electrochemical DNA sensor to identify breast cancer susceptibility. *Microchem. J.* **2017**, *133*, 37–42. [CrossRef]

17. Singh, N.K.; Arya, S.K.; Estrela, P.; Goswami, P. Capacitive malaria aptasensor using Plasmodium falciparum glutamate dehydrogenase as target antigen in undiluted human serum. *Biosens. Bioelectron.* **2018**, *117*, 246–252. [CrossRef] [PubMed]

18. Yang, Z.; Castrignanò, E.; Estrela, P.; Frost, C.G.; Kasprzyk-Hordern, B. Community Sewage Sensors towards Evaluation of Drug Use Trends: Detection of Cocaine in Wastewater with DNA-Directed Immobilization Aptamer Sensors. *Sci. Rep.* **2016**, *6*, 21024. [CrossRef]

19. Cardiovascular Diseases (CVDs). WHO Fact Sheet 2017. Available online: https://www.who.int/en/news-room/fact-sheets/detail/cardiovascular-diseases-(cvds) (accessed on 24 September 2020).

20. Taylor, C.J.; Ordóñez-Mena, J.M.; Roalfe, A.K.; Lay-Flurrie, S.; Jones, N.R.; Marshall, T.; Hobbs, F.D.R. Trends in survival after a diagnosis of heart failure in the United Kingdom 2000-2017: Population based cohort study. *BMJ* **2019**, *364*, l223. [CrossRef]

21. Dörner, K. *Klinische Chemie und Hämatologie*, 7th ed.; Georg Thieme Verlag: Stuttgart, Germany, 2009.

22. Luppa, P.B.; Junker, R. (Eds.) *Point-of-Care Testing: Principles and Applications*, 1st ed.; Springer: Berlin, Germany, 2018.

23. Rapp, B.E.; Voigt, A.; Dirschka, M.; Länge, K. Deposition of ultrathin parylene C films in the range of 18 nm to 142 nm: Controlling the layer thickness and assessing the closeness of the deposited films. *Thin Solid Films* **2012**, *520*, 4884–4888. [CrossRef]

24. Länge, K.; Gruhl, F.J.; Rapp, M. Surface acoustic wave (SAW) biosensors: Coupling of sensing layers and measurement. *Methods Mol. Biol.* **2013**, *949*, 491–505.

25. Länge, K.; Gruhl, F.J.; Rapp, M. Influence of preparative carboxylation steps on the analyte response of an acoustic bio-sensor. *IEEE Sens. J.* **2009**, *9*, 2033–2034. [CrossRef]

26. Kestell, J.; Abuflaha, R.; Garvey, M.; Tysoe, W.T. Self-Assembled Oligomeric Structures from 1,4-Benzenedithiol on Au(111) and the Formation of Conductive Linkers between Gold Nanoparticles. *J. Phys. Chem. C* **2015**, *119*, 23042–23051. [CrossRef]

27. Olson, D.; Hopper, N.; Tysoe, W.T. Surface structure of 1,4-benzenedithiol on Au(111). *Surf. Sci.* **2020**, *702*, 121717. [CrossRef]

28. Lauková, L.; Konečná, B.; Janovičová, Ľ.; Vlková, B.; Celec, P. Deoxyribonucleases and Their Applications in Biomedicine. *Biomolecules* **2020**, *10*, 1036. [CrossRef]

29. Kavsak, P.A.; Roy, C.; Malinowski, P.; Clark, L.; Lamers, S.; Bamford, K.; Hill, S.; Worster, A.; Jaffe, A.S. Sample matrix and high-sensitivity cardiac troponin I assays. *Clin. Chem. Lab. Med.* **2019**, *57*, 745–751. [CrossRef] [PubMed]

30. Ilies, M.; Iuga, C.A.; Loghin, F.; Dhople, V.M.; Thiele, T.; Völker, U.; Hammer, E. Impact of blood sample collection methods on blood protein profiling studies. *Clin. Chim. Acta* **2017**, *471*, 128–134. [CrossRef] [PubMed]

 *biosensors* 

Perspective

Space Biology Research and Biosensor Technologies: Past, Present, and Future [†]

Ada Kanapskyte [1,2], Elizabeth M. Hawkins [1,3,4], Lauren C. Liddell [5,6], Shilpa R. Bhardwaj [5,7], Diana Gentry [5] and Sergio R. Santa Maria [5,8,*]

[1]   Space Life Sciences Training Program, NASA Ames Research Center, Moffett Field, CA 94035, USA; akanapskyte@gmail.com (A.K.); e.hawkins742@gmail.com (E.M.H.)
[2]   Biomedical Engineering Department, The Ohio State University, Columbus, OH 43210, USA
[3]   KBR Wyle, Moffett Field, CA 94035, USA
[4]   Mammoth Biosciences, Inc., South San Francisco, CA 94080, USA
[5]   NASA Ames Research Center, Moffett Field, CA 94035, USA; lauren.c.liddell@nasa.gov (L.C.L.); shilpa.r.shankar@gmail.com (S.R.B.); diana.gentry@nasa.gov (D.G.)
[6]   Logyx, LLC, Mountain View, CA 94043, USA
[7]   The Bionetics Corporation, Yorktown, VA 23693, USA
[8]   COSMIAC Research Institute, University of New Mexico, Albuquerque, NM 87131, USA
[*]   Correspondence: sergio.santamaria@nasa.gov; Tel.: +1-650-604-1411
[†]   Presented at the 1st International Electronic Conference on Biosensors, 2–17 November 2020; Available online: https://iecb2020.sciforum.net/.

Abstract: In light of future missions beyond low Earth orbit (LEO) and the potential establishment of bases on the Moon and Mars, the effects of the deep space environment on biology need to be examined in order to develop protective countermeasures. Although many biological experiments have been performed in space since the 1960s, most have occurred in LEO and for only short periods of time. These LEO missions have studied many biological phenomena in a variety of model organisms, and have utilized a broad range of technologies. However, given the constraints of the deep space environment, upcoming deep space biological missions will be largely limited to microbial organisms and plant seeds using miniaturized technologies. Small satellites such as CubeSats are capable of querying relevant space environments using novel, miniaturized instruments and biosensors. CubeSats also provide a low-cost alternative to larger, more complex missions, and require minimal crew support, if any. Several have been deployed in LEO, but the next iterations of biological CubeSats will travel beyond LEO. They will utilize biosensors that can better elucidate the effects of the space environment on biology, allowing humanity to return safely to deep space, venturing farther than ever before.

Keywords: space biology; deep space; biosensors; space radiation; microgravity; CubeSats

Citation: Kanapskyte, A.; Hawkins, E.M.; Liddell, L.C.; Bhardwaj, S.R.; Gentry, D.; Santa Maria, S.R. Space Biology Research and Biosensor Technologies: Past, Present, and Future [†]. *Biosensors* **2021**, *11*, 38. https://doi.org/10.3390/bios11020038

Received: 31 December 2020
Accepted: 27 January 2021
Published: 29 January 2021

Publisher's Note: MDPI stays neutral with regard to jurisdictional claims in published maps and institutional affiliations.

1. Introduction

NASA currently has plans to return humans to the Moon and eventually land crewed missions on Mars. This goal is unachievable unless we can ensure the safety and health of the astronaut crew and other terrestrial biology on those missions. The goal of this *Perspective* is to provide a brief introduction to examples of past and current technologies in space biology research, and how they influence the development of biosensor technologies for future missions to deep space. The last time NASA performed space biology experiments beyond low Earth orbit (LEO) was during the Apollo 17 mission in 1972. Since then, long-duration missions have been confined to LEO, such as those to the International Space Station (ISS).

The deep space environment is characterized by ionizing radiation and reduced gravity, both of which can have detrimental effects on biology. Beyond the Earth's magnetosphere, biology will be exposed to a constant, low-flux shower of high-energy ionizing

357

radiation, such as that from galactic cosmic rays (GCRs) and solar particle events (SPEs). Ionizing radiation causes damage to biology through several means, including direct DNA damage like double-strand breaks and indirect damage such as that caused by reactive oxygen species [1]. Microgravity also induces health risks such as muscle atrophy and bone density loss in humans. Reduced gravity can have effects at the subcellular level as well, affecting gene expression and cell growth pathways [2]. For example, in plants, a cellular-level phenomenon called gravitropism causes roots to grow downward, but in space, their roots grow randomly [3]. Additionally, many bacteria have been shown to display increased virulence and antibiotic resistance when exposed to the space environment [4].

Unfortunately, it is nearly impossible to mimic the complex conditions of space using facilities on Earth. Attempts to model the space environment are limited to particle accelerators and single-element radiation sources to simulate cosmic radiation, and rotating wall vessels or similar instruments to simulate microgravity. However, even facilities that model GCRs by consecutively exposing biological samples to single, high-energy particles cannot overlap both radiation and microgravity to mirror the conditions of space. Thus, flight missions are crucial for gaining essential insight into how biology will fare in such a unique and hostile environment.

It is critical for the future of space exploration that more biological studies be conducted querying the deep space environment; however, it is expensive and dangerous to send humans to space. A method of simplifying biological experiments is by using model organisms, including microbes, plants, invertebrates, and rodents. Since 1972, NASA has performed many missions within LEO that have utilized model organisms to understand the biological impacts of the space environment. Higher-order multicellular eukaryotes like rodents or primates yield more human-relevant information. However, they often require complicated and bulky technology and are resource-intensive to maintain. While many space experiments have been performed utilizing higher eukaryotes, currently planned missions beyond LEO only include microbes and plant seeds [5,6]. The NASA Artemis-1 vehicle will carry five biological payloads beyond LEO; four will be inside the Orion multicrew capsule carrying model organisms such as fungi, algae, and plant seeds, and the BioSentinel satellite will carry the budding yeast *Saccharomyces cerevisiae* [7]. These model organisms were selected not only because they share similarities with human cells, but also because they can remain viable in stasis for long periods of time. Current launch schedules require payload integration up to a year or more before the projected launch date. Once integrated, experiments will be without life support, exposed to the ambient temperature and humidity of the storage facility, until mission start. Thus, the limiting factor preventing mammalian cells from being used in current CubeSat platforms is the current prelaunch conditions, not technology constraints. Additional benefits of using microbes as model organisms, in comparison to higher eukaryotes, include that they require minimal care and interaction and that relevant biological and biochemical assays can be performed using small, low-cost instruments.

By combining microbes with the miniaturization and automation of new technologies, it is possible to perform highly sophisticated experiments. Biological research in space requires very specialized hardware, such as microfluidics and detection sensors, as well as reliable automation and data handling. Small satellites known as CubeSats are platforms that can accommodate these requirements, and can be used to answer questions about the effects of the space environment on biology. In recent years, microbial-derived biosensors aboard CubeSats have been used, for example, to investigate the effect of microgravity on antibiotic resistance in pathogenic bacteria and to study the effect of a fungicide on yeast cells [8,9]. These recent studies have been built on a foundation of decades of space biology research.

2. Past and Current Technologies

To fully understand the role of biosensors in space research, it is helpful to reflect upon a timeline of NASA's life science programs (Figure 1). In 1966, NASA launched the first of three uncrewed satellites through its Biosatellite program. The aim was to assess the effects of spaceflight on living organisms, ranging from microorganisms to a pigtail monkey. The program was ambitious and unfortunately incurred several failures; however, it provided valuable lessons for future life science missions [10]. Seven years later, the United States' first space station, Skylab, was launched. The goal of Skylab (1973–1974) was to serve as a laboratory environment for a variety of experiments spanning the fields of solar physics, Earth sciences, medicine, materials processing, and biology [11]. Many of the biology experiments focused on crew health and human physiology, that is, validating instruments for measurements of mass in the absence of gravity and performing cytogenetic studies of blood [11]. However, there was also an early interest in microbial studies. NASA partnered with the National Science Teachers' Association to involve high school students in Skylab experiments, which resulted in the Skylab Student Project. Out of approximately 4000 applications, 25 student projects were selected for flight [12]. Among these projects was ED31, a study investigating the viability, growth rates, and morphology of dormant microbes and spores in microgravity. The design and hardware for the experiment were simplistic; dormant bacterial and spore samples were immobilized onto sterile filter discs, wrapped in aluminum foil, and loaded beside corresponding agar plates into a larger cylindrical capsule to be inoculated in space [12]. Although rudimentary in its use of hardware, this experiment paved the way for future capabilities and advances in the technology of microbial studies in space.

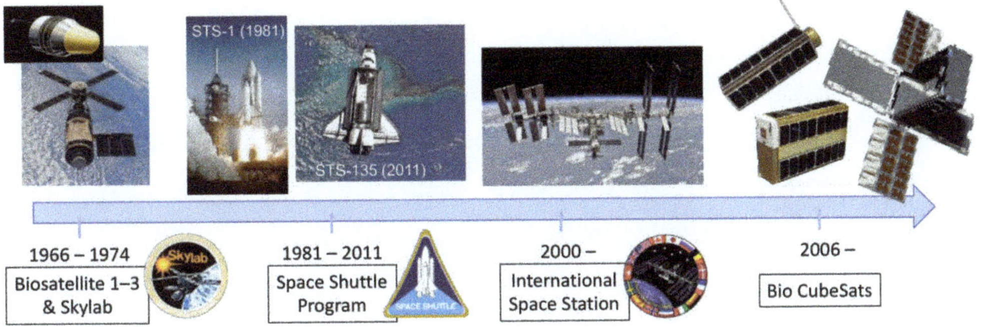

Figure 1. NASA's life science programs.

Another iteration of life science missions came with the birth of the Space Shuttle Program in the 1980s. Aboard the Columbia shuttle launch in 1996 was the Life and Microgravity Spacelab, containing 16 life science experiments, with a primary focus on human life sciences and animal models [13]. Although not particularly advancing technologies for investigations of microorganisms, the Spacelab missions were fundamental in setting up the infrastructure for the International Space Station (ISS), which is now a key resource for space microbiology research and associated technologies.

The ISS, over the course of its lifespan, has implemented over 40 facilities providing capabilities for life sciences research. As defined by NASA, an ISS facility is an internal or external structure or device on the ISS used for various investigations. Commonly, these facilities have attachment points for additional research investigations and equipment [14]. Among the ISS facilities are key technologies for conducting studies on microorganisms. In particular, there is an increasing prevalence of semi- or fully automated systems, such as the Advanced Biological Research System (ABRS), the BioCulture System, and the Mobile Spacelab, among others highlighted in Table 1. Although the ISS contains many facilities supporting physical science, advancing technology, and human research, examples listed in Table 1 focus on facilities supporting biology and biotechnology research, with an emphasis on microbe, mammalian cell, and tissue experiments. This list is not exhaustive, but instead aims to highlight some of the automated technologies for conducting such research, including microfluidics, various microscopy techniques, bioreactors, and multi-sample collection systems, all of which are crucial for biological experiments [15–17].

Table 1. Examples of International Space Station (ISS) facilities employing automated technologies for biological experiments.

ISS Facility	Description	Automated Technologies
Advanced Biological Research System (ABRS)	Single system with two independent growth chambers for plants, microorganisms, insects, and spiders [18,19]	Illumination via LEDs, temperature, CO_2 level controls; green fluorescent protein imaging system; data downlinking [18,19]
ADvanced Space Experiment Processor (ADSEP)	Single unit with thermal control for three independent experiments [18,20]	Programmable internal computer for temperature control in each cassette-based experiment; up to 44 individual experiments in each cassette [18,20]
BioChip SpaceLab (subcomponent of Mobile SpaceLab)	Cell and tissue culture platform with imaging capabilities [21]	Microfluidics for delivery of media, reagents, fluorescent particles; bright field and fluorescence time-lapse imaging; $1 \times g$ centrifuge [16,21]
BioCulture System	Cell, microbe, and tissue culture platform [18]	Hollow fiber bioreactor for medium delivery and waste removal; sample collection, protocol additions (i.e., growth factors); 10 independently controlled experiments [14,18]
Cell Biology Experiment Facility (CBEF)	Incubator with microgravity compartment and $1 \times g$ compartment with centrifuge [22]	Telemetry-controlled or pre-programmed experimental parameters [23]
Commercial Generic Bioprocessing Apparatus (CGBA)	Cold storage or incubation unit [14,22]	Programmable and accurate temperature control from -10 to 37 °C; can be fitted with bioprocessing inserts for automated sampling [14,24]
European Modular Cultivation System (EMCS)	Incubator with controllable, multi-gravity environment ($0.001–2 \times g$); two independent rotors [19]	Autonomous run of pre-defined programs for event-triggered or time-based day/night cycles, imaging sessions, or gravity thresholds [19]
Fluid Processing Cassette (FPC)	Insert placed into ADSEP; contains feeding and fixation bags for microbe cultivation [25]	Automated sampling and sample fixation [25]
Multiple Orbital Bioreactor with Instrumentation and Automated Sampling (MOBIAS)	Bioprocessing insert for CGBA made of stackable trays and used for sample processing [18]	Automated sampling [18]

Importantly, the automation of biology and biotechnology experiments onboard the ISS saves precious astronaut crew time and resources that can be devoted to maintaining life support systems and other critical tasks. Automation is also a prerequisite to deep space biological missions. The next key step to enabling missions beyond LEO is miniaturizing and converting these automated technologies to systems independent of a larger facility like the ISS. The development of autonomous biological CubeSats, described in the next section, aids in accomplishing this step.

3. Biological CubeSat Missions

In 2006, the NASA Ames Research Center pioneered a new era of biological studies and technology development in space with the advent of biological CubeSats. CubeSats are miniature satellites that are made up of one or more 10-cm cube modules or units (1 unit = 1U = 10-cm cube). GeneSat-1 was the first fully automated and self-contained biological CubeSat to go to space. GeneSat-1 employed some of the fundamental capabilities of the ISS facilities discussed previously—microfluidics and cell growth detection systems—contained within a free-flying, 3U platform to study gene expression in LEO [26]. From there, NASA Ames developed five additional free-flying biological CubeSats, each building on the previous CubeSat's infrastructure. An overview of these small satellites can be seen in Table 2. PharmaSat launched in 2009 and utilized a three-LED optical sensor to monitor microbial activity, this time testing yeast cells and their response to a fungicide in microgravity [27]. In 2010, Organism/Organic Exposure to Orbital Stresses (O/OREOS) successfully integrated two independent astrobiology studies in one Cube-Sat [28]. O/OREOS Space Environment Survivability of Live Organisms (SESLO) studied the ability of bacteria to adapt to the stresses of the space environment. O/OREOS Space Environment Viability of Organics (SEVO) monitored the stability and changes in different organic molecules [28]. Four years later, SporeSat was launched, employing unique lab-on-a-chip devices termed biology compact discs (bioCDs). These devices utilized ion-sensitive electrodes to measure concentrations of calcium in fern spores, and were rotated to simulate artificial gravity using miniaturized centrifuges, validating novel CubeSat technologies for biological experiments in space [29]. EcAMSat launched in 2017 as the largest biological satellite thus far, and was the first 6U CubeSat to be deployed from the ISS [8]. Its main objective was to study antibiotic resistance in a pathogenic bacterium. The microfluidics and infrastructure used for all these LEO missions would set up the technological framework for the next and most recent NASA mission.

Table 2. NASA's biological CubeSat missions.

CubeSat Mission (Size; Launch)	Biological Organism	Research Investigation	Technology Development
GeneSat-1 (3U; 2006)	*Escherichia coli* (bacterium)	Microgravity effects on gene expression	12-well fluidic card with LED optical detection
PharmaSat (3U; 2009)	*Saccharomyces cerevisiae* (yeast)	Microgravity effects on antifungal response	48-well fluidic card with 3-LED optical sensors
O/OREOS SESLO (3U; 2010)	*Bacillus subtilis* (bacterium)	Microgravity and LEO radiation effects	3-LED optical sensor; multiple-time-point activation
SporeSat (3U; 2014)	*Ceratopteris richardii* (fern spores)	Microgravity effects on calcium transport	Artificial gravity; lab-on-a-chip devices
EcAMSat (6U; 2017)	*Escherichia coli* (uropathogenic)	Microgravity effects on antibiotic response	48-well card; 3-LED optical sensors; variable dose delivery
BioSentinel (6U; 2021/2022)	*Saccharomyces cerevisiae*	Deep space radiation effects	18 fluidic cards (288 wells); LET spectrometer; ISS control experiment

Aside from NASA-based missions, other LEO CubeSats of interest include the SpacePharma DIDO-2 (launched 2017) and DIDO-3 (launched 2020) 3U missions that investigated enzymatic reactions and antibiotic resistance in bacteria under microgravity, among other experiments [30]. There are several biological CubeSats under development, including India's 2U RVSAT-1 and Poland's 3U LabSat, which will study the survival of microorganisms in extreme conditions [31,32]. Lastly, although significantly larger than a CubeSat, it is also worth mentioning the Bion-M2 mission, the newest iteration in the series of uncrewed, recoverable Bion satellites first launched in 1973 for a multi-week study of biological organisms in LEO [10]. Led by Roscosmos, the Bion-M2 mission will travel to the inner Van Allen radiation belt, providing a radiation and microgravity environment for potential space biology investigations [33].

The newest biological satellite in the succession of NASA Ames' biological CubeSat program is BioSentinel. BioSentinel is the first CubeSat designed to perform biological experiments in interplanetary deep space and is planned to launch as the sole biological secondary payload on NASA's Artemis-1 rocket [5,7,34]. After deployment and a lunar fly-by, BioSentinel will reach a stable heliocentric orbit and perform experiments for a minimum of six months. The primary goal of the mission is to investigate the DNA damage response to the deep space environment in the budding yeast *S. cerevisiae*. BioSentinel is a highly sophisticated and autonomous 6U CubeSat equipped with a series of subsystems designed and developed for the deep space environment, including solar panel arrays, batteries, star tracker and micro-propulsion navigation systems, transponder, antennas, and command and data handling systems. These systems occupy approximately 2U of the spacecraft [7]. The remaining 4U volume is occupied by the BioSensor payload, which contains all the instruments required to autonomously support the biological experiments. The BioSensor also contains a Timepix-based linear energy transfer (LET) spectrometer for radiation dose measurements and particle characterization. The microfluidics system is composed of 18 fluidic cards with 16 microwells per card (total of 288 microwells) [7]. Once in space, desiccated yeast cells will be rehydrated by injection of a mixture of growth medium and a metabolic indicator dye. Cell growth and metabolic activity will be monitored using an optical detection system consisting of three different LED lights and a light-to-voltage optical converter per well [34]. Each fluidic card also has a dedicated thermal control system that allows it to maintain the yeast cells in a benign cold environment until activation at a higher temperature. All the data will be telemetered back to Earth via the Deep Space Network (DSN). In addition to the deep space mission, an identical copy of the BioSensor payload will be flown on the ISS, allowing for biological comparisons in deep space and LEO.

A number of improvements have been made to CubeSat flight heritage technology with BioSentinel. These include, but are not limited to, the use of biocompatible materials and sterilization techniques, the low-cost fabrication of custom microfluidics, off-the-shelf high-precision microfluidic parts (e.g., fluidic valves, pumps, and bubble traps), an onboard LET spectrometer to enable the comparison of biological responses to space radiation to actual physical dosimetry, the inclusion of independent calibration cells for optical data normalization, the ability to store and return optical time series data, a profound increase in the capacity of sample size (288 wells compared to 48 previously), the ability to activate experiments at multiple time points and distances from Earth, and the long-term preservation of biological samples and reagents before experiment activation. These improvements are accomplished while maintaining a compact volume and relatively low cost [5,7]. Additionally, BioSentinel provides a new avenue for the space research community to conduct future missions using a variety of organisms and instruments in different space platforms, which are discussed in the next section.

4. Future Technologies and Conclusions

As highlighted in the preceding sections, technology continues to evolve as humanity once again prepares to embark upon deep space missions. Automated technologies like those used in the aforementioned ISS facilities and CubeSat missions enable more biological experiments to be performed with minimal human interaction. They also set the framework for biological missions beyond LEO—to the Moon, the Lunar Gateway, and Mars—all of which will be even more restrictive in budget, size, and available crew time. NASA's current objective to return to the Moon is being carried out by planned Artemis missions, with projects like the Lunar Gateway, Commercial Lunar Payload Services (CLPS), the Human Landing System (HLS), and others. For example, the Lunar Gateway—located outside of Earth's protective Van Allen radiation belts—will operate autonomously and create a unique platform for studying space radiation [35]. It will be an opportunity to adapt the technological infrastructure of previous CubeSats for future lunar missions.

In particular, microfluidics systems flown in previous biological CubeSats are adaptable frameworks for a variety of future space biology missions to deep space [5,7]. They can be used to deliver antibiotics, metabolic dyes, or selective growth media to better understand the biological response beyond LEO. With multiple independently activated fluidic cards and dedicated thermal control for each card now available in CubeSat platforms, we could potentially perform different, simultaneous experiments to answer separate research questions within the same payload. One particular area of interest is the potential acquisition of adaptive beneficial mutations in a reduced gravity environment together with the constant presence of high-energy ionizing particles. When microbes are subjected to environmental stress, natural selection favors genetic changes that give cells an advantage in that adverse environment [36]. The significance of exploring microbial adaptation in space has important implications. For example, adaptations that lead to abnormal cell growth and physiology—especially in pathogens—could be detrimental to astronaut health, especially considering NASA's upcoming long-term missions to outer space. On the other hand, altered microbial growth in space could be advantageous for the production of high-value products, including medicines, vitamins, and food. Such experiments could be performed with currently available fluidic and optical detection systems, for example, by using specialized growth media to select for the acquisition of genetic markers in microbes.

Automation technologies will allow experiments currently only suitable for the benchtop, or ISS facilities with astronaut intervention, to be adapted for stand-alone payloads. For example, developing technologies such as miniaturized PCR instruments or commercial DNA/RNA sequencers like the MinIon^{TM} could be integrated into small satellite platforms for mutagenesis and gene expression studies, together with advanced microfluidic delivery, sample processing, and detection systems. In addition, novel biosensors which are currently under development hold the potential to allow new experimental assays and designs, expanding the range of biological data that can be taken in space. One example is dielectric spectroscopy. This method takes advantage of a common technique used in industrial fermentation processes to measure changes in cell physiology. It utilizes the understanding that cells can be polarized when exposed to an electric field, and their ability to be polarized changes the overall capacitance of the cell suspension, which can then be measured at a range of different frequencies. Capacitance changes as the cells undergo growth, replication, protein synthesis, increases in cell membrane size, and changes in cell shape [37]. Dielectric spectroscopy correlates such cellular changes to capacitance measurements, and is one of many methods of measuring biologically relevant data in space that can be miniaturized and automated. This type of biosensor technology could be potentially implemented onboard an existing CubeSat foundation, like BioSentinel, to assay the effects of the space environment on biology. Moreover, it can advance current optical detection systems to allow the study of transient changes in vivo, independent of metabolic indicator dyes.

Another promising avenue for microfluidics devices in space is the development of organ-on-chip devices. The use of these devices provides new platforms for combining microbial research with systems more closely resembling human physiology (e.g., pathogen infection processes and changes in the human microbiome), thus more directly translating research to human applications [38,39]. Recently, the National Center for Advancing Translational Sciences (NCATS) at the National Institutes of Health (NIH) partnered with the ISS to launch the NIH Tissue Chips in Space initiative [40]. Many projects, all aimed to investigate human disease and potential therapeutics, employ microphysiological systems (MPS) or organ-on-chip devices to study tissues that are affected by the space environment, such as muscle, lung, and bone marrow [40]. These MPS technologies are prime examples of opportunities for scientific investigation that are enabled by the refinement of available hardware and the use of automated systems to answer a full spectrum of biological questions about different space environments. However, even though technologies suitable to maintain mammalian cells or organ-on-chip devices onboard small satellites like CubeSats already exist, it is currently not possible to ensure proper conditions to maintain the viability of these cells during pre-launch and launch activities. Therefore, CubeSat experiments will be mostly limited to microbial organisms due to the long pre-launch periods and constraints associated with the upcoming missions to deep space. Nevertheless, future missions might allow for the loading of biological payloads closer to launch and provide life support during launch and deployment, thus opening new possibilities for more human-relevant studies.

The previously mentioned fluidics and biosensor technologies focus on space biology research to study the effects of space exposure on terrestrial organisms. However, similar instruments are being developed for life detection and to search for signs of extraterrestrial life. Therefore, the resulting sample input, fluidic processing, and sensor and analysis technologies in astrobiology may also be useful for future space biology applications. Two examples of such miniaturized fluidics processors developed by NASA are SPLIce (Sample Processor for Life on Icy Worlds) and MICA (Microfluidic Icy-world Chemistry Analyzer). SPLIce is a microfluidic processing-and-handling system that can take in microliter volume samples and prepare them for a wide range of functions, including pH and conductivity measurements, multi-year storage of dehydrated reagents, and the retention of samples for microscopy [41]. MICA builds upon the Wet Chemistry Laboratory flown on the Phoenix Mars mission and employs an electrochemistry sensor array to quantitatively measure key chemical properties of Europa's surface materials and provide sample context in the search for evidence of potential biosignatures [42]. Other fluidics-based life detection instruments currently under development include Spain's Centro de Astrobiología's CMOLD (Complex Molecules Detector) and SOLID-LDChip (Signs of Life Detector—Life Detector Chip), JPL's Chemical Laptop, and UC Berkeley's EOA (Enceladus Organic Analyzer) [43–46].

Although space biology research has been conducted for decades, there is still much to do. By building upon the heritage and technologies of the past and present, planned and future missions beyond LEO will make it possible to move forward confidently and safely into the next era of human space exploration.

Author Contributions: A.K. and E.M.H. contributed equally to this work. A.K., E.M.H., and S.R.S.M. conceptualized this work. A.K., E.M.H., L.C.L., S.R.B., D.G., and S.R.S.M. wrote the paper. All authors have read and agreed to the published version of the manuscript.

Funding: The BioSentinel mission is supported by the Advanced Exploration Systems Program Office at NASA Headquarters.

Institutional Review Board Statement: Not applicable.

Informed Consent Statement: Not applicable.

Data Availability Statement: Data sharing not applicable.

Acknowledgments: We thank past and present members of the NASA BioSentinel mission and the Space Life Sciences Training Program (SLSTP).

Conflicts of Interest: The authors declare no conflict of interest.

References

1. Furukawa, S.; Nagamatsu, A.; Nenoi, M.; Fujimori, A.; Kakinuma, S.; Katsube, T.; Wang, B.; Tsuruoka, C.; Shirai, T.; Nakamura, A.J.; et al. Space Radiation Biology for "Living in Space". *BioMed Res. Int.* **2020**, *2020*. [CrossRef]
2. Bizzarri, M.; Monici, M.; van Loon, J.J.W.A. How Microgravity Affects the Biology of Living Systems. *BioMed Res. Int.* **2015**, *2015*, 1–4. [CrossRef] [PubMed]
3. Ferl, R.J.; Paul, A.-L. The effect of spaceflight on the gravity-sensing auxin gradient of roots: GFP reporter gene microscopy on orbit. *NPJ Microgravity* **2016**, *2*. [CrossRef] [PubMed]
4. Taylor, P.W. Impact of space flight on bacterial virulence and antibiotic susceptibility. *Infect. Drug Resist.* **2015**, *8*, 249–262. [CrossRef] [PubMed]
5. Massaro Tieze, S.; Liddell, L.C.; Santa Maria, S.R.; Bhattacharya, S. BioSentinel: A Biological CubeSat for Deep Space Exploration. *Astrobiology* **2020**, *20*. [CrossRef] [PubMed]
6. NASA Research Announcement for Appendix A: Orion Exploration Mission-1 Research Pathfinder for Beyond Low Earth Orbit Space Biology Investigations. Available online: https://www.nasa.gov/feature/small-samples-with-big-mission-on-first-orion-flight-around-the-moon (accessed on 29 December 2020).
7. Ricco, A.J.; Santa Maria, S.R.; Hanel, R.P.; Bhattacharya, S. BioSentinel: A 6U Nanosatellite for Deep-Space Biological Science. *IEEE Aerosp. Electron. Syst. Mag.* **2020**, *35*, 6–18. [CrossRef]
8. Padgen, M.R.; Chinn, T.N.; Friedericks, C.R.; Lera, M.P.; Chin, M.; Parra, M.P.; Piccini, M.E.; Ricco, A.J.; Spremo, S.M. The EcAMSat fluidic system to study antibiotic resistance in low earth orbit: Development and lessons learned from space flight. *Acta Astronaut.* **2020**, *173*, 449–459. [CrossRef]
9. Diaz-Aguado, M.F.; Ghassemieh, S.; Van Outryve, C.; Beasley, C.; Schooley, A. Small Class-D spacecraft thermal design, test and analysis—PharmaSat biological experiment. In Proceedings of the IEEE Aerospace Conference, Big Sky, MT, USA, 7–14 March 2009. [CrossRef]

10. Souza, K.A.; Hogan, R.; Ballard, R. *Life into Space: Space Life Sciences Experiments, NASA Ames Research Center 1965–1990*; National Aeronautics and Space Administration, Ames Research Center: Moffett Field, CA, USA, 1995.

11. Skylab: A chronology. Available online: https://ntrs.nasa.gov/citations/19780017172 (accessed on 14 October 2020).

12. MSFC Skylab Student Project Report. Available online: https://ntrs.nasa.gov/citations/19740025164 (accessed on 14 October 2020).

13. Life and Microgravity Spacelab (LMS). Available online: https://ntrs.nasa.gov/citations/19980206462 (accessed on 14 October 2020).

14. Neigut, J.S.; Tate-Brown, J.M. *International Space Station Facilities Research in Space 2017 and Beyond*; NASA ISS Program Science Office: Houston, TX, USA, 2017.

15. Mobile SpaceLab. Available online: https://www.nasa.gov/mission_pages/station/research/experiments/explorer/Facility.html?#id=7692 (accessed on 14 October 2020).

16. BioChip SpaceLab. Available online: https://www.nasa.gov/mission_pages/station/research/experiments/explorer/Facility.html?#id=7666 (accessed on 14 October 2020).

17. Cell Culturing. Available online: https://www.nasa.gov/mission_pages/station/research/experiments/explorer/Facility.html?#id=377 (accessed on 14 October 2020).

18. Mains, R.; Reynolds, S.; Baker, T.; Sato, K. *A Researcher's Guide to: International Space Station Cellular Biology*; NASA ISS Program Science Office: Houston, TX, USA, 2015.

19. Zabel, P.; Bamsey, M.; Schubert, D.; Tajmar, M. Review and analysis of over 40 years of space plant growth systems. *Life Sci. Space Res.* **2016**, *10*, 1–16. [CrossRef] [PubMed]

20. Advanced Space Experiment Processor. Available online: https://www.nasa.gov/mission_pages/station/research/experiments/explorer/Facility.html?#id=369 (accessed on 29 December 2020).

21. CASIS (Center for the Advancement of Science in Space) and the International Space Station National Laboratory: Research in Space for Earth Benefits. Available online: https://iee.ucsb.edu/sites/default/files/docs/ucsb_presentation_nov6_16b.pdf (accessed on 28 January 2021).

22. Barker, D.C.; Costello, K.A.; Ruttley, T.M.; Ham, D.L. *International Space Station Facilities Research in Space 2013 and Beyond*; NASA ISS Program Science Office: Houston, TX, USA, 2013.

23. Cell Biology Experiment Facility. Available online: https://www.nasa.gov/mission_pages/station/research/experiments/explorer/Facility.html?#id=333 (accessed on 29 December 2020).

24. Commercial Generic Bioprocessing Apparatus. Available online: https://www.nasa.gov/mission_pages/station/research/experiments/explorer/Facility.html?#id=329 (accessed on 29 December 2020).

25. Fluid Processing Cassette. Available online: https://www.nasa.gov/mission_pages/station/research/experiments/explorer/Facility.html?#id=378 (accessed on 29 December 2020).

26. Ricco, A.J.; Hines, J.W.; Piccini, M.; Parra, M.; Timucin, L.; Barker, V.; Storment, C.; Friedericks, C.; Agasid, E.; Beasley, C.; et al. Autonomous genetic analysis system to study space effects on microorganisms: Results from orbit. In Proceedings of the TRANSDUCERS 2007—2007 International Solid-State Sensors, Actuators and Microsystems Conference, Lyon, France, 10–14 June 2007. [CrossRef]

27. Ricco, A.J.; Parra, M.; Niesel, D.; Piccini, M.; Ly, D.; McGinnis, M.; Kudlicki, A.; Hines, J.W.; Timucin, L.; Beasley, C.; et al. PharmaSat: Drug dose response in microgravity from a free-flying integrated biofluidic/optical culture-and-analysis satellite. In Proceedings of the SPIE 7929, Microfluidics, BioMEMS, and Medical Microsystems IX, San Francisco, CA, USA, 23–25 January 2011. [CrossRef]

28. Nicholson, W.L.; Ricco, A.J.; Agasid, E.; Beasley, C.; Diaz-Aguado, M.; Ehrenfreund, P.; Friedericks, C.; Ghassemieh, S.; Henschke, M.; Hines, J.W.; et al. The O/OREOS mission: First science data from the Space Environment Survivability of Living Organisms (SESLO) payload. *Astrobiology* **2011**, *11*, 951–958. [CrossRef] [PubMed]

29. Salim, W.W.A.W.; Park, J.; Rickus, J.L.; Rademacher, A.; Ricco, A.J.; Schooley, A.; Benton, J.; Wickizer, B.; Martinez, A.; Mai, N.; et al. SporeSat: A nanosatellite platform lab-on-a-chip system for investigating gravity threshold of fern-spore single-cell calcium ion currents. In Proceedings of the Solid-State Sensors, Actuators and Microsystems Workshop, Hilton Head Island, SC, USA, 8–12 June 2014. [CrossRef]

30. SpacePharma. Available online: https://www.space4p.com/missions (accessed on 23 January 2021).

31. Hegde, K.M.; Abhilash, C.R.; Anirudh, K.; Kashyap, P. Design and Development Of RVSAT-1, A Student Nano-satellite With Biological Payload. In Proceedings of the IEEE Aerospace Conference, Big Sky, MT, USA, 2–9 March 2019. [CrossRef]

32. SatRevolution. Available online: https://satrevolution.com/missions/labsat/ (accessed on 23 January 2021).

33. NASA Research Announcement for Solicitation of Proposals for Possible Inclusion in a Russian Bion-M2 Mission. Available online: https://www.nasa.gov/feature/nasa-selects-space-biology-experiments-to-study-living-organisms-on-russian-bion-m2-mission (accessed on 23 January 2021).

34. Santa Maria, S.R.; Marina, D.B.; Massaro Tieze, S.; Liddell, L.C.; Bhattacharya, S. BioSentinel: Long-Term Saccharomyces cerevisiae Preservation for a Deep Space Biosensor Mission. *Astrobiology* **2020**, *20*. [CrossRef] [PubMed]

35. Artemis Plan: NASA's Lunar Exploration Program Overview (September 2020). Available online: https://www.nasa.gov/sites/default/files/atoms/files/artemis_plan-20200921.pdf (accessed on 28 January 2021).

36. Nickerson, C.A.; Ott, C.M.; Wilson, J.W.; Ramamurthy, R.; Pierson, D.L. Microbial Responses to Microgravity and Other Low-Shear Environments. *Microbiol. Mol. Biol. Rev.* **2004**, *68*, 345–361. [CrossRef] [PubMed]

37. Al Ahmad, M.; Al Natour, Z.; Attoub, S.; Hassan, A.H. Monitoring of the Budding Yeast Cell Cycle Using Electrical Parameters. *IEEE Access* **2018**, *6*, 19231–19237. [CrossRef]

38. Chin, W.H.; Barr, J.J. Phage research in 'organ-on-chip' devices. *Microbiol. Aust.* **2019**, *40*, 28–32. [CrossRef]

39. Bein, A.; Shin, W.; Jalili-Firoozinezhad, S.; Park, M.H.; Sontheimer-Phelps, A.; Tovaglieri, A.; Chalkiadaki, A.; Kim, H.J.; Ingber, D.E. Microfluidic Organ-on-a-Chip Models of Human Intestine. *Cell. Mol. Gastroenterol. Hepatol.* **2018**, *5*, 659–668. [CrossRef] [PubMed]

40. Low, L.A.; Giulianotti, M.A. Tissue Chips in Space: Modeling Human Diseases in Microgravity. *Pharm. Res.* **2019**, *37*, 8. [CrossRef] [PubMed]

41. Chinn, T.N.; Lee, A.K.; Boone, T.D.; Tan, M.X.; Chin, M.M.; Mccutcheon, G.C.; Horne, M.F.; Padgen, M.R.; Blaich, J.T.; Forgione, J.B.; et al. Sample Processor for Life on Icy Worlds (SPLIce): Design and Test Results. In Proceedings of the International Conference on Miniaturized Systems for Chemistry and Life Sciences (MicroTAS 2017), Savannah, GA, USA, 22–26 October 2017.

42. Noell, A.C.; Jaramilllo, E.A.; Kounaves, S.P.; Hecht, M.H.; Harrison, D.J.; Quinn, R.C.; Forgione, J.; Koehne, J.; Ricco, A.J. MICA: Microfluidic Icy-World Chemistry Analyzer. In Proceedings of the AbSciCon, Bellevue, WA, USA, 24–28 June 2019.

43. Fairén, A.G.; Gómez-Elvira, J.; Briones, C.; Prieto-Ballesteros, O.; Rodríguez-Manfredi, J.A.; López Heredero, R.; Belenguer, T.; Moral, A.G.; Moreno-Paz, M.; Parro, V. The Complex Molecules Detector (CMOLD): A Fluidic-Based Instrument Suite to Search for (Bio)chemical Complexity on Mars and Icy Moons. *Astrobiology* **2020**, *20*, 1076–1096. [CrossRef] [PubMed]

44. García-Descalzo, L.; Parro, V.; García-Villadangos, M.; Cockell, C.S.; Moissl-Eichinger, C.; Perras, A.; Rettberg, P.; Beblo-Vranesevic, K.; Bohmeier, M.; Rabbow, E.; et al. Microbial Markers Profile in Anaerobic Mars Analogue Environments Using the LDChip (Life Detector Chip) Antibody Microarray Core of the SOLID (Signs of Life Detector) Platform. *Microorganisms* **2019**, *7*, 365. [CrossRef] [PubMed]

45. Mora, M.F.; Kehl, F.; Tavares da Costa, E.; Bramall, N.; Willis, P.A. Fully Automated Microchip Electrophoresis Analyzer for Potential Life Detection Missions. *Anal. Chem.* **2020**, *92*, 12959–12966. [CrossRef] [PubMed]

46. Mathies, R.A.; Razu, M.E.; Kim, J.; Stockton, A.M.; Turin, P.; Butterworth, A. Feasibility of Detecting Bioorganic Compounds in Enceladus Plumes with the Enceladus Organic Analyzer. *Astrobiology* **2017**, *17*, 902–912. [CrossRef] [PubMed]

MDPI

St. Alban-Anlage 66

4052 Basel

Switzerland

Tel. +41 61 683 77 34

Fax +41 61 302 89 18

www.mdpi.com

Biosensors Editorial Office

E-mail: biosensors@mdpi.com

www.mdpi.com/journal/biosensors

www.ingramcontent.com/pod-product-compliance
Lightning Source LLC
LaVergne TN
LVHW071602170726
843515LV00008B/2322